THE LUNAR MEN

Jenny Uglow grew up in Cumbria, and has worked in publishing since leaving Oxford; she is also Honorary Visiting Professor at the University of Warwick. She has written widely on eighteenth- and nineteenth-century literature and culture, including the acclaimed biographies of Elizabeth Gaskell and William Hogarth, studies of George Eliot and Henry Fielding, and *Cultural Babbage: Technology, Time and Invention* (with Francis Spufford). She lives in Canterbury.

Widely praised as 'book of the year' by other writers and reviewers, *The Lunar Men* was the winner of the Hessel-Tiltman Prize for History and was shortlisted for the James Tait Black Memorial Prize for Biography, the Chrysalis Biography of the Year in the British Book Awards, and the South Bank Show literature award.

Praise for *The Lunar Men*

'In this spectacular, epic book, Jenny Uglow shows how childlike day-dreams and Heath-Robinson contraptions gave way to some of the greatest inventions of mankind . . . Never has the eighteenth century come so much to life in the twenty-first.' Gaby Wood, *Observer*

'Jenny Uglow's book represents an astonishing feat of research, inquiry and fact-collecting. It is also a superbly original idea . . . *The Lunar Men* is a considerable historical achievement and through its examination of these men, adds significantly to our knowledge of our world, which they did so much to create.' *Literary Review*

'It is Uglow's achievement to give these men – "whose interests leapt from subject to subject like grasshoppers in summer" – the charming and energetic treatment they deserve.' Arthur Herman, *The Scotsman*

'Some biographies are as dry as chalk dust, but this book is packed with bizarre, entertaining and downright hilarious anecdotes . . . The story is filled with the sparks and flashes of electricity, the hammering of pistons, the hissing of steam engines, and the not infrequent angry mumblings of various violent crowds that characterised the era.' *Focus*

'What Uglow did for eighteenth-century London in her biography of Hogarth, she has now achieved for another city . . . This is an exhilarating book, then, filled with wonders. Jenny Uglow is the most perfect historian imaginable, alive to detail and alert to general patterns.'
Peter Ackroyd, *The Times*

'A magnificently accomplished and enjoyable book. And if it highlights the Lunar Men's prodigious learning, it also demonstrates the exceptional abilities of at least one Lunar Woman. For in the sheer range of Jenny Uglow's erudition, her obvious delight in the inquisitiveness of her savants, and in her skills as a writer, she has proved herself a worthy member of that distinguished club.' John Adamson, *Sunday Telegraph*

'Each member of the Lunar Society deserves a biography. Combining their lives – emotions as well as achievements – in a single, coherent volume is a task that few writers could have accomplished with anything like success. Jenny Uglow manages the near impossible by writing with an ingenuity which the Lunar Men themselves would have admired.'
Roy Hattersley, *Independent*

'Jenny Uglow packs every page of her wonderful book with riveting information about this group of titans who together precipitated eighteenth-century Britain into the modern world . . . Uglow recounts their story with tremendous narrative drive and gives a vividly realised impression of the times as backdrop to the main action.' Penelope Hughes-Hallett, *Daily Telegraph*

'An absolute wonder of a book, huge in its span and close in its detail, nothing less than a snapshot of what and who was best about Britain and its intellectual life in the middle of the eighteenth century.' *Economist*

'Jenny Uglow presents a delightful and vivid account of the Industrial Revolution's often packed and complex history . . . Meticulously researched, sympathetic and lively, Uglow has written one of the most compelling books on the period in ages.' *Evening Standard*

'Jenny Uglow escapes into the past with the skill of a master storyteller in this beautifully written book . . . As well as providing a classic example of how to write a history book, Jenny Uglow's captivating narrative offers a rare account of an amazing collection of visionaries.' *Birmingham Post*

'Here is a fine read, and essential for all who would understand the modern world.' *New Statesman*

'Jenny Uglow's book is an absorbing and rich account of the dreams and determination of the engineers of the first Industrial Revolution.'
Times Literary Supplement

'This plump book is like a treasure galleon out of Acapulco, stuffed to the poop with riches and marvels.' *The Oldie*

'The best single explanation of the genesis of the industrial revolution.'
Ian Bell, *Sunday Herald*, Books of the Year

'An extraordinarily gripping account.' Richard Holmes, *Guardian*, Books of the Year

'The best book I've read this year – full of unexpected information, amazing characters and the real sense that scientific curiosity is as exciting as any "artistic pursuit".' A. S. Byatt, *Guardian*, Books of the Year

'This collective biography is a revelation from start to finish.'
Michael Kerrigan, *The Scotsman*, Books of the Year

'A brilliantly orchestrated account . . . the best example I know of that difficult art, the collective biography.' Michael Holroyd, *Guardian*, Books of the Year *by the same author*

GEORGE ELIOT
ELIZABETH GASKELL
HENRY FIELDING
HOGARTH
MACMILLAN DICTIONARY OF WOMEN'S BIOGRAPHY
(editor)
CULTURAL BABBAGE: TECHNOLOGY, TIME AND INVENTION
(edited with Francis Spufford)

THE LUNAR MEN

The Friends who made the Future

1730–1810

JENNY UGLOW

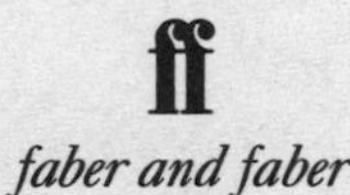

faber and faber

To Desmond King-Hele and Shena Mason

First published in 2002
by Faber and Faber Limited
3 Queen Square London WC1N 3AU
This paperback edition published in 2003

Typeset by Agnesi Text Hadleigh
Printed in England by Mackays of Chatham, plc

A CIP record for this book
is available from the British Library

ISBN 0–571–21610–2

This book is set in a version of Baskerville,
based on the original typeface designed
by John Baskerville in the 1750s.

2 4 6 8 10 9 7 5 3 1

CONTENTS

THIRD QUARTER

FOURTH QUARTER

WANING

THE PRINCIPAL LUNAR MEN IN ORDER OF AGE

John Whitehurst (1713–1788)

Matthew Boulton (1728–1809)

Josiah Wedgwood (1730–1795)

Erasmus Darwin (1731–1802)

Joseph Priestley (1733–1804)

William Small (1734–1775)

James Keir (1735–1820)

James Watt (1736–1819)

William Withering (1741–1799)

Richard Lovell Edgeworth (1744–1817)

Thomas Day (1748–1789)

Samuel Galton (1753–1832)

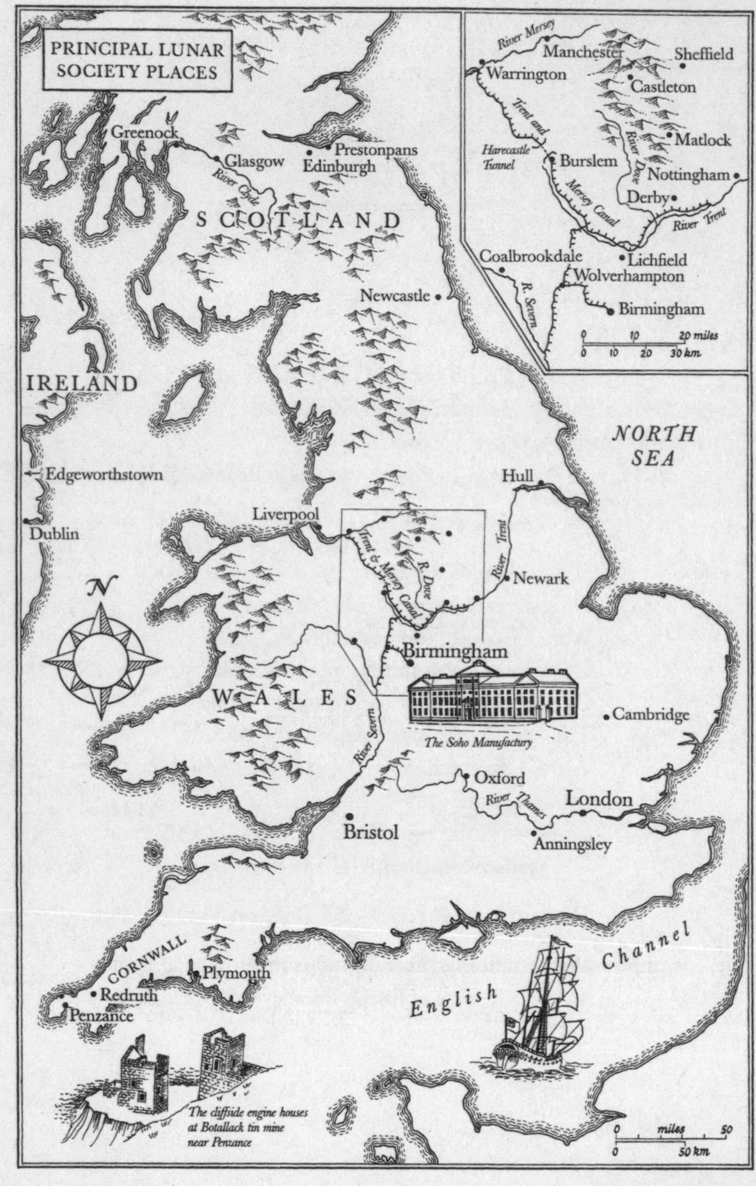

The Soho Manufactory
The cliffside engine houses at Botallack tin mine near Penzance

WAXING

I shall never forget Mr Boulton's expression to me: 'I sell here, Sir, what all the world desires to have – Power.'

James Boswell, 1774

'Twas in truth an hour
Of universal ferment; mildest men
Were agitated; and commotions, strife
Of passion and opinion fill'd the walls
Of peaceful houses with unquiet sounds.
The soil of common life was at that time
Too hot to tread upon; oft said I then,
And not then only, 'what a mockery this
Of history; the past and that to come!'

William Wordsworth, *The Prelude*, 1805

There is universally something presumptuous in provincial genius.

Review of Joseph Priestley's *Memoirs*, 1806

'O! pray! move on, Sir, said she, this is amazingly fine: I fancy myself travelling along with that little earth in its course round the gilded Sun . . .'

PROLOGUE : 'SURPRISE THE WORLD'

The earth turns and the curving shadow sweeps round the globe. The sun sets, the moon rises, and all that is familiar feels suddenly strange. In an age before street lights, link-boys carry torches to see city-dwellers home, while in the countryside starlight and moonlight are the only guides. The footpads are out, a darker blackness against shadow, so for safety's sake men walk together when they roll back from the coffee-house, the tavern and the club. And in the eighteenth century clubs are everywhere: clubs for singing, clubs for drinking, clubs for farting; clubs of poets and pudding-makers and politicians. One such gathering of like-minded men is the Lunar Society of Birmingham. They are a small, informal bunch who simply try to meet at each other's houses on the Monday nearest the full moon, to have light to ride home (hence the name) and like other clubs they drink and laugh and argue into the night. But the Lunar men are different – together they nudge their whole society and culture over the threshold of the modern, tilting it irrevocably away from old patterns of life towards the world we know today. That is why I wanted to write about them.

Amid fields and hills the Lunar men build factories, plan canals, make steam-engines thunder. They discover new gases, new minerals and new medicines and propose unsettling new ideas. They create objects of beauty and poetry of bizarre allure. They sail on the crest of the new. Yet their powerhouse of invention is not made up of aristocrats or statesmen or

'The Grand Orrery', and a quotation from John Harris, *Astronomical Dialogues*, 1719 scholars but of provincial manufacturers, professional men and gifted amateurs – friends who meet almost by accident and whose lives overlap until they die.

So who are they?

First to enter is Erasmus Darwin, doctor, inventor, poet and – half a century before his grandson Charles – pioneer of evolution. (Enormously gifted and enormously fat, eventually he has to cut a semi-circle in his dining table to fit his stomach.) Then comes Matthew Boulton, flamboyant chief of the first great 'manufactory' at Soho, just outside Birmingham, followed by his anxious Scottish partner James Watt, of steam-engine fame. Another member is the ambitious young potter Josiah Wedgwood, and eventually, in 1780, Joseph Priestley arrives, the preacher with the stuttering voice and flowing pen, the chemist who isolates oxygen and becomes the visionary leader of Rational Dissent.

This quintet forms the core. But around them weave other stories, a string of names that take on shape as they turn up in their top-coats and breeches, driving newfangled carriages, talking of freedom, of riots and reform, love and laughing-gas. Among them are the Scots chemist James Keir, reliable as a rock; the clockmaker John Whitehurst, who works with minutes but dreams of millennia, the age of the earth itself. Then come the doctors: the diplomatic William Small who seals their early friendships, and the austere William Withering, who brings digitalis into mainstream medicine. And a wilder note sounds with the arrival of two young, idealistic followers of Rousseau, Richard Lovell Edgeworth and Thomas Day.

Ten of these men became Fellows of the Royal Society but only a few had a university education and most were Nonconformists or free-thinkers. This placed them outside the Establishment – an apparent disadvantage which proved a real strength, since they were unhampered by old traditions of deference and stuffy institutions. They came from varied backgrounds but when they edged towards rows they agreed to differ, turning back to the things they shared. 'We had nothing to do with the *religious* or *political* principles of each other,' wrote Priestley. 'We were united by a common love of *science*, which we thought sufficient to bring together persons of all distinctions, Christians, Jews, Mohametans, and Heathens, Monarchists and Republicans.'[1] Like a living unit, the group stretched to encompass the awkward and odd: only rarely was there an absolute impasse. Their passionate common exchange and endeavour was of a type that would never be possible again – until today, with the fast, collaborative intimacy of the Internet.

To begin with they came together simply through the pleasure of playing with experiments, what Darwin called 'a little philosophical laughing'.[2] They caught at discoveries with delight, sure that every find could help them to crack the elusive codes of nature. And Nature, on every hand, offered herself for investigation. The great vogue for collecting that had grown through the previous century now reached new peaks. Sometimes the collections were 'evidence' in an argument, like the unsurpassed collection of minerals and fossils amassed at the start of the century by geologist John Woodward, to prove the revolutionary thesis that fossils were indeed the remains of ancient organisms, not patterns in rocks, or mysterious designs placed there by God.[3] At other times, the whole of the natural world suddenly became 'collectible', as if knowledge were conveyed directly, visibly, tangibly by the objects in a cabinet of curiosities. When Peter the Great asked the philosopher Leibniz in 1708 what he should collect, the answer, it seemed, was 'everything':

> Such a cabinet should contain all significant things and rarities created by nature and man. Particularly needed are stones, metals, minerals, wild plants, and their artificial copies, animals both stuffed and preserved . . . Foreign works to be acquired should include diverse books, instruments, curiosities and rarities . . . In short, all that could enlighten and please the eye.[4]

However, Peter's daughter-in-law Catherine the Great (another great collector) disparaged this old, baroque style of freakish accretion: 'I often quarrelled with him', she wrote, 'about his wish to enclose Nature in a cabinet – even a huge palace could not hold Her.'[5]

Nature would not be confined. In the mid-eighteenth century, across Europe, in Britain and in America, ordering the vast and complex riches of Nature became a priority. This was the age of great scientific expeditions. When the naturalists Joseph Banks and Daniel Solander travelled with Captain Cook on his voyage to the South Seas from 1768 to 1771, they brought back 1,000 new species of plants, 500 fish, 500 bird skins, numberless insects and hundreds of drawings. It was against this background that Erasmus Darwin translated Linnaeus, wrote his epic poem

The Botanic Garden and developed his own controversial theories of evolution.

In exploring such matters Darwin and his friends were part of the great spread of interest in science that extended from the King and the Royal Society to country clergymen and cotton-spinners. When people talk of eighteenth-century culture this is the swathe that is often missed out: the smart crowds thronging to electrical demonstrations; the squires fussing over rainfall gauges; the duchesses collecting shells and the boys making fire-balloons; the mothers teaching their children from the new encyclopaedias with their marvellous engraved plates of strange animals and birds and plants.

The Kentish Hop Merchant and a Lecture on Optics, satirical engraving

Science was popular because it was 'gentlemanly' and cultured, and like all crazes it produced its share of jokes. But it was also a great spur to industry, helping Britain to surge ahead of other European nations.[6] As professors and savants brought their improved mathematics and theoretical knowledge of chemistry, minerals, heat or hydraulics to bear on the *ad hoc* wisdom of old crafts, so the artisans developed new processes and technologies at an astonishingly accelerated rate. The manufacturers among the Lunar men pounced on the new findings. Their ambitions were unbounded: 'I hate piddling, you know,' wrote Wedgwood, who also declared that he would 'surprise *the World* with wonders'.[7]

But the idealists among them, particularly Priestley, wanted to surprise the world in a different way. Their technocratic fix, they thought, could bring paradise on earth: just as chemists could make 'pure' air to cure diseases, so knowledge could light the fuse of democratic change. Anything seemed possible – steamships, manned flight, diving bells. Darwin speculated quite seriously about changing the windflow over Britain, and suggested that European governments, 'instead of destroying their seamen and exhausting their strength in unnecessary wars', should use their navies to tow icebergs to the Equator to cool the tropics and ease the northern winters.[8]

There was no man-made Georgian global warming – but what happened in Britain was dramatic. In two generations, roughly from 1730 to 1800, the country changed from a mainly agricultural nation into an emerging industrial force. By the time these friends died, iron and coal and cotton were king and the provinces no longer looked automatically to London to lead the way.[9] The 'universal ferment' that accompanied this shift was as potent as any political revolution, affecting the lives of millions, opening the way to the factory age, the railway, the forging of empire. Although there was no sudden, sharply datable 'industrial revolution', for all the makers and merchants the late eighteenth century was a cluttered, cut-throat world, different to that which their fathers had known. They now had to appeal to the affluent 'middling classes' who were rushing to buy new domestic goods: clocks, prints, earthenware, curtains and cutlery and carpets.[10] The country was driven to rethink the whole relationship of 'luxury' to culture and such issues were argued over not only by philosophers but also by smart consumers such as Lady Caroline Lennox, who declared stoutly that shopping was not only fun but a 'rational exercise, a commitment to the civilising powers of trade'.[11]

Caroline's word 'rational' is the key. When she was growing up the nation prided itself on its open, rational outlook. In the early years of the century, Continental philosophers such as Voltaire saw Britain as a model of freedom, with its balanced constitution and religious tolerance and its openness to public discussion. Many thinkers were convinced that the light of reason would dispel the shadows of superstition.[12] Yet here too, change was slow: when the murrain decimated herds of Midlands cattle in the 1740s, an educated boy still prayed, 'God grant that the people of the land may turn away the wrath of God by true repentance, and that we may *sin no more lest a worst thing come upon us*.'[13] And new discoveries themselves often seemed to defy reason – the idea that seas of fire rolled beneath the solid earth, or that chalk contained gas 'fixed' into it, which could be freed into the air like a genie from a lamp.

Contradictions abound. The age of progress was also one of retrospection, in which people hunted endlessly for 'origins'. The age of reason was also one of sensibility, whose gurus stressed the power of the passions and senses as much as the mind. Science itself was intensely physical: medicine was a saga of bleeding and blisters; chemistry a matter of green fumes and red fumes, of the tang of acid on the tongue, of sneezing and choking and watering eyes. And this sensual bias was embedded in the terms they used: as chemical substances proved mysteriously choosy, reacting with some substances and repelled by others, so chemists hunted for 'affinities', patterns of union as binding (and baffling) as choices in love. The language of science rippled with the suggestiveness of sex and the human body itself became a source of fascination. Was it a machine or a bundle of vibrating nerves? How did we feel sensations? How did we register them in our minds?

These were key questions not only for medicine but for education and artistic taste. In the time of the Lunar men science and art were not separated: you could be an inventor and designer, an experimenter and a poet, a dreamer and an entrepreneur all at once without anyone raising an eyebrow. In 1772, when the young British Museum bought the first great collection of antiquities belonging to Sir William Hamilton, ancient bronzes and vases and specimens of natural history all found a place together, in the 'Department of Natural and Artificial Productions'. Constantly the different realms overlapped. As botanists listed plants, so flowers bloomed across teapots and plates. As scholars compiled tables of minerals, so manufacturers printed catalogues and grouped their goods in families and types. As geologists argued about rock formations and volcanoes, so artists and poets began to show wild regions not as deformed but as 'sublime'. At the same time, factories joined ruins on the tourist trail. In 1781, the Hon. John Byng advised that the best way to enjoy Tintern Abbey was 'to bring wines, cold meat, with corn for the horses':

> Spread your table in the ruins; and possibly a Welsh Harper may be obtained from Chepstow. I next visited several of the iron works up the stream, and with wonder observed the gradations of the iron from the smallest wire to a large cannon.[14]

Yet if optimists like Priestley genuinely believed a peaceful millennium was on the way there was always a darker side. Enclosures emptied villages. Factories and machines could turn workers into cogs. Quaker anti-slavery campaigners sold guns to Africa. Some people already suspected that progress might create a hell instead of a heaven on earth and a counter voice fought to make itself heard, through writers such as Kit Smart and William Blake, who asked us not to trust in 'Reason' but to look within for the divine, the springs of creation.

This book smells of sweat and chemicals and oil, and resounds to the thud of pistons, the tick of clocks, the clinking of cash, the blasts of furnaces and the wheeze and snort of engines but it also speaks of bodies, courtships, children, paintings and poetry. The excitement of science and manufacturing went side by side with experiments in living which aroused horror in the icy evangelical respectability that followed. (When Charles Darwin wrote his grandfather's biography, his daughter Henrietta took the proofs and firmly scored out any hint of Erasmus's shocking 'atheism'.) The Lunar men shared the praise and the abuse together, and although over the years the dynamic of their friendships changed, they remained remarkably close and influential. We know so little about their work together in comparison, say, with the Romantics – yet once we do know them it is impossible to read Romantic poetry in quite the same way. As their glow fades, so the cloudy moon of Coleridge and Shelley sails into the sky, both reflecting and rejecting the old Lunar ideals.

I realize that I am looking at these men through the spectacles of my own time and interests, and have shaped the facts according to my own guiding images: the Lunar tides driving up the beach, with each wave curling back to let the next one break; or the ancient elements – Earth, Water, Air and Fire. The elements both offer a map of the group's preoccupations and suggest their story's form: the physical and intellectual earth from which they grow; the way that their lives flow together; their airy ascent, and their final, fiery, revolutionary years. And the old Aristotelian names also signal the slow but profound change in scientific thinking over the century, until in the late 1780s the new 'French chemistry', with its new terms, began to hold sway.

From then on we spoke a new language. Feeling some disloyalty to the Lunar men, I have often used modern terms such as 'sulphuric acid' or 'hydrogen', because they are easier to understand than 'vitriol' or 'inflammable air'. And I have called people 'scientists' because that expresses what they were doing in our modern terms, although the word was not even coined until the 1830s. Yet such translation marks the mental gap between our time and theirs, since language is a key to a whole way of thinking. At the time, 'science' meant knowledge; interest in the material world was 'natural philosophy'. And when people spoke of the 'arts', they did not mean only the fine arts but also the 'mechanic arts', the skills and techniques in agriculture, say, or printing. So the relationship of philosophy to the arts could mean the usefulness of natural knowledge to industry – almost the opposite of what we mean today.

We have to wrench our minds round, abandoning divisions, to think back into this age, but for me that mind-shift has been revealing. I now marvel at the way the history of technology underpins the simplest things in our lives, such as the coins in our pocket, the plate on the breakfast table and the newspaper beside it, let alone the toaster or kettle. And science has given us the great modern narrative whose stories mutate like the variants of myths. Driven by curiosity, we build and rebuild explanations for mysteries we cannot fully understand – from the spinning of the cosmos to the growth of a cell. No wonder the Lunar men seemed so powerfully seductive in their day, and so dangerous to the entrenched *status quo*. Fallible and extraordinary at once, they were, without a doubt, men who changed the world.

FIRST QUARTER

For MATTER is the dust of the Earth, every atom of which is the life.
For MOTION is as the quantity of life direct, & that which hath not motion, is resistance . . .
For the EARTH which is an intelligence hath a voice and a propensity to speak in all her parts.

Christopher Smart, *Rejoice in the Lamb* (*Jubilate Agno*), 1759–63

1: EARTH, ELSTON & ELECTRICITY

On 12 December 1731, Erasmus Darwin was born at the Old Hall in Elston, about ten miles north-east of Nottingham, the sturdy seventh child of Robert and Elizabeth Darwin. His baptism was celebrated in style with a feast for the tenants and a special beer bottled in his honour – two unopened bottles, and the menu, survive. He grew up in a house noisy with children, beneath a steep-pitched roof with crumbling Elizabethan chimneys, shaded by trees and set amid flat fields of corn and cattle.

Darwin, Boulton and Wedgwood were all born in the heart of England, all descended from 'yeomen', small landowners and farmers. They came from different sides of the Midlands, where the counties curve around the Derbyshire Peak. And in the winter of 1739–40, when Darwin was eight, all across their region the earth froze. Snow fell on New Year's Day and lay until March. Post-boys perched on coaches died of exposure in the cutting wind. In Birmingham steam from the forges clouded in ice-crystals around men's heads; in Staffordshire the potters' clay formed rigid crags; in Derby, women woke to find their breath frozen on the sheets. The trunks of ash trees split from top to toe; fish became slivers of steel and small birds fell dead from the trees, so that for three or four years their flocks were diminished. After the frost, 'there came such a cold dry, stern, cutting & backward spring, as can hardly be parallel'd'.[1] The harvest was poor and the price of grain spiralled; even the roots of the furze bushes, the fuel of the poor, froze in the iron ground.

'The Leyden Experiment'

Many families suffered – Joseph Priestley's mother died this winter, after the birth of his youngest brother.

Darwin's family, however, was well padded against the long chill, warm by their great log fires. Erasmus was a youngest child, tagging after his elders, longing to make them take notice; large and big-boned, he seemed sunny, impetuous and confident, yet he stammered all his life. His older brother Robert had several random memories of him as a small boy – mostly of disasters.[2] He had a lock of white hair after a 'blow from a maid servant by accident' when he was five. He nearly drowned when they went fishing and his brothers stuck him in a sack with only his feet poking out, then twirled him round on the bank so that he walked straight into the river. He did once catch a hare with his brother John (celebrated in a 'fol-de-rol' song by the fourteen-year-old Robert),[3] but on the whole he disliked exercise and country sports. Instead he preferred poetry and experiments, although at school he and a friend, Lord George Cavendish, had a nasty scare with gunpowder. 'These things', decided Robert, 'made a deep impression & fixt habits of precaution on a bold temper.'

In the long holidays the Darwin boys hunted and fished, lazed and read. Their nearest town was Newark, where they could watch the laden barges on the Trent, carrying coal and lead, barley and malt, cheese and pots, to the wharves of Gainsborough and Hull. There was history here too, as well as commerce. Guarding one of the main crossings over the Trent, Newark had been a great Royalist stronghold, resisting Cromwell's army to the last despite plague and siege. Nearly a hundred years later, just as its ruined castle still loomed above the river so the Civil Wars still cast their shadow. Thousands of lives had been lost in the conflict and many families divided, and after the Restoration the tension continued, as Nonconformist ministers were ejected from their parishes, Presbyterian congregations were purged and Quakers were tried and imprisoned. Yet however strongly the Tory gentry and churchmen clung to their power, from Nottingham to Birmingham the Dissenters formed tight, independent communities and in the coming generations many became leaders of industry, banking and trade.

In the Old Hall at Elston there was no thought of trade: the sons of the gentry were destined for higher things. In the autumn of 1741 Erasmus joined his older brothers at school in Chesterfield, twenty miles north on the edge of the Pennines close to the Yorkshire border. Until this generation of the 1730s the Darwin family had not produced noticeable scholars. Their forebears were Lincolnshire landowners who held minor posts under James I and Charles I.[4] Robert, Erasmus's father, had given up his law practice at Lincoln's Inn when he married at the age of forty-two; his wife Elizabeth, a Lincolnshire girl, was twenty years his junior and they had seven children in as many years.

When Erasmus was born his father was nearly fifty, still working on his law even when no business came his way. In his son's memory, 'He was frugal, but not covetous; very tender to his children, but still kept them at an awful kind of distance.'[5] Two generations later, Charles Darwin mused on the portrait of this lawyerly great-grandfather and thought he looked, 'with his great wig and bands, like a dignified doctor of divinity'.[6] Charles also suggested rather hopefully that he might have had some taste for science since he was a member of the well-known Gentleman's Society of Spalding, in the south of Lincolnshire. Indeed Robert won a mention in the Royal Society's *Philosophical Transactions* as a 'Person of Curiosity' when he gave his fellow member William Stukeley an account of a skeleton 'impressed in Stone', a rare marvel, 'the like whereof has not been observed in this island, to my knowledge'.[7] Stukeley thought it was a prehistoric crocodile, but, as if with some uncanny premonition of the evolutionary interests of future Darwins, the bones found in the rectory garden across the road from the hall turned out to be the first fossil plesiosaur found in Britain.

Robert's own interests were more antiquarian than scientific. The Spalding Society had been founded in 1710, and was unusually distinguished, with close links to the London Society of Antiquaries and members including Isaac Newton and Sir Hans Sloane.[8] Their rules stated that the Chairman of the day should have the seat by the fire, and that there should be plenty of coffee, a pot of Bohea Tea, '12 clean pipes and an Ounce of Best Tobacco', a Latin Dictionary and Greek Lexicon, and a chamber pot. Robert Darwin was not one of the most intellectual members; Erasmus described him as 'a man of more sense than learning'.[9] By contrast, his wife Elizabeth was remembered as 'a very learned lady'. Full of spirit and humour, she lived to be ninety-five and 'to the last day of her life got up to feed the pigeons'.[10]

Erasmus's move to school meant parting from the hens, and from the leather-covered books in the Elston library. With its famous twisted spire, Chesterfield was a handsome, busy place, surrounded by high moors. In the old grammar school, in this town of wealthy tanners, shoemakers and iron-masters, he knuckled down to six years of solid classical education. What excitements there were – apart from the experiments with gunpowder – came from outside. In 1745, Chesterfield and Elston, like the rest of the nation, were caught up in the panic of the Jacobite invasion. Charles Edward Stewart, the Young Pretender, landed in the Hebrides in July and took Edinburgh virtually unopposed in September. With the Government at Westminster divided and ill-prepared the Jacobite army cut swiftly through northern England and reached Derby by December. Here, while Londoners fled from the expected attack, Charles dithered. Eventually he turned back, and his army straggled north through the winter storms to be butchered at Culloden in April 1746. The ten-year-old James Watt saw his father's workshop searched, amid rumours that Bonny Prince Charlie lay concealed at Greenock.

As the Jacobites marched south, the men of the Potteries buried their money and hid their cattle in the gorse. In Derby, the family of the painter Joseph Wright, like many others, fled as the Jacobites approached. But even when the threat was near, to young men it often seemed less important than immediate things: work, clothes, love. Although William Hutton – future historian of Birmingham – was then actually living in Derby, he treated the invasion merely as an interesting aside: 'The Rebellion broke out, which provided sufficient matter for inquiry and conversation.'[11] And for Birmingham lads like Boulton the blood of Culloden meant a night on the town, with 'Bonfires, Fireworks, giving great Quantities of Ale to the Populace, and Illumination of Windows throughout the whole Town'.[12] Yet for all the future Lunar men, as for so many of the coming generation, the Rebellion was a key point in the forging of a stout, Protestant, Hanoverian nationalism, whose patriotic rhetoric would ring loud in future projects.

For Darwin the waters of daily life soon closed over the '45. His school week was full of Greek and Latin translations and exercises. It was 'tedious and insipid', he declared off-handedly to his favourite sister Susannah, already celebrated in a boyish scribble:

My dearest Sue
Of lovely hue
No sugar can be sweeter;
You do as far
Excel Su-gar
As sugar does saltpetre.[13]

In February 1749 Sue wrote to him of family and friends, cramming her current diary neatly on to the back of her letter:

Thursday, call'd up to Prayers, by my Larum; spun till Eight, collected the Hens' Eggs; breakfasted on Oat Cake, and Balm Tea; then dress'd and spun till One, Pease Porrage, Pottatoes and Apple Pye; then turned over a few pages in Scriblerus; eat an Apple and got to my work . . . red in the Tatlar and at Ten withdrew to Prayers; slept sound . . .[14]

Sue also set her brother a puzzle for Lent. A 'learned Divine' had told her that hog's flesh was fish, and had been so 'ever since the Devil entered into them and they ran into the Sea': so could she eat the meat when the family pig was killed?

A fortnight later Erasmus scrawled his reply.[15] Of course he agreed, but the story of the Gadarene swine meant pork was a 'devillish sort of fish'. On the other hand, he himself had happily eaten 'roast beef, mutton, veal, goose, fowl, &c for what are all these? All flesh is grass!' He then burst into a mock invocation to Temperance, imagining all the 'Whimsical Tribe of Phisitians' cheated of their fees. Without doctors, he thought men would still live to be a hundred; fever would be 'banished from our Streets, limping Gout would fly the land . . . and death himself be slain'.

When Darwin became a physician himself, temperance would be his key prescription. But not at seventeen. And even in adulthood his restraint did not apply to food. To Susannah's postscript, 'Excuse hast, being very cold', he responded, 'Excuse Hast, supper being called, very Hungry.' Food figured large, too, in a Pope-style Christmas verse letter to his school-friend Samuel Pegge:

Thus spoke the dying Pigg, 'Let all abroad
The bright Black-pudding smoak upon his Board:
While snaky Sausages their volumes roll,
And hiss and spit before the burning coal'.[16]

*

Darwin enjoyed his verse. But although he never forsook his Muse, by now he had set his sights on a different career, as a doctor. Medicine was at the forefront of change, as *Chambers Cyclopedia* declared resoundingly:

Medicine is become free of the tyranny of any sect, and is improved by sure discoveries in anatomy, chymistry, physics, botany, mechanics &c. See MECHANICS.[17]

This would suit Darwin. His brother Robert remembered how when young they often corresponded in verse, 'viz in Enigmas and other trivial matters; & he has often told me, if it had not been for me, he shou'd never have been a Poet'.[18] Then he added:

– he was also always fond of Mechanicks. I remember him when he was very young making an ingenious alarum for his watch; he used also to show little experiments in electricity with a rude apparatus he then invented with a bottle.

In Darwin's youth an interest in natural philosophy was an accepted attribute in polite society, like a taste for art, or music, or collecting curiosities. The vogue had grown first in court circles after the founding of the Royal Society in 1660, with its emphasis on experiment and on making new discoveries public. Slowly the interest spread and by the 1720s scientific lecturers were gathering admiring crowds in London, and soon in the provinces. Such an interest could certainly be reconciled with poetry. Brought up on the collected volumes of the *Tatler* and *Spectator*, the Darwin children were familiar with Joseph Addison's acclaim of Newton as 'the Miracle of the Modern Age'.[19] The Newtonian view of the universe, the earth and the planets filled men, said Addison, 'with a pleasing astonishment, to see so many worlds, hanging one above another, and sliding round their axles in such an amazing pomp and solemnity', amid the 'wild fields of ether' extending to infinity.[20] When Newton died in 1727, four years before Darwin was born, he was mourned as a national hero, and Alexander Pope's famous couplet hailed him as Britain's gift from God:

Nature and Nature's Laws lay hid in Night.
GOD said *Let Newton be!* and all was Light.[21]

The language of Darwin's own *Botanic Garden*, written at the end of his life, still carries echoes of these raptures.

Yet was God a clockmaker or a chemist? Few people really grasped the mathematics or even read Newton's great *Principia* of 1687, although, said Voltaire, everybody talked about him. Most people probably turned to simplified primers, yet his *Opticks* of 1704, with its speculative 'Quaeries' at the end (to which Newton kept adding until 1717), set the agenda for experimental philosophy until the mid-century. And Newton's insistence on method – on inductive reasoning, drawing general conclusions from experiment and observation – became the universal lore, while his model of gravitational order was quickly transferred to all realms of life. Whig thinkers applied it neatly, for example, to the constitution, with monarch and ministers bound by 'natural' laws of attraction:

> What made the planets in such Order move,
> He said, was harmony and mutual Love.
> The Musick of his Spheres did represent
> That ancient Harmony of Government.[22]

But was the order as stable as this verse implied? Or was life a matter of perpetual flux, of shape-shifting, marvellously varied change and alteration, as Darwin came increasingly to think?

Children learned from the world around them just as much as from books. Some of Darwin's own early mechanical experiments – like the alarm for his fob watch or clock that Robert remembered – were probably inspired like this. Derby, twenty-five miles away, was renowned for the clock-making trade and for its many scientific interests. Lectures in natural philosophy were held here from the 1740s.[23] This was a vigorously commercial town, using the local metals and minerals in instrument-making, ironwork and gem-cutting – it was the home of John Flamsteed, first Astronomer Royal, and of George Sorocold, the brilliant designer of Sir Thomas Lombe's silk mill in 1718, the first great landmark of the industrial age. The designs were based on drawings smuggled home by Lombe's half-brother John from Italy (where he was said to have been poisoned by jealous Italian workmen), and all the machinery in the mill was driven by a huge water-wheel, 23 feet in diameter. Sorocold was a friend of the steam pioneer Thomas Savery, and by 1731 the factory had its own 'fire-engine', housed in a huge block five storeys high, keeping warm air flowing to stop the silk filaments snapping.[24]

One of Darwin's future friends, John Whitehurst, the son of a watch- and clock-maker from Congleton, Cheshire, had moved to Derby in 1736 when he was twenty-three. He was already becoming known for his ingenious wind-vanes, barometers, pyrometers and especially for his clocks. And clocks had glamour. Britain was transfixed by the competition to find an accurate method of estimating longitude and in 1735, after seven years' work, John Harrison perfected the first of his famous chronometers.

Mechanical wonders entranced many children. Just as Robert recalled Erasmus's fascination, so Richard Wright remembered how his brother Joseph, the future painter,

> being of an active mind, would frequently spend his vacant time from school in going to different shops to see the men work & when he returned home would imitate their works and compleat them in a masterly manner such as joiners goods, chests of drawers, clocks, spinning wheels, guns &c.[25]

A few years later Richard Lovell Edgeworth, a son of the Anglo-Irish gentry, fell under the same spell. His passion too began in childhood, when a Dublin acquaintance brought an electrical machine to treat his mother, who was partially paralysed after a stroke. Invited to his workshop, the seven-year-old Edgeworth was thrilled by the tools and machines and globes. In this Aladdin's cave, he wrote, the 'good natured philosopher' showed him 'a syphon, and the parts of a clock; he melted some metal for me in a crucible; he explained to me the bellows, and construction of an organ'. From that moment, Edgeworth was 'irrecoverably a mechanic'.[26]

But if clocks and globes and bellows were absorbing, then electricity, another feature of Darwin's childhood experiments, was even more magical. In the 1740s the educated public learned of new discoveries in the *Gentleman's Magazine* and flocked to the spark-filled demonstrations of the public lecturers, or 'electricians'. A generation before, in 1706, Francis Hauksbee, Newton's assistant and Curator of experiments at the Royal Society, had built a strange machine, with a great wheel which twirled a whirling ball of glass, rubbed to produce the 'electrical force': after some time the inside of the glass globe shone with a strange purple, blue-green glow, and lines of light crackled like lightning within it. It must have been extraordinary – thrilling and startling – to people who had never seen

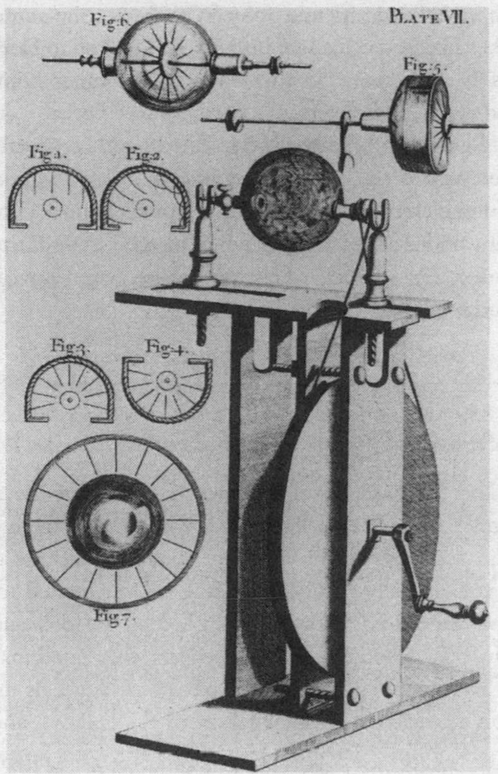

Hauksbee's electrical machine, from *Physico-Mechanical Experiments*, 1714

such a thing before.[27] From now on, discovery followed discovery. In the 1730s the British experiments of Stephen Gray on conduction and insulation, 'electric' and 'non-electric' bodies, attraction and repulsion, became widely known. In particular Gray had shown that electricity could be communicated from the rubbed glass tube, through a rod or a long wire, so that a charge could be carried for astoundingly long distances. Almost anything, it seemed, could either be a transmitter – silk, hair, glass and resin; or a receiver – ivory, metal and vegetables, 'soap bubbles, water, a map of the world, an umbrella'.[28] Most dramatically, the human body itself proved a spectacular conductor: in 1730, when Gray suspended a charity boy from a frame and touched him with his rubbed glass tube, the shivering leaves of metal placed on plates beneath shot upwards and clung to his body. From this point on, there were many famous public displays in Britain and on the Continent.

Soon the German professor G. M. Bose built a powerful machine whose massive wheel could generate static of a force unknown before. News came of one German demonstration in which a man kissed an electrified woman. 'Fire flashed from her lips in such abundance', wrote an English observer, Henry Baker, 'that they were both heartily frightened and also felt some Pain.'[29] Baker was sure that when the device reached London 'our own Country-Women will be found to have as much Fire in their Lips as well as in their Eyes as any of their Sex in *Germany*'. A report of Bose's fiery sparks reached Philadelphia in late 1745, and set Benjamin Franklin on the track of yet more electrical findings. It was held almost as irrefutable that electricity – like heat and light – was an 'imponderable', a physical but mysteriously weightless stream of minute bodies or corpuscles, shot out with great speed: a fluid, or 'effluvium'. And if it was a fluid, the real problem was how to catch it, hold it, make it portable. The German Hauksbee-type engines were massive and expensive. They were also phenomenally clumsy, with their huge wheels, and glass globes spun against the hand (if you were French), or against a leather pad (if you were British) – or against the foot, if you were a suspended German boy.

The breakthrough came by accident in 1746. Pieter van Musschenbroek, a Leyden professor, was trying to obtain 'electrical fire' from water electrified in a glass jar by a wire running into it from a conductor, which was a gun barrel hung from silk thread and charged by the rapidly spinning globe near by. The jar was carefully placed on an insulated stand, since once you had managed to lead the 'fluid' charge into any sort of container, it was thought that you had to insulate it as a precaution against it being 'leaked'. A lawyer friend of Musschenbroek, Andreas Cunaeus, tried to repeat the experiment at home: but instead of putting the jar on any base he simply held it in one hand and touched the electrified gun barrel with the other. With a world-shaking shock, he drew the charge straight into himself.[30] Cunaeus survived, but when the intrigued Musschenbroek repeated this he too found his hand 'struck with such force that my whole body quivered just like someone hit by lightning'.[31]

Abbé Nollet's experiment with the electrified boy, from *Recherches sur les causes particulières des phénomènes éléctriques*, 1749

He vowed never to attempt the experiment again. Others immediately tried the jar, issuing reports of nose-bleeds, paralysis and jolts that felt as if their arms and legs were being struck off. The little bottle upset all accepted theories; it condensed the weak electric static into a powerful shock: it could be carried from room to room, and it delivered its shocks until the charge was spent. Soon jars were ranged in a 'battery', piling up the power.

At the same time, the electricians worked out that if one man held the jar and another the conductor, they would jump with shock when they touched. The eminently serious and influential Jean Antoine Nollet, lecturer and *Académicien*, who taught natural philosophy to the French royal family, entertained the court by sending a charge through a set of guardsmen holding hands – and then topped this by placing several hundred Carthusian monks in a long line and electrifying the lot: they were said to give a 'sudden spring' when the contact was completed. Such fun apart, the simplicity of the Leyden jar meant that experiments were no longer the property of the royal and the rich. The jar was cheap, and easy to charge with a spinning globe or rubbed rod, and instrument-makers sold a range of devices, plus 'directions for gentlemen who have electrical machines, how to proceed in making their experiments'.[32] Everywhere, people tried electric shocks on themselves and their friends. (One man heard that the painter and experimenter Benjamin Wilson had made *all* his kitchen utensils 'into Leyden bottles. If so, I should not much care to dine with him.')[33]

The interest in electricity went beyond the thrill of experiment. Indeed, it aroused hot arguments on the propriety of demonstrating in public at all. Was it right to reveal these 'marvellous' effects to gaping crowds? Was electricity a material or divine emanation, 'the Soul of the World'? Were the demonstrations a manifestation of 'nature' or a new class of conjuring trick?[34] A controversy arose, dividing on party lines, with progressive-minded Whigs championing the demonstrators, and high-church Tories claiming it was blasphemy to expose God's secrets to an ignorant populace. Darwin sided with the first group. Leaving school and heading for Cambridge in October 1750, he was sure that asking questions could be nothing but good, and that the only way to find general 'truths' was from experiment. This might not be infallible, Newton had said, 'yet it is the best way of arguing which the nature of things admits of'.[35] It was the way that Darwin, for one, would argue throughout his life.

2 : TOYS

The rows over electricity made the spectacular public shows even more popular, in the provinces as well as in London. In early 1747 in Northampton you could see an electrical orrery, demonstrating the movements of the planets; in Birmingham, audiences were offered their own personal shocks from the Leyden jar.[1] And maybe, people thought, electricity was even the energy within us all. A little later Erasmus Darwin wrote to a student friend, Albert Reimarus, about a host of subjects, including wild speculations 'on the resemblance between the action of the human souls and that of electricity'.[2] Even a layman such as Matthew Boulton, scribbling comments on something he had read, wondered if electricity 'is that animal Spirit wch is secreted by the Brain & is the source of Motion and Sensation'.[3] A practical bloke, Boulton plumped soundly for material interpretations: however subtle electricity was, it had nothing to do with the soul: 'we know tis matter & there tis wrong to call it Spirit'.

Boulton briskly cast aside such 'Cymoras of each others Brain' which muffled truth. He shone with the contemporary confidence in observation. Thanks to recent work, he thought:

> we are much better enabled to say what Electy. is to know its uses & understand its Laws & propertys than the Philosophers of any preceding Age for we can both hear it see it smell it & feel it . . . We have it as much in our power as any of the other Elements we are acquainted with to experiment upon therefor let us consider it just as it appears to our senses.

Sword-hilts, from Boulton & Fothergill pattern-book, 1760s

For his part, he added, he loved electrical experiments 'and should have a great pleasure in contributing my mite to the Science but am an absolute Sceptic in it'.

When Darwin was at university in 1755, Boulton, three years older, was making a note of the books he had bought to set up his study. His list was a model in miniature of the gentleman's library, the kind of books that the Darwin children had grown up with. They included four collected volumes of the *Tatler* (with gilded spine) and eight of the *Spectator* (with frontispieces), English, Italian and French dictionaries and the complete works of Pope, Swift, Shakespeare and Locke. In the middle came more practical works such as 'Clare's Introduction to Book-keeping'. With touching pride and pleasure he was going at this full tilt, noting down everything, including 'sett of Locks for my desk' and 'Hodgkins for making my desk and wood and brickwork'. He marked the prices neatly against every item apart from a little bunch at the foot which he had obviously had for some time:

> I have on Electricity ye underneath Books: 3 Vols of Franklin's containing in all 154pp, Freke (640), Benjamin Wilson's treatise (242), Hoadly and Wilson, Simon Lovett, Benj. Martin (40), Lectures by M. l'Abbe Nollet (278), Gowin Knight on Attraction etc. (95).[4]

For Boulton, as for Darwin, these were the books that stirred his imagination – far more than the gilded *Tatler*, or the complete works of Pope.

When he made this list, Boulton was twenty-seven and had already worked in his father's business for ten years. He had no manor house, Spalding Society or classical education behind him. His technical interest arose from his trade, while his curiosity and ambition made him determined to understand every advance, even if he could not exploit it directly – and especially if he could.

He was born on 3 September 1728 in the family house in Whitehalls Lane (now Steelhouse Lane) on the northern fringes of Birmingham and named Matthew after his father. This name had originally been given to the first-born son, who had died at the age of two in 1726. He was thus a double namesake, and although he had a brother, John, and two sisters, as he grew up it was he who carried his father's hopes.

Boulton senior was a 'toy-maker', but his 'toys' were not for children – this was the general name for the wealth of small metal goods for which Birmingham was already famous. The makers provided luxury goods for the rich, but also wooed the lesser purchasers, whose spending power increased as the century rolled on, who could now afford new buttons and silver buckles, brass candlesticks and snuffers, a clock on the mantelpiece, an enamelled snuff-box in the pocket. Small traders and their wives left these prized objects in their wills, along with their new calicoes and silks and muslins. In the later eighteenth century toy-making grew into a great trade, its wealth and variety summed up by the definition of 'Toy Makers', in *Sketchley's Birmingham Directory* of 1767:

> An infinite variety of Articles that come under this denomination are made here; and it would be endless to attempt to give a list of the whole, but for the information of Strangers we shall here observe that these Artists are divided into several branches as the Gold and Silver Toy Makers, who make Trinkets, Seals, Tweezer and Tooth Pick cases, Smelling Bottles, Snuff Boxes, and Filligree Work, such as Toilets, Tea Chests, Inkstands &c &c. The Tortoishell Toy maker, makes a beautiful variety of the above and other Articles; as does also the Steel; who make Cork screws, Buckles, Draw and other Boxes: Snuffers, Watch Chains, Stay Hooks, Sugar knippers &c. and almost all these are likewise made in various metals.[5]

Birmingham was the place for a man to make a fortune. Many men, said the bookseller William Hutton, came on foot and left in chariots. (Even in the 1970s a local saying held that 'Any fool can make money in Birmingham.') Boulton senior had come here from the cathedral town of Lichfield, fourteen miles to the north, and in 1724 he married Christiana Piers from Chester and settled down to build his trade and raise a family. It was a good move. In the 1720s the town was already booming. Perched on its sandstone bluff, the new church of St Philip (with a dome modestly modelled on St Paul's Cathedral) looked down over crowded streets to the fields beyond the river Rea. Forges and workshops clustered in low-lying areas around the alleys of Digbeth and Deritend, Well Street and Corn Cheaping, while the better-off citizens moved uphill to drier land and clearer air, building new, modern houses and squares.

The metal trades depended on the plentiful supplies of coal in South Staffordshire and Warwickshire. Iron, too, had been worked in the Midlands since Tudor times in small charcoal blast furnaces high on the

An Iron Forge, engraving by Richard Earlom, 1773, after the painting by Joseph Wright of Derby

hills. In 1709 the Quaker iron-master Abraham Darby began smelting iron with coke instead of charcoal at Coalbrookdale in Shropshire but until the wood for charcoal grew scarce, ironworking was slow to change. Years later, in the early 1770s, Joseph Wright would paint nostalgic scenes of the old forges: sparks flaring from a blacksmith's anvil; a forge in a barn with a great tilt-hammer, and the owner in his striped waistcoat, folding his arms and looking on. In another painting, though, the owner is now a smart gentleman industrialist and the gap between him and the workmen is far sharper.[6] Time is moving on.

The ironworkers of the hills were a fierce, separate community, but the metalworkers they supplied congregated in the small towns around the coalfields, specializing in nails or locks, scythes or buckles or guns. Birmingham had long rung with the sound of anvils; in 1538, John Leland had written of 'many smithes in the towne that use to make knives and all mannour of cutting tooles, and many lorimers that make bittes, and a great many naylors, soe that a great part of the towne is maintained by smithes who have their iron and sea-cole out of Staffordshire'.[7] The red sandstone was good for grinding edges; the streams flowing down from the ridge turned the many water-wheels. After the Civil Wars came new trades – coining, minting and gun-making. Brass-working arrived and all the businesses were boosted by the boom in novelties and trinkets that followed the Restoration, and by protective legislation that banned the importation of buttons. In 1683 there were two hundred forges here. Six years later the French visitor Alexander Missen wrote that although he had seen fine swords, cane-heads, snuff-boxes and works of steel in Milan, they 'can be had cheaper and better in Birmingham'.[8]

The copper and brass and steel were worked into thin sheets elsewhere, and brought into the town to be 'hammered' into goods, whose variety increased as new tools were introduced: stamps and dies, the turning lathe and the drawbench. Being far from the great rivers, the town's workers concentrated on small, valuable items that could be carried cheaply across the land, especially buckles and buttons. The glittering buttons and delicate buckles, so fashionable for hats, shoes and knee breeches, could be made of iron, brass, copper or polished steel, cut into fine shapes, or covered with fabric, silver or gilding.

A city of makers and traders, Birmingham almost seemed itself to be 'in the making', always looking forward. Many people claimed that part of the reason for its growth was its freedom from rules. It had no charter to shackle it, and no ancient craft guilds to block enterprise with strict apprenticeship and trading rules. The town had supported Cromwell in the Civil Wars and many Nonconformists settled here, especially after the 1660s, when the punitive laws of the Clarendon Code banned them from worshipping in the chartered towns.[9] The Test Acts of 1673 excluded Dissenters from public office, teaching and the universities, but after the Toleration Act of 1689 allowed some public worship, meeting houses and chapels sprang up in many streets. Strong-minded and determined, the Baptists, Presbyterians and Quakers infused the place with energy.

'Freedom' was built in to Birmingham's self-image, and into Matthew Boulton's. Its citizens boasted of its industry, its independence, its bustle and its power. There were no great ironworks or manufactories: this was chiefly a town of small independent masters, rarely employing more than

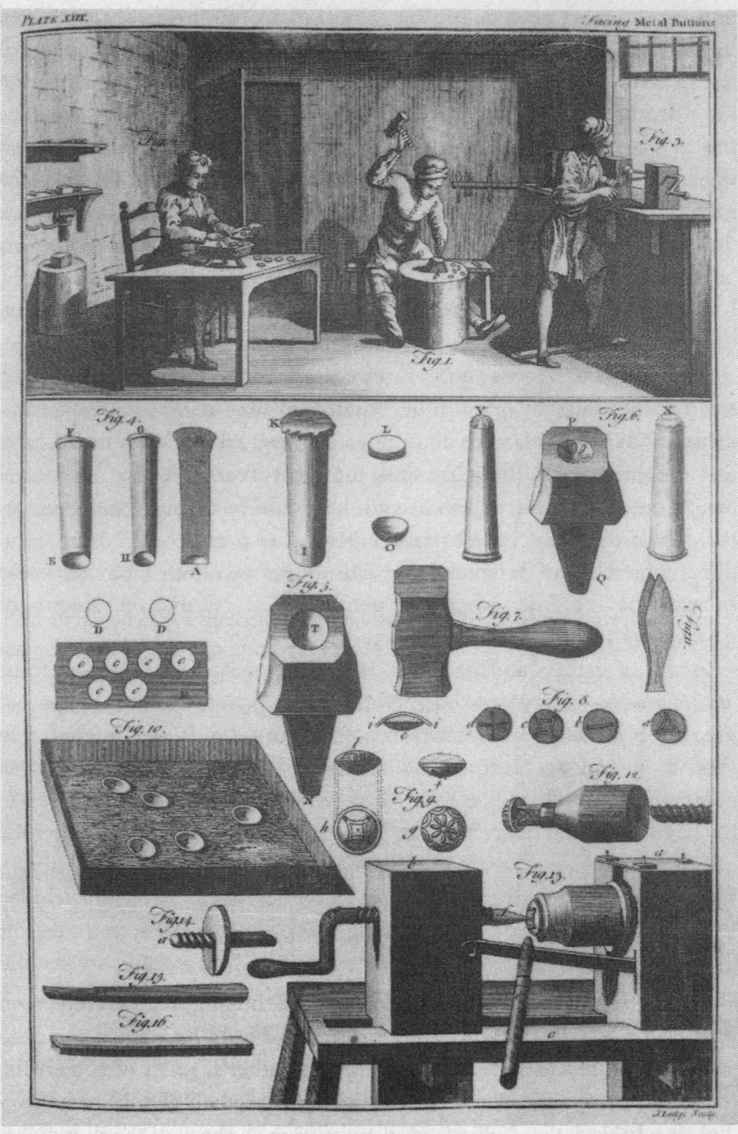

Button-making, from Croker's *Complete Dictionary of the Arts and Sciences*, 1764–66

twenty men and girls, with most of the work done by family or apprentices.[10] When William Hutton first came here as a runaway apprentice in 1740 he thought the people in the streets seemed more alert, more awake, than any he had seen. But, he wrote, 'I could not avoid remarking that if the people of Birmingham did not suffer themselves to sleep in the street, they did not suffer others to sleep in their beds; for I was, each morning by three o'clock, saluted with a circle of hammers.'[11] In countless poems and broadsheets, Vulcan – not St Philip – is the patron of Birmingham, an artist and artificer, but also a thief, and a dangerous, powerful, pagan force.

In Boulton's lifetime Birmingham would grow from a small craft town to a major manufacturing centre, its population doubling in each generation: from 15,000 in 1730 to 35,000 by 1760, reaching 70,000 by 1800.[12] And to those who watched it, the rise of Birmingham had a romance of its own. Writing his history of the town in 1785, Hutton drew breath as he reached the brink of the eighteenth century. So far, he wrote, his readers had seen the town in its infancy, growing slowly through the centuries, 'comparatively small in her size, homely in her person, and coarse in her dress. Her ornaments wholly of iron, from her own forge'. But now:

> Her growth will be amazing, her expansion rapid, perhaps not to be paralleled in history. We shall see her rise in all the beauty of youth, of grace, of elegance, and attract the notice of the commercial world. She will add to her iron ornaments the lustre of every metal that the whole earth can produce, with all their illustrious race of compounds, heightened by fancy and garnished with jewels. She will draw from the fossil, and the vegetable kingdoms; press the ocean, for shell, skin and coral; she will tax the animal, for horn, bone and ivory, and she will decorate the whole with the touches of her pencil.[13]

In reality Birmingham was noisy, dirty and chaotic. Yet its exuberant individualism and inventiveness made it just as much 'a City of the Enlightenment' as Bath, or Edinburgh, or Bordeaux.

The Boultons' business specialized in buckles, buying steel on credit from one of the big ironmongers who ran the trade. As he walked to school Matthew passed many workshops like his father's, often with a casting shop and stamping house as well as the old workshop with its hearth and bellows. In slack times, the goods were piled in the back of the shop; when overseas trade was brisk there was a rush to fill orders before the boats sailed. Birmingham makers still rode out themselves, carrying their clean linen in their bag, and a pair of pistols in their holsters, to get orders and fix prices and deal with agents from London and elsewhere. Their goods reached across England to the Continent and the colonies. As early as 1720 English brassware was sent to Holland, France, Italy, Germany, Poland and Russia, and Birmingham toys even penetrated the illustrious court of France.

On the road, the hardwaremen might join up with Manchester manufacturers, Sheffield cutlers or Staffordshire potters, travelling together as protection against highwaymen, stopping at taverns and gradually building up a complex net of friendships, deals, shared knowledge. Within the town the links were closer still. The Boultons' friends included the influential button- and hardware-maker Samuel Garbett and his partner John Roebuck, a pioneering industrial chemist, and the great printer John Baskerville. All three were important influences on Matt; all were independent-minded men, risk-takers pursuing their ends with dogged perseverance.

Garbett fought all his life for the interests of Birmingham to be represented in London. His partner Roebuck was the son of a Sheffield cutler, educated at the Dissenting Academy at Northampton and trained as a doctor in Edinburgh and Leyden. In Birmingham in the mid-1740s he worked on new methods of smelting and on producing the acids used in the trade. His laboratory became the town's first refinery, recovering gold and silver from scrap, stripping them from the base metals that covered them and Roebuck's stroke of genius was to make this on a large scale using great lead chambers. Vitriol (sulphuric acid, made with nitre and sulphur or iron pyrites) was central to this process. In 1746 he and Garbett opened a factory in Steelhouse Lane, and soon they started a bigger factory in Prestonpans, east of Edinburgh, where glassworks and salt-pans and potteries clustered along the Firth of Forth. Here the partners engaged in still more ventures, culminating in 1760 in Scotland's first major ironworks by the river Carron in Stirlingshire.

Garbett taught Boulton how to finance ambitious projects and manipulate patronage; Roebuck showed him that science could pay. His third mentor, Baskerville, demonstrated that art could be combined with experiment. He had allegedly been footman to a clergyman, employed to teach the parish boys to write, and came to Birmingham in the 1720s where he used his skill in calligraphy to design epitaphs on gravestones and to teach writing in a small school. In 1738, inheriting his father's estate, he taught himself japanning – covering metalware with layers of varnish, often decorated with pictures – and set up a workshop in Moor Street.[14] He was a good friend to Matt as he grew up, almost a second father. Caustic and witty, self-taught and obstinately independent, he was (like Boulton) also markedly 'fond of shew'. A short man, he 'delighted to adorn that figure with gold lace', and 'although constructed with the light timbers of a frigate, his movement was solemn as a ship of the line'.[15] Hutton's daughter Catherine remembered his 'cream-coloured horses, and his painted chariot, each pannel a picture, fresh from his own manufactory of japanned tea-boards'.[16] (He stuck in her memory too, 'by the token that he once took me up in his arms and kissed me'.) He succeeded without truckling to convention, living openly with his companion Sarah Eaves – whose husband had left her – braving disapproval and gossip. And he was equally disdainful of religious convention, pouring scorn on 'revelation' and shrugging off the barbs of 'the ignorant and bigoted', who were bamboozled into professing belief in 'absurd doctrines about which they have no more conception than a horse'.[17]

Matthew Boulton senior was altogether more conventional. As his toy-making prospered the family moved to Snow Hill, then a country lane running down through orchards on the north side of the city. The houses here were new, built around 1720, set back a little from the road with unusually large chimney stacks, which could allow metalwork to be done on the hearths.[18] About half a mile away a tree-lined lane led to the old manor of New Hall, the seat of the Colmore family, hidden behind iron gates at the end of an avenue of elms. New Hall's pools and trees, with their larks and cowslips, would soon be swallowed by houses, but in Boulton's childhood the parkland stretched from Snow Hill to Paradise Street, with views over open country all around.

Matt walked to school at the other side of the town, not to the grammar school, which had fallen into decline, but to an academy in Deritend run by the Reverend John Hausted, chaplain of the old St John's Chapel. His route took him past the newly laid-out churchyard of St Philip and the fine houses in Temple Row, occupied by wealthy businessmen, lawyers and professionals. Then he could run down towards the steep, narrow

High Street, scene of several fatal accidents when loaded wagons overturned on the tight corner. All around, the lanes and courts were packed with the workshops of jewellers and instrument-makers, glass-cutters and toy-makers. Markets were everywhere: butchers' stalls crammed one street, flowers and shrubs another; a double range of stalls clogged the Shambles; there was one market for cattle and another for pigs, sheep and horses. At the foot of the hill, corn and garden produce were sold in the Bull Ring in front of the old houses that ringed the parish church of St Martin's. 'Beds of earthenware lay in the middle of the footways,' remembered Hutton; while 'fruit fowls and butter were sold at the Old Cross; nay it is difficult to mention a place where they were not.'[19]

Several bookshops lay along this route. (Indeed when Matthew was seven, and still at dame school, Samuel Johnson was staying with his friend Edmund Hector at the house of Birmingham's leading bookseller, Mr Warren, struggling to translate Lobo's *Voyage to Abyssinia*, and making eyes at his future wife, Tetty Porter.) But learning was obscured by the glint of cash: the projecting eaves swung with shop signs, emblems and tavern posts. And on the low ground across the river, the road through Digbeth was lined with open forges, where sweating men and boys, and sometimes women, swung their hammers on the glowing iron bars. There were many distractions, from cock-fighting to bowling-greens. Spectacle abounded, such as the waxworks of the Royal Family which were on show in 1746, or the theatres in New Street and Moor Street where '*A tragedy called "Hamlet Prince of Denmark"*' played in 1747. There were scientific lectures, given by men such as Benjamin Martin or the Northampton engineer Thomas Yeoman, who promised the people of Birmingham that they would be 'agreeably entertained with a Variety of surprising Experiments in ELECTRICITY (that branch of Philosophy which engrosses so much Conversation everywhere, and is the Subject of so many learned debates)'.[20] And there were exhibitions of mechanical marvels such as the 'curious and unparallel'd Musical Clock' at the Wheatsheaf Inn, and a 'Grand, curious and splendid representation of the Temple of Apollo at Delphos, in Greece', displayed in a 'Machine' twelve feet high and nine wide, 'and not seen through any glass'.[21] The town and the workshop taught Boulton quite as much as the classroom. A smattering of classics stayed with him but by fifteen he had left school. At seventeen he had already developed the technique of inlaying steel buckles with enamel: these became so fashionable that they were 'exported in large quantities to France, from whence they were brought back to England and sold as the most recent productions of French ingenuity'.[22]

Matt Boulton was neat and dark and dapper, with curly brown hair, keen eyes and a broad grin. Frank and humorous, always with an eye to the main chance, he was a man on the make, like his town. But a business needed capital, and if love and money went together, so much the better; and better still if it was all kept in the family. When it came to finding a wife, it was to his family that Boulton looked. Throughout his childhood he had visited Lichfield where his maternal grandmother Elizabeth lived until her death in 1746. He had a web of relations here, many belonging to the Babingtons and the Dyotts, powerful local families. One Royalist forebear is still commemorated in a plaque in Dam Street, where Lord Brooke, General of the parliamentary forces besieging the Cathedral Close in March 1643, died 'by a shot in the forehead from M. R. Dyott, a gentleman who had placed himself on the battlements of the great steeple to annoy the besiegers'.

Now it was Boulton's turn to lay siege. On 9 February 1749, at St Mary's Church in Lichfield, he married his distant cousin Mary, the daughter of Luke Robinson, a wealthy mercer with a farm at Whittington, three miles outside the city. In one move he cleverly reconnected himself to the lost, grand side of his family (his great-grandmother, and Mary's grandfather and mother were all Babingtons) and scooped a great deal of money. At ten, Mary had inherited a substantial estate from her godmother and an additional £3,000 on her father's death in 1750 – quite enough to buy a business or a small estate. After their marriage the couple lived briefly in Lichfield with her mother, before returning to Birmingham. When Matthew was twenty-one his father made him a partner. Flying on his own optimism and the security of Mary's fortune, he now set out to make his name.

3: SCOTLAND

Matthew Boulton's career seemed dictated by his family and the trades of his home. Far to the north, the same was true of James Watt, seven years younger, growing up among the planes and lathes and wood-shavings of his father's business at Greenock, where the salt air rang with the whistle of wind in rigging. By the mid-eighteenth century, Britain was on the verge of a boom and the captains of the ships on the Clyde were eager to share it. For three generations now, adding to the old exchange with northern Europe and the Mediterranean, colonial trade had been growing. Fierce navigation laws, passed almost a century before, required that all colonial goods be sent to Britain, in British ships, before being re-exported to Europe, while the colonists could buy only from the mother country. From the southern American seaboard and the West Indies came sugar and rum, tobacco and rice and mahogany; from New England and Newfoundland came timber and fish and furs. On the back of this trade, Greenock grew from a small fishing village into a busy port. The deep-bottomed boats could sail no further up the sandbank-filled estuary of the Clyde and the triangular trade that made Glasgow rich was based here. The ships left Scotland laden with local goods for the colonies – everything from cradles to coffins – returning with tobacco, which was then re-exported to Europe and exchanged for more goods and raw materials.

When Watt was born on 19 January 1736, his father was a substantial figure, a general merchant, builder, shipwright, carpenter and cabinet-

The Old College of Glasgow and Blackfriars Chapel

maker, and part owner of several vessels. He made the first crane in Greenock for unloading the heavy, scented bales of tobacco, and into his workshop the captains brought their instruments for repair. This was the trade Watt set his heart on. Instrument-makers were the unsung heroes of the scientific revolution. The sixteenth-century burst of exploration had fostered the mathematics of navigation and the improvement of astrolabes, quadrants and compasses, while on land surveying instruments were vital to map new territories.[1] Meanwhile the clock- and watchmakers were developing their craft, and the spectacle-makers and glass-grinders were working on new optical instruments, telescopes and microscopes. Yet the theoretical aspects of their work had little status: in Cambridge in the 1630s, 'Mathematicks . . . were scarce looked upon as Academical Studies, but rather mechanical, as the business of Traders, Seamen, Carpenters, Surveyors of Land, or the like.'[2]

The ground shifted in the late seventeenth century with the new fashion for demonstration and experiment. Mathematics achieved dignity, and Newton set a new tone with his title *The Mathematical Principles of Natural Philosophy*. Precision instruments now commanded more respect, especially after Newton's disciple Willem Jacob 'sGravesande developed a magnificent range of new instruments for his lectures in Leyden. Demand soared and in London a world-famous trade grew up. Meteorologists wanted barometers and thermometers, chemists more accurate balances, surgeons more delicate forceps, lecturers more spectacular models. Rich, aristocratic collectors requested orreries or armillary spheres, beautifully made, often in brass and silver, models of the solar system in which the planets rotated round the sun within a ring engraved with signs of the zodiac.

Watt knew all about the practical application of instruments: his grandfather, who had come west from Aberdeenshire to be a 'bailie' or agent for the local landowner, was described in the Burial Register as a 'teacher of navigation', while his uncle John, a lecturer in mathematics, astronomy and surveying, was responsible for the first survey of the Clyde in 1734.[3] Among the few possessions handed down to Watt were his uncle's notebooks, and two portraits, of Newton and of John Napier, the inventor of logarithms. He himself studied astronomy and botany, and pored over his father's copy of 'sGravesande's *Elements of Natural Philosophy*, translated by Desaguliers.

As a child James was constantly ill and the cosseting by his mother Agnes, who had already lost three children in infancy, might have helped to set him on his lifelong track of hypochondria. He stayed at home until he was ten or eleven, and when he did go to school he was miserable: only when he moved to the grammar school at thirteen did he begin to shine, especially in mathematics. The workshop was Watt's retreat. His father gave him a workbench, tools and a small forge, where he made models and miniatures – tiny working cranes, pulleys and pumps, a barrel organ, a punch ladle hammered from a silver penny.[4] But in 1753, when he was seventeen, his secure Greenock home was struck a double blow: his mother died and his father's business suffered a series of losses, including a shipwreck. The following year he left for Glasgow to pursue his own trade, listing the tools and clothes he took with him in a solemn round hand: his jack and chisels and files, his ribbed stockings and ruffled shirts, his holland night cap and tartan waistcoat, his leather apron and his hat with its crêpe mourning band.[5]

Until recently, Glasgow had been a small cathedral city surrounded by hills and woods and nursery gardens but by now the town was dominated by the legendary 'tobacco lords' with their scarlet cloaks and gold-topped canes. As well as the old linen-weaving industry, there was an iron foundry and a rolling mill, a fine printing press established by the Foulis brothers in 1741 and a pottery started by two Dutch brothers in 1748. In the columned arcades beneath the Trongate's tall buildings shoemakers, silversmiths and haberdashers ran their shops, while clubs such as the Hodge-Podge, the Accidental, the Grog, the Pig, the What-You-Please, met in the taverns. Life was lively, if hardly sophisticated. James Wolfe, stationed here in 1753, wrote home: 'We have plays, concerts and balls, public and private, with dinners and suppers of the most execrable food on earth, and wine that approaches to poison. The men drink till they are excessively drunk.'[6]

Watt stayed with relations of his mother, the family of George Muirhead, Professor of Humanity at the University. Passing through the first courtyard of the 'Old College', and under the sandstone arch of the tower, you reached College Green, a meadow stretching down to a brook where new houses for the professors were being built in a row facing the church. Scottish Calvinism could inspire bigotry but it also encouraged a self-sufficient, questioning approach, and learning was seen as the key to progress. Scotland had five universities to England's two, and was proud of its up-to-date, specialized courses.

Glasgow University was quietly progressive and determinedly practical. Its Political Economy Club linked merchants, gentry and academics, while its professors often gave public lectures and acted as consultants for patrons or for the 'Board of Trustees for Fisheries, Manufactures and Improvements in Scotland'. In 1757 Francis Home, the Professor of *Materia Medica*, published the pioneering *Principles of Agriculture and Vegetation*, while William Cullen, Professor of the Practice of Medicine since 1751, and an inspired chemist, advised on bleaching, salt-boiling and alkali-making.[7] A noted character, 'known everywhere by his strange pendulous lips, huge peruke, bigger hat, big coat-flaps sticking out and huge sand-glass to measure patient's pulses',[8] Cullen was an influential teacher whose network of students would spread his ideas throughout Britain.

Among the university men whom the young Watt met and impressed was Robert Dick, Professor of Natural Philosophy, who asked him to help set up a new batch of astronomical teaching instruments. By now, his move to Glasgow had begun to seem an error, since no one there was qualified to teach him instrument-making and the 'optician' he worked for knew less than he did. Instead, Dick persuaded him to go to London, promising that if Watt's father agreed he would provide him with introductions. It seemed the only course. Preparing to leave Scotland for the first time, Watt sent his trunk ahead by sea from Leith and on 7 June 1755 he set off south with his friend John Marr, a naval instructor due to join his ship on the Thames. His father gave him two guineas, noting it carefully in his memorandum book, together with the carriage for the chest. For two weeks Watt and Marr journeyed slowly down the Great North Road, carrying their Bibles, refusing, like staunch men of the kirk, to travel on the sabbath, and tut-tutting at the ceremonies and the chattering of the clergy in York Cathedral. Finally, after crossing the Trent at Newark, near Erasmus Darwin's home, they reached London.

Many newcomers were daunted when they breasted Highgate Hill and gazed down across the smoke and the spires to the forest of masts on the Thames. Watt made his way into the maze, carrying Dick's letter of introduction to James Short, a highly regarded Scottish instrument-maker with a business on the Strand. But, unlike Birmingham, the capital was ruled by a rigid guild system and as Watt had served no formal apprenticeship, Short would not take him on, nor would the other makers. As he explained wretchedly to his father, instrument-makers were controlled by the Worshipful Company of Clock-makers, whose rules decreed that they must not employ any non-Londoners who were not already Freemen of the Company of Clock-makers, or apprenticed to one.[9] It took a month of despairing visits before he found a place with John Morgan, of Finch Lane, Cornhill.

Morgan was a master craftsman and a fine mathematician who had written a paper on the sand-glass and longitude, and had made a telescope for the King of Spain in 1752 which cost an astounding £1,200.[10] But his terms were hard: instead of receiving pay, Watt paid a fee of twenty guineas and promised full use of his services. It seemed worth it, since he hoped to pack into this single year a training that usually took four, and 'though he works chiefly in the brass way', he told his father, 'yet he can teach me most branches of the business, such as rules, scales, quadrants &c.'.[11] Watt learned fast. He moved swiftly from making rules and dividers to brass scales and quadrants and theodolites. Existing on eight shillings a week, he worked from early morning until nine in the evening, fitting in extra tasks at night until his hands shook from working. He pined for home and felt uneasy in the vast city. War with France broke out in 1756 and as he sweated in Finch Lane, lacking any official guild status, Watt was terrified of being press-ganged into the Navy, kidnapped by the East India Company for their army, or shipped off to West Indies plantations:

They now press anybody they can get, landsmen as well as seamen, except it be in the liberties of the City, where they are obliged to carry them before my Lord Mayor first, and unless one be either a 'prentice or a creditable tradesman, there is scarce any getting off again. And if I was carried before my Lord Mayor, I durst not avow I wrought in the City, it being against their laws for any unfreeman to work, even as a journeyman within the Liberties.[12]

As the winter passed his spirits rose. In April he wrote, 'I think I shall be able to get my bread anywhere, as I am now able to work as well as most journeymen, though I am not so quick as many.'[13] Three months later, despite coughs and backache and exhaustion, he felt he had mastered the craft and could even 'make a brass sector with a French joint'.[14] In July his year was up. Back in Greenock he unpacked his materials and tools and his translation of Bion's *Construction and Use of Mathematical Instruments*.

Soon his training paid off. At the end of September, when Watt was in Glasgow on business, he sent a note to his father: 'wd have come down today but there is some instruments that are come from Jamaica that Dr Dick desired that I would help to unpack.'[15] The University had been debating for some time how to pay for an observatory, and the chance came with these astronomical instruments, bequeathed by a wealthy former student, Alexander Macfarlane, who had built his own observatory in Jamaica. Many had been damaged on the voyage and Watt was asked to repair them. He was given a fee of £5 and the use of a room near the Department of Natural Philosophy in the University.

The Macfarlane collection included many fine instruments and among the curious who came to Watt's room were Professor Joseph Black and the eighteen-year-old student John Robison. Robison remembered how he thought himself a 'pretty good proficient' in mathematics and mechanics and

> was rather mortifyd at finding Mr Watt so much my superior. But his own high relish for these things made him pleased with the Chat of any person who had the same tastes with himself . . . I loung'd much about him, and, I doubt not, was frequently teazing him. Thus our acquaintance began –[16]

He reminded Watt, too, how he had met Black in Watt's rooms:

> where you was rubbing up McFarlane's instruments. Dr Black used to come in, and, standing with his back to us, amuse himself with Bird's Quadrant, whistling softly to himself in a manner that thrilled me to the heart – I tryed to imitate him.

Watt, Black and Robison became lasting friends. In 1755 William Cullen had moved to Edinburgh and, at twenty-eight, Black succeeded to his old teacher's chair. He came from a very different milieu from Watt. His father was an Ulster wine merchant in Bordeaux, his brothers were Belfast manufacturers, and his friends included enlightened aristocrats such as Lord Kames. He was tall and fair-skinned, with large dark eyes, and in later life always dressed elegantly in black with silver-buckled shoes and carried a cane or green silk umbrella. With his learning went immense

Joseph Black, from John Kay, *A Descriptive Catalogue of Original Portraits*, 1836

charm. 'The wildest boy respected Black,' wrote Lord Cockburn. 'No lad could be irreverent towards a man so pale, so gentle, so elegant and illustrious.'[17]

Black's Edinburgh MD dissertation had included the first of his historic discoveries, the isolation of 'fixed air' (carbon dioxide) from limestone. He realized that this gas – which could also be obtained by using acid on chalk, magnesia, soda and potash – was actually 'fixed in', part of their chemical make-up. He noticed, too, that it softened the harshness of caustic alkalis: when quicklime absorbed fixed air, for example, it became chalk. Robison later recalled the astounding new vistas revealed by the discovery

> that a cubic inch of marble consisted of about half its weight of pure lime and as much air as would fill a vessel holding six wine gallons . . . What could be more singular than to find so subtile a substance as air existing in the form of hard stone, and its presence accompanied by such a change in the properties of the stone?[18]

When Black took over Cullen's chair, he began to focus on heat, one of the great topics of chemistry in this century. The whisky distillers had asked him for advice on cost-cutting, and he set about investigating the heat involved in changes of state, starting with basic questions: exactly *how* did water absorb heat? Why doesn't ice melt straight away on a sunny day? He found that when ice is heated, its temperature increases to freezing and stays there until all the ice has melted; similarly, if you boil water, its temperature stays the same until it has all evaporated – though in both cases you need to add heat to keep the process going. (In today's terms: 'The heat gives the water molecules enough extra kinetic energy to escape from the surface of the water.'[19]) To Black it seemed that heat – then thought of as a chemical substance itself – was actually combining with the ice or the water, and that a definite quantity was needed to make the transformation into melt-water or steam. This 'lost' or 'hidden' heat he called 'latent heat': a formula with great implications for the future.

Watt's vision broadened in the company of Black and Robison. He taught himself German to read Leupold's *Theatrum Machinarum* and Italian for other sources. Glasgow had several flourishing student societies and he joined the Anderston Club, a discussion club which included Black, Cullen, Adam Smith (who taught at Glasgow from 1751 to 1764), the radical lawyer John Millar and the argumentative John Anderson, 'Jolly Jack Phosphorus', now Professor of Natural Philosophy, who encouraged workmen and mechanics to attend his lectures without charge. 'Our conversations then,' Watt remembered, 'besides the usual subjects with young men, turned principally on literary topics, religions, belles-lettres, &c.; and to those conversations my mind owed its first bias towards such subjects, I never having attended a college, and being then but a mechanic.'[20]

That linking of science to philosophy and literature was typical of the ethos of contemporary Scotland. Only a few years before, the mountains had run with blood in the wake of the Jacobite uprising of 1745, and, as if desperate to put this behind them, the Protestant Scots of the cities and the Lowlands turned their backs on the Highlands and the past, and looked to the south and to the future: Scotland must take the lead in the modern age. By the mid-1750s journals, books and newspapers poured from the presses and discussion clubs such as those of Glasgow flourished across the country. And while some thinkers were investigating the physical world, others were subjecting the make-up of man and society to equally fierce scrutiny. In 1739 the philosopher David Hume, exhilaratingly bold, had described his *Treatise of Human Nature* as an attempt to bring experimental methods to bear on moral subjects and 'extend our conquests over all those sciences, which more intimately concern human life'.[21] To his distress, Hume's book 'fell dead born from the press' but over the next few years his *Essays* and *Political Discourses* built his reputation and since 1751 (looking more like 'the Idea of a Turtle-eating Alderman than of a refined Philosopher') he had been Keeper of the Advocates Library in Edinburgh, the national copyright library.[22]

Among Hume's friends was the thirty-year-old Adam Smith, currently working on the lectures that would be published as the *Theory of Moral Sentiments* in 1759. Smith too appealed to underlying laws of 'Nature': the idea that prices naturally 'gravitated' to certain levels and that regulation damaged commerce, which could bring greater freedom and betterment to all. Both he and Hume set sensibility and the passions at the heart of their theories, and believed that self-interest and appetite drove economic and social growth.[23] The ideas of such men, the concerns of the Edinburgh clubs and the varied interests of Watt's circle in Glasgow would permeate the culture of the Lunar men. So much so, indeed, that at times it would seem as though Birmingham itself was an intellectual colony of Scotland.

4 : THE DOCTOR'S BAG

In these early years the lives of the Lunar men crossed like cotton threaded between pins on a map. By 1753 Erasmus Darwin was in Edinburgh, encountering many of the same men whom James Watt knew in Glasgow. Darwin, like Watt, loved machines, while Watt said he might have become a surgeon if he had not been so squeamish, as an interest in the body's mechanism was natural to an engineer.

When he left school in 1750 Darwin went with his brother John to St John's College, Cambridge. Although he won a scholarship of £16 per annum these were lean years, and he would tell his second wife that if she cut the heel out of a stocking he could put a new one in 'without missing a stitch'.[1] Still, he learned more than darning at Cambridge. He enjoyed his classics and won a name as a poet in 1751 with a flowery elegy on the death of Frederick, Prince of Wales. Ever keen on short cuts, he learned shorthand, making 170 pages of lecture notes, under headings such as 'the fossil and animal kingdoms' and 'waters, earths, metals stones; insects, fish, birds, quadruped and man'.[2] He also made a careful longhand copy of manuscripts on physic and on the pulse which had been left at the college by a former teacher, the great physician William Heberden. Fellow students borrowed this eagerly, scrawling their ironic comments on the cover: 'Damn you Darwin you have spelt a thousand words wrong, you son of a whore.'[3]

Cambridge, however, offered little at the cutting edge of medical knowledge. For that he had to go elsewhere. In early 1753 he took lodgings in

'I cure all'; the Doctor, from a contemporary broadsheet of professions and trades

London so that he could go to the anatomy lectures of William Hunter and Noah Thomas's two alluring-sounding courses at St Thomas's Hospital on salivation (the mercury and 'sweating' cure for venereal disease) and on 'acrimonious' and 'narcotic' poisons. Thomas was a good teacher but Hunter was the star, and the steeply raked benches in his class were a crush of broadcloth coats, powdered wigs, gold-topped canes. The seventh child of a farmer from East Kilbride, and one of Cullen's brilliant protégés, he had opened his anatomy school in Covent Garden in 1746. He ran courses six days a week, using a breathtaking collection of models and specimens.[4] He and his brother John eventually became the most distinguished surgeons of their day.

Later that year Darwin rode north, with his eldest brother Robert as his travelling companion, to finish his studies in Edinburgh, 'the hotbed of genius'. The Luckenbooths, where Erasmus lodged, 'at Miss Ogston's in Goldielocks Land', was a row of massive sixteenth-century buildings facing St Giles Church. All classes bumped here on the narrow common stair: sweeps and messenger boys in the cellars; merchants on the ground floor; a countess or a judge on the first; shopkeepers, dancing masters and clerks above; and artisans in the attics. Student lodgings were low-ceilinged, stuffy and cold. Oliver Goldsmith, who arrived the winter before as a medical student from Ireland, wrote mournfully, 'I have hardly any society but a Folio book a skeleton my cat and my meagre landlady.'[5]

Darwin however found plenty of society. There were cheap taverns such as Johnnie Davie's, known for its ale and its toasted ham, herring and whiting; and there were assemblies and dances, societies and drinking clubs. Among the many friendships he made here, one in particular, with James Keir, would last a lifetime; a decade later Keir would move south, to become a central figure in Lunar life, 'the wit, the man of the world, the finished gentleman who gave life and animation to the party'.[6] Four years Darwin's junior, Keir too was a youngest child, the last of eighteen. He came from a well-connected family and after his father died when he was eight his education was supervised by his uncles, the Linds, one of whom was Lord Provost and MP for Edinburgh, the other Sheriff of the County. Tall and broad-shouldered, Keir was invariably good humoured, with a gravity that could suddenly crack to reveal unexpected wit and feeling. He was amused by Darwin's jovial extravagance and remembered how he stood out among the students with his poetry, his wit, his Cambridge ways and classical background. Yet, like any student, Erasmus drank, pursued women (always his weakness) and staggered home down the dark streets before dawn. Typically, he left one bizarre, semi-scientific memory: discussing phosphorescence, he noted that the Edinburgh folk often threw their fish heads in the streets, and 'I have on a dark night easily seen the hour by holding one of them to my watch.'[7]

Edinburgh had its grander intellectual side, as well as the folk who threw fish heads. 'Here I stand, at what is called the cross of Edinburgh,' wrote one ecstatic visitor, 'and within a few minutes take fifty men of genius by the hand.'[8] Among the gowned lawyers and black-suited elders who strode through the city streets, the university men were a conspicuous group. The Scottish universities were open to Dissenters (unlike Oxford and Cambridge) and the Town Council had long ago seen that the sale of learning was a way to enrich its city. Under the brilliant mathematician Colin Maclaurin Edinburgh had become a hub of Newtonian mathematics and astronomy, and other schools flourished too, including the medical school, founded in 1726.[9] Darwin's courses included medical practice, theory and chemistry (which he and Keir studied with the well-known teacher Andrew Plummer), and clinical medicine, a very modern subject which was taught at a special ward opened at the Royal Infirmary in 1740.[10] Here the professor, John Rutherford, impressed on his students the importance of observation, noting all the signs of the patient's face – the reddened eyes, sore gums, pale skin. But diagnosis was a matter of deduction: the examinations rarely went further and looking at the rest of the body, let alone touching it, was out of the question.

This limitation was typical and indeed the progressive ideas of the Edinburgh medical school were constantly held back by the crudeness of contemporary medicine itself. In surgery, there was no anaesthetic and operations were often performed at lightning speed. As for the *materia medica* – the doctor's bag of remedies – the ingredients listed in the university's *Pharmacopoeia* contained spider's webs, Spanish flies, pigeon's blood, hoofs of elks, eggs of ants, spawn of frogs, dung of horse, pig and peacock, human skulls and mummies.[11] Until the forward-looking Cullen revised the *Pharmacopoeia* in the next decade, students still had to learn elaborate concoctions simmered in everything from Rhenish wine to treacle.

Yet if the cures were medieval the theories were firmly post-Newtonian. It was in Edinburgh that Darwin and many others first grappled with the new ideas of the body, of perception and human understanding that would mark enlightened British thought with its distinctive blend of belief both in reason and in 'sensibility'. The first Edinburgh professors had all been taught in Leyden by the powerful Hermann Boerhaave, who had applied Newtonian physics to the body, explaining health and sickness in terms of forces, weights and hydrostatic pressures; health was a matter of achieving equilibrium, balancing the pressures of internal fluids. In this teaching, a doctor does indeed sound more like an engineer, as Watt saw, or even a plumber. Keir put it well, years later, when he thought back to the narrow Boerhaavian system,

> . . . in which man was considered as an hydraulic machine, whose pipes were filled with fluids capable of chemical fermentation, while the pipes themselves were liable to stoppages or obstructions (to which obstructions and fermentations all diseases were imputed).[12]

How did Darwin, Keir wondered, manage to put this behind him and move on 'to the more enlarged consideration of man as a *living being*, which affects the phenomena of health and disease more than his merely mechanical and chemical properties'? In fact, as Keir acknowledged, even when they were students some Edinburgh teachers were beginning to 'throw off the Boerhaavian yoke'. Influenced by the Swiss physiologist Albrecht von Haller, critics were focusing less on the vascular and more on the nervous system, looking at irritability, sensibility, excitability and reflexes.[13] On the Continent, anatomy and physiology were already blending into one science, 'living anatomy', and in Edinburgh, one of the most notable proponents of this was Cullen, who also fostered an interest in fevers, epidemics and nosology – the classification of disease. From now on, Edinburgh-trained doctors would be at the forefront in the new, radical linking of health and the environment.[14]

But if the body was not in the end 'a machine that winds its own springs', as the French theorist La Mettrie put it in his *L'homme machine* of 1748, where did the life-giving force come from? Was the 'soul' a God-given facility, separate from the body, as the animists and religious dualists held? Was it a natural function of the nervous system as the 'vitalist' camp believed? Or an unconscious but active entity controlling the body, as another Edinburgh professor, Robert Whytt, declared?[15]

Darwin was fiercely interested in such arguments. Everything he learned at Edinburgh concentrated his mind on the physical and the material. Inevitably he rejected conventional religion, with its mystical Trinity, its promise of salvation and store of miracles. He himself was what one might call a Deist, developing his ideas with his close friend, Albert Reimarus, son of a noted Dutch Deist philosopher. He was still ready to accept the notion of some controlling force – a First Cause, a Being of Beings or 'Ens Entium' – but he refused to believe that this remote God had much to do with everyday life. When his father died in November 1754 he wrote to an old Cambridge friend:

> That there exists a superior ENS ENTIUM, which formed these wonderful creatures, is mathematical demonstration. That HE influences things by a particular providence, is not so evident. The probability, according to my notion, is against it, since general laws seem sufficient for that end.[16]

General laws were quite enough to 'roll this planet round the sun'. And as to life after death, no one could know: 'The light of Nature affords us not a single argument for a future state; this is the only one – that it is possible with God; since he who made us out of nothing can surely recreate us.'

In June 1755, a few months after he penned these thoughts, there came a parting of the ways. In a long Latin letter, Keir broke the news that he was leaving Edinburgh without graduating; burning to see foreign lands, he would join the Army. In the same month, Darwin took his Bachelor of Medicine in Cambridge and the following summer, after another year in Edinburgh, he presented his thesis. At twenty-four he could call himself Dr Darwin.

His entry into the world of work was markedly unsteady. He dawdled through the high-summer months at Elston until the corn was cut. Then at last he set off for Nottingham, where he took lodgings with an upholsterer and declared his practice open. It seemed a good choice as the town was busy and prosperous (and he could easily slip off to the races), but Darwin had no connections here and no introductions. The autumn wore on and still no paying patients came. To make matters worse, he became embroiled in a row with a London surgeon over an operation for a young Nottingham worker, which Darwin – who never charged his poor patients – had assumed would be free, but was not.[17]

Nottingham was a financial disaster. Darwin's sole case seems to have been the treatment of an infection after a stabbing in a brawl between two shoemakers. He prescribed medicines with alarming enthusiasm (a lasting habit) but, after rallying briefly, the unhappy cobbler expired: 'Convulsions of the muscles of the Face. Death. Dissection.'[18] Not until the post-mortem did Darwin realize that the man's stomach had been pierced. But how could he tell? he lamented. There was no vomiting, and 'what evidence had we, except the pulse'? Dead patients are not a good advertisement. In November, a month before his twenty-fifth birthday, Darwin moved on to Lichfield.

Lichfield, 'the mother of the Midlands', was the most important of an arc of old Midland towns flanking the woods and heaths of the Birmingham plateau. The cathedral stood on a long sandstone ridge to the north of the town, its three spires reflected in the series of shallow pools that divided the close from the borough, with its Guildhall and market square, parish church and fine houses of russet-coloured brick. Between city and close ran a muddy causeway. Lichfield was an important staging post on the London–Holyhead route but more important to Darwin it was the heart of a web of county families, and the cultural centre of the region. Darwin – soon comfortably ensconced with his sister Susannah as housekeeper – found that Lichfield also had a good cathedral lending library, several printers and booksellers and a strong intellectual tradition. Addison had been the son of a Lichfield Dean; David Garrick, whose brother Peter still lived near the cathedral, was now blazing across the London stage; Samuel Johnson, whose *Dictionary* was published in 1755, was the son of a town bookseller. 'Sir,' Johnson told Boswell in 1776, 'we are a city of philosophers; we work with our heads, and make the boobies of Birmingham work for us with their hands.'[19]

This time Darwin brought the vital introductions, one to a local dowager, Lady Gresley, the other to the Reverend Thomas Seward, Canon Residentiary of the Cathedral (a fellow graduate of St John's). Since the bishop lived elsewhere, Seward lived in the Bishop's Palace and

Darwin immediately found himself part of his circle. The Canon, like Darwin, had poetic pretensions and had contributed to Dodsley's *Miscellany* in 1746, including a poem on 'The Female Right to Literature', written in the persona of 'A young Lady'. He undoubtedly had in mind his own daughter Anna, a prodigy who could recite Milton at the age of three and was now writing poetry herself, with her father's encouragement.

Anna Seward was nearly fourteen when Darwin arrived. Much later, after his death, when her youthful fondness for him had long soured, she recalled her first impressions. What did he look like? He was tall, 'his form athletic, and inclined to corpulence; his limbs too heavy for exact proportion'. His pock-marked face lit up in company, but his features were heavy and his stooping shoulders and full-bottomed doctor's wig made him look twice his age. Yet he swept all before him:

> Florid health, and the earnest of good humour, a sunny smile, on entering a room, and on first accosting his friends, rendered, in his youth, that exterior agreeable, to which beauty and symmetry had not been propitious. He stammered extremely; but whatever he said, whether gravely or in jest, was always worth waiting for, though the inevitable impression it made might not always be pleasant to individual self-love.[20]

So – he could be charming but he could also be arrogant, capable of wounding irony, and always revenging opposition, 'by sarcasm of very keen edge'.

Most Lichfield people took kindly to this large, loud, generous, big-souled man, and within a few weeks – by sheer luck – he also made his name as a physician, when he cured a well-connected Staffordshire man, whose own doctor had prophesied a speedy death. Thanks to Darwin's 'reverse and entirely novel course of treatment' (its details still a mystery) the patient recovered, later to flourish as a local magistrate. At the end of the year, Darwin happily noted down his fees, adding up to £18 7s 6d for the seven weeks since his arrival.[21] As the Lichfield matrons totted up his rising income, he began to seem quite a catch. Darwin, however, eluded their nets and fell in love as impulsively as he did everything. Across the Close from the Sewards lived the widowed solicitor Charles Howard, a schoolfriend of Samuel Johnson and now Proctor in the Ecclesiastical Court, an affable, efficient man with a son Charles and a seventeen-year-old daughter Mary, known as Polly. Darwin was soon showering her with verse, giving her a copy of Dodsley's *Miscellany* embellished with a long poem of his own packed with hints of 'chequer'd Passions', 'serious Pains' and 'soft-eddying Pleasures'.[22] By late 1757 they were engaged.

On Christmas Eve, staying with friends in an old manor house, Darwin poured his whimsical passion into a letter, pretending he had found a dusty old recipe book, full of receipts for piecrust and tarts – and 'To make Love'. This, he thought, he must send to Miss Howard next post.[23] Another recipe was: 'To make a good Wife'. 'Pshaw,' Darwin continued, 'an acquaintance of mine, a young Lady of Lichfield, knows how to make this Dish better than any other Person in the World, and she has promised to treat me with it sometime.' And thus 'in a Pett', he said, he threw down the book. So he rattled on. But with the wedding imminent, he suddenly panicked. He would be home in five days, but could Mr Howard order a licence now? Then they could have the ceremony discreetly next morning at eight o'clock, 'as the Voice of Fame makes such quick Dispatch with any News in so small a Place as Lichfield'.

> Matrimony, my dear Girl, is undoubtedly a serious affair (if any Thing be such), because it is an affair for Life. But, as we have deliberately determin'd do not let us be *frighted* about this Change of Life; or, however, not let any breathing Creature perceive that we have either Fears or Pleasures upon this Occasion.[24]

Whatever the pre-wedding nerves, on the morning of 30 December 1757, Mary and Erasmus were married in St Mary's Church. Soon they moved into an old half-timbered house at the western end of the Close, the lease being taken by Darwin's patron, the formidable Lady Gresley ('Greasly', as Darwin ungratefully called her: 'Lady Blackwig' to others[25]). On 3 September 1758, eight months after the wedding, their first son was born, named Charles after Mary's father and brother. That summer Mary stayed with old Mrs Darwin in Elston, and Erasmus wrote to his 'dear Pollakin' of deaths and of a hearse rattling past the door, of gossip and love affairs, of his queer-coloured rabbits and bantam hen.[26]

Next year, the pattern was the same. He was altering the house to suit their status, turning it entirely around. The old house looked east into the Close but Darwin added a new front facing west across the road towards open fields. The new Georgian façade had fine Venetian-style windows, letting in the afternoon sun. Across the yard were stables and outhouses – space both for children and experiments. Between the new front and the road ran a deep dell, overgrown with briars, the remains of the old semicircular moat around the Close. Across this, Anna Seward remembered, Darwin 'flung a broad bridge of shallow steps with chinese paling, descending from his hall-door to the pavement'.[27] He cleared the bottom to 'lawny smoothness', made a terrace, and planted lilac and roses.

Anna was keeping a sharp eye on her new neighbours. She was immensely fond of Mary, and Darwin was still her 'poetic preceptor'. All her critical knowledge, she said, came from him, although a tactless burst of enthusiasm when he declared her verses better than her father's had led to the jealous Canon's disapproval. Her father withdrew his support, for fear, thought Walter Scott, of producing 'that dreaded phenomenon, a learned lady'.[28] From now on, Anna's writing was done in secret: in public she confined herself to music and sewing. From this point too, as she watched Mary wearied by childbirth before she was twenty, her admiration of Darwin faltered. But Darwin himself was untouched by doubt.

Darwin's house in Lichfield, drawn by Rosemary Thomas

Each year he worked out his earnings – 'The Profits of my Business' – and each year they rose: £192 in 1757; £305 in 1758; £460 in 1759; £544 in 1760.

There were some mad moments in this steady career. One, recorded with possible exaggeration by Anna forty years later, was a midsummer outing on the Trent. The day was hot and sultry and, according to the story, his party had a picnic and lashings of wine, promoting 'a high state of vinous exhilaration'. Without warning he leapt into the river, clambered up the bank and strode coolly across the meadows towards Nottingham. A dripping river-god, he was found standing on a tub holding forth to the crowd – without a trace of a stammer. – on the virtues of industry and fresh air.[29] On the whole, he steered clear of such public displays. Occasionally he did try more serious lectures, even venturing into anatomy, a subject that crystallized all the superstitions that fuelled distrust of science: at Tyburn relatives would cling to the feet of hanged men to prevent them being carried to the surgeon's hall. Executions were rare around Lichfield, but he was ready when one came, as this advertisement suggests:

> October 23rd, 1762 – the body of the malefactor, who is order'd to be executed at Lichfield on Monday the 25th instant, will be afterwards conveyed to the House of Dr Darwin, who will begin a Course of Anatomical Lectures, at four o'clock on Tuesday evening, and continue them every Day as long as the Body can be preserved; and shall be glad to be favoured with the Company of any who profess Medicine or surgery, or whom the Love of Science may induce.[30]

The love of science attracted unlikely folk – in 1768 two clergymen friends asked if he could mount another show.

Looking at Darwin as a young man, one gets the feeling that the fires were damped down, that concern for his reputation stifled his smouldering originality. Yet wherever he went, his lodgings were cluttered with tubes and wires and chemicals. The dearth of patients in Nottingham in 1756 had at least given him time to play. He made friends among clock-makers, passed on scandal about plagiarisms in John Hill's new *British Herbal* and wrote to Albert Reimarus in a tangle of English and bad Latin about the soul, about Egyptian mummies, about using spring rims on coach wheels as shock absorbers, about Benjamin Franklin and electricity. Franklin had now been working on this, among many other things, for ten years, and was best known for his work on lightning and 'pointed conductors'. Trying to understand how clouds charge themselves with lightning, he had suggested that the 'fire' was collected from the friction of salt and water in the ocean, held on the surface of clouds and discharged as a shock when the cloud met a mountain, a steeple, a tree, a ship's mast.[31] Then in May 1752, in the small village of Marly, French electricians had proved that lightning could indeed be drawn safely from the clouds by a metal rod and 'grounded'. From Poland to Portugal, church towers sprouted their new conductors.

All this intrigued Darwin, and the debates about electricity inspired his first scientific paper, which was published in the Royal Society's *Transactions* in early 1757, a few months before his marriage.[32] In this he set out to demolish a recently proposed theory that vapours, like clouds, rise and stay up because they are electrically charged. Instead Darwin argued that solar heat, cooling and expansion were enough to explain the behaviour of clouds and described an ingenious experiment to prove it.

And around this time he did meet someone who shared his interests: Matthew Boulton. The Robinsons, Mary Boulton's family, were among Darwin's patients and the two men were also admirers of John Baskerville, by now an innovative printer, to whose beautiful edition of *Virgil* they both subscribed, and friends of John Michell, an inspired natural philosopher and astronomer, whom Darwin had known in Cambridge and Boulton entertained in Birmingham.[33] But the chief bond between them was the love of invention and experiment. Very quickly they realized how they could complement each other: Darwin the university-educated theorist, Boulton the man with the technical know-how. Equally outspoken, energetic and ebullient, they were two sides of a coin. The first Lunar link had been forged.

5 : POTS

In a sketch map of Burslem in Staffordshire drawn in 1750 a broad straggle of houses runs along the top of the hill, with the church in the valley below.[1] The shops of a cobbler, a barber and a couple of butchers are dotted amid clusters of potworks and inns: the Turk's Head, the Jolly Potters, the Court House, the Bear. Between the potteries lie triangles of common land and lanes twisting outwards to meadows and crofts. A maypole is raised on the village green and on a hill named 'The Jenkins' a windmill turns. It seems a rural spot, yet the landscape was more like the moon than the English countryside, gouged with pits and humped with mounds of drying clay and towering shard-rucks of spoiled pots. And on every side great bottle-shaped kilns curved and smoked against the sky.

Wedgwoods had been potters here since the early seventeenth century, and in 1758, while Darwin was building his practice and Boulton his business, the young Josiah Wedgwood was planning to start his first pottery. Much later, looking back at the notebook he kept in early 1759, he wrote, 'I saw the field was spacious, and the soil so good as to promise an ample recompense to any one who should labour diligently in cultivation.'[2]

The agricultural image was apt. Throughout the eighteenth century the land still dictated most people's lives. Great estates, such as that owned by Lord Gower at Trentham on the edge of the Potteries, straddled the Midland counties, and if such men owned great reserves of coal, or timber,

Creamware shapes from a Wedgwood catalogue

or stone they had no snobbery about making money from them. The rich soils gave farmers a fat living and the unenclosed woodlands and heaths supported smallholders and commoners. The earth also fed the region's industry: the deposits of iron ore and lead, the limestone and flint of the hills, the brown and yellow clays of the Potteries. There were many Midland trades. In the late sixteenth century the glass-makers came, often immigrants from France, settling around the Potteries and in the Stour valley west of Birmingham. A century later, there was brewing in Burton-on-Trent; silk-weaving and ribbon-making near Coventry; framework knitting around Nottingham. From Cheshire rock salt was sent down-river to the Mersey where the flow of the tide dissolved it into brine in great lead cisterns, clouding the horizon with briny smoke.

One of the oldest of the local crafts was pottery. For two hundred years, in a handful of small towns and villages dotted along the low hills of North Staffordshire, potters had been making marbled and mottled ware from local clay, selling butterpots and pitchers and patterned plates. There were six villages in this cluster, running from north to south – Tunstall, Burslem, Hanley, Stoke, Longton and Fenton – bafflingly known later as the 'Five Towns' and now united in the rambling Stoke-on-Trent.

Ceramics had always been a mix of science, design and skill, and every good potter was in a sense an experimental chemist, trying out new mixes and glazes, alert to the impact of temperatures and the plasticity of clay. Years of trials had preceded the great breakthrough in Dresden in 1708, when the formula for porcelain was discovered, a secret held in China for a thousand years. The Royal Saxon Porcelain Company at Meissen spurred rival royal factories across Europe but porcelain from the East was still a treasure and the tons of blue-and-white ware ballasting the ships of the East India Company were never enough. The same ships brought new luxuries – tea, coffee and chocolate – which in turn increased demand for fine teacups and pots. (Both pottery and porcelain alike soon became loosely known as 'china'.) To meet the fashion, Dutch potters, particularly at Delft, had long ago begun decorating their tin-glazed pottery in the much admired blue-and-white designs, and soon English workshops followed suit. Then in 1745 the first English soft-paste porcelain (whose ingredients were more like those for glass) was made at Chelsea, and then in Bow and Derby, Staffordshire and Worcester. But only in Bristol and Plymouth was the true 'Chinese' hard-paste porcelain made, after the discovery of kaolin, the vital ingredient, in Cornwall.

Sensing the wealth to be made, a new kind of entrepreneur moved in. Apothecaries and doctors with chemical knowledge, or jewellers and goldsmiths with skills in design and modelling and a feel for the luxury trades, set up partnership with potters.[3] But fortunes did not come easily: there were huge technical and financial problems in producing both sorts of porcelain, soft and hard, and the losses in firing were tremendous. Factories foundered and bankruptcies multiplied. Yet the fashion inspired a search for something similar but cheaper and less risky, a fine white earthenware to suit 'polite' tables. There were basically two kinds of pottery. The first was earthenware, which could be made from local clay and fired at quite low temperatures (around 1000 °C), but which remained porous and easily breakable, and had to be glazed a second time before it could hold water. The other type was stoneware, in which the clay was mixed with flint and fired at a far higher heat, so that the ingredients vitrified and it became glassy and non-porous, a great advance.

Wedgwood cherished the history of his trade, the way Staffordshire potters had learned, stage by stage.[4] Each step forward gave potters a competitive edge, and they held their secrets tight. Folktale stories of chance or cunning often cloaked industrial espionage. The story of salt-glazing is typical: although the Dutch had long used soda for glazing (a technique known since the Greeks), in the mid-seventeenth century Burslem men still used the old powdered lead-glaze melted in charcoal 'hearths' high on the moors. And then in 1680, so they said, a happy 'accident' occurred: a servant on a farm near Burslem was making a salt-ley for curing pork; it boiled over, tumbling down the sides of the pot, and when it cooled it left a clear, glassy glaze: the farmer told his neighbour, a potter, and soon salt was arriving from Cheshire by the ton. In fact salt-glazing had spread slowly from Europe to London and the North. Its advantage was that only a single firing was needed, although the heat had to be intense.

New kilns were built, wide and high, able to hold a mass of ware packed into saggars, the rough clay boxes that protected the pots against uneven heat and flame and toxic fumes. Standing on scaffolding on the outside, 'firemen' emptied sacks of salt into holes high in the kilns and as the fumes circulated in the great vertical flues a glassy silicate formed on the pots, giving them a lasting sheen. Each Saturday morning a heavy white cloud hung over the countryside round Burslem, so thick 'as to cause persons often to run into each other, travellers to mistake the road; and strangers have mentioned it as extremely disagreeable and not unlike the smoke of Etna or Vesuvious'.[5]

Burslem also produced red and black earthenware. In the late seventeenth century the Dutch potter J. P. Elers settled here, working a newly discovered seam to imitate fashionable red teapots of Oriental porcelain. (So secretive was he, people said, that he laid underground pipes to detect the sound of strangers' footsteps and employed an idiot to turn his wheel.) Elers also made a strong 'Egyptian black', the precursor of Wedgwood's blackware, and other local potters often blackened their wares with 'car' found in the drainage from the coalmines, full of heavy iron oxide. The trade moved on. In the 1720s some potworks began using blue and white 'ball clay' from Dorset and Devon to make a fine white earthenware, while others made salt-glazed white stoneware, using ground flint. Then around 1740 a new process was developed, in which earthenware was fired to 'biscuit', then glazed and refired to produce a lustrous 'creamware'. With this new body – the basis of Wedgwood's later ware – the Staffordshire potters could finally attack the fashionable markets.

Josiah had to learn all the stages of his craft, beginning with the long preparation before the clay even reached the wheel. Once dug, it was weathered in piles in the fields and yards for two or three years to make it more plastic. Then it was mixed with water, traditionally in a huge, shallow outdoor pit called a sun kiln, but more often now in great 'slip-houses' which gave protection from the weather. Men with poles and paddles mixed or 'blunged' the cloudy liquid so that gravel and pebbles and heavy grains sank to the bottom. Then it was sieved and poured into heaps to dry, building up layer by layer until it could be cut into blocks and stored in a damp shelter.[6] Finally the clay was 'wedged', pounded and kneaded like bread to get rid of any air pockets which might make it fracture in the kiln. Only then was it fit for the wheel.

Most of Wedgwood's close family were involved in this trade. Two of his uncles, Thomas and 'Long John' – one an expert thrower, and the other the best 'fireman' in the district – were leading makers of salt-glazed stoneware and by the early 1740s had made a fortune, building the first house in Burslem to be roofed with slates, a square-cut three-storey building standing out on the hillside, simply known as 'The Big House'. Both seemed confirmed bachelors, and their relations were staggered and amused when first John married, at the age of fifty-three (and had six children), and then Thomas took the plunge at sixty-two. 'A VOICE this moment breaks in upon me with – NEWS, NEWS, NEWS,' wrote Wedgwood in the middle of a letter to his friend Thomas Bentley, '& what do you think it is? why truly, the Marriage writeings are making between my Unckle Thomas, and my Cousin Molly, both of venerable memory, this may serve as *a Choice drop of Comfort to Old maids & Batchelors*.'[7]

Another uncle, Richard, left the district to become a cheese-factor in Cheshire, growing wealthy enough to act as a private banker. Meanwhile, Josiah's father, a maker of 'moulded ware', inherited the Churchyard Works which had belonged to his grandfather and father before him. Josiah, baptized on 12 July 1730 in the church next door, was the youngest of thirteen children, of whom seven survived. At six, he joined his older brothers and sisters at a school in nearby Newcastle-under-Lyme run by Thomas Blunt, an ascetic man with some classical knowledge, interested in mathematics and chemistry. Anecdotes selected (or invented) with hindsight stress Josiah's dexterity and eye for design, such as his schoolmates' memories of how he borrowed his sisters' scissors to 'cut out the most surprising things' like 'an army at combat, a fleet at sea, a house and a garden, or a whole pot-work, and the shape of the ware made in it'.[8]

For three years Josiah walked to school, seven miles a day there and back, and cut out his paper armies. Then in June 1739, when he was nine, his father died. In a hasty last-minute will Thomas left the works to his eldest son, another Thomas, and £20 each to his six younger children: Margaret, John, Aaron, Richard, Catherine and Josiah. But this was wishful thinking. Old Thomas Wedgwood had been a good potter but a bad businessman, and his last small legacy was not paid until the 1770s, by Josiah himself.

It was said (though it is hard to prove) that on his father's death Josiah began straight away in the family pottery where he and his brother Richard, three years his senior, 'sat at work at the respective corners of a small room'.[9] Old stories also claimed that when he was almost twelve the smallpox struck and that he was severely ill, the infection penetrating his joints, particularly his right knee, leaving him with a marked limp. Certainly he bore the marks of smallpox into middle age, as did Erasmus Darwin, whose face was pitted with 'the traces of a severe smallpox'.[10]

On 11 November 1744, a year before the Jacobite army passed near the Potteries, Wedgwood was bound for five years as apprentice to his eldest brother. Thomas contracted to teach him 'the Art, Mistery, Occupation or Imployment of Throweing and Handleing', a sign that he was marked down to be a master-potter, and to provide him with board and lodging and clothing, 'Linen and Woollen and all other Necessaries,

The Churchyard Works, Burslem

both in Sickness and in health'.[11] In return he agreed to serve Thomas faithfully, not divulge his 'secrits' or embezzle his goods. And, as was standard, the Indenture dictated his leisure: 'at Cards Dice or any unlawful Games he shall not Play, Taverns or Ale Houses he shall not haunt or frequent, Fornication he shall not Commit – Matrimony he shall not Contract.'

The potter's wheel was usually turned by kicking a treadle, and perhaps because of the pain in his knee at fifteen he abandoned hopes of becoming a thrower and turned instead to moulding, design and experiment. He was already deeply interested in chemistry, at least judging by one notebook, annotated in later years by his secretary Alexander Chisholm as 'Experiments done before 1750'.[12] This shows that he was reading widely, copying down experiments he had come across, relating not just to pottery but also glass and metalwork and other fields. When his apprenticeship ended in 1749, he worked for his brother for three years and then left to make his own way: perhaps the Churchyard Works was too small to support two partners, or the household was disrupted when Thomas remarried – his first wife had died in 1750. Whatever the reason, Wedgwood now joined Harrison & Alders, a minor pottery in Stoke.

Josiah was too clever and too ambitious to stay with Harrison & Alders for long. In 1754, aged only twenty-four, he obtained a partnership at Fenton Vivian near Stoke, with Thomas Whieldon, eleven years his senior and a pivotal figure in pottery history. Whieldon had begun modestly by making ceramic bases for snuff-boxes for the Birmingham metal trade but was soon employing several journeymen and apprentices (one of whom, in the late 1740s, was Josiah Spode), and now owned a second factory as well as Fenton Hall and its flint mill.[13] Many of the innovations credited to Wedgwood, such as the division of labour, with men employed on different tasks – throwing, turning, handling, decorating, mixing slip – were started by Whieldon. He was the first man in the neighbourhood to rent accommodation to his workers, as Wedgwood did later, and he paid his men well, sometimes adding a shirt or a pair of shoes to make up small bills, all neatly entered in his account books: 'I am to give him a old pr. stockins, or somthing.'[14] Better still, he was a great pioneer, keen on all technical advances and a leading maker of creamware.

Each year, thousands of crates of pottery were sent down the river Weaver to Liverpool, or the Trent to Hull. Dealers in the Strand and shops across the country advertised their stock, as Mrs Ley did in Birmingham:

> . . . a choice selection of Staffordshire ware, viz., white cased stone plates and dishes both round and oval, carved sauce boats of various sorts and sizes, Dutch pudding cups, mellons, scalloped shell-shaped flummery and artichoke cups, custard cups, and all other curiosities that are made.[15]

Advertisements for Whieldon's wares appeared as far away as Boston, New York and Philadelphia.[16] Orders poured in for square sugar boxes and curvaceous ewers, tea-caddies with birds and fruit, dishes with scalloped edges, fat striped teapots and punch-pots and tall pear-shaped coffee-pots. Many were mottled wares with beautiful, semi-transparent coloured glazes; others were a deep, buttery cream, crusted with reliefs of flowers, painted with Oriental scenes or enamelled with pastoral vignettes.

Wedgwood had already won a name for new glazes and designs, and Whieldon encouraged him with a partnership agreement that let him pursue his research without divulging his methods.[17] He worked on new glazes, yellow and green. He experimented with tortoiseshell and marbled ware, where the cream body was dusted with crystals of metallic oxides such as copper, iron or manganese, which dissolved during the firing to leave streaks of blue, brown and green. He tried new forms of 'Agate', where red, yellow and white clays were wedged together or mixed in the surface slip, and by the time he left, he wrote, 'I had already made an imitation Agate which was esteemed beautiful, and made a considerable improvement.'[18] In early 1759 he began to keep notes of these trials and much later he added a note explaining that 'this suite of experiments' was begun at Fenton Hall:

> . . . for the improvement of our manufacture of earthenware, which at that time stood in great need of it, the demand for our good decreasing daily, and the trade universally complained of as being bad and in a declining condition.

There follows the first of nearly five thousand carefully recorded trials, carried out over the next thirty-five years.

Although a slump had followed the poor harvests of the early 1750s and the war with France, it was not sales that worried Wedgwood so much as staleness. The low price of white stoneware, he wrote, meant that potters could not pay attention to 'Elegance of Form'; the country was tired of tortoiseshell and even his fine agate fell victim to the weariness with variegated colours. He had to find something new. He was learning all the time and one of his mentors was Warner Edwards, a Shelton potter known for his chemical knowledge, whose 'secret partner' was a local Nonconformist minister. Similarly, Wedgwood apparently took lessons from the Reverend William Willet, the Unitarian minister at Newcastle-under-Lyme, 'a man of great mechanical ingenuity' who married Josiah's youngest sister Catherine in 1754.[19] (Wedgwood's maternal grandfather had himself been a Dissenting minister and this connection would be a central thread all his life.)

He was becoming restless at Whieldon's and only needed a push to make him move on. In 1756 he might have hoped for something from his rich, eccentric aunt Katherine Wedgwood Egerton; heiress to a large property and further enriched by three marriages, Katherine could neither read nor write but she was an extremely shrewd businesswoman. Yet she bequeathed her considerable estate, right down to the featherbed she was lying in, to Josiah's older brother Thomas, apart from small individual legacies, including £10 to Josiah himself. That £10, however small, was one spur to independence. The other was his cousin Sarah – daughter of Richard Wedgwood, his Cheshire merchant uncle. At twenty-two Sally (as Wedgwood always called her) was not beautiful, but striking, tall and slim, with pale skin, reddish hair and grey-green eyes. They had known each other since childhood but she was better educated and far wealthier, and her father was wary, supposedly telling Josiah that he would not consent until he could match Sarah's dowry of £4,000, 'guinea for guinea'.[20]

Wedgwood set out to do this. At the end of 1758 he arranged to employ his cousin Thomas, who was four years his junior and had worked in a porcelain factory at Worcester. On May Day, when Thomas's contract began, Wedgwood rented the Ivy House Works in Burslem from his uncle John of The Big House. When he reached thirty in 1760, he was employing fifteen men and boys and already laying claim to being a master-potter.

Much of his ware used the rich green and yellow glazes he had developed at Whieldon's, now applied to his popular 'greengrocery' in the shape of cauliflowers and pineapples, artichokes and melons. He also made plain creamware, some of which was painted locally or sent to be enamelled by the firm of Rhodes in Leeds. And in the early 1760s he took a more significant step, arranging for his pots to be decorated by the new transfer printing. In this revolutionary process, the craftsmen took prints from engraved copper plates, made on paper or on sheets of glue using ceramic colour, and pressed them on to the glaze. The origins of the technique are disputed but it seems to have been developed at Bow and

Chelsea and was taken up by Sadler and Green in Liverpool in the mid-1750s for use on tiles.[21] From 1761 Wedgwood was placing orders here, and from now on he and other potters brought British tables and dressers to vivid life, with flowers and sentimental love scenes, landscapes and exotic birds, monarchs and heroes and emblems.[22]

Earthenware teapot cast in the form of a cauliflower, with green glaze

It was on one of his trips to Liverpool in 1762 that Wedgwood met the man who would become his closest friend and partner. On his way, Wedgwood was forced off the road by a carriage near the Mersey bridge at Warrington and injured his weak leg in the accident. He was taken to the Golden Lion in Liverpool, where he was treated by the surgeon Matthew Turner, a keen experimental chemist. Soon Turner introduced him to a local friend, Thomas Bentley, a general merchant in the town. Bentley was exactly his age, with similar Nonconformist connections, but he was far more accomplished and confident and could open Wedgwood's eyes to new vistas. He was classically educated, had travelled on the Continent, spoke French and Italian and knew about ancient and Renaissance art; he had an elegant house in Paradise Street and was energetically involved in local life as the Liverpool Trustee for the Dissenting Academy at Warrington, co-founder of the new Octagon Chapel and later of the Public Library and the Academy of Art.[23] Brave in his beliefs, he was also a dogged opponent of the slave trade on which much of Liverpool's new wealth depended.

As soon as he was back in Burslem Wedgwood wrote rapturously to thank Bentley for his kindness, taking the liberty, he said, of addressing him as 'My much esteemed friend'; if Bentley did not think the address too free, he added, 'I shall not care how Quakerish or otherwise antique it may sound, as it perfectly corresponds with the sentiments I wish to continue towards you.'[24] Much of their talk during Wedgwood's convalescence had been about chemistry and he reported that he had since 'found time to make an experiment or two upon the Aether', and was now assaulting Turner with 'a tedious Account of Acids & Alcalies, Precipitation, Saturation &c.'

They talked on other subjects too. Indeed from the start they could discuss everything under the sun: when they met, Bentley was writing on female education, and Wedgwood asked his opinion of Rousseau's *Emile*, published that year. The firm of Bentley and Boardman soon became Wedgwood's Liverpool agents, and in years to come, Bentley would keep Wedgwood up to date with new fashions and artistic taste, as well as wooing wealthy customers.[25] Bentley gave ballast to Wedgwood's mercurial, questing mind – made him slow down and think things through – and his letters were a window on the world: 'The very feel of them, even before the seal is broke, cheers my heart and does me good,' Wedgwood wrote. 'They inspire me with taste, emulation and everything that is necessary for the production of fine things.'[26] (Sarah Wedgwood, not surprisingly, sometimes became impatient as he eyed his mail over the breakfast table.)

Wedgwood's meeting with Bentley and Turner gave him confidence to speak out on poetry and politics as well as pots. Praising James Thomson's poem 'Liberty', he commented, 'Happy would it be for this island, were his three virtues the foundation of British liberty – independent life – integrity in office & a passion for the common weal more strictly adhered to amongst us.'[27] But to create that independent life, Wedgwood looked for guidance not to the men of culture and ideas but to the practical visionaries – like the rising star of Birmingham, the toymaker Matthew Boulton.

6 : HEADING FOR SOHO

Boulton always signed his business letters 'from father and self' but by 1757 it was already clear who was the driving force. That year his father retired to the relative peace of Sarehold Farm, leased from relations of Baskerville's companion Sarah Eaves. Their trade was expanding and Boulton now paid rates for workshops on both sides of Slaney Street behind the old house on Snow Hill. He was fired by each new invention, quick to snap it up. At the start of this year, when he placed an order with Benjamin Huntsman, the Sheffield inventor of fine 'crucible steel' so perfect for watch springs, pendulums and metal-cutting tools, he added, 'I hope thy Philosophic Spirit still laboureth within thee, and may it soon bring forth Fruit useful to mankind, but more particularly to thyself, is the sincere wish of Thy Obliged Friend.'[1]

His own philosophic spirit was unbounded. From the start of the 1750s he kept a notebook, its headings ranging widely: notes on precipitation; on the temperatures of different liquids; the freezing and boiling points of mercury; the expansion and contraction of different types of cords, flax and wire and their different behaviour when wet or dry; on people's pulse rates at different ages; on sunbeams; on the movements of the planets; on how to make phosphorus and sealing wax and even 'To write in a secret manner' – a recipe for disappearing ink.[2]

Many memoranda jotted down on odd scraps of paper were concerned with his own trade, with entries on alloying copper and silver and

Soho from the Heath, by John Phillp, 1790

hardening tin, or the recipe for semilor, a gold-hued metal finish used in button gilding.[3] He was also searching for accurate measurements of heat, a universal problem at the time. Thermometers were the trickiest of all instruments and up to now most people had been content to note changes in temperature without quantifying them, but in the 1720s and 1730s three new systems were put forward, those of Réaumur, Fahrenheit and Celsius. Boulton tested both Réaumur's and 'Ferenheath's' scales, and tried making his own thermometers. In 1762 Darwin told him jokingly that a mutual friend, Dr Petit, 'desires I would use my Interest with your Worship, to procure him a Thermometer or two – now why won't you sell these Thermometers, for I want one also myself'.[4]

But Boulton was interested in such things for their own sake as well as their commercial value. As Keir wrote after his death:

> Mr B is proof of how much scientific knowledge may be acquired without much regular study, by means of a quick & just apprehension, much practical application, and nice mechanical feelings. He had very correct notions of the several branches of natural philosophy, was master of every metallic art, & possessed all the chemistry that had any relation to the objects of his various manufactures. Electricity and astronomy were at one time among his favourite amusements.[5]

Like Watt in Glasgow and Wedgwood in Liverpool, he learned most of all through friendship, as Keir acknowledged: 'It cannot be doubted that he was indebted for much of his knowledge to the best preceptor, the conversation of eminent men.'

Soon Boulton was in close contact with one of these eminent men, the Derby clockmaker John Whitehurst. As young men in their twenties, keen for knowledge and seeking mentors, both Boulton and Darwin learned much about instrumentation and invention from Whitehurst – who seemed impressively old and experienced in 1758, aged forty-five. That January, Whitehurst sent Boulton a bill for some work, adding a note announcing that he had finally built a pyrometer that pleased him: 'It has all the perfection I could wish for, and will, I think, ascertain the expansion of Metals with more exactness than any machine extant.' He would come over to Birmingham soon, he said, 'and hope to spend one day with you in trying all necessary experiments'.[6] He was full of excitement at all the subjects they were tackling, from barometers and heat expansion to the problems of stopping 'the vibration of bells in chime music'.

Boulton found such experiments irresistible. And in 1758 his hero Franklin – who also knew Whitehurst and had worked on clock design with him – came to Birmingham. The previous year, aged fifty-one, Franklin had settled in London as the agent of the Pennsylvania Assembly and had demonstrated his electrical experiments at the Royal Society. He was now touring the Midlands, partly, he said, 'to recover my health, and partly to improve and increase Acquaintance among Persons of Influence'.[7] Having made his money as a printer in Pennsylvania, he was keen to meet Baskerville, to whose *Virgil* he too had subscribed, and was tracking down his innumerable family connections, including several Birmingham buttonmakers. One was a prosperous master, another a lively old lady who looked just like his daughter, with 'exactly the same little blue Birmingham eyes'.[8] All were 'industrious, ingenious working people and think themselves vastly happy that they live in dear old England'.

Unstuffy and enthusiastic, Franklin had great appetites and varied interests; his inventions included fire-grates and water-closets as well as lightning rods; like Darwin (who admired Franklin greatly) he believed in the virtues of fresh air, advocating flinging wide your casement and baring all to the breezes. In Birmingham, his party dined at different houses and went 'continually on foot, from one manufactory to another and were highly entertained in seeing all the curious machines and expeditious ways of working'.[9] John Michell had sent a note to Boulton, introducing 'the best Philosopher of America, whom you are already very well acquainted with though you don't know him personally'.[10] With infectious excitement, Boulton swept the great man round his friends.

Boulton was now one of the electrical brotherhood. Just after Franklin's visit, the poet William Shenstone wrote from his nearby estate at the Leasowes, thanking him for introducing them. He also asked if Boulton could 'procure an electrical apparatus' for a friend, adding, 'He wishes it to be such as may effectually exhibit all the common experiments in electricity; & *This* he leaves entirely to your Judgement who are so much better vers'd in it.'[11] Within a month, Shenstone, who swore he had no technical interests, was confessing, 'I have burnt my fingers with electricity already; having told ye story of Mr Franklyn's bottling up ye Lightning, till I am thought as great a lyar as a Popish legendary.'[12] Soon he was commissioning gilded pineapple ornaments for his carriage, shoe and knee buckles and 'enamel'd stars for Horses fore-heads'.[13] Boulton quickly saw that in business his philosophic contacts could, most definitely, be put to good use.

But while Boulton's public reputation was rising he faced grief and anxiety at home. The house at Snow Hill was a sad one: in the early years of the Boultons' marriage, before 1753, three babies were born, Dorothea, Anne and Maria: all died in infancy and now Mary herself was far from well.[14] She is the silent woman in this male story: we have no picture, no letters, no reminiscences. And in 1759 she vanishes altogether. In August that year she died suddenly and was buried in the family vault at Whittington.

There is no hint as to whether Mary died in childbirth or from a fever, although some suggestive sheets lie among Boulton's early papers, covered with notes on 'Hystericks'.[15] One sheet gives the symptoms, beginning with creeping coldness and continuing to headaches and spasms: 'Hence it is that Hysterickal Women feel constriction in the throte as if strangled.' A list of cures ends in a desperate sequence of anti-epileptic remedies:

> Roots of Peiony, leaves of Lily of the Valley, Seeds of rue, misseltoe of ye Oak, or hazle, Box wood, Spt. of Black Cherry, Spt. of human blood . . . human Cranium, tooth of Sea Horse, Castoroum, peacocks dung, Camphor & ye Salt & oyl of amber. To recover a person in a fit tobaco smoak, the flowers & spts. of Sal Armoniac is good. Elks Claws is good.

But sea-horses and elk's claws were hard to find in Birmingham: nothing could help Mary. Before her funeral Boulton wrote a poem to put in the coffin, grasping at formal phrases, praising her sincerity, compassion, duty to her family, and love for him:

> If bearing many Children, & enduring many pains & illnesses,
> with patience under his Eye,
> If preserving fair Virtue around her unpolluted Bed,
> If passing through Life without one Black Spot upon her Fame,
> If these things can endear a Wife to a Husband
> Thou wert dear to me.[16]

He added a prayer for himself and for her. 'Farewell,' he ended. 'Farewell.' Mary's ghost is there in the chiselled phrases; forthright, plain-spoken, tender-hearted and stoical. Yet the tone is circumscribed: she is the one who loves to excess, while her sterling qualities only 'endear' her to Matthew.

In 1759 his father also died, and from now on, running the business on his own, he became even more ambitious, travelling to London in search of customers for his silver buckles, sword-hilts and snuff-boxes. In December he and his brother John were staying at the George in Aldersgate, 'a Noisy Hurrying Inn', preparing their assault on the Quality.[17] Boulton arranged for a friend to present a sword-hilt to Prince Edward, who wore it to the play; the Prince of Wales then ordered one and a delighted Boulton thought he could take some time off and 'indulge one day Now with some of my Philosophical Friends'.[18]

Back in Lichfield Dorothy Robinson was clearly devoted to her handsome widowed son-in-law, to the growing irritation of her own son, Luke. Within months Boulton was writing chatty letters to his 'dear Mamma' as if Mary's death had never happened. On the way back from London, he wrote of how pleased he was at getting to his journey's end 'but should feel much more so could I peep through a little hole & see you pert or could by any means be satisfied that you & your Fireside were chearfull & well . . . I shall not forget the China or Fiddle Strings or drops or any other commissions'.[19] In the margin of one effusive letter he scrawled, 'excuse my manner of writing, for you know that I write always in that Style which Floats uppermost'.[20]

Even more startling than the buoyant jollity so soon after his wife's death was a new and insistent theme – Mary's younger sister Anne. He begged Anne to write to him in London and when she did his reply showed more than brotherly affection: 'be assured that my Love and friendship for thee is as steady as the very Foundations of the Earth'.[21] A little later she was 'the source of all the real joy & happiness that does or ever will attend thy very sincere & loving Swain, Matt^w. Boulton'. Once he had a goal, Boulton fixed on it, in love as in business. As winter thawed into spring his letters flew faster and his tone grew ever warmer: 'My dear Heart', 'My Dear Charmer', 'My dear Angel'. Obstacles made him only more determined. By chance, among the gossip Darwin had passed on to

Polly in 1758 was that 'Lamb pays close addresses to Miss Robinson so that Sudal [a rival suitor] is quite turn'd off.'[22] Yet in early 1760 Boulton was writing to Anne anxiously about this same Sudal, having assured her mother 'that I was very certain you would never have him'.[23]

This spring he was in London with other Birmingham worthies, including Baskerville, giving evidence on behalf of local buckle-makers who were supporting a Bill to prohibit the export of buckle-chapes – the tongue that fastens the buckles to the strap. On 22 April 1760 he appeared before the House of Commons Committee. His stand was entirely self-interested: if the chape-makers continued to export, then foreign buckle-makers would benefit and his own trade would suffer. He was clear and to the point, arguing unashamedly for this ruthless restriction on fellow manufacturers.[24] All the time, though, he was brooding on marriage. He and Anne had a serious problem: under ecclesiastical law (though not common law) marriage with a dead wife's sister was forbidden. Anne was daunted and Matthew feverishly begged her not to torture him, swearing he would be torn to pieces by wolves before his own resolution faded, protesting that he would never live to see her in the arms of another, 'for if I do my brain will be overturned with madness, & I should do some desperate deed'.[25] As ammunition he seized on a recent pamphlet, 'Fry on Marriage', which argued that such a marriage was quite 'fit and convenient being opposed neither to law nor morals'.[26] On 22 April, the very day he was appearing before the Commons Committee, he ordered 180 copies of this tract from a Fleet Street bookseller (shrewdly getting a 50 per cent discount on his massive order). As soon as the chape-makers' Bill was dropped, he rushed back to the Midlands, his bags filled with presents and pamphlets, which he scattered like confetti.

One person, however, spurned his gifts and his tract. As Luke Robinson watched both his sisters and his mother succumb to Boulton's charm, he began, understandably, to fear that all their wealth might be sucked into Boulton's speculations. And this amounted to quite a sum. By a deed of gift, both Robinson girls stood to inherit £14,000 each from the family estate, and on Mary's death her share had passed to her sister. Anne was now heiress to £28,000 – somewhere between £1½ million and £2 million in today's money. No wonder Luke was suspicious. Dorothy Robinson was now seriously ill and Luke's hostility was fuelled by rumours. 'I find it is told thy Brother,' Boulton wrote, 'that you and I are laying our Heads together to persuade thy Mamma to give all she has from him unto us which is a most Vilinous, Enveyous & Malicious insinuation.'[27] He had to work hard in protesting that his interest in Anne was romantic, not financial. And when her mother died at the end of May his chief ally was lost. Swiftly, he sent Anne to London, where no one who knew them was around to oppose the banns. On 25 June 1760 – witnessed by the clerk and by Boulton's elder sister – they were married at St Mary's Church, among the warehouses and docks of Rotherhithe. A month later came a dressmaker's bill for a white satin négligé, with Persian sleeve linings.[28]

Twenty years after this Richard Lovell Edgeworth found himself in the same situation, hoping to marry his dead wife's sister. He appealed to Boulton, who promptly recommended he read Fry. And if he did decide to go ahead, Boulton added:

> I advise you to say nothing of your intentions but go quickly and snugly to Scotland or some obscure corner in London, suppose Wapping, and there take Lodgings to make yourself a parishioner. When the month is expired and the Law fulfilled, Live and be happy. The propriety of such a marriage is too obvious to men who think for themselves to need my comments . . . I recommend Silence, Secrecy & Scotland.[29]

After their marriage Boulton and Anne settled in Snow Hill, where he fitted up the house, finding servants and cookmaids and organizing upholsterers, painters and cabinet-makers. Anne took easily to being the wife of a leading manufacturer, ordering striped silk gowns, dresses of Italian crêpe and black silk petticoats.[30] Like her husband, she made good use of credit, staving off this bill for five years.

Boulton had need of credit. The 1750s had been fruitful for trade: Britain's population was increasing and, with rising markets at home and in the colonies, the economy started to flourish. The towns grew and spread. Old market towns and cathedral cities such as Lichfield, with their assemblies and concerts and race-meetings, were rivalled by upstarts such as Birmingham. There was ample work for builders and decorators and furnishers, and the windows of the shops were crammed with goods. For the first time people were buying on impulse, eager to be ahead of fashion.

In the newspapers advertisements blossomed like flowers for china and silver, clothes and toys, while exotic goods from the colonies now came within reach of all but the poor – silk, coffee, tobacco and tea. Stairs were carpeted for the first time and instead of looking to Turkey and France, purchasers could simply buy from the new English workshops at Wilton and Kidderminster.

The toy-makers thrived in this climate of consumption, providing everything from buttons and snuff-boxes down to the tiniest hooks for pinning jewellery to dresses, bonnets and wigs. And while the quality of its goods was often questioned, Birmingham was already known for its efficiency and cheapness. In 1759, the leading manufacturers, John Taylor and Samuel Garbett, told Parliament that 20,000 people worked in the toy trade in Birmingham and the surrounding districts. The annual output of the trade could be valued at £600,000, of which five-sixths was exported. This was a cosmopolitan business, alert to fashion and design. The town already had 'two or three drawing schools for the instruction of youth in the arts of designing and drawing', Taylor said, where 'thirty or forty Frenchemen and Germans' were working.[31]

Taylor was a role model for budding manufacturers. A real pioneer, he had broken with the old practice of 'putting-out' work, whereby a manufacturer received an order and then delegated the work to different craftsmen – to one man who made the button 'shells', another who made the thread rings, a third who made the decorated tops. Instead he had set up a 'factory', not a huge building on one site, but a system of linked, specialized workshops, and was already employing five hundred people. In 1765 he would join with Charles Lloyd, another rich local manufacturer, in founding the Birmingham Bank, which funded many local projects and became the ancestor of Lloyds Bank.

For old-established merchants such as Taylor and for young bucks such as Boulton, these were promising times. Taxes were minimal; trade drove prosperity, and commerce became a badge of patriotism. In particular the old trading war with France revived sharply in the reaction against all things Catholic and French after the '45. And while France held the lead in luxury goods, Britain was showing the way in new machines and technologies. Increasingly, French attention was concentrated on acquiring these – the French Government declaring openly that it was foolish to waste time reinventing what had already been discovered across the Channel.[32] In Britain patriotic clubs sprang up, such as the Society for the Promotion of Arts, Manufactures and Commerce, founded in 1754, which gave prizes and premiums for designs, inventions and improvements.[33]

Even when trade plummeted at the start of the Seven Years War in 1756, Boulton's business had kept growing. His letter books were packed with orders and his evidence to Parliament in 1759 already showed a knowledge of the trade in Holland, Spain and Portugal. He shared the buoyant mood of the nation at large, dazzled by the victories of this year – Pitt's *annus mirabilis*. In India Clive had driven the French down the Coromandel Coast; in the Caribbean British gains included the rich sugar island of Guadeloupe; at sea the French fleets were routed and Britain controlled the Mediterranean; in North America the fortresses on the Great Lakes were finally taken, and Wolfe's daring, fatal attack on Quebec secured the conquest of French Canada. When George II died in October 1760 and the eighteen-year-old George III took the throne – the first Hanoverian monarch to be born here and to speak English as his first language – patriotic fervour knew no bounds. Huge crowds welcomed his bride, the young Charlotte of Mecklenburg, and cheered the couple at their coronation. Briefly, helped by a fine summer with good harvests and orchards heavy with fruit, the nation was at ease: the moment felt right for new projects, new adventures.

Armed with the security of Anne's money and his own inheritance from his father, Boulton made plans. The workshop and warehouse on Snow Hill were too small and he dreamed of a site big enough to have stores of raw materials, drawing and design rooms, workshops for all stages and products, and a warehouse for finished goods. He also wanted a mill to drive machines for basic operations such as rolling and polishing and possibly lathe-turning. He got in touch with the engineer John Smeaton, an expert on watermills, and began to look for a mill of his own. Soon he approached Edward Ruston and John Eaves who held a ninety-nine-year lease on thirteen acres at Handsworth, a mile and a half north of Birmingham, just over the Staffordshire border. Their plot ran down a steep slope to the valley below, and by making a half-mile 'cut' and damming the Hockley Brook as it tumbled downhill, they had created a wide millpond, with a steep drop for the water to rush down to power their new Soho Mill. Meanwhile, on the hilltop, they replaced an old cottage with the new Soho House. In 1761 Boulton bought the lease and all the buildings for £1,000 – just about what they had spent on the improvements alone.

In the past Soho had been no more than a scrubby heath covered with heather and gorse, with nothing but rabbit warrens and the old warrener's hut. But although Boulton always romanticized his venture as creating wealth out of a desert, this waste was common land. Even though most of the Soho site was already enclosed, and his own enclosures took in only a relatively small amount of common, this could still be seen as robbing the poor of their rights. In his sixties, writing with the careless intolerance of a proud merchant prince, he dismissed this as nonsense:

> I speak from experience; for I founded my manufactory upon one of the most barren commons in England, where there existed but a few miserable huts filled with idle beggarly people, who by the help of the common land and a little thieving made shift to live without working. The scene is now entirely changed. I have employed a thousand men, women and children, in my aforesaid manufactory, for nearly thirty years past.[34]

The tumbledown cottages had gone, and the area now boasted 'hundreds of clean comfortable cheerful houses'.

This was true, but it was also a classic justification, used by a whole class. The sweeping change from the old rural economy to the 'wage-slave' economy demanded a certain iron in the blood and in 1761 Boulton firmly closed his ears to any disquiet. In February, apologizing for delay in fulfilling an order, he simply explained to the London merchant Timothy Hollis that he had just bought 'the most convenient water-mill in England for my business, which I shall not work before Lady day'.[35] When it was working it would help him to send some new patterns 'upon such terms as I presume will greatly increase our dealings'.

He set about finishing the house, so that his mother and sister Mary could move in. Then he laid out a kitchen garden and plantation of firs and built a warehouse, workshops and workmen's dwellings.[36] Most of his energy, however, went on the mill itself, which he thought inefficient and promptly demolished and rebuilt.[37] The work dragged on into the winter when he built his own brick kiln and dug a clay mine, planning to make bricks for the work in the spring and sell the surplus. All this time he still kept his house and workshops on Snow Hill, rushing between there and Soho, and as the new year came he acknowledged that he needed help. In the icy January of 1762, on a business trip to London, he was approached by John Fothergill, a Russian-born Birmingham merchant who had just separated from his employer; the partnership agreement was signed on Midsummer Day and Fothergill replaced Boulton's sister and mother in Soho House.[38]

Fothergill had been apprentice to a cousin in Konigsberg and his mercantile experience on the Continent proved invaluable, but he was a cautious man and, unlike his partner, tended to worry about money. Even before the partnership began he wrote from London to warn Boulton of the cut-throat competition. One jeweller said that Taylor and others had already been down 'like so many wolves for orders', offering him 'any encouragement he would accept of for the sake of a little business'.[39] But Boulton was sure he could beat the pack. Although several large workshops were started in Birmingham around this time, making light metal and japanned wares, in terms of innovation Soho led the way: within two years Boulton introduced fusion-plating or 'Sheffield plate' and remained its sole maker in the district for many years. Finance, however, was always the problem. He operated on a knife edge, pouring all funds back into the business and rarely having cash in hand. This was typical of the era, where the credit chain ran from manufacturers buying raw materials to wholesalers and exporters and retailers and sometimes even labourers; many rich merchants, like Taylor and Lloyd, moved gradually into banking.

To build up the business at the new Soho site Boulton needed still more cash. And once again it came through his wife. In September 1764, having suffered delirium and fits, her brother Luke Robinson died. Immediately there was a row about his will. A tattered notebook is filled with depositions of local people, among them his doctor – none other than Erasmus Darwin – who found Luke's 'pulse, body & mind' so agitated by the anxiety that Darwin 'was almost in a passion with his servants for suffering him to be disturbed'.[40] But the will was confirmed, and on Luke's death Anne inherited the whole Robinson estate. Almost at once Boulton began siphoning his in-laws' old farm-based wealth straight into his factory. With this security he could raise yet more funds: soon

Soho Manufactory

Baskerville arranged a hefty loan of £3,500 (later increased to £10,000) from the London publisher Jacob Tonson.

As soon as the money was secured, or promised, Boulton decided to build a new warehouse and workshops. Everything was done on a grand scale. The warehouse, which was known as 'the principal building', was nineteen bays wide, on three floors, with a Palladian front, a clock tower and a carriage drive worthy of a stately home. Clerks and managers and their families lived on the upper floors, and on each side were wings enclosing a great yard. The partners had their own rolling-mill below this yard, and later, when their needs increased, they rented another at Holford Mill, a couple of miles away on the river Tame.[41] The production was arranged in different workshops according to the objects made or the techniques needed.[42]

As soon as Boulton's factory opened it was cried up as a marvel of the new industrial scene. In 1765 Fothergill estimated that it could hold four hundred workers – although Boulton, in flamboyant mood, upped this to seven or eight hundred.[43] Money was invested in every possible machine for working metals and alloys, stone and glass, enamel and tortoiseshell. But if the works were magnificent, so were the bills. The cost of the principal building turned out to be £10,000 instead of the estimated £2,000: however high the turnover, it could never compensate for these costly fixed assets – a difficulty made worse by the partners' hopeless inability to cost and price their products properly.[44] By 1764, Boulton & Fothergill had losses of over £3,000. Both partners poured more money in and their constant borrowing built up a huge 'Bill Account', a major headache for years. And as the loans piled up so did the plaintive appeals from creditors: Boulton became a dab hand at persuasive, Micawberish letters.

Gradually he brought all the production under one roof. He had never really intended this, he said later, but the out-work across Birmingham was making his clerks despair and at the Christmas stock-taking in 1765, 'all our assistants declared that it was impossable almost to guard against the Losses we were so exposed to by haveing our patterns goods & materials scattered about in so many different Streets & places'.[45] He also felt that the partners had to trust their workmen more than was prudent '& therby were obliged to give em greater prices than we ought or should do if it was more under our Eyes & immediate management'.

He now decided to move to Soho House himself, ousting Fothergill, who was far from pleased at the announcement that he would be more useful overseeing the warehouse in town, meeting foreign agents and taking orders. Boulton, who was something of a man for lists, numbered his reasons indignantly over a sheaf of pages: 'As Fothergill is not of the least use in the Manufactory, if he will not live near a warehouse in Town Query of what use will he be?' (He had not even, Boulton argued, put 'one load of muck' on the garden.) By contrast, Soho needed 'a master of some resolution some knowledge of human nature & great skill & ingenuity in all mechanick arts both in theory & practice'.[46] He must live on the spot, since issues would arise hourly. But where would they find such a master?

> shall we find him among the unlettered Birmingham handycraft men or shall we find him amongst the speculative theorists whose knowledge have been drawn from Books no neither will do: but he must be one that is both . . . if B hath any claim to that character let it be him.

Having thoroughly convinced himself that he was a man of both books and business and therefore the only person fit to run things, Boulton never looked back. But the wonderful 'manufactory', which looked so fine, nearly ruined both partners: the search for new sources of income, new contacts and new ideas became even more vital.

7: INGENIOUS PHILOSOPHERS

Even when he was scurrying to sort out his factory, Boulton kept up his experiments. In the late 1750s he had been reading Réaumur on incubating eggs; Louis XV had patronized him and liked helping the chickens escape from their shells. What was good enough for a king would do for him. He tested equipment, commandeered his friends' hens, and noted when the eggs hatched.

> I placed my thermometer under Lee's Cocks wing & it rose to 103
> I then placed it under his Hens wing & it rose to 103 also
> I then placed it under the Wing of Harris's Hen & it rose to 99
> but when placed among the Eggs it rose to 96.
> June 19th, Monday, 3 o'clock placed my Thermometer in the Heat of Sun wch. rose to 121 at wch. time it stood at 72 which 49 Difference . . .[1]

Really, it was the thermometers that excited him. But the eggs were good too: the Earl of Hopetown told him that eggs were the most nutritious food of all, 'so that he thinks two with Salt & Bread only are sufficient for a Meal'.[2] Three and a half minutes, Boulton thought, should do the trick.

Once Soho was up and running, he had less time for dabbling with hens. Yet the interest in natural philosophy was like a fever; once caught there was no cure. The 1760s were alive with discoveries and as circles of friends spread in Birmingham and Scotland, Liverpool and Warrington, the sense of individuals working on their own began to give way to a feeling

Benjamin Franklin

of shared ground. Although Boulton and Wedgwood were brimming with a passion for modern methods, they also knew that innovation depended on exchange. They began to link up with others: with the men of the Dissenting academies, such as Joseph Priestley, and the 'electricians' and idealists who clustered around Benjamin Franklin in London.

These men were interested in experiment, but sometimes for very different reasons. To Joseph Priestley, for example – a member of the Dissenting circle that included Matthew Turner and Thomas Bentley in Liverpool – exploring electricity, magnetism or chemistry meant uncovering the workings of Providence in nature. Such learning opened men's eyes, allowing them to question authority and cast off the chains of the past. Knowledge was a light and a guide, a right and a weapon. And as James Thomson had shown in *The Seasons*, or Mark Akenside in his *Hymn to Science*, the divine and the scientific could share the same language:

> Science! thou fair effusive ray
> From the great source of mental day,
> Free, generous and refined!
> Descend with all thy treasures fraught
> Illumine each bewilder'd thought
> And bless my labouring mind.[3]

Priestley was always in search of this all-illuminating light. Slight and eager, with a long nose, bulging eyes and gentle mouth, he was a comical figure at first glance. He had a disconcertingly lopsided face so that his two profiles looked like two different people; he walked with 'a kind of disjointed, bird-like trot'; he talked non-stop at a rattling speed, the flow chopped into jerky waves by a terrible stammer. (He eventually tamed this by reading 'very loud and very slow every day'.[4])

In 1762 Priestley was teaching at the Academy on the banks of the Mersey at Warrington. 'I bless God,' he once stoutly declared, 'that I was born a dissenter, not manacled by the chains of so debasing a system as that of the Church of England, and that I was not educated at Oxford or Cambridge.'[5] Ironically, the Nonconformists' exclusion from the old universities, where the classics-based courses had hardly changed since Tudor times, was a great spur to British culture. Many Dissenters went to Europe or to Scotland, returning to work in the Dissenting academies, which positively welcomed the new, teaching modern languages, modern history, politics, mathematics and natural philosophy.

Priestley was born in March 1733, the son of a Yorkshire cloth-dresser, and after his mother died when he was six he was brought up by his aunt, an open-minded Presbyterian who welcomed all neighbouring preachers, however heretical, if she thought them 'honest and good men'.[6] Like Josiah Wedgwood and Erasmus Darwin, he played with experiments – his brother Timothy remembered him shutting spiders in bottles to see how long they could live. Also like Wedgwood, he was helped by a local minister, who had studied at Edinburgh and introduced him to Locke, Newton and 'sGravesande.[7] A natural scholar, he studied Latin, Greek and Hebrew and seemed destined for the priesthood, but next turned towards commerce, learning French, German and Italian and acting as translator for a merchant uncle. As a teenager he suffered consumption and his family was about to send him off to Lisbon, to work in a counting-house and breathe warmer air, but at the critical moment – suddenly frozen by visions of physical extinction and spiritual damnation – he swerved back to the idea of medicine or the ministry and enrolled at Daventry Academy. His sense of doom vanished in this liberal school, where work was sweetened by fun; blind man's buff, kissing and skating, bowling and singing, and flirtation with 'the cuddliest creature I ever beheld'.[8]

He arrived in Warrington in 1761 after a first, miserable, ill-paid job at Needham Market in Suffolk (where he was virtually ostracized as a 'furious unbeliever'), and then four happier years as minister at Nantwich in the Cheshire salt district. Here he founded a successful school, creating a lively new curriculum for girls as well as boys and sneaking in a science course which enabled him to 'purchase a few books, and some philosophical instruments, as a small airpump, an electrical machine, &c.'[9] He taught his older students to care for these and to entertain their parents and friends with experiments, with the result, he confessed, that 'I considerably extended the reputation of my school: though I had no other object originally than gratifying my own taste'.

In Warrington it was still easier to gratify these tastes. The atmosphere was lively and progressive, and the school was patronized by the local iron-masters, glass-makers, linen-weavers and rich Merseyside merchants. He made many friends: the rector John Seddon; the former student

Thomas Percival, a doctor and campaigner for public health; and John Aikin, the theological tutor. The Aikins were a prime example of the scientific and literary families found everywhere in Dissenting circles: their son John became a physician, while their daughter Anna Laetitia won fame as a poet under her married name of Barbauld. In later years her writings for children became standard fare for a whole generation, but in 1760 she was a slender, witty young woman with startling blue eyes and, although there were no official women students, 'we have a fine knot of lassies', she told a friend, 'as merry, blithe and gay as you could wish; and very smart and clever'.[10]

There were dances, card games and parties, and classes took the form of open discussion. Priestley was obviously a natural teacher – his *Chart of Biography* and *New Chart of History* were popular for decades. He was clear and direct, unless his stammer got the better of him. He asked his students round, lent them his lectures to read at home, and was delighted when they challenged him. 'I do not recollect', wrote one former student,

> that he ever shewed the least displeasure at the strongest objections that were made to what he delivered; but distinctly remember the smile of approbation with which he usually received them, nor did he fail to point out in a very encouraging manner the ingenuity or force of any remarks that were made, when they merited these characters. His object . . . was to engage the students to examine and decide for themselves, uninfluenced by the sentiments of any other person.[11]

Priestley's attractiveness came both from his openness to questioning and his deep-held visions of a better future. Growing up in a Calvinist household he had been terrified by ideas of predestination, which divided humanity into saved and damned, and he could find no 'sign' that assured him he was not destined for hell – it was now that his stutter began. 'I felt occasionally such distress of mind,' he remembered, 'as it is not in my power to describe, and which I still look back upon with horror.'[12] Slowly, he threw off this grim doctrine and looked for a more rational religion – the start of his later attacks on all dark, tyrannical authority.

One significant step was his reading at Daventry of David Hartley's *Observations on Man, his Frame, his Duty and his Expectations* (1749), a book that also influenced Darwin, Godwin and many others.[13] Hartley, a doctor and the son of an Anglican clergyman, combined some dubious physiology (based on a mechanical theory of the mind) with the ideas of Locke and

Hume – that far from being born into original sin, the child's mind was a blank, inscribed through sensation and perception and the association of ideas. At base Hartley hoped to prove that both our 'inner' and social lives were governed by Newtonian laws – a sort of moral push–pull mechanism. Pleasure and pain were the key: attraction to good and sympathy for others created happiness while their rejection caused misery: 'self-interest' was thus synonymous with benevolence. Social change would spring from 'the diffusion of knowledge to all ranks and orders of men, to all nations, kindred, tongues and people', setting humanity on a trajectory, 'which cannot now be stopped, but proceeds ever with an accelerated velocity'.[14]

Whereas Darwin was intrigued by Hartley's physiological ideas of the nervous system, Priestley responded passionately to his missionary view of learning. If people were moulded by circumstances then they could change, become 'perfected' – finally reaching the happiness intended by Providence. This necessitarian confidence resembled that of Hume and Adam Smith, who wrote in his *Theory of Moral Sentiments* that 'The happiness of mankind, as well as of all other rational creatures, seems to have been the original purpose intended by the Author of nature, when he brought them into existence.'[15] In Priestley's mind this goal merged with wider political battles for liberty and reform. The ex-Calvinist was always, in some way, looking for a new millennium: 'The morning is upon us', he wrote, 'and we cannot doubt that the light will increase, and extend itself more and more into the perfect day.'[16]

Priestley confessed that he was a materialist. As he would explain it later, mankind did not consist of two opposing principles of matter and spirit or 'soul'; the mental powers were part of the organic structure of the brain itself. When the body died that was the end. (He did manage to hang on to a belief in Judgement Day, but declared that the dead were still thoroughly dead; they would be reconstituted only by a process 'derived from the scheme of revelation'.[17]) There was certainly a God, whose plan could be mistily discerned in the workings of the universe, but no 'Holy Spirit' or divine Christ. Yet, if he was a materialist, to Priestley matter itself was far from inert. Not even molecules or atoms were solid or hard. Instead, inspired partly by his work on electricity and partly by the Jesuit theorist Boscovich, he saw matter as made up of points, centres of force in perpetual motion governed by attraction and repulsion – our feeling of 'solidity' was simply resistance to force.[18] In this eighteenth-century pre-vision of atomic physics, all matter was energy, and thus in some sense spiritual. Religion and politics and science swept together. Around all his experiments and ideas hovered a shimmering aura of transformation.

Sometimes Priestley can sound like Voltaire's Pangloss in his belief that all is for the best. His landlord taught him to play the flute and although he was never any good, he enjoyed it for years, recommending it to all 'studious persons' especially if they had no ear for music, since they would be 'more easily pleased, and be less apt to be offended when the performances they hear are but indifferent'.[19] Life in Warrington did seem to offer all the cards to this eternal optimist. Among his former Nantwich pupils was William Wilkinson, the son of a Wrexham iron-master (he and his brother John would themselves, in time, become the wealthiest iron-masters of the age). They had a younger sister, Mary, now seventeen and one of Anna Laetitia Aikin's close friends among the 'knot of lassies'. In June 1762, just before Priestley's thirtieth birthday, he and Mary were married. Like her fierce-tempered brothers, she spoke her mind, had a penetrating glance and startling smile: she was also well read, generous spirited and a fiendish organizer. With a happy sigh, Priestley admitted that 'greatly excelling in every thing related to household affairs, she entirely relieved me of all concern of that kind', so he could give all his time to his studies and 'other duties of my station'.[20]

These duties apparently embraced long visits to friends, and he often visited Bentley in Paradise Street, sitting up late to talk. Here he met Wedgwood and Turner, and helped to arrange for Turner to teach chemistry and anatomy at the Academy (later admitting that he knew almost no chemistry before he went to Turner's lectures). Riding a new wave of enthusiasm he now decided that experimental philosophy needed a written history, to show how the discoveries of nature were linked to social progress. Why shouldn't he write one himself? Or at least begin one – if it was beyond him, others could take it up. With the same confidence that Boulton and Wedgwood applied to business empires, Priestley attacked science – nothing was impossible. He would begin with electricity, about which he knew most. In 1765, he launched himself on London, armed with a letter of introduction from the rector, John Seddon. Here he found a second support network, the 'electricians', who were linked to what

Benjamin Franklin called 'the Club of Honest Whigs', a philosophical and political club that met at St Paul's Coffee-House. Priestley soon became close to several members of this group: Franklin, the veteran electrician William Watson, John Canton, and the mathematician and theologian Richard Price, who took him to the Royal Society. On his return (telling Canton that the visit seemed like 'a pleasing dream'), he tackled his book.[21]

He tried out disputed experiments before he wrote about them, revving the handles of his electrical machines for hour after hour until he was completely 'fatigued with the incessant charging of the electrical battery and stunned with the frequent report of the explosion'.[22] His brother Timothy – a carpenter as well as a Calvinist minister – came over from Manchester to help him build equipment, including a large kite for bringing electricity down from thunder clouds. This was six feet wide, said Timothy, but 'Joseph could put the whole thing in his pocket, for the frame would take to pieces, and he could walk with it as if he had no more in his hands than a fishing rod'.[23] The kite worked fine. Priestley added a clanking chain to earth it (which protected the experimenter but proved fatal to a curious goose), and his London friends sent a thermometer, hygrometer and barometer so that he could make meteorological soundings when he flew it. In June 1766, to help the reception of his book, they proposed his election to the Royal Society.

Priestley pressed on. In September 1767 he moved from Warrington to Leeds, to become minister at Mill-Hill Chapel, and that autumn the *History and Present State of Electricity* was published. Its success was immediate. Priestley's style was vivid, his account was accurate, his information was fresh. Some of his findings had important consequences, especially his linking of electricity and chemistry when he discovered that charcoal was an excellent conductor. And although mathematics and theory were not his strongest points, he had some shrewd hunches, including his suggestion that electrical attraction worked according to the same law as gravity.[24]

His style was almost as significant as his content. His whole approach sprang from his belief that knowledge could be a democratic weapon to fight obscurantism and tyranny and he put this forcefully in his stirring preface to *Experiments and Observations* in 1772:

The rapid progress of knowledge, which like the progress of a wave of the sea, or of light from the sun, extends itself not in this way or that way only, but in all directions, will, I doubt not, be the means, under God, of extirpating all terror and prejudice, and of putting an end to all undue and usurped authority in the business of religion as well of science; and all the efforts of the interested friends of corrupt establishments of all kinds, will be ineffectual for their support in this enlightened age.[25]

Priestley chose to write a 'history' rather than a treatise because he felt this was a form that showed knowledge forever advancing, telling a broad, onward-driving story that would arouse 'sublime' emotions.[26] In writing his book he supplied his own neat drawings and described his own experiments, failures or not. Although he realized the admission of failure was 'less calculated to do an author honour', he hoped, he said, that by showing that no special genius was required for experiments, others might be encouraged to have a go. Even if he looked a fool he could 'contribute more to make other persons philosophers, which is a thing of much more consequence to the public'. And his political agenda also underlay his choice of form. Whereas works of fiction were like mechanical demonstrations 'such as globes and orreries', offering limited models of human ingenuity:

real history resembles experiments by the air pump, condensing engine and electrical machine which exhibit the operation of nature, and the God of nature himself. The English hierarchy (if there be anything unsound in its constitution) . . . has equal reason to tremble even at an air-pump, or an electrical machine.[27]

Bentley and Wedgwood shared Priestley's democratic ideals, and Bentley also helped with his experiments, telling Wedgwood at one point that perhaps electricity might be used in decorating pottery. Wedgwood replied that he would be pleased to assist in any way towards rendering Priestley's experiments 'more extensively usefull', but, he teased:

What daring mortals ye are! to rob the Thunderer of his Bolts, – and for what? – no doubt to blast the oppressors of the poor and needy, or to execute some public piece of justice in the most tremendous and conspicuous manner, that shall make the great ones of the Earth tremble! . . . But peace to ye mortals! – no harm is intended – Heaven's once dreaded bolt is now called down to amuse your wives and daughters – to decorate your tea-boards and baubles.[28]

He would be happy to work with them, and wished all success to the Doctor's 'delightfull and ingenious researches into the secrets of nature'.

*

In Birmingham – with no theological baggage of any kind – Boulton had also been tinkering with electricity. In 1760 he worked with Franklin, describing their efforts as an attempt to seal the Leyden jar to prevent leakage of electrical 'fluid' (although electricity cannot 'leak'), and soon he had two electrical machines and was even receiving enquiries about using electricity to cure rheumatism.[29] Franklin also showed off his pet invention, his famous 'glassychord', a mechanized version of musical glasses[30], and, from time to time, Boulton would supply him with new glass.

Boulton basked in vicarious fame. He met experimenters from across Britain, like the Scottish astronomer James Ferguson, an associate of both Franklin and Whitehurst, who lectured in Birmingham in 1761. The town saw plenty of such lectures – four years later a set of twelve talks covered everything from electricity to lenses, hydraulics to magnets.[31] Almost every second man, it seemed, was working on new processes or tools, leading to a rash of patents.[32] And the most famous Birmingham scientist–industrialist of them all was Samuel Garbett's partner, John Roebuck. The sulphuric acid plant at Prestonpans was flourishing and in 1760 Roebuck and Garbett opened their ironworks at Carron. Skilled workmen came up from Coalbrookdale, Smeaton designed machinery, and in the first year they melted 1,500 tons of iron. In Birmingham around this time Roebuck and Boulton were working together on thermometers. 'Mr Bolton & I have spent many Pounds in the structure of thermometers but have not yet finished any which we are thoroughly satisfied with,' Roebuck told John Seddon of Warrington Academy (Roebuck was the Birmingham Trustee for the Academy, as Bentley was for Liverpool) 'as soon as ever we finish any we shall remember our promise to yourself and also return your own.'[33]

Another elder statesman among Birmingham experimenters, of course, was Boulton's old friend Baskerville. In 1760 he was fifty-four and Boulton thirty-two, yet they shared a cast of mind that crossed the generations, common to many innovators of the time – Wedgwood is another example. They were 'gaffers', masters who could do all the tasks their men did but were quick to buy in skills they needed. They combined an equal passion for design and techniques, and were imaginative experimenters who persisted and persisted in their drive for perfection.

Perhaps it was because they were self-taught that such men seemed to think that anything they set their heart on was achievable. Baskerville had worked for years, challenging a notoriously conservative craft, before he printed his ground-breaking *Virgil* of 1757. He not only designed new typefaces but cast the type and set it and improved printing-press design, paper-making and ink-making. Many of his experiments had the lateral-thinking quality that marked the Lunar circle – an *ad hoc*, quick readiness to seize the potential of things near to hand, to test the properties of everything they stumbled across, whether it be rocks, metals, acids or tools. Baskerville's lustrous, oily, near-purple ink gained its unique colour from being mixed with 'fine-black', soot collected from the glass-pinchers' and solderers' lamps; his paper's prized glaze came from 'hot-pressing', a mysterious process probably based on a technique from his japanning work.[34]

Despite phases of depression, Baskerville won through. So did another flamboyant pioneer, Henry Clay, an apprentice of Baskerville's who would make a fortune in the 1770s by patenting hard papier mâché, varnished for use in panels, screens, tables, even the roofs of coaches. Other inventors were less lucky. One of Boulton's neighbours in Snow Hill was John Wyatt, the eldest of eight brothers, fathers of a branching dynasty of inventors, sculptors and architects. A mechanical and mathematical genius, Wyatt saw the future – from the 1730s he worked on machines to replace manual labour – but he was absolutely hopeless at business. First he invented a machine to cut files (sold to a Birmingham gunsmith who went bankrupt); then, with Lewis Paul, he developed the first machine for spinning cotton yarn, using rollers revolving at different speeds. So fast did this run that it devoured the slivers of cotton before the carders could supply them by hand, so Paul went on to invent the first carding machine. With the support of two unlikely projectors, Johnson's friend the bookseller Thomas Warren and Edward Cave of the *Gentleman's Magazine*, they set up a mill in Birmingham and then another in Northampton, employing fifty people – often called the first textile factory. This too failed, although twenty-five years later, in 1769, Richard Arkwright developed their ideas, and made a fortune. Wyatt also designed the first suspension bridge (dismissed as an airy fantasy), drew plans for piped water supplies, ball-bearings and new types of harpoon, and his weigh-bridge for loaded carts was snapped up by corporations keen to stop fraud at markets. A fraction of the royalties on this would have made him rich. Yet in February 1760, the father of six children, he wrote sadly to Boulton, 'I am on the brink of Ruin . . . I am sorry to give you this trouble but if I attempt to speak to this purpose the subject chokes me.'[35] Boulton made him his foreman until his death in 1766, and later took on his two sons, John and Charles. But if he proved a generous friend to the family, he also bought a good mind very cheap.

Outside the smoky rush of Birmingham, Boulton's strongest friendship was still with Darwin. In the calm shade of Lichfield Cathedral, Darwin was prospering. He finished his house, with its airy rooms and curving staircase. He shone in the Seward circle. He took inordinate pride in his sons – Charles was four in 1762, and Erasmus a sturdy three-year-old. The only shadow was Polly's poor health, which he briskly prescribed for. And his reputation as a doctor was still growing. In October 1762 he was confidently advertising his 'Anatomical Lectures' in the *Birmingham Gazette*, the corpse dwindling, presumably, after each gory session.[36] In the same month he sent Boulton Dr Petit's request for some thermometers, adding that Petit 'desires you'll write a Paper and become Member of the R.S.'[37] He himself had written a second paper for the Royal Society's *Transactions* in 1760, about a patient who woke in the night spitting blood. His Boerhaavian diagnosis was that while the man slept his lungs were not 'sufficiently sensible' to drive the circulation, so the blood gathered, putting pressure on the blood vessels. His proposed remedy was simply to wake the patient in the middle of the night, before the tiny veins in his lungs ruptured.[38] On the strength of his two papers Darwin became a Fellow of the Royal Society in 1761. He liked the cachet, but never signed the Charter Book and did not visit London for another twenty years.

Boulton was often in Lichfield and Darwin occasionally came to Birmingham. In 1763 both men may have seen the 'Microcosm' which toured the Midland towns, an embellished orrery which showed the planets, 'with a wonderful variety of Moving Figures, Landscapes, &c, &c'.[39] Darwin watched Boulton's business grow with amusement and some envy. In July 1763 he wrote wryly, 'As you are now become a sober plodding Man of Business, I scarcely dare trouble you to do me a Favour in the nicknachatory, alias philosophical way.' This time, he needed Boulton's practical help in making some equipment, having spent all day 'twisting the necks of Florence-Flasks – in vain!'

Now if you like Florence Wine I beg leave to make you a present of one Bottle, or two, if the first does not answer, to drink Success to Philosophy and Trade, upon condition that you will procure me one of their Necks to be twisted into a little Hook . . . it must be truly hermetically seal'd, air-tight, otherwise it will not answer my End at all.[40]

'I am extremely anxious for this new Play-Thing!' he added. A little drawing showed exactly what he needed. Using a balance and a glass box, he planned to make a hygrometer to test the humidity of air. Letting his ideas free-wheel, Darwin tossed out theories that foreshadowed discoveries made decades later.[41]

But while Darwin, Boulton, Wedgwood and Priestley made their experiments, built their careers and raised their families, they were always aware of the political swings that could affect their lives. The honeymoon glow of George III's accession had not lasted: soon people started to count the cost of the French war, while the Whig politicians were dismayed at the new King's support of their Tory rivals and his reliance on his former tutor, the hated Earl of Bute. In March 1763, Wedgwood was writing excitedly from London about the ferocious debates against Bute's plan to extend the excise laws – 'It gives universal disgust here & is the general topic of every Political Club in Town' – and noting that 'Mr Wilks ye Author of ye North Briton is gone to France but I cannot learn on what account'.[42] In April Bute resigned as Secretary of State but in the same month his opponent Wilkes was tried for seditious libel for the outspoken article in No. 45 of his paper the *North Briton* and London heaved with chants of 'Wilkes and Liberty'. By the end of the year Wilkes had fled to France, four years later he would return in triumph as a popular hero, elected MP for Middlesex.

Although Wilkesite riots filled the papers, work and new discoveries were of even more interest to these slowly converging groups of friends in Liverpool and the Potteries, Birmingham and Lichfield. And in May 1765 a new figure arrived who would become a linchpin of their circle over the next decade. This was the thirty-year-old Scottish doctor William Small, who

William Small, drawing by unknown artist

landed on the doorstep at Soho clutching a letter from Franklin that proclaimed him 'both an ingenious Philosopher & a most worthy honest man'.[43]

For the past seven years Small had been Professor of Natural Philosophy and Mathematics at the College of William and Mary in Williamsburg, Virginia, an oasis of white clapboard in a region of rivers and swamps. His lectures fired a generation of students, among them the future President Thomas Jefferson, who said that meeting Small, with his liberal views and 'happy talent of communication', was 'my great good fortune, and what probably fixed the destinies of my life . . . and from his conversation I got my first views of the expansion of science & of the system of things in which we are placed'.[44] Not everyone was so impressed. Small had graduated from Marischal College in Aberdeen, where he was a protégé of Dr John Gregory, the friend of Hume, Kames and Black, member of the Royal Society and later the King's physician: it was probably through his influence that Small landed the Williamsburg job, answering an appeal sent through the Bishop of London. He was only twenty-three when he arrived and to the jealous, enclosed faculty, almost entirely composed of Anglican clergymen, he seemed a dangerous young man. His father and brother were Presbyterian ministers: his liberalism was threatening, his ideas were new, his closest friends were students. The whole college was in turmoil with the faculty violently at odds with the Board of Visitors, local gentry who included many of Small's cronies. There were drunken rows and sudden dismissals and, despite his inexperience, he found himself having to teach humanities as well as natural philosophy, and to write a completely new curriculum, replacing rote-learning with lectures, adding *belles-lettres*, ethics and modern languages.[45]

Small lived well here, boosting his salary by working as a physician (and building up a fat bill at the wine merchants), but the backbiting and the climate defeated him. Giving advice to Stephen Hawtrey, a young ex-Etonian about to sail for Williamsburg, he warned of the severe winters and blasting summers, 'for the heat is beyond conception', and advised him to take large linen waistcoats, 'that they mayn't stick to your hide when you perspire', and 'Callico shirts, as they suck up the moisture'.[46] As soon as he was offered the chance to come to England to buy scientific instruments for the College, he seized it. In late June 1764 he packed his bags. Although the Williamsburg archives show that he did in fact send back the instruments, jokes are still made there about 'that rascal who ran off with our money'. He had also refused to take an oath that would allow peremptory dismissal of professors, and part of his brief in England was to find out if such a power was valid. The rows and politics of Virginia dogged him: 'I find he is rather inclined to party & Opposition', wrote Hawtrey's brother.[47]

Small soon gave up any idea of going back. He spent much of the autumn of 1764 in Edinburgh and Glasgow, before returning to London where he met Benjamin Franklin, whom he had known in America.[48] The London clubs and societies were friendly and welcoming, like the Monday Club at the famous George and Vulture tavern in the City, 'The Club of Honest Whigs'.[49] Franklin and Small later lived a few streets apart in the Charing Cross district, and in January 1765 Franklin took Small to the Royal Society – as he would soon do with Priestley – and introduced him to his friends. For a short time Small toyed with the notion of lecturing to medical students in the capital, but when he heard that a practice was vacant in Birmingham, he realized that this would suit him perfectly. Equipped with an Aberdeen MD fixed by his old friend Dr Gregory, he set off for the Midlands. Within months he was so settled that he turned down an offer to go to work in Russia, and soon he was sharing a house and opening a clinic with John Ash, the chief campaigner for the new Birmingham infirmary. With his easy, dry manner, Small quickly became Boulton's confidant, doctor and unofficial secretary. In December this year, when Baskerville suggested taking Anne to London to meet Boulton, who was on his way home from a trip, he added that Dr Small thought it a good idea: 'He observed that it would be as useful as a Journey to Bath, & the Consequence no doubt a Son & Heir, at which she laughed heartily, & said, then she would not go.'[50]

In many ways Small sounds an ideal doctor yet he often joked that medicine was 'a prison'. His real interests were in mathematics, mechanics and chemistry but, like Darwin, he was reluctant to step forward as a 'crankish' inventor, so he had no empire of his own to guard. He was one of those people of whom everybody becomes increasingly fond, with a charm that is hard to pin down. Slight and delicate, he was completely unthreatening and absolutely open to ideas, able to pick them up and play with them, bringing to bear his own crisp intellect. He carried no heavy philosophical baggage or commercial ambition and was blissfully non-judgemental. To his 'extensive, various and accurate knowledge', wrote James Keir, he added 'engaging manners, a most exact conduct, a liberality of sentiment, and an enlightened humanity'.[51] In short, he was a perfect addition to any network – unassuming in himself yet accelerating the flow of information between others. Even more valuable, he proved an instinctive diplomat who could ease potential conflicts while somehow managing not to betray confidences. Franklin – who was keen on encouraging groups that would work together for knowledge and reform, openly or secretly – had been clever to direct his steps here.[52]

Boulton soon introduced him to Darwin. His new closeness to Small involved a slight adjustment to their relationship, but although there is a tinge of hurt pride in one or two of Darwin's letters, the correspondence soon became three way. In March 1766, when Darwin heard a new doctor had arrived in town, he hoped this would bring no uneasiness to

> our ingenious Friend Dr Small, from whom and from you, when I was last at Birmingham, I received Ideas, that for many Days occured to me at the Intervals of the common Business of Life, with inexpressible Pleasure.[53]

This total permeation, the way 'pure' and 'applied' scientific pursuits fused together, characterizes the nature of the group. For Darwin, Boulton and Small – as for Wedgwood, Bentley and Priestley – 'philosophical friendship' would increasingly become one of the great pleasures, not separate from, but inextricably linked with, the common business of life.

QUEEN's WARE and ORNAMENTAL VASES, manuſactured by Joſiah Wedgwood, Potter to her Majeſty, are ſold at his Warehouſe, the Queen's Arms, the Corner of Great Newport Street, Long Acre, where, and at his Works at Burſlim in Staffordſhire, Orders are executed on the ſhorteſt Notice.

As he now ſells for ready Money only, he delivers the Goods ſafe, and Carriage free to London.

☞ His Manufacture ſtands the Lamp for Stewing, &c. without any Danger of breaking, and is ſold at no other Place in Town.

8 : REACHING OUT

Soon Wedgwood too joined this fellowship of ingenious philosophers, and like the circles of inventors, his flexible web of commerce also spun out from the Midlands to the wider world. Wedgwood was now a master and a married man: in 1763 he had finally decided it was time to ask his uncle for Sally's hand. He turned to Bentley, whose wife Hannah had died in childbirth in 1759, and who was immensely understanding about Wedgwood's lengthy wrangling with Sally's father. 'You will be sensible how I am mortify'd', Wedgwood wrote,

> when I tell you I have gone through a long series of bargain makeing – of settlements Reversions – Provisions &c:&c: Gone through it did I say: would to Hymen I had. No, I am still in the Attorney's hands, from which I hope it is no harm to pray '*good Lord deliver me*' Miss W: & I are perfectly agreed, & could settle the whole affair in three lines & so many minutes, but our Pappa, over carefull of his Daughter's interest, would by some demands which I cannot comply with, go near to separate us if we were not better determin'd.[1]

When all was settled Josiah persuaded Sally to name the day, 'the blissful day! when she will reward all my faithfull services, & take me to her Arms! to her Nuptial bed! to – Pleasures which I am yet ignorant of'.[2] They were married at the end of January 1764. On their wedding day Wedgwood asked Bentley to 'think it no sin to wash your Philosophic evening pipe with a glass or two extraordinary, to hail your friend & wish him good speed into the realms of Matrimony'.[3] By May he was sending

Advertisement for Queensware, 1769

him the best respects of 'two married Lovers, happy as this world can make them'.[4]

Bentley remained his chief adviser, but another was his brother John, nine years older, a dandified, depressive, effervescent character, partner in the London warehousing firm of Wedgwood & Bliss. John belonged to the group that gathered around Ralph Griffiths, a Staffordshire watchmaker turned bookseller who had made a pile from publishing John Cleland's scandalous novel *Fanny Hill* in 1750, and then founded the *Monthly Review*, which he edited for fifty-four years. Writing to John in the early 1760s Wedgwood presented himself meekly as an aspiring provincial potter, pressing his face to the windows of Griffiths's mansion at Turnham Green, envying his brother for being part of 'the meeting & collision of such Geniuses', and wishing for 'a pair of wings, & a learners seat amongst them'.[5]

In fact Wedgwood was learning fast, and his modesty, though genuine, was also a useful tactic for winning allies. But it was true that the provinces were his base: his family, his home and his pottery were always closest to his heart. When his first child, Susannah, was born on 3 January 1765, John sent lobsters from London to celebrate her christening. 'Sukey is a fine sprightly lass', Wedgwood told him, 'and will bear a good deal of dandleing, and you can sing – lulaby Baby, whilst I rock the cradle.'[6] From now on her teething, her schooling, and her mishaps would all figure large in her father's letters. Wedgwood was very much a family man, and this was very much a family business, involving uncles and cousins and brothers. Boulton too used his family connections[7] but Sally Wedgwood took far more interest in the business than Anne Boulton, being a useful guide to the growing female market in giving her opinion on all new designs and learning to read her husband's shorthand to keep up his experiment book. When Sukey was three months old, Wedgwood was deep in experiments, explaining that 'Sally is my chief helpmate in this as well as other things, and that she may not be hurried by having too many *Irons in the fire* as the phrase is I have ordered the spinning wheel into the Lumber room.'[8]

Overseeing firings, designing new models and testing new clay bodies and glazes, Wedgwood turned out his pots. Two years earlier, at the start of 1763, he had leased the Brick House Works in Burslem (known as the Bell House after he installed a bell to call his workers, rather than using the traditional horn). This was an elegant house with a large garden, with several two-storey workshops attached, and several more behind. For the past few years his main effort had been concentrated on improving his cream-coloured ware, aiming at a fine, uniformly coloured body whose glaze would not crack or craze. By 1763 he had developed 'a species of earthenware for the table quite new in its appearance, covered with a rich and brilliant glaze bearing sudden alterations of heat and cold, manufactured with ease and expedition, and consequently cheap having every requisite for the purpose intended'.[9] The new ware was an astonishing advance and was quickly copied throughout the Potteries, especially since Wedgwood asked local potters to supply it to his specification when orders came too fast.

The next step, as soon as he married in 1764, was to put Sally's dowry to use.[10] He rented more workhouses from his uncle and began to take a lead in local affairs. More family members were brought in: in the spring of 1765, when John sold his partnership in the warehouse for a healthy profit, he agreed to become his London agent. Meanwhile, their nephew Tom Byerley was taken on to help with accounts and French correspondence. Tom caused Wedgwood much gnashing of teeth – wanting first to be an author, then a strolling actor, and next sailing for America, where he overspent and got gaoled – but eventually he returned to Burslem, a loyal manager for many years. Beyond the family, waves of contacts rippled out through local patrons, and in 1765 Wedgwood heard to his delight that Lord Gower had talked of his creamware at dinner and had 'said that nothing could exceed them for a fine glaze'.[11]

Word spread. In early 1765, through Lady Chetwynd, a lady-in-waiting with Staffordshire connections, an order came for a tea-set for Queen Charlotte, complete with candlesticks and fruit baskets, 'with a gold ground, and raised flowers upon it in green'.[12] Wedgwood shrugged off the honour by claiming it came only 'because nobody else would undertake it', and he may have been right that the gilding was too costly and difficult for most potters to risk it. But he set to work at once, asking John to buy the gold powder made only by a certain Mr Shenton, to an exotic-sounding recipe in which gold leaf was ground in honey then steeped in water, so that when the honey dissolved the gold was left at the bottom. 'Pray put on *the best suit of Cloaths you ever had in your life*,' he told his brother, 'and take the first opportunity of going to Court.' Wedgwood then went himself, in *his* best clothes: new blue surcoat, scarlet waistcoat with lace, 'a lite brown Dress Bobwig', and a sword bought at the Sign of the Flaming Sword in Great Newport Street.[13]

Making the service was not easy – 'I am just teazed of my life with dilatory, drunken, Idle, worthless workmen,' he fumed – but the royal commission sealed his reputation.[14] He followed it up by sending a box of patterns and vases to the Queen and soon won permission to style himself 'Potter to Her Majesty' while his creamware was granted the name of 'Queen's Ware'. A second service was ordered by the King, to a simpler design which became known as 'the Royal Pattern'. Two years later he was still stunned by his success:

> The demand for this said *Creamcolour*, Alias *Queens Ware*, alias *Ivory* still increases. It is really amazing how rapidly the use of it has spread allmost over the whole Globe, and how universally it is liked. How much of this general use and estimation is owing to the mode of its introduction – and how much to its real utility and beauty? are questions in which we may be a good deal interested for the government of our future Conduct . . .[15]

If a royal or aristocratic introduction was 'as necessary to the sale of an Article of Luxury as real elegance and beauty', then the manufacturer should bestow as much pains, 'and expence too if necessary', in gaining such patronage as he would on the design itself.

From this time on Wedgwood realized that the future, as far as finding ideas and marketing products went, lay in London. Boulton knew this too, however much he moaned that he still had a hundred and fifty people to see at all corners of town, and was 'heartily tired of London & want sadly to be home, nor shall I have any repose until I am got into my old green bed'.[16] But while Boulton dealt through London agents, Wedgwood was beginning to think of setting up a permanent base there. In August 1765 he told John, 'I have just had the honour of the D. of Marlbro., Ld Gower, Ld Spencer, & others at my works – They have bought some things & seem'd much entertain'd & pleas'd.'[17] His visitors, he added, had been surprised there was no London warehouse to show off his patterns. Potters traditionally sold their wares through the big London dealers, or 'chinamen', but some porcelain manufacturers already sold from warehouses in the city and now Wedgwood followed their lead.[18] Within a year he had rented space at 5 Charles Street, off Grosvenor Square, sending his Burslem accountant William Cox down to look after it and taking on a new man, Peter Swift, in Cox's place.

In 1766 he made his cousin Thomas a full partner, responsible for all the 'useful' ware (later defined as 'such vessels as are *made use of at meals*').[19] And he had three other things firmly in mind: to start a grander 'ornamental' line, to fix a partnership with Bentley and to build a new factory. From late 1765 onwards, when Boulton was developing Soho, Wedgwood was negotiating to buy the Ridgehouse Estate, 350 acres lying between the villages of Hanley, Burslem and Newcastle. The stubborn proprietor, Mrs Ashenhurst, grumbled and delayed but Wedgwood persevered, though 'she scolds and huffs away at a hard rate, and seems to be in a good way for making me a hard bargain'.[20]

A factory alone was not enough, even for the Queen's potter, and Britain alone was not enough of a market. In 1763 the Seven Years War had finally ended, bringing new dominions – Canada, Nova Scotia, Dominica, Grenada and Tobago. And although trade had never stopped during the war, the Continent itself was now more open to merchants and manufacturers. Boulton had long traded with agents in Europe and once his partnership with Fothergill got under way this was one of the areas he developed most keenly. Over the next few years letters flew from Soho not only to London or Sheffield, but to Aix en Provence, Altona, Augsburg, Berlin, Bordeaux, Breslau, Cadiz, Dresden, Frankfurt, Geneva, Ghent, The Hague, Hamburg, Hanover, Holland, Iserlohn, Königsberg, Leghorn, Leipzig, Liège, Lyons, Marseilles, Nancy, Naples, Nuremberg, Orléans, Riga, Rotterdam, Smyrna, Vienna, Wesel.[21] Some of the correspondents have code names, such as 'Montreal', 'Niagara', 'Guadelupe', 'Senegal'; many later ones are in French, and even more in German as so much business was done with the German states. The names of the cities were strung like beads on the thread of Boulton's ambition. Trade crossed all boundaries and legal restrictions were merely obstacles to be overcome. One Birmingham export salesman sent some typical advice from Lyons:

> The inlaid buttons must be wrapped in Paper Parcels, as if they were Links, and have a card of Links on the outside, and placed in the second row near the bottom of the Cask, about the middle – that is to say, all round near the wood there must

be other goods, being as in the Custom House they frequently cut the Casks in one or two places, and take out the Parcels that come first, so you will please to disguise well and place carefully for all your friends in France, the Gilt, mettal, inlaid, and lacquered coat and breast buttons, all of them being prohibited in France . . .[22]

The cheerful skulduggery was manageable. No rules would block the way.

Boulton made skilful use of his foreign contacts. One of his silent partners was a German merchant; another was a Dutch financier; a third was the founder of the Royal Danish Guinea Company. All these men were sources of information as well as capital.[23] But the plundering of skills went both ways. One Birmingham hardwareman, Michael Alcock, had already skipped the country in 1755, and with French Government support he set up a foundry in La Charité sur Loire. Boulton himself was wooed by offers to establish an iron foundry in Sweden, where he bought much of his steel. He toyed quite seriously with the idea, so much so that the Earl of Halifax, Secretary of State, effectively asked Roebuck to spy on him to find out the truth.[24]

He was far keener, however, on pinching ideas from the Continent than on selling his own. In late November 1765 he was on his way to France, crossing the Channel in a 'hurricane' with four shabby Frenchmen and an old Dutch woman and her family. Everyone, beginning with himself, was violently sick. Pale but exhilarated, he smuggled his patterns and samples into Calais 'by the help of wide breeches and pockets' and from there he launched himself on a Paris choked by winter frost and fog.[25] His aim was to win orders and spy out new techniques, but also, secretly and illicitly, to get his hands on various Boulton & Fothergill goods held by a recently bankrupted agent. 'Silence at my being here & concerning Oppenheim answer nothing,' he warned Anne. He succeeded, in part, but the real gain was his first-hand experience of high French style: 'Fine painting, very fine sculpture in marble, porphyry & Brass – Water works & curious workmanship without end or number.'[26] He mingled with the English merchants, took coffee with the Ambassador, the Duke of Richmond, and investigated French silvering and brassfounding.

Although he complained half-heartedly that it was taxing to have to dress up, he was in his element. He headed his letters 'Paris' in huge swirling copperplate. He enjoyed the flirtations in the Tuileries Gardens, the smart dinners, the playhouse and the singers, and he bought the sweetest and most fashionable French silk for Anne. Paris, he found, was well stocked with Birmingham goods. Much to his delight he even met the hero of the day, 'that exiled patriot & friend to English Liberty', John Wilkes. It all made a great impression. In January the architect William Wyatt told Boulton that the builder had a sketch of his new house at Soho, 'but tells me the manner of it will not please you & thinks your journey to France has alterd your Tast in regard to the alterations there, I did not know that France was very famous in Architecture –'.[27] Calming down, instead of a mock Versailles, Boulton settled for a solid Georgian mansion looking down through spreading parklands to his new works.

It was good to feel part of Europe and to mingle with high society in London and sell to the court. But Boulton and Wedgwood found it maddening being linked to the ports and the capital only by long, tough journeys. Pack-horse trains and overloaded carriages struggled across quagmires in winter and rock-hard ruts in summer: often a third of Wedgwood's pots were smashed when they reached their destination. Gradually, mile by mile, stretches of road came under the control of the turnpike trusts, set up by Acts of Parliament to raise money for repairs, repaid by tolls. A petition for a turnpike in the Potteries in 1763 claimed that there were now a hundred and fifty potteries around Burslem employing seven thousand people, exporting 'in vast quantities' to America, the West Indies and Europe from Bristol and elsewhere, while imported flintstones and West Country clay were brought first by sea and then down the rivers and finally overland by carrier to Burslem.[28] Wedgwood – articulate and charming and coolly persistent – became the potters' spokesman. He rode to countless meetings, soothed nearby turnpike towns who feared new roads would cut their trade, and campaigned in London.

Crucial to his effort, and providing a useful lesson in lobbying, was the support of the local landowner Lord Gower, former MP and Lord of the Admiralty and now a highly placed official at court. Gower ensured the presentation of the turnpike petition to Parliament in 1763, which brought a new road to Burslem. But even the turnpikes were badly maintained. When Arthur Young toured the country in 1770 he filled his notebook with curses:

Newport Pagnall to Bedford so narrow that it was at the peril of our necks to drive on . . .

Rotheram to Sheffield excruciably bad, very stony and excessively full of holes . . .

To Lancaster (turnpike) and to Preston (turnpike) and again to Wigan (turnpike) very bad with ruts which measured 4 foot deep and floating with mud only from a wet summer . . .

From Billericay to Tilbury of all the accursed roads that ever disgraced this Kingdom none can equal this.[29]

There must be a smoother, cheaper way for manufacturers to reach their suppliers and their markets. Surely water was the answer – for all the land-bound makers who sent their goods to the coast, and reached out eagerly to the wide shores beyond.

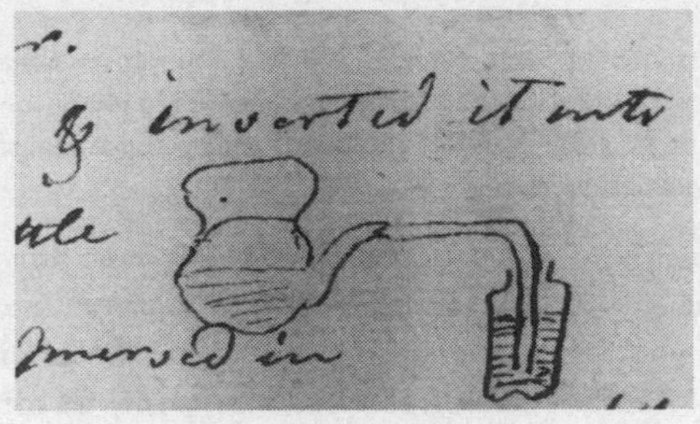

Watt's kettle, from a letter to William Small

9 : STEAM

Erasmus Darwin, like Wedgwood, was a man who knew his roads. Many of his best, or wildest, ideas came as he dashed from patient to patient across country: in 1766 he calculated that he travelled ten thousand miles a year, and he once allegedly received a letter simply addressed 'To Doctor Darwin – On the Road'. When the lanes were too narrow, cut deep beneath overhanging trees, he abandoned his carriage and took to his horse, Doctor, returning tired, muddy and stiff, clattering up to the Close in the dusk. Even in his carriage, bumping over ruts and holes and bashing against hedges, his journeys were a bone-shaking business. Not surprisingly he now became immensely keen on designing improvements to carriages, getting local coachbuilders to adapt the wheels, improve the turning circle, soften the jolting of the axles with springs. Sometimes, in a combination of despair and inventive passion, his mind leapt ahead to a different form of transport altogether. 'As I was riding Home yesterday,' he told Boulton one day around 1764, 'I concider'd the Scheme of the fiery Chariot, and the longer I contemplated this favourite Idea, the [more] practicable it appear'd to me.'[1]

Darwin scrawled on, the pages of his long letter spattered with ink from his quill. He wanted to put down his ideas, 'crude and undigested as they appeared to me', to provoke Boulton's own thoughts, but to keep the idea secret from clever inventors such as John Wyatt.

And as I am quite mad of this Scheme, I begg you will not mention it, or shew this paper, to Wyot or any Body.

These things are required. First a Rotatory motion. 2 easyly altering its Direction to any other Direction. 3 To be accelerated, retarded, destroy'd, revived instantly and easyly. 4 the Bulk, Weight and Expence of the Machine as small as possible in proportion to its use.

With admirable clarity, Darwin had laid down the basic principles for any powered vehicle. But should his steam-powered carriage have three wheels or four? And which wheel should the power act on? With four wheels, he could not get over the problem of transferring power in cornering so he opted for three, drawing scratchy diagrams showing a disembodied hand manœuvring a large 'rudder wheel'. He decided he would have a steam-engine with two cylinders and a single boiler. If the steam cocks were opened 'gradually and not with a Jerk' the power could be pretty smooth and adjusting the cocks would allow him to accelerate, brake and stop. 'And if this answers in Practice as it does in theory, the Machine can not fail of Success! Eureka!'

It was far harder to develop this into any machine that could possibly work. He hoped to get rotary power – to make steam turn a wheel – by using a small beam-engine, tilting back and forth to wind and unwind chains around a split back axle. As the beam rose and fell and the cog wheel with the chain moved round, so the wheels would roll forward. If Boulton could tell him how much power he would need, and 'the Expence of coals of a common Fire-engine, and the Weight of Water it draws' he would draw up an estimate, build his machine and take out a patent. He had hoped Boulton would be a partner but he was sceptical and had enough expense anyway at Soho, so Darwin's design fell by the wayside, so to speak. In 1769, the world's first successful steam-carriage was built in France, by Nicolas Cugnot. (Cugnot's car managed only two and a half miles an hour, and even then it had to stop every twenty minutes to get up steam. It was also rather risky: his first model flattened a stone wall and his second overturned dramatically in a narrow Paris street: Cugnot – and his machine – were imprisoned, for fear of more damage.[2])

Although Boulton did not take to the fiery chariot, steam-engines were increasingly on the minds of the Birmingham friends – not as novelties for the road, but as big, thundering beam-engines. At Soho, Boulton's watermill turned his lathes and machines, but summer droughts or work higher upstream often lowered the level and he began to consider using a steam-engine to pump the water back up into the millpond to keep the levels high. He asked Smeaton, Franklin, Small and Darwin for advice. He and Small even built a model and sent it to London for inspection. Darwin was desperate at missing it: 'This Model I am so impatient to see . . . that I am determined to spend a Day with you, the first vacant day that occurs to me: and shall trust to the Stars for meeting with you at Home.' [3] If Boulton was out, perhaps Anne could be asked to show it to him?

They had corresponded about this for some months, and Darwin's habit of taking nothing for granted had already raised basic questions – did evaporation, for example, take place on the surface of boiling water where you saw the steam, or at the bottom, near the fire? In a flask, you could see bubbles rising, condensed again by the cold water above, 'and their Sides clapping together make that Noise call'd *Simmering*'. And when you boiled a kettle, 'what a Quantity of Steam? – is this only at the upper surface? I humbly conceive not.' Although scientifically interesting, this did not get Boulton much further. As if admitting as much Darwin swooped into whimsy. As a man's 'Momentum' depends on the 'vital Steam rising into the Brain from his boiling Blood', so the amount of steam in the steam-engine's boiler, he declared, 'is a Question of the utmost Importance in the Creation of this Animal'.[4]

Unknown, as yet, to the Midland men, in Scotland James Watt was also working on steam, the great source of energy that would bring him to Birmingham in years to come. His interest had begun in 1757, when his friend John Robison developed a sudden passion for the steam-engine, 'a machine', Watt admitted, of which he was 'very ignorant'.[5] Robison claimed boldly in the *Universal Magazine* that steam could run wheel-carriages and asked Watt to try to make a model to prove it. Their first attempt had two cylinders acting alternately by 'rack motions' on pinions fixed to the wheels – a design very like Darwin's. It did not work, and, 'neither of us having any idea of the true principles of the machine', the scheme was dropped.

Aggravated by this failure, Watt began to chase up the work of the early experimenters, Papin, Savery and Newcomen, but his reading was squeezed into brief leisure hours and his experiments hampered by his need to make a living. By now he too had established his own business. In the summer of 1757 he was appointed Mathematical Instrument Maker to the University and given his own shop in the quadrangle. His mail was now directed to College Green and, anticipating visitors, he asked his father to send not tools but '½ Doz afternoon China tea cups a stone teapot not too small a sugar Box & Slop Bowl as soon as possible'.[6] Over the next two years, notes sped down the Clyde to his father and his brother Jock as he sent instruments they had ordered ('Capt. Rowells glass price 3/–'), asked for the quadrants he was making to be 'approved by some experienced seaman', requested barometer tubes he had left behind and worried about money or crowed about a West Indies order. 'I have a Good deal of things to do,' he wrote, '& can get no lads here to help me.' Did their friends know 'any lads that can file (Brass) tolerably well'?

In 1759, that year of victories, when Wedgwood was moving to the Ivy House and Boulton was buying his land at Soho, Watt opened a shop in Glasgow's Saltmarket, financed by an architect partner, John Craig. Four years later he moved to the Trongate in the centre of the city, selling 'all sorts of mathematical and musical instruments, with variety of Toys and other goods'.[7] One of the best-selling items was his newly invented apparatus for drawing in perspective, based on a pantograph system, cleverly folding away into a pocket-sized box. Apparently he also repaired and made bagpipes, guitars, fifes and flutes – perhaps even passing some of the latter off as the work of a top French maker.[8] But if his musical instruments showed that Watt sailed near to forgery, they also showed the theoretical scientist at work. 'Tho' we all knew he could not tell one musical note from another,' Robison remembered, to his friends' astonishment Watt agreed to build an organ for a Masonic Lodge in Glasgow.[9] He had already learned a bit from repairing one, which had intrigued and amused him, and he now prepared for this daunting commission by building a small organ for Joseph Black. Starting from scratch, he went back to basics, 'noticing a thousand things that no Organ builder would have dreamt of', such as adjusting the strength of the air-blast to the pipes and improving the regulators. Robison ransacked the university library and Watt immersed himself in the theory of harmonics, especially the new science of tuning keys to 'equal temperament' (an age-old mathematical problem, brilliantly solved in Bach's *Well-Tempered Clavier*, of 1726). Watt's calculations put him ahead of the best mathematicians in Europe and his Masonic organ worked perfectly.

He also bought a share in the Delftfield Pottery Company and soon moved into a house near by. (Wedgwood liked to call him a 'fellow-potter' and in the mid-1760s they were both experimenting in the chemistry of clays and glazes.) By now, Watt's work also had to support others. In July 1764, he had married his lively, down-to-earth cousin, Margaret Miller, always called Peggy. The only person who ever made Watt genuinely light-hearted, Peggy eased his life; his headaches lessened, his panics and depressions calmed. In letter after letter, she told him to look after himself, to keep warm, not to worry, to come home soon, and she kept the shop while he was away, taking orders and dealing with clients. Their first son, John, born in 1765, died in infancy, but he was quickly followed by two healthy children, Margaret and James.

Watt's family made him even more nervous of risking a sound income for a life of experiment. While Boulton, Wedgwood and Priestley snapped up opportunities, everything Watt did seemed agonizingly slow. Before and after his marriage he continued to brood on steam. Around 1761 he got hold of a Papin 'digester' – a round pot with a narrow valve in the neck through which steam burst at terrific pressure, like a sort of primitive pressure cooker. To test this he fixed an ordinary apothecary's syringe to the valve, and put a little piston inside it with a rod pointing out of the top, on which he balanced some weights. Between digester and syringe he fitted a steam cock which he could turn so that the steam either filled the syringe or escaped. This Heath-Robinson contraption worked devastatingly well. When the steam whooshed in, his tiny piston was forced up the syringe so hard that it could lift a weight of fifteen pounds.

The experiment was simple and dramatic – almost too dramatic, making Watt afraid of bursting the boiler. After this he put his inquiries into the expansive power of steam aside and refused to look at it again. (Half a century later it would be exploited by the Cornishman Richard Trevithick, who built high-pressure engines whose booming, cracking exhaust could be heard five miles away.) Watt was often overly cautious, and it is interesting in this respect that, unlike Boulton, Darwin and Priestley, he was never very interested in electricity. One feels that perhaps it was too sudden and explosive for him – he wanted to harness the simmer, not the lightning.

Watt was a craftsman, a perfectionist. He wanted power that was efficient, steady and manageable and so he backed the 'atmospheric' principle used by Thomas Newcomen at the start of the century. Newcomen's engines exploited basic atmospheric pressure, building on the way that air had been found to rush into a vacuum. A vacuum could be created by sucking air out of a closed vessel with a pump, as Boyle had shown, but it could also, as Papin had demonstrated, be created by using steam. If you filled a cylinder with steam this drove out the air, but as the steam cooled it condensed into water which took up only a fraction of the space, so that the rest was almost a vacuum. When a piston was fitted into the top of the cylinder, the ordinary air pressure on its surface would drive it down into this vacuum. Pump steam in again and it forced the piston up – cool it and down it came.

The 'fire-engines' that Newcomen and his successors built had a great cast-iron cylinder above a domed brick boiler, and a moving piston attached by chains to a huge rocking beam. At the other end of the beam were heavy pump rods. As the steam forced the piston up, the beam tilted and down went the rods. As the steam condensed – cooled by a jet of cold water sprayed into the cylinder – the air pressure drove the piston down, pulling down the beam and lifting the pump rods. Steaming and cooling, pushing and pulling, creaking and thumping, stroke after lumbering stroke, the engines drained the mines.

Watt had never seen a full-size Newcomen engine – there was still only one in Scotland – but Glasgow University had a model, made by Jonathan Sisson of London. This had been sent back to Sisson for repair and in June 1760 Watt's friend Professor Anderson was charged to get it back. When it arrived it was still not right, and Watt was asked to make it work. He did so, though fuming at how badly it was built. Nothing maddened him more than bad workmanship, and even when the model was mended he was still bothered by the immense amount of fuel it consumed in proportion to its tiny cylinder, only two inches in diameter. From this point, obsession took hold.

As he watched his small model Watt could hardly believe how much steam he had to create to keep it going: the engine could manage only a

A Newcomen atmospheric engine, engraving by Henry Beighton, 1717

few strokes before it boiled dry. Nothing he tried affected this – changing the type of boiler, moving the fire, even making a wooden cylinder-cover so it would lose less heat. Slowly, doggedly, he set to work to find out why, helped by Black and Robison, who returned to Glasgow in 1763 after four years at sea. In his years away (besides adventures such as accompanying Wolfe to Quebec on the fatal night he scaled the heights), Robison had learned surveying and had been employed by the Admiralty to make observations on Harrison's chronometer in the West Indies. He felt more on a par with Watt now, he said, and found him 'good and kind as ever', but 'continually striking into untrodden paths, where I was always obliged to be a follower'. He also remembered how Watt's wide range of knowledge and insatiable curiosity made him popular at Glasgow:

> All the young Lads of our little place that were any way remarkable for scientific predilection were Acquaintances of Mr. Watt; and his parlour was a rendezvous for all of this description – Whenever any puzzle came in the way of any of us, we went to Mr. Watt. He needed only to be prompted – everything became to him the subject of a new and serious study; and we knew that he would not quit it till he had either discovered its insignificancy or had made something of it.[10]

Those china tea-cups must have come in handy.

Watt loved puzzles, and the steam-engine presented plenty. He had a craft background, and was not afraid of getting his hands dirty, but he was no on-site experimenter, working by trial and error. He was concerned with the principles of his subject, the laws of hydraulics and hydrostatics, the findings on variable temperatures and pressures, the application of mathematical theory. The first questions he asked were very like those Darwin had put to Boulton, about the nature of heat and steam. He experimented with the capacities for heat of different liquids, apart from water. He worked out how much water could be evaporated in a particular boiler by a pound of coal; how much steam was needed for each stroke of Newcomen engines of different sizes; how much cold water was required to condense the steam.

This last was a very real problem, since Watt and others discovered that water boils at a far lower temperature in a vacuum, so the danger was that instead of cooling the steam the opposite might happen – the steam might make the cold water boil. Investigating this Watt turned to the most obvious domestic example, a battered copper kettle. He took a bent glass tube, inserted one end in the kettle's spout and put the other into a marked flask of cold water. As the steam poured into the flask it gradually heated the water to boiling point, and Watt saw that the water level had risen by a sixth: this was the steam that had condensed before the whole flask bubbled and boiled. So steam could raise six times its own volume of water to boiling point.

When he turned this into measures of temperature and calculation of heat units it appeared to Watt that water in the form of steam could contain more heat than it could as pure water:

Being struck with this remarkable fact, and not understanding the reason of it, I mentioned it to my friend Dr Black, who then explained to me his doctrine of latent heat, which he had taught for some time before this period, but having myself been occupied with the pursuit of business, if I had heard of it, I had not attended to it, when I thus stumbled upon one of the material facts by which that beautiful theory is supported.[11]

Black then explained his theory of 'latent heat' – as the water continues to boil, although its temperature doesn't rise any further, the process of making steam continues to absorb heat. When the steam cools, this heat is released.

Watt now saw that the great drawback with the Newcomen engine was the loss of this extra heat through the alternate heating and cooling of the cylinder. Although Black's theory, he said, did not *suggest* his improvements to the engine, his knowledge and method – 'the correct modes of reasoning, and of making experiments of which he set me an example' – helped his work immeasurably.[12] But however much he fiddled with different ways of injecting the water or letting in the steam, he could never get round the insuperable problem: the cylinder had to be kept hot to make it efficient, yet the steam had to be cooled to create a vacuum. It was impossible to do both.

For months this worried him, and then in the spring of 1765 the answer came, so dazzling in its simplicity that the moment stayed sunlit and sharp in his memory. It was on the College Green:

I had gone to take a walk on a fine Sabbath afternoon. I had entered the Green by the Gate at the foot of Charlotte Street – had passed the old washing-house. I was thinking upon the engine at the time and had gone as far as the Herd's house when the idea came into my mind, that as steam was an elastic body it would rush into a vacuum, and if a communication was made between the cylinder and an exhausted vessel, it would rush into it, and might be there condensed without cooling the cylinder . . . I had not walked further than the Golf-house when the whole thing was arranged in my mind.[13]

Standing on the Green, which on weekdays was white with linen laid out to bleach, the realization 'flashed on his mind at once, and filled him with rapture'.[14] But it was the Sabbath, and no good Presbyterian could work. The grass was bare of cloth and Watt had to wait. On Monday, in his workshop in a courtyard behind the Beef Market, he built a model. This

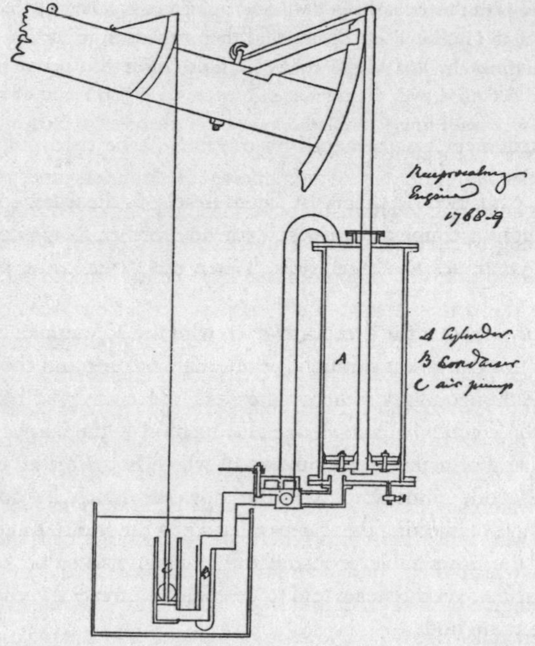

Drawing showing Watt's separate condenser, prepared for the Patent Specification, 1769

was not a beam-engine, but an inverted engine, pulling weights up by a hook. The cylinder, once again, was simply a large brass syringe, borrowed from a friend. Next to the cylinder he fixed an air pump, and instead of using a jet to cool the steam, he made a 'surface condenser' by soldering a rectangular tank on to the bottom and filling it with cold water. With the steam rushing out of the cylinder into the separate condenser, and the cylinder kept hot by a special insulating jacket, his makeshift model worked.

He was now sure that his calculations were accurate and convinced that a full-size engine on these lines would prove more powerful and use less fuel than its predecessors. At the end of April 1765 he wrote exuberantly to James Lind:

and if there is not some devil in the hedge mine ought to raise Water to 44 feet with the same Quantity [of] steam that theirs does to 32 . . . in short, I can think of nothing else but this Machine. I hope to have the decisive tryal before I see you.[15]

His excitement was balanced by caution. As he tried out valves and steam-cocks and different arrangements of cylinders and surface condensers – soldering together tin-plate into little tanks, with tubes or cells for the water to circulate – he became extremely cagey. When the enthusiastic Robison dashed unannounced into Watt's parlour he found him sitting by the fire, gazing at a little tin cistern on his knee. As Robison talked – 'something about Steam' – Watt gently put the cistern on the floor. When pressed, 'at last he looked at me and said briskly, "You need not *fash* yourself any more about that Man; I have now made an Engine that shall not waste a particle of Steam. It shall all be boiling hot, aye and hot water injected, if I please."'[16] He looked down at his cistern, saw Robison watching, and shoved it under the table with his foot.

There was, however, a vast difference between an instrument-maker's model in which every part was lovingly tailored to fit, and a massive working engine built of brick and iron. The engine-builders and designers were often millwrights (as Newcomen himself had been), used to huge structures, heavy wheels and gearing and beams made of tree-trunks – not accurate measurements and precision workmanship. Years of frustrating technical problems lay ahead. Watt had no money for large-scale models and although Black generously lent him as much as he could, he too was short of funds and was anyway leaving to take up a professorship in Edinburgh, handing his Glasgow chair on to Robison. Looking for other ways to finance Watt's work, Black thought of John Roebuck. So far, everything Roebuck had touched had turned to gold but he had recently leased the estate of the Duke of Hamilton at Kinneil, near Barrowstoness, with its collieries and salt-works. To pay for the mines Roebuck threw in his own capital, borrowed more money from Garbett, and took out loans against Prestonpans and Carron, but the deep coal seams flooded so fast that his Newcomen engines could not drain them. Knowing this, Black told him of Watt's engine. It was a neat idea: Roebuck needed the power; Watt needed the backer. And in contrast to Watt, so quickly depressed by setbacks and unforeseen costs – 'modest, timid, easily frightened by rubs and misgivings, and too apt to despond' – Roebuck was always 'ardent and sanguine in the pursuit of his undertakings'.[17] His bullish energy, let alone his cash, might finally get the engine into production.

By September 1765, after initial scepticism, Roebuck was keenly urging Watt on. But he did not pay Watt directly so that he could be free of other work, and as Watt made new and bigger models, now with brass cylinders, so he encountered new and bigger problems. His design made it impossible to use a layer of water as a seal above the piston, as Newcomen engines did, and so it proved a nightmare to keep the piston steam-tight, however much he tried sealing it with oil and animal fat, or packing it with oiled paper, pasteboard and rags. And when the work for a large experimental engine was put in hand, the iron cylinder, specially made at Carron, was so clumsily made that it proved unworkable.

Disillusioned, Watt turned to other schemes, planning an improbable 'circular engine', a wheel with a hollow rim which could be filled with steam, and three steam-pipes acting as spokes; with a careful use of valves, the wheel could rotate. This design cul-de-sac diverted his energies over months and even years. And all the time his money troubles were increasing, especially when his old partner John Craig died and he had to repay his capital to his heirs. In the summer of 1766 he sold the instrument-making shop and turned to surveying, opening an office in King Street. The engine at Kinneil lay rusting, in pieces.

SECOND QUARTER

For WATER is not of solid constituents, but is dissolved from precious stones above.
For the life remains in its dissolvent state, and that in great power.

Christopher Smart, *Rejoice in the Lamb*

10: THEY BUILD CANALS

In their frustration with the roads and their eagerness for profit, the entrepreneurs and landowners, the surveyors and engineers, the visionaries such as Darwin and the manufacturers such as Boulton and Wedgwood, all turned their minds more and more to dreams of inland waterways.

The main arteries were still the Severn and the Trent, but increasingly the Mersey gained in importance and, as time went by, the rivers flowing into all these were deepened and straightened. Since the start of the century people had dreamt of linking the rivers from east to west and north to south but such visions were rejected as too costly, or technically impossible.[1]

In the end, British canal work was triggered by private enterprise, not national plans. Between 1755 and 1761 the little Sankey Brook in Lancashire was turned into a canal, taking cheap coal to Liverpool. A few miles away lay Worsley Hall, owned by the twenty-one-year-old Duke of Bridgewater. A budding experimental philosopher, intrigued by science and engineering, on his Grand Tour the Duke had been impressed by the Canal du Midi and now, with the model of Sankey before him, he planned a massive private speculation – a canal to Manchester direct from his Worsley collieries.

As surveyor, Bridgewater employed James Brindley, whose sturdy face heads this chapter. The son of a crofter from the Derbyshire peaks, Brindley was now forty-two and a millwright in Staffordshire (where he rented a millwright's shop from Wedgwood's uncles, building the flint mill

James Brindley, engraving, 1771

The Duke of Bridgewater displaying his canal and aqueduct

for them on the Jenkins in Burslem – when the sails blew off on its first day, he firmly put this down to 'bad louck').[2] Despite his lack of education Brindley was famed for his knowledge of mills and water and his interest in steam and engineering. Known as 'the Schemer', he was asked for advice by many landowners: we can still hear his forthright voice in his spelling as he makes an 'occhilor survey or a ricconitoring' for the Duke of Bridgewater. In 1759 the Duke won his Act and within two years the first stretch was finished, provoking widespread wonder and amazement.[3] In one place, barges crossed the Irwell on an aqueduct as high as the tree-tops, hanging in the air; at Worsley tiers of canals drove deep into the mine itself; at Manchester there were docks and wharves and a tunnel taking the coal straight up to Deansgate, the heart of the city. Standing at the prow of his pleasure boat in his dark-brown coat with gold buttons, his hair tied back in a black rosette, the Duke showed off the marvels to his friends on summer evenings.

Across Britain, projectors were jolted into action; canal schemes were canvassed in coffee-houses and councils, on street corners and in stately homes, in Parliament and in the press. An era of fierce competition opened, a battle to the death between the new entrepreneurs and the old companies who monopolized river transport. In 1758, even before he worked on the Bridgewater Canal, Brindley had been asked by two Staffordshire landowners and coal-owners, Lord Gower of Trentham (Bridgewater's brother-in-law) and Thomas Anson of Shugborough, to survey a possible route connecting the Trent and the Severn. Another idea was to link the Trent to the Mersey, joining Hull to Liverpool. Although the scheme was dropped because of cost it was not forgotten. In 1760, Brindley and John Smeaton published a revised plan of a section from Burslem to Wilden Ferry, south of Derby – including a branch from the old Minster pool at Lichfield, just down the hill from Darwin's house.

Erasmus Darwin was not a man to let such an opportunity slip. Having watched Boulton's business grow he quite fancied becoming an entrepreneur himself. In the early 1760s he and Samuel Garbett, with Darwin's good friend the paper-maker John Bage and a local banker, John Barker, planned a new iron-rolling and slitting mill at Wychnor, north-east of Lichfield.[4] This was where the turnpike crossed the Trent and where Brindley's canal would join the river for a while. If they dug part of the route in advance – a cutting from the Trent, about half a mile long – they could sell it to the canal company later. Meanwhile, the drop of water from their own canal back into the river was about seven feet, and this would provide water power for their iron mill.

They first had to appease the Wychnor corn-miller, whose water supply would be affected, and after testy negotiations they bought him another mill, with a profitable eel fishery attached. Between patients Darwin rode across to meetings, or dashed off letters to Barker, about lawyers, engineering problems, expenses and rents and compensation for landowners. It was exhausting and expensive; before they even started digging, Darwin complained, they had spent £2,000. In the end the cut was made (and eventually sold as planned to the Trent and Mersey Company) and the rolling-mill was opened in 1765. It was run by Barker until his death in 1781 but it was never very profitable: poor Bage lost so much money – around £1,500 – that he turned for consolation to literature: he became a widely acclaimed novelist, best known for his radical-feminist fantasy *Hermsprong*, published in 1796.[5] Darwin was not driven to writing (not yet, at least) but he never engaged in a similar venture again. Instead, with regard to canals, he pitched in as adviser, consultant, shareholder, irritant – the grit in the oyster.

In 1765, Wedgwood too joined the canal entrepreneurs. He was busy and happy: Susannah had just been born, his potworks were flourishing, his turnpike activities were bearing fruit. Now he turned his mind to navigation. The Trent–Mersey scheme had never quite faded, and in January he discussed a revised plan with Brindley and a group of potters. Just as they were thinking of taking this further, he heard that a slightly different proposal was being concocted by Samuel Garbett and Thomas Gilbert (Lord Gower's agent, now MP for Newcastle-under-Lyme), and that they were now due to meet Gower in Lichfield. From the start canal-building was a matter of assaults and triumphs: 'Astonish'd, confounded and vexed', Wedgwood and Brindley dashed to Lichfield. Although alarmed at gatecrashing a cabal where he knew he was not wanted, a passionate Wedgwood knocked down all Garbett's arguments:

> in 10 minutes time he found his *baseless fabrick* tumbling down to the ground & deserted it immediately without one attempt to rear it up again. I am afraid you

will think me rather too figurative, but remember I am not reasoning now I am only huzzaing and singing Io' after a conquest.[6]

That evening he persuaded Gower to back him, and next day his scheme was adopted by a big public meeting.

The potters were understandably keen. 'Our gentlemen seem very warm in setting this matter on foot again,' he told his brother John in early March, 'and I could scarcely withstand the pressing solicitations I had from all present to undertake a journey or two for that purpose.'[7] Once he had made a decision, Wedgwood always acted fast. If it was to succeed the 'Grand Trunk' had to build a coalition of interests: West Midlands iron-masters and potters, landowners in Derbyshire, Staffordshire and Warwickshire, and the Duke of Bridgewater himself, keen to protect his canals in Lancashire and Cheshire. By the start of April Wedgwood had already enlisted support from Birmingham – including that 'very intelligent Gentleman' Garbett, who noted helpfully that by using the canal the Baltic merchants might save up to fifteen shillings a ton importing iron and flax, and began persuading his government contacts, even gaining a nod of approval from the King. In Cheshire Wedgwood won over the salt manufacturers, tired of high tolls on the river Weaver. In Liverpool Bentley tackled the merchants, corporation and mayor. And in Lichfield Wedgwood met Erasmus Darwin. Darwin, of course, was unabashedly enthusiastic and Wedgwood – who was always slightly over-impressed by educated, literary men – was ready to enrol him completely, telling him he had 'public spirit enough to be Generalissimo in this affair'.[8] Darwin graciously declined on grounds of time, but for the rest of the year they were in constant touch.

If Darwin and his partners had faced a difficult mill-owner and suspicious landowners in Lichfield, the promoters of a great scheme such as the Grand Trunk met these problems a hundredfold. The first thing Wedgwood had to do was sweeten or trounce the opposition. Landowners had to be approached to sell their land; mill-owners needed to be reassured that reservoirs would be built so that they would not lose water power. Rival companies, such as the 'Cheshire gentlemen' of the Weaver Navigation, and the owners of the Trent Navigation, had to be brought in or beaten down. The canal supporters themselves had to settle two vital points: whether the canal would be run by a trust, like the turnpikes, or as a joint-stock company (the accepted option); and whether the Weaver Navigation or the Duke of Bridgewater would control the Mersey end.

Briskly, Wedgwood roped in support. And in the middle of all this came the order from Queen Charlotte. Between canal work he was experimenting on white bodies and 'making powder gold' and cursing 'drunken, Idle, worthless workmen' who held up the royal commission. At the start of the year he had written, 'I scarcely know without a good deal of recollection whether I am a Landed gentleman, an Engineer or a Potter, for indeed I am all three by turns, pray heaven I may settle to something earnest at last.'[9] This was even truer now.

The idea of the canal system inspired him. It was a vision, something so huge that it might transform British industry and society for ever. Canals involved planning, engineering and co-ordination on an unprecedented scale. There were risks, but also huge profits to be made. Financiers in London and the provinces watched the canals keenly. It was new, daring – a young man's game. There was something 'joyful', to use Wedgwood's own word, about the way he attacked every aspect of his life at this time. As he said of the canals, his '*heart* was in it': arguments and actions came easily and his spirits soared as his different interests meshed effortlessly together. In the heat of midsummer, for example, when he and a colleague spent the day talking navigation plans at Worsley Hall, Bridgewater placed an order for 'the completest table service of Cream Colour that I could make', and showed them a Roman urn, said to be fifteen hundred years old, which his workmen had dug up near by. As the sun set, Wedgwood stepped into the Duke's gondola and sailed for nine miles along his canal, 'through a most delightful vale to Manchester'.

The wooing was not all one-sided. The Duke was now sorely short of money to extend his own canal and the expensive tea-service shows how keen he was to join Wedgwood's scheme. From spring onwards articles about the proposed canal – critical, satirical and supportive – flooded the press. By June Wedgwood had decided that to counter their opponents they needed a new, detailed pamphlet. Canal advocates feverishly collected information on the financial benefits and Darwin became unstoppable in his requests for information: 'I have no more facts for you,' Wedgwood eventually replied. 'If you are short you must create some.'[10] Soon Darwin sent the first two parts of a three-part essay that would eventually stretch to sixty-seven handwritten pages. (His surviving papers are full of drafts and redrafts, notes and loose pages, scribbles and crossings out.[11]) Wedgwood thanked him eagerly for the 'very ingenious and most acceptable paquet', which he was sure would do the trick when made public, 'the arguments are strong, conclusive and easy to be understood, and the dedication in my humble opinion is exceedingly clever'.[12] (The Dedication was to the Queen, humbly entreating her patronage and permission to call the canal 'The River Charlotte'.)

Wedgwood wanted Darwin to acknowledge authorship: 'Unless you break out from behind the cloud, what shall we do for a name for our pamphlet?' But Erasmus shied away, fearful of his standing as a doctor and anxious not to alienate local bigwigs who opposed the canal. In the end Bentley edited the pamphlet, incorporating Wedgwood's hard facts and toning down Darwin's flourishes (including the prized dedication). Wedgwood approved – apart from asking him to cut a section on the Potteries, 'lest our Governors should think it worth taxing'. Then he gave the edited draft to Darwin. This was an error – Wedgwood drily passed on a torrent of criticism. 'In general,' he told Bentley, 'he thinks the style rather too flat or tame, but be it remembered that he is himself a Poetical Genius.'[13] Next day, when he bumped into Darwin by chance at an inn at Uttoxeter, the doctor had even more comments at the ready: 'Dr Darwin says *nobody* writes *Grace*, & Rt. honourable, but Taylors & such like folks, so do as seemeth good in your own eyes.'[14] This small, snappy exchange encapsulates the delicate, sometimes jarring social differences. Here is a gentleman, telling manufacturers that elaborate courtesies towards the quality mark you out as sycophant, an oily shopkeeper trying to sell them something. And Wedgwood had the wit to see that Darwin's social know-how was useful, even if his assumption of superior status was galling.

In the spat between Darwin and Bentley all Wedgwood's humour and tact was called for. With Bentley he joked about the 'long & critical epistle from our ingenious & poetical friend', which doubtless 'afforded you entertainment & shook your diaphragm for you, whatever it may have done respecting your Pamphlet on navigation'.[15] To Darwin, on the other hand, he explained that the pamphlet had, alas, to be straightforward; suspicious folk would see rhetoric as a sign of deceit, and it was best 'to disappoint them with the manly Simplicity of Truth'.[16]

Equable though he was, Wedgwood found it maddening to be held up by literary squabbles. He appealed desperately to Bentley in mid-October not to let the doctor's comments postpone things:

Must the Uniting of seas & distant countrys depend upon the choice of a phrase or a monosyllable? Away with such hypercriticisms, & let the press go on, a Pamphlet we must have, or our Design will be defeated, so make the best of the present, & correct, refine, and sublimate if you please in the next edition.[17]

On 18 November the great pamphlet was finally printed. At this point, not surprisingly, Wedgwood collapsed. Darwin came over to Burslem and sat up all night with him, leaving a prescription which, Wedgwood reported, 'will strengthen the machinery and set it all to rights again'. Wedgwood now wrapped the whole pamphlet row up lightly, telling Bentley, 'The Dr. acknowleged he had wrote you two or three very rude letters, & said you had drub'd him genteely in return, which he seemed to take very cordially, & to be very well pleas'd with his treatment.'[18] He remained devoted to both his writer friends and, for his part, Darwin relished the whole affair.

Meanwhile Wedgwood took Sally to London. Amid canal business they went to the theatre and spent time with Ralph Griffiths and his circle at Turnham Green, and he met artists and designers and aristocratic patrons, 'taking patterns from a set of French China at the Duke of Bedford's worth at least £1500, the most elegant things I ever saw'.[19] Wherever he went, Wedgwood carefully left behind a copy of the canal pamphlet. But the Grand Trunk fight was not over. The tussle with the Cheshire Gentlemen continued: first they lowered their rates, then threatened to build a separate canal of their own. Bentley now suggested they avoid the Weaver by running the canal straight to Liverpool, an idea that fell in with Bridgewater's own plan to take his canal past Runcorn. Wedgwood and Bentley, rightly as it turned out, harboured serious doubts as to whether the Duke genuinely meant to push on with his route 'to serve the Public', or simply wanted to control the Mersey end. Pragmatically, they decided that the best policy was 'whatever be his *motives*, we endeavour to make his *actions* contribute to the execution of our plan'.[20]

They could make Bridgewater commit himself, and defeat the Cheshire men and win over recalcitrant landowners, only if powerful interests were *seen* to be on their side. Their last hope lay in Lord Gower. Now came a spell of cliff-hanging tension, with Wedgwood even refusing to use the mail because of spies. Scribbling a note to Bentley headed 'Monday morning before sun rising', he noted, 'I am ordered when our battle waxeth hot, which it soon will do, to keep 4 or 5 running footmen at my elbow & trust nothing of consequence to the post.'[21] Gower finally agreed to come down 'PUBLICKLY' to the Potteries as open patron, taking the chair at a key meeting on 30 December at Wolsely Bridge. Here Brindley presented his plans and a vote was passed to apply for a Bill in the next parliamentary session. That night there were joyous bonfires in Burslem marketplace.

Wedgwood had hardly a minute to call his own, telling Darwin, 'Indeed my dear friend, I am so hurryed up and down and allmost off my life with this Navigation that I have not time to write, speak or think out of that Magic circle.'[22] He was lobbying Parliament, giving evidence and combating counter-petitions from the Weaver and the Trent Navigations who fought the Bill right up to Committee stage. 'It is impossible to express the joy that appeared throughout the Potteries,' declared the *Birmingham Gazette* when the Bill passed the Commons. 'For nothing but an Inland Navigation can ever put their Manufactory on an Equality with their Foreign Competitors.'[23] In May the Act received the Royal Assent and in early June the Proprietors of the Trent and Mersey Canal made Wedgwood their Treasurer and decided on a motto, *pro patria populoque fluit* – 'It flows for country and people.'

On 26 July 1766, every pottery closed for the day and the local gentry and masters and workmen and their families, all dressed in their best, poured on to the low-lying land beneath the Brownhills in Burslem. At noon Wedgwood dug the first sod of the new canal and Brindley ceremoniously wheeled the earth away in a wheelbarrow. Speeches were made and toasts were drunk in Staffordshire ale, followed by feasts in all the Burslem inns. As evening drew on, a sheep was roasted for the poor and more bonfires were lit: 'a *feu de joie* was fired in front of Mr Wedgwood's house, and within a very large company assembled to partake of the bounteous hospitality which Mrs Wedgwood's skill as a housewife had prepared'.[24] Next day work began in earnest.

Although he worked as Treasurer for free, Wedgwood was hardly disinterested. From the beginning he had a particular, private interest in the Trent and Mersey. The Ridgehouse estate, where he wanted to build his new factory, was situated exactly where the planned canal route would go. In December 1767, when Mrs Ashenhurst died and the purchase was confirmed, Wedgwood moved like lightning. By the middle of the month he had been to Derby to discuss his plans with the architect Joseph Pickford, and had marked out the course of the canal. His long, low factory, like Soho, would have a fine Palladian front, and needed a graceful landscape to go with it, a 'serpentine line' as current taste decreed. But the fields were so flat that the canal could run straight, and after long arguments with Brindley's assistant, Hugh Henshall, he reported glumly, 'I could not prevail upon the inflexible Vandal to give me one *line of Grace* – He must go the nearest, & the best way, or Mr Brindley would go mad.'[25]

On the same day as the Trent and Mersey Act was passed, a separate group in Wolverhampton won an Act for a connection to the Severn: now Liverpool, Hull *and* Bristol would be linked by water. But building these canals involved staggering costs. The Staffs. and Worcs. had an authorized capital of £70,000, and the Trent and Mersey £130,000 (supplemented by two further amounts of £70,000 each, in 1770 and 1775). The money was raised in £200 shares, with no one allowed to hold more than twenty, and tolls were set at 1½d per mile. The routes, too, were complex. Brindley's canals were rarely straight – the rigid lines came with later engineers such as Rennie and Telford. Instead they curved around inclines, looped aside to serve collieries, and cut branches and tributaries to small towns and to factories. The great dipping arc of the Grand Trunk was 93 miles long, the Staffs. and Worcs. 46: both were 4 feet deep and 15 feet wide, allowing two 7-foot narrow boats to pass. On the two canals together, there were 120 locks, 4 aqueducts and several tunnels. Brickworks had to be set up, stone brought from quarries, stacks of timber supplied.

In midsummer 1766 gangs of skilled cutters and unskilled labourers began work with picks and shovels. Some were local farmworkers driven from the land by enclosures, some were men from the Fens, used to ditching and banking, others were vagrants and discharged soldiers or immigrants from Scotland and Ireland. The canals indeed saw the birth of the social phenomenon of mass, mobile labour. And as they moved from place to place, they took the *spectacle* of the industrial revolution with them wherever they went: this was sixty years before the building of the railways, for which the canals provided the precedent, the model, and even the name for the labourers – navvies, navigators. With the unskilled labourers came the specialists. Masons worked on the locks and aqueducts, while miners dug shafts from the surface and then blasted their way through the rock, setting gunpowder charges by candlelight, alert for sparks or floods. When the canal was finally dug the bottom was lined with 'puddle', clay kneaded with water to form a thick skin – for this hard, long task cattle were sometimes herded up and down to pound the clay, but elsewhere barefoot labourers painfully tramped the line for weeks.

Slowly the canals gouged their way through the landscape, and the gangs moved on from village to village. Brindley became a national hero, a model of how practical genius could triumph over low birth and near illiteracy. It was said that he was 'as plain a looking man as one of the boors of the Peake, or one of his own carters, but when he speaks, all ears listen, and every mind is filled with wonder, at the things he pronounces to be practicable'.[26] Although he could infuriate his grand clients by not delegating responsibility or by dodging meetings he was on site in rain and snow, surveying routes in all parts of the country. In 1767 Wedgwood wrote:

> I am afraid he will do too much, & leave us before his vast designs are executed; he is so incessantly harrassed on every side, that he hath no rest, either for his mind, or Body, & will not be prevailed upon to have proper care for his health . . .
>
> I think Mr. Brindley – the *Great*, the *fortunate, money-getting* Brindley, an object of Pity! . . . he may get a few thousands, but what does he give in exchange? His *Health*, & I fear his *Life* too, unless he grows wiser, and takes the advice of his freinds before it is too late.[27]

Wedgwood was right. Brindley was only fifty-six when he died in 1772, the year that the stretch of canal from the Trent to the Potteries, and the Staffs. and Worcs. Canal, were both finished. The northern line of his astounding Grand Trunk took another five years, slowed down by the immense engineering feat of the Harecastle tunnel, 2,880 yards long. But once it was opened his canal changed the region. The freight costs to and from the Potteries fell from 10d to 1½d a mile.[28] Lime and manure for the fields were often carried toll-free, new markets were opened and local labourers and farmers made money from construction work or even took to the water, running their own boats. There was an end to local grain monopolies, allowing cheaper flour and bread, and cheap coal for fires, even in rural areas, where theft of timber and destruction of hedgerows were crimes often born of desperation.

The years between 1768 and 1776 saw the first great canal boom. Progress then slowed – dampened by rising interest rates and the impact of the American Wars – until the 1790s saw a new canal mania. From the start, a nationwide web seemed a real possibility. Brindley was full of plans. In 1768 came an Act for the Coventry Canal, to join the Trent and Mersey near Lichfield, and a year later another authorized an extension to the Thames at Oxford (and thence, of course, to London). A further Act of 1768 paved the way for the Birmingham Canal, linking the town to the Staffs. and Worcs. and thus to the Severn and the Trent.

The Birmingham Navigation Committee contained familiar names, including Garbett, Small and Boulton, although the latter was too busy to be active. In late 1765 Darwin had tried to involve him in the battle against the Weaver Navigation. 'I desire you & Dr Small will take this Infection', he wrote, 'as you have given me the Infection of Steam-Enginry: for it is well worthy your attention, who are Friends of Mankind, & of the Ingenious Arts.'[29] Boulton casually signed the petition and was elected to the Committee of the Trent and Mersey, but he backed away hastily when he was fined for missing a meeting. He was busy moving his whole workforce and family to Soho, he explained to Wedgwood; 'I have now & have had all this summer about 70 Masons Carpenters & Labourers at Work.'[30]

While the metal trades would undoubtedly gain, the most interested parties in Birmingham were the local coal-owners. Within a year £55,000 in £100 shares had been raised, to build a curving canal with cuts running off to the collieries. (By contrast, the campaign for the new infirmary came to a halt for several years for lack of subscriptions.) Once the committee realized the potential profits from coal they rushed to build boats and urged the builders on, regardless of expense. Almost as soon as the first stretch of canal from the collieries to the new basin at Paradise Street was opened in November 1769 the price of coal fell from over 15 shillings to 4 shillings a ton, steadying around 7 shillings. The company controlled the whole chain: buying the coal, transporting it, loading it and landing it day and night and even distributing it in their own carts.

Ironically, within two months the town was almost choked. And although the profit margin was kept low and offices were set up to distribute coal to the poor, canal shares shot up from £100 to £140 within a year and cries of monopoly were heard.

Boulton was having his own disputes over the canal. Only when work started did it dawn on him that the planned extension to New Hall, the old Colmore estate whose lands ran down towards Soho, posed a real danger to his own water supplies – as Garbett had warned him long ago.[31] Acting as Boulton's lieutenant, Small negotiated behind the scenes. The New Hall extension mysteriously disappeared from the plans, due to an 'unforeseen difficulty with the tunnel'.[32] Charles Colmore was enraged and took the canal company to the King's Bench and then to Parliament. Small worked even harder, dashing to London, sending weekly – even daily – express letters, while carefully keeping Boulton's name out of it, telling their lawyer, 'you are allowed to read Mr Boulton's letter but not to speak of it'.[33] Garbett, noted Small, was 'subpoeaned into Mr C's camp', and the engineers joined battle on opposing sides, Brindley speaking for the committee, Smeaton for Colmore. (Boulton scornfully rejected the evidence of this 'whole tribe of Jobbing ditchers'.[34]) The case dragged on until Parliament supported Colmore and the Newhall Ring was built.

By September 1772 all was forgotten. The canal was finished: a great flight of locks carried the narrow boats to Wolverhampton and a complete new port grew up at Stourport. Boulton wrote proudly to the Earl of Warwick, 'Our navigation goes on prosperously . . . we already sail from Birmingham to Bristol and to Hull.'[35] Over the next forty years a filigree of waterways threaded across the district.

Almost all the Lunar men owned canal shares, and Boulton had a lucrative sideline supplying metal parts, locks, bolts, brass valves for pistons, copper boxes, taps and rings.[36] Wedgwood was the king, but Small's interest was also substantial as he gradually bought transfers from other shareholders. He and Boulton even began to dream of working canal boats by steam. And he carefully kept abreast of developments elsewhere through a new friendship, with James Watt.

Canals were being promoted with equal fervour in Scotland, and in late 1766 Watt was joint surveyor of a route from the Forth to the Clyde,

A perspective view of Stour Port, engraved by Peter Mazell after a drawing by James Sherriff, 1776

backed by Roebuck's Carron Ironworks. In the spring of 1767 he went to London to promote a Bill to authorize this. His mission failed, since another route was chosen instead, but on the way back he visited the Bridgewater Canal and before that stopped in Birmingham to meet Roebuck's partner, Garbett. Boulton was away but Small took him round Soho, and he also stayed at Lichfield with Darwin.

Darwin, Small and Watt took to each other at once, and Small's correspondence over the next few years would sustain Watt in many bouts of depression. At this point it looked as though his future might even lie in civil engineering: he surveyed the harbour for Port Glasgow, worked on the canal from the Monkland coalfields, made improvements to Ayr harbour and planned a fresh water supply for Greenock. Then in 1773 he set off to the Highlands, to see if a canal could be built through the wild, lake-dotted country between Inverness and Fort William, joining the North Sea to the West Coast.

At one stage Small hoped to lure Watt south as surveyor to the floundering Coventry Canal. Replying to Small's queries Watt sent meticulous details of depths and plans and locks and costs and methods of saving water.[37] But he really never enjoyed this work: he was out in all weathers, working in near-impossible conditions, and found it difficult dealing with demanding employers and hard-pressed labourers. In September 1770 he admitted:

> Nothing is more contrary to my disposition than bustling & bargaining with mankind, yett that is the life I now constantly lead . . . I am also in a constant fear that my want of experience may betray me into some scrape, or that I shall be imposed upon by the workman, bothe which I take all the care my nature allows to prevent. I have been tolerably lucky yett; I have cut some more than a mile of canal besides a most confounded gash in a hill & made a bridge & some tunnels for all which I think I am within the estimate, notwithstanding the soil has been of the hardest, being a black or red clay engrained with stones.[38]

In the surveying and building of canals, locks and tunnels the stock of knowledge about hydraulics and engineering was immeasurably increased. Watt's letters were full of his improvements to instruments, such as the delicate adjustments to his surveying level, using a telescope, creating his 'cross-hair micrometer' for measuring distances (known later as a telemeter). 'Have you invented any new gimcracks lately?' Small used to tease him. In 1777 even Darwin designed a revolutionary canal lift. Ten years ahead of its time, the lift was intended to avoid locks, or to link a canal with a river below – raising and lowering the boats in two massive water-filled wooden boxes hung on chains, one acting as counterweight to the other.[39]

The drive for canals caught up all sorts of men, from the pessimistic Watt to the super-confident Wedgwood and Darwin the indefatigable inventor. Through canals such men literally made their mark on the land but they also unearthed its secrets, from the clay of the Monkland to the fossils of the Midlands. Their canal work involved them in the practical hurly-burly of the early industrial world, the intricacies of finance and lobbying and marketing. But they were still natural philosophers, and in building canals – even while transforming whole regions – they could still seek after hidden knowledge.

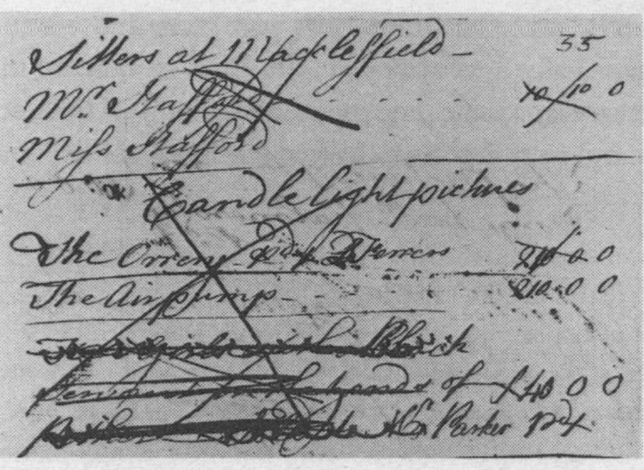

Mr. Stafford
Miss Stafford
Candle light pictures
The Orrery
The Air pump

11: PAINTING THE LIGHT

The canals that cut across the land, meeting, joining, intersecting, brought together Wedgwood and Darwin, Boulton, Small and Whitehurst. So far, when a couple of friends had corresponded, there was only a two-way run of information. Now they could turn for help to three, four, five people. There was a real sense of bartering, tit-for-tat swapping of knowledge – an exhilarating, companionable sharing of interests.

The alluring sociability of this culture can be felt in paintings from this time by Joseph Wright of Derby. Wright was never a formal member of the Lunar group but he was certainly a Lunar man. He had grown up in Derby as a close neighbour and friend of John Whitehurst and around 1766 he became a patient of Darwin, who treated him for years for the lethargy and depression that alternated with his spells of creativity. His interest in light, in the details of apparatus, in the technicalities (and market realities) of his art, and in the long tradition of natural philosophy, shown in paintings such as *The Alchymist*, gave them much common ground.

In the 1760s Wright painted a remarkable series of nocturnes, 'candle-lit studies', one of which, showing three young men admiring a brilliantly lit model of the 'Borghese gladiator', was greatly admired at the Society of Artists in 1765. But the following year he caused an even greater stir when the centrepiece in his new painting was not a classical statue but a beautifully made orrery, demonstrating the movement of the planets round the sun. Science can fire the polite imagination, Wright implies, just

Account for 'Candle-Light pictures', Joseph Wright's Account Book as strongly as classical art. In *A Philosopher giving that lecture on the Orrery, in which a lamp is put in place of the Sun*, the lamp is there so that the lecturer can explain an eclipse: the sudden darkness that was once so terrifying but that science can now explain. The scene is domestic, but the suggestion is that natural philosophy is open to all – to men and women, old and young – from the white-haired lecturer and the young man taking notes to the solemn girl in her ribbon-trimmed hat and the little girl with her arm round her brother, smiling as she points out the orbit of Saturn's moons. The light catches the angles of cheeks and foreheads, as the sun does the peaks and planes of the earth.

The *Orrery* may have been an opportunistic work, painted in the hope that it would be bought – as indeed it was – by Earl Ferrers, the passionate amateur astronomer with whom Wright's friend Peter Perez Burdett stayed while making a topographical survey of Derbyshire. Ferrers was one of the Derby 'philosophers'; he had built an expensive orrery, and had been elected to the Royal Society in 1761 for his paper on the transit of Venus. There is something of Whitehurst, crossed with Newton, in the grey-haired, benign philosopher, while the man taking notes is almost certainly Burdett.[1] Whitehurst was already working on his theory of the formation of the earth, of whirling particles drawn together by gravity in space. And although the painting has no direct connection with the Lunar Society, it catches one aspect of their hopeful view of the world, its progress defined by solid geometry – orderly, harmonious, serene.

The opposite seems true in the companion picture, *An experiment on a bird in the Air Pump*, exhibited in 1768. A rare white cockatoo flutters in panic as the air is sucked from the glass that contains it: although we can see that the tranced, bard-like philosopher is just about to flood back the life-giving air, the outcome is uncertain, the moment still shocking, and we recoil, as the two girls do. The painting is crammed with allusions: to the bird as emblem of love and to the arrogance and potential cruelty of experiment itself in the figure of the boy at the back, which echoes a boy pointing to a skeleton in Hogarth's violent scene of dissection, the last of his *Four Stages of Cruelty*. This vision of the active force of nature, the perilous, precious breath of life, is in dramatic contrast to the serene heavens of the *Orrery*, gestured at by the sailing moon, glimpsed through the window.

Wright's paintings were enormously popular when they were engraved as prints. He showed no new science – the orrery and the air pump had been familiar to lecture audiences for years. Instead he created brilliant, evocative and complicated depictions of two different sides of science, the theoretical and the experimental, the orderly Newtonian mechanics and the volatile experiments with air.[2] These are domestic pictures, with no worries about keeping children away from disturbing experiments or costly apparatus. Much the same feeling comes from an anecdote about Richard Lovell Edgeworth, who horrified the visiting French philosopher Pictet by passing his precious pocket sextant around among even the smallest children.[3] In these balanced studies of intimate, awe-struck investigation of the laws of the universe and the forces of nature, Wright marvellously suggests the absorption, the frisson of danger, and the shared, profound pleasure of the experimental life the Lunar men so enjoyed.

In 1766 their Midland group was augmented by the arrival of the young Irish inventor Edgeworth (who soon introduced his eccentric, idealistic friend Thomas Day), and Darwin's old Edinburgh friend James Keir, who had decided to leave the army and find a job and a wife, perhaps 'some Lichfield *fair* that has more money and love than wit'.[4]

The circle met often. When Watt's friend Robison passed through Lichfield in July 1768 (being 'quite charmed with Dr Darwin's unaffected ease and civility') he found Small there too, and 'your very warm friend Mr. Boulton'.[5] Their meetings were as much for pleasure as business. As the years rolled on, they tried to dine at two o'clock, and usually planned to stay until at least eight.[6] The wine flowed (despite Darwin's later advocacy of temperance) and the tables were heavy with fish and capons, Cheddar and Stilton, pies and syllabubs. At dinner the wives sometimes joined the men and the children dashed in and out. But when the meal was cleared away, out came the instruments, the plans and the models, the minerals and machines. In the house and in the workshops they talked long into the evening. Often the 'philosophical feast' swept on to the next day, and even longer if they were working on a particular project. (In the early years, they regularly stayed overnight when they visited Darwin in Lichfield.) They developed their own cryptic, playful language and Darwin, in particular, liked to phrase things as puzzles – like the charades and poetic word games people used to play. Even though they were down-to-earth champions of reason, a part of the delight was to feel they were unlocking esoteric secrets, exploring transmutations like alchemists of old. Boulton must tell him, Darwin demanded teasingly years later, had they 'discover'd the longitude and philosophers stone. And above all if you keep yourself and friends in good spirits, and happyness.'[7]

They were sharing an exuberant spread of knowledge. The mood was typified perhaps by the publication, between 1764 and 1766, of a three-volume work edited by Henry Temple Croker, which pinched plates from Diderot's *Encyclopédie* and other sources, and bore the modest title of *The Complete Dictionary of Arts and Sciences. In which the whole circle of human learning is explained, and the difficulties attending the acquisition of every art, whether liberal or mechanical, are removed.* As they worked and talked, they felt part of a wider community linked to the Royal Society, the Society of Arts and other clubs. In October 1766 Boulton wrote to Anne from London, 'Having nothing to do I am (and have during this week) so intensly engaged with Phylosophers & Artists that I cannot get to write any other particulars in this post than that I am in good health.'[8] Keir soon introduced Edgeworth into 'a society of literary and scientific men' who met at Slaughter's Coffee House in St Martin's Lane, its members over the years including the surgeon John Hunter, the botanists Joseph Banks and Daniel Solander, John Smeaton the engineer and the instrument-maker Jesse Ramsden. Wedgwood, Bentley and probably Small and Boulton were also members, and Edgeworth vividly described the quick talk of 'the first hints of discoveries, the current observations, and the mutual collision of ideas'.[9] Arguments were fierce and fools were jeered, but as 'the knowledge of each member of such a society becomes in time disseminated through the whole body', so the 'talents of numbers' could forward the ideas of a single person.

In the lives of these men, the light of knowledge shone like the candle, or lamp, or glowing furnace – or even the glittering moon – at the heart of so many of Joseph Wright's paintings. The Lunar men were joined in a shared hunt for illumination, a quest that would bind them ever more closely together as the decades passed.

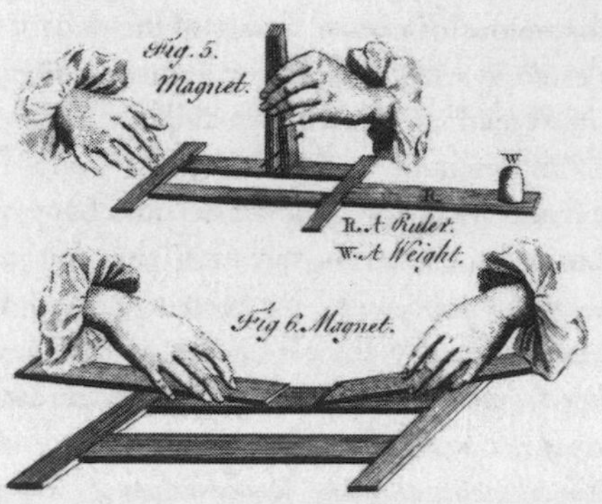

12: MAGIC & MECHANICS

In the summer of 1766, as the navvies began digging the canal at Burslem, Darwin wrote jubilantly to Boulton:

I have got with me a mechanical Friend, Mr Edgeworth from Oxfordshire – The greatest Conjuror I ever saw – G–d send fair Weather, and pray come to my assistance, and prevail on Dr Small and Mrs Boulton to attend you, tomorrow Morning, and we will reconvoy you to Birmingham on Monday, if the D–l permit, adieu
E Darwin

He has the principles of Nature in his Palm and moulds them as He pleases.

Can take away Polarity or give it to the Needle by rubbing it thrice on the Palm of his Hand.

And can see through two solid Oak Boards without Glasses, wonderful! astonishing! diabolical!!! Pray tell Dr Small he must come to see these Miracles.[1]

Here is Darwin the showman, a style he liked and would often use again, whipping back the curtain with a ringing 'Roll up! Roll up!' Here too is the Lunar fascination with power over Nature, the surprise and slightly blasphemous glamour of scientific 'miracles'. Darwin loved the lure of magnetism, the old power of the loadstone, so long linked with magic. Years later he presented his friend John Michell (who developed techniques for making artificial magnets) as a sorcerer controlling the mystic powers of earth:

John Canton's method of making artificial magnets

Last MICHELL's hands with touch of potent charm
The polish'd rods with powers magnetic arm . . .
The obedient Steel with living instinct moves,
And veers for ever to the pole it loves.[2]

The Birmingham men were intrigued by the sound of Edgeworth. On the back of Darwin's note Boulton scrawled a reply: he and Anne were coming over on Sunday to dine with the local MP and landowner John Levett and 'should be glad to place Dr Small in the middle of the Chaise and drop him at Litchd. where he may enjoy a philosophical feast and return with Boulton and Wife on Monday Noon to Birmingham'.[3] Boulton was particularly interested since he was developing a line in artificial magnets for export.[4] He asked Edgeworth over and sent him round Soho and other workshops. In a few hours, Edgeworth said, he 'became intimately acquainted with many parts of practical mechanicks, which I could not otherwise have learned in many months'.[5] But much as he admired Boulton, it was Small – 'a man esteemed by all who knew him, and by all who were admitted to his friendship beloved with no common enthusiasm' – whom he saw as holding together this group of men, of very different characters, but all devoted to literature and science. Their intimacy, he wrote touchingly in later life, 'has never been broken but by death'.[6]

In 1766, Edgeworth was twenty-two – a good twelve years younger than Darwin – a tall, dark-haired, talkative, arm-waving Irishman. His conjuring was a side interest, taken up with the rake Sir Francis Delaval with whom he put on entertainments based on his working out of the magnets and mechanisms used by the great French conjuror Comus. (Comus was all the rage at the time: Watt saw and admired him the following spring, telling Peggy that 'Comus the conjuror is really a surprising fellow and very inexplicable.'[7]) Delaval also allegedly sabotaged a show at the Haymarket Theatre – though this is a bit hard to believe – by wielding his own powerful magnet from the audience and thus rendering the conjuror, Philip Breslau, totally unable to control his magnetically driven 'learned little swan, which was supposed to pick out letters from a bowl in response to questions'.[8] As with electricity, huge rows blew up between 'performers' and 'philosophers' about the right to display natural forces.

Edgeworth's own pulling power was potent. He tended to go at everything headlong, from mechanics to marriages, eventually notching up four wives and twenty-two children. As a student at Trinity College, Dublin, he lived at such a drunken pace that his alarmed father sent him to Oxford, lodging him with an old friend, a down-at-heel lawyer, Paul Elers (the son of J. P. Elers, one of the potters who had worked in Burslem at the start of the century). The ruse backfired: within a year Edgeworth had to marry one of the Elers daughters, Anna Maria, and their first son was born the following year. Hauled back to Ireland, Edgeworth spent a bored year reading 'some law and more science' and making a wooden orrery.

By late 1765, having come of age, he escaped and took the boat to England. Arriving in Chester, he saw the Microcosm, the mechanical exhibition of the planets, and he went back so often that he made friends with the showman, who mentioned that at Lichfield he had met another inquisitive spectator, Erasmus Darwin. He then described a carriage Darwin had invented, which could turn in a small circle without overturning and 'without the encumbrance of a crane-necked perch'.[9] Edgeworth immediately seized on this and at his new house in Hare Hatch, Hertfordshire, he designed a phaeton on Darwin's principles and had it built by a local coachmaker. He then took it to London: soon Sir Francis Delaval had one like it, and the design received official approval from the Society of Arts.

He had told the Society that this was Darwin's idea and now wrote to Darwin, who invited him to Lichfield. When he arrived, Polly was alone, and they talked together until supper finished, when there was a loud rapping at the door and a bustle in the hall. Hearing Polly exclaim that they were bringing in a dead man, Edgeworth dashed out to find Darwin directing his servants, who were lugging in an apparently lifeless body.

> 'He is not dead,' said Doctor Darwin. 'He is only dead drunk. I found him', continued the Doctor, 'nearly suffocated in a ditch; I had him lifted into my carriage, and brought hither, that we might take care of him tonight.'
>
> Candles came, and what was the surprise of the Doctor, and of Mrs Darwin, to find that the person whom he had saved was Mrs Darwin's brother! who, for the first time in his life, as I was assured, had been intoxicated in this manner . . .[10]

(That sly 'as I was assured', is very Edgeworth.)

Darwin seemed surprised that the visitor and his wife had spent the evening together until Edgeworth's classical allusions explained all: Darwin had assumed from his letter that he was 'only a coachmaker' – a good example of the currents of class beneath the surface democracy of natural philosophy. Once recognized as a gentleman, Edgeworth was quickly introduced to Lichfield society, dining with the Sewards, flirting with the poetic Anna (until Polly Darwin ironically raised her glass to 'Mrs Edgeworth'), and spending evening after evening 'in different agreeable companies'.

He was in some ways a younger version of Darwin: classically educated, socially at ease. Unlike Darwin, however, Edgeworth did not work. In Oxfordshire, reading in a desultory way for the Bar, he lived on an allowance from his father, played cards with the gentry and befriended other amateur inventors, such as the clergyman Humphry Gainsborough (brother to the artist), an expert in hydraulics who was working on improving the steam-engine. To fill his empty hours, Edgeworth piled scheme upon scheme. Many of his ideas were ahead of their time: to win a bet with the gamblers and racing men surrounding Delaval, he invented a crude telegraph system to carry news from Newmarket races to London, a system he returned to and improved over many years.

Among the 'various contrivances' he planned in the workshop behind his house were a 'sailing carriage' (a kind of go-cart with sails, highly disconcerting to travellers on the Reading road), a machine for cutting turnips; an umbrella for covering haystacks; a cart on rollers to protect road surfaces. For months he worked on a huge hollow wheel containing a barrel in which a man could stand, pushing it forward faster and faster by walking, like a mouse on a treadmill. This came to a cataclysmic end when some local boys launched it on a slope running down to a chalk pit: the wheel gathered speed, the boy inside jumped clear; the barrel vanished. Returning from London, 'to my no small disappointment', Edgeworth found his wheel at the bottom of the chalk pit 'broken into a thousand pieces'.[11]

This was hardly rational mechanics. In fact many of Edgeworth's ideas have a touch of fantasy, a visionary, almost Gothic quality as if they came to him in a dream, like the giant wooden horse, which he still remembered fondly in his old age:

> I was riding one day in a country, that was enclosed by walls of an uncommon height; and upon its being asserted, that it would be impossible for a person to leap such walls, I offered a wager to produce a wooden horse, that should carry

me safely over the highest wall in the country. It struck me, that, if a machine were made with eight legs, four only of which should stand upon the ground at one time; if the remaining body were divided into two parts, sliding, or rather rolling, on cylinders, one of the parts, and the legs belonging to it, might in two efforts be projected over the wall by a person in the machine; and the legs belonging to this part might be let down to the ground, and then the other half of the machine might have its legs drawn up and be projected over the wall, and so on, alternately. This idea by degrees developed itself in my mind so as to make me perceive, that as one half of the machine was always a road for the other half, and that such a machine never rolled upon the ground, a carriage might be made which should carry a road for itself.[12]

Even after forty years of experiments and a hundred models, this never worked. But it was not all nonsense; it looked forward to the caterpillar track, the tank and the train. 'It is already certain', Edgeworth added, 'that a carriage moving on an iron rail-way may be drawn with a fourth part of the force requisite to draw it upon a common road.'

His interest in transport was very much of its time, even if he took it to extremes. The new turnpikes had made record-breaking and road races the talk of the town. Flashy models were like sports cars and advanced carriage design was a subject of lasting appeal to almost all the Lunar men. In 1767, Watt was poring over sketches of Darwin's plans for axles and steering. Five years later, Wedgwood described a super-fast journey to London in a new carriage, made by the Lichfield coachmaker to Darwin's design, 'with *patent spring wheels*':

every spoke being a spring by which means the Vis inertia of all sudden obstructions it meets with upon the road are overcome without any Jolt to the rider, or what is more with perfect ease to the poor horses. Another advantage these carriages have over the common ones is their *stillness*, and as every spoke in the wheel is in the form of *the line of beauty*, I am told they have the most elegant appearance.[13]

As late as 1780 Boulton was still filling notebooks with sketches of carriage designs, adding lift-up paper flaps to show how the carriage looked with its hood up or down. In the cause of science, he worked out the speed and pulling power of horses and took detailed measurements of all his friends' conveyances, including 'Fothergill's coach', 'Mr Watt's one Horse Chaise', 'Mr Edgeworth's phaeton 4 wheels', and 'Dr Darwin's post Chaise'.

Darwin was undeniably the leader here. He spent much of every day on the road, and had a vested interest in improving his transport. His brightly painted carriage was known around the county, crammed with books, medicine bags and hampers of food to keep him going. After Edgeworth's first letter, the Society of Arts wrote to Darwin asking for details of his high-wheeled design, and in reply he said that he had already built a phaeton and a post-chaise, in each of which he reckoned he travelled about ten thousand miles a year. In the previous six years he had added springs, worked on his new steering principles and mechanism (developing a revolutionary system not used again until early modern motor cars). He allowed Edgeworth to develop these improvements and take the credit, in the form of a gold medal from the Society in 1769.[14]

Edgeworth even took up Darwin's plans for a steam-carriage, a subject that roused a flush of competition in Watt. In June 1768 Darwin told Wedgwood that Edgeworth had 'nearly completed a Waggon drawn by Fire, and a walking Table that can carry 40 men'.[15] Two months later Small was warning Watt that the Irishman ('young and mechanical and indefatigable') had 'taken a resolution of moving land and water carriages by steam, and has made considerable progress . . . He knows nothing of your peculiar improvements.'[16] Edgeworth forged ahead and Small used his work, and that of other rivals such as 'Linen-draper Moore', to prod Watt into action. 'Come to London with all possible speed,' he urged:

> At this moment how I could scold you for negligence. However if you will come hither soon I will be very civil, & buy a steam-chaise of you & not of Moore. And yet it vexes me abominably to see a man of your superior genius neglect to avail himself properly of his great talents.[17]

In the end, however, both Edgeworth and Watt put the steam-carriage aside, and although the idea would be taken up by others in the 1780s (much to Watt's chagrin), this too was an invention that could not be realized until technology improved. What sometimes seemed quirky, fantastic, hypothetical was in fact brilliantly ahead of its times.

During these years, all the members of the group were involved in making things as well as in experiments, and the interest helped to forge close links between them. In May 1767, Wedgwood visited Soho, where he was extremely impressed by an improved lathe, fitted with a rosette, a guide tool for intricate work, which he hoped to persuade Boulton to part with.[18] He was equally taken by Boulton himself, whom he thought 'the first – or most complete manufacturer in England, in metal. He is very ingenious, Philosophical & Agreeable.'

It was later that summer, on his way back from London on canal business, that Watt visited Birmingham and Lichfield for the first time. By the end of August, Darwin was already addressing him warmly:

> Now my dear new Friend, I first hope you are well, and less hypochondriacal; and that Mrs Watt and your Child are well. The plan of your Steam-Improvements I have religiously kept secret, but begin myself to see some difficulties in the Execution that did not strike me when you was here . . . I wish the Lord would send you to pass a week with me, and Mrs Watt along with you, – a Week! – a Month, a Year.[19]

Watt's possessiveness about any form of steam-engine development was intense, yet his own work crawled at a snail's pace. Many things intervened, including his canal work, and it was not until April 1768 that he felt he had made real progress with the second experimental engine built at Kinneil. He had found a better way to pack the piston ('tho this new apparatus is not perfectly air tight') and was relatively happy with his trials: 'I found the engine could easily make 20 strokes pr minute & snift properly when the steam was middling strong.' At last Roebuck was persuaded to back him financially, agreeing to pay off Watt's debts to Black and his late partner, Craig, in return for a two-thirds share in the invention. A few days later, it was agreed that Watt should apply for a patent, and in July he set off for London. On his way back he stayed at Soho for a fortnight: it seemed a mechanic's paradise, a promised land, a wonder of modernity with Boulton its chief sorcerer.

This was just the impression Boulton wanted to create. By now well aware of Watt's work, he was keen to lure him south and on this visit was already discussing a possible partnership. The sight of Soho made the situation even more agonizing for Watt and the letter he wrote when he arrived back in Glasgow in October was almost despairing: Roebuck was so busy at Bowness that he could not spare the time to organize the production of parts, and so 'the greatest part of it must devolve on me, who am from my natural inactivity and want of health and resolution, incapable of it'.[20] For his part, Roebuck was becoming more and more frustrated with his engineer: 'You are letting the most active part of your life insensibly glide away. A Day, a Moment, ought not to be lost.'[21]

Early the following year, pushed by Watt, Roebuck offered Boulton the right to manufacture engines for three counties only, Staffordshire, Warwickshire and Derbyshire. But Boulton wanted all or nothing. 'I was excited by two motives to offer you my assistance,' he told Watt, 'which were love of you and love of a money-getting, ingenious project.' He had seen that the engine would need investment, accurate workmanship and extensive research, and wanted to keep it out of the hands of ignorant 'empirical engineers':

> To remedy which and produce the most profit, my idea was to settle a manufactory near to my own, by the side of our canal, where I would erect all the conveniences necessary for the completion of engines, and from which manufactory we would serve all the world with engines of all sizes. By these means and your assistance we could engage and instruct some excellent workmen (with more excellent tools than would be worth any man's while to procure for one single engine) could execute the invention 20 per cent. cheaper than it would be otherwise executed and with as great a difference of accuracy as there is between the blacksmith and the mathematical instrument maker.[22]

Placing Watt on a high hill, tempting him with the advantages of scale and the promise of dreams fulfilled, Boulton signed off: 'It would not be worth my while to make for three counties only; but I find it very well worth my while to make for all the world.'

He could vary his rhetoric, his tone of appeal. He had hinted at a partnership at their Soho meeting, he said, because he sensed that Watt 'wanted a midwife to ease you of your burthen, and to introduce your brat into the world'. It would take a while yet for that mechanical brat to be born, but as it happened (and as Boulton may well have guessed), he was writing only two days after the birth of Watt's son James. With another mouth to feed, the ever anxious Watt welcomed the sympathy as much as the promise of money and technical support. 'Pray remember,' Boulton urged him, 'when you come, always to put up at the *Hotel d'Amitié sur Handsworth Heath*, where you will always find a friendly reception.'

Boulton was the most ambitious, but the needs of the growing industries and the thought of a 'money-getting, ingenious project' spurred them all. Darwin, for example, designed a horizontal windmill for Wedgwood to grind colours ('if it should happen to grind anything',[23] Josiah noted sceptically in the margin of Darwin's letter explaining its principles). The windmill sail was placed horizontally at the top of an octagonal tower with slatted boards – when the wind, from any direction, hit the boards, they lifted, carrying the air up to the sail. Darwin claimed it had a third more power than the normal, vertical windmill and built a three-foot working model, appealing to the Society of Arts for help in constructing a full-scale version. When he saw the model, Wedgwood was won over. But although he wanted the windmill at once, Darwin now advised him to hold back in view of Watt's work on steam, which would be 'a Power so much more convenient than that of Wind'.[24] Watt, of course, was slow. In 1771 Darwin wrote wistfully, 'your windmill sleeps at my house'.[25]

Some of the mechanical marvels they made were huge – like the steam-engine or the windmill – but others were delicate and intricate. Around 1771 Small became preoccupied with clocks – not the elaborate astronomical clocks, with their globes, calendars and models of the planet, like those Whitehurst and James Ferguson made, but small timepieces with radically simplified mechanisms, employing one wheel or even (hard to visualize) no wheel at all. Small patented his design in 1773, hoping, he said, that his watches might become 'an article of commerce' and inevitably Boulton flirted briefly with clockmaking at Soho.

Watt and Small also both experimented with telescopes: 'I propose when you & I meet to spend some of our Leisure time on trying some Experiments &c on Glass for Dollands,' wrote Watt in January 1770.[26] Darwin too became interested in telescopes, working out a complex system of mirrors to improve the brightness of the image. On the theoretical side, helped by John Michell, who lived nearby in Yorkshire, Joseph Priestley had also now moved from electricity to optics. His *History and Present State of Discoveries relating to Vision, Light and Colours* continued a series that he hoped would cover the whole of experimental philosophy. It was published in 1772, giving a full account of optical instruments, and questioning the nature of light itself. The book was not a commercial success, and Priestley abandoned his ambitious series, but it had the unexpected outcome of an invitation to join Cook's second voyage of discovery as astronomer. Hesitating – thinking of his family – he consulted Mary and

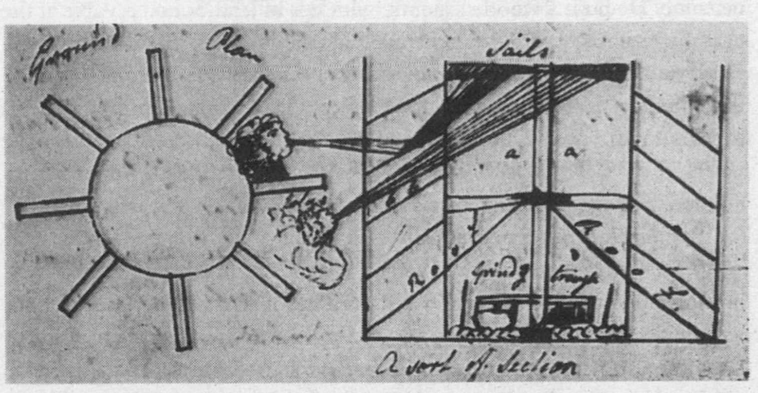

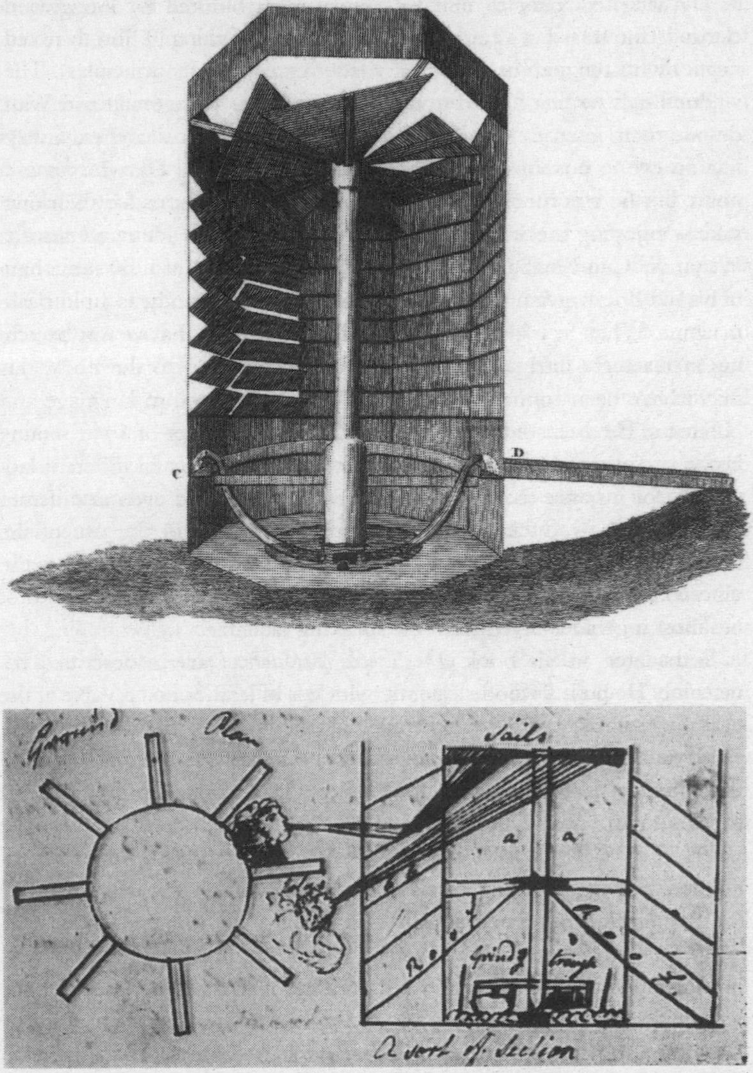

Darwin's horizontal windmill, from *Phytologia*, 1800, and Wedgwood's sketch in his letter to Bentley, 1768

finally accepted, only to find his appointment blocked by intransigent clerics in the Board of Longitude, whom Joseph Banks told him were outraged by his religious and political views.[27]

Priestley's writing helped to add to his income, while Small and Watt, despite their intense, absorbed and often frustrating researches, always had an eye to possible money-making schemes. So did Darwin, up to a point, but he was more in love with speculative inventions for their own sake – enjoying the whole process of thinking up an idea and asking: Why not? Can I make it work? As a doctor, retaining at least something of his old Boerhaavian training, he could even see the body as an intricate machine. When he told his old friend Albert Reimarus that he was becoming increasingly intrigued by the relation of the mind to the body, this might have been connected to another current interest, in language and phonetics. He discussed this with Franklin in the summer of 1771, noting how the consonants and vowels were formed differently in different languages, for instance the 'W' in Welsh and German, and even in different dialects, such as Cockney or Northumbrian. How did the use of the mouth and jaw and throat affect sound?[28] Patiently, he analysed the vibrations of air for each sound, and this in turn led to one of his oddest but most ingenious inventions: the speaking machine.

Years later, in his book *The Temple of Nature*, Darwin described his machine. He built a wooden mouth, with lips of leather and a valve at the back for nostrils, which could be opened and closed by the fingers. A silk ribbon was stretched from this between two hollowed pieces of wood, so that when he blew on it with some bellows it vibrated, creating a sound like a human voice. His head could produce the sounds *p*, *b*, *m* and the vowel *a*, so well 'as to deceive all who heard it unseen, when it pronounced the words mama, papa, map and pam; and had a most plaintive tone, when the lips were gradually closed'.[29]

Typically, Darwin never got round to finishing his model, although he imagined it singing to the piano and even believed that if a really gigantic version was built, it 'might speak so loud as to command an army or instruct a crowd'. It was a fantastically complicated project since the pitch differentiation in speech depends largely on shortening and lengthening the entire vocal system by moving the larynx, widening and narrowing the cheeks, and by the dramatic shape changes made by the tongue.[30] But even unfinished, his talking head caused a sensation. At the end of the century, Edgeworth heard a new French machine which he said was not a patch on Darwin's, and remembered how he had 'placed one of your mouths in a room near some people in 1770, who actually thought I had a child with me'.[31] In September 1771 Boulton endowed this wonder of mechanics with the businessman's ultimate accolade, a contract, duly witnessed by Small and Keir. It was a joke, but not altogether – he had, after all, nothing to lose:

> I promise to pay to Dr. Darwin, of Lichfield, one thousand pounds upon his delivering to me (within two years from the date hereof) an Instrument called an organ that is capable of pronouncing the Lord's Prayer, the Creed and Ten Commandments in the Vulgar Tongue, and his ceding to me, and me only, the property of the said invention with all the advantages thereto appertaining.[32]

Like the Chester Microcosm, or Comus's tricks with machinery and magnets, or Edgeworth's eight-legged, wall-jumping wooden horse, Darwin's famous speaking machine proved once again that mechanics could have all the allure of magic. Even human speech, from the miraculous babbling of a baby to the crowd-pulling power of a dictator, could be reduced to theory and reproduced mechanically. As Darwin himself had said of Edgeworth's conjuring, it was 'wonderful! astonishing! diabolical!!!'

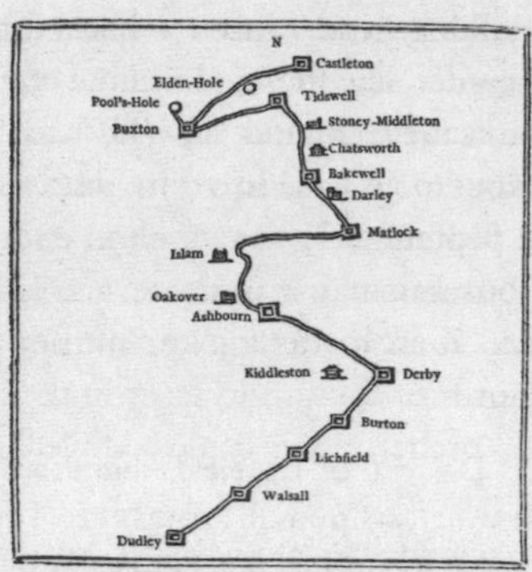

13: DERBYSHIRE EXPLORATA

From the end of the seventeenth century visitors came to the Peak District of Derbyshire, winding up from Lichfield to Derby, then up the Derwent valley to Matlock, taking detours to the mansions of Kedleston and Chatsworth before climbing the moors to Buxton and the caves. They followed special itineraries of 'The Seven Wonders of the Peak', including 'Two Founts, two Caves, one Palace [Chatsworth], a Mount and a Pit'.[1] These early tourists were both enticed and appalled by the 'ugliness' and 'deformities' – later hallowed as sublime or picturesque – of the stony moorland and jagged cliffs, the rivers that vanished and suddenly leapt into the light and the vast cave systems with their Stygian lakes, stalagmites and sparkling minerals. Around Poole's Hole near Buxton and the Peak Cavern near Castleton, colonies of beggars sold specimens while special shops opened in Matlock and Derby.

One anecdote, passed on by an old lady whose father had mounted cameos in Birmingham, told how Wedgwood's father turned some shelves in his work-shed into a sort of museum 'decorated with fossil shells and other curiosities', brought by the packhorse men from nearby coal mines.[2] Here the local potter was a man of his day. In the collecting of minerals and fossils, natural philosophy, tourism and taste overlapped. During the eighteenth century the rocks and caves of Derbyshire aroused intense curiosity, inscribing the history of the earth in a code as yet unsolved. Respect was matched by desecration – in three minutes, stalactites that

Itinerary for the Peak District, from Warner's *Tour through the Northern Counties*, 1802 had taken 3 million years to develop were hacked untimely from their place to please the dilettantes, like those adorning Pope's grotto at Twickenham. Decades later Elizabeth Bennet in *Pride and Prejudice* could still console herself for having to go to Derbyshire instead of the Lake District by thinking she might pick up 'a few petrified spars'.[3]

For the Midland-born Lunar men this was their playground and all their lives it was a place to escape to and explore. As a young man in the 1750s Boulton rode north to Buxton to relax in the moorland air and in his old age he was still orchestrating trips, complete with boat rides and romping country dances. Even in the 1790s, Anne Boulton's friend, young Patty Fothergill, could still find Matlock 'one of the most romantic places', declare Bonsal Dale 'one of the finest rides in England' and name the carved wood at Chatsworth 'the most beautiful thing in the World'.[4] Wedgwood often rode east from Burslem up into the hills and he too organized Derbyshire tours, meeting up with the Bentleys, Whitehurst and other friends. As a boy, Whitehurst had set out from Cheshire to roam the Peak; as a man he mapped and plotted its landforms season after season, year after year. Joseph Wright, his Derby neighbour, would paint some of his most remarkable landscapes here. But even Derbyshire was changing: the new survey, engraved and published in 1767 by Wright's friend Burdett, concentrated more on turnpikes, forges and mines than on caves and stately homes. Burdett's map, sponsored by the Society of Arts, brought Derbyshire into the modern age of 'patriotic progress': all the latest instruments were used; every place was plotted according to a universal system of co-ordinates; all positions were fixed by trigonometrical survey.[5] Now the county had an added allure, connected with the region's manufacturing life; from mines and quarries, Derby silk mill and Spode china, to the startling five-storey cotton mill opened at Cromford by Richard Arkwright in 1771.

The limestone massif, crossed and re-crossed, became built into a new kind of tour, combining old curiosities with new industrial wonders. In May 1771, for example, Franklin set off with John Canton, the chemist Dr Ingenhousz and his young friend Jonathan Williams. They travelled via Matlock and Bakewell (visiting a marble mill, examining the water-driven saws and polishers), and then on to Castleton, where they explored the Peak Cavern – then known as the Devil's Arse – travelling

The Devils Arse *near Castleton.*

A the Devils Arse. B. Houses within the Arch where many poor peo-ple live. C. the first Water. D. the second Water. E. the third and last Water. where the Rook and the Water Closes and you can pass no farther.

'The Devil's Arse' at Castleton, from *The Genuine Works of Francis Cotton*, 1715 by boat, lying on their backs when the cave roof loomed close, or being carried on the backs of guides. They visited Manchester, and saw the Bridgewater Canal and mines before doubling back across the Pennines to Leeds to call on Priestley. In Yorkshire they took in silver-plating at Sheffield and iron-smelting at Rotherham, and on their way south they toured Chatsworth and saw the silk mill at Derby. From here, where Franklin stayed with Whitehurst, they headed for Burton-on-Trent ('remarkable for good Ale') then journeyed on to Darwin at Lichfield, and from thence to Birmingham and Boulton and the marvels of Soho. A packed itinerary.

The variety of rocks and the curious arrangement of strata crammed into the Peak District never failed to dazzle foreign geologists.[6] For the Lunar men in the late 1760s it became the site of an investigation into minerals and a search for raw materials, and also a canvas on which they could draw different narratives of the past. At a straightforward level, the Peak was a key site for the hunt for ingredients for pottery. Wedgwood was still trying to make an earthenware more like porcelain and was searching for a clay that was truly white, free of the iron traces that browned it in firing. For this he looked far afield, and indeed the most promising source seemed to be deposits on Cherokee land across the Atlantic in South Carolina.[7] In July 1767 – wary of official government channels – he commissioned Ralph Griffiths's brother Thomas, a landowner in America, to sail for Carolina as his confidential envoy. Griffiths returned a year later, after adventures with gales and pirates at sea and hair-raising escapes in the interior, bringing back five tons of Cherokee clay, at a cost of £500, a price that stopped Wedgwood in his tracks.

Wedgwood kept searching, receiving samples from all over Britain, from Germany and even from China. But he needed other ingredients too, and for these he looked to Derbyshire. Porcelain is made by mixing white clay with a fusible stone, or spar – what Wedgwood called '*spath fusible*' – and over the next five years he experimented with moor-stone or 'china-stone', a granite feldspar resembling the famous Chinese spar, *petuntse* (*pai-tun-tzu*). (He was not alone in the group in his anxiety to find this: Boulton was also considering making pottery and Watt was testing spar for his own Glasgow potworks.)

It was a frustrating quest to find the perfect moor-stone since every find was different and the Protean nature of his samples drove Wedgwood mad. 'Moor stone and Spaith fusible are the two articles I want,' he would write impatiently in 1774:

They have plagued me sadly of late. At one time the body is white and fine as it should be, the next we make, having used a different lump of the Spaith, is a Cinamon colour. One time it is melted to a Glass, another time as dry as a Tob:Pipe – And this way it has led me a dance since I came home, without my knowing why 'till I tried each separate lump by itself.[8]

Since the mid-1760s, his friends had been collecting rocks and clay for him wherever they went. Whitehurst, who had interests in the Derbyshire lead mines, asked his miners to put by samples of earth and clay: in return Wedgwood promised to tell him about any 'curious products, or facts' found in digging the canals.[9] At the same time, Brindley reported that the workmen at the lead mines near Matlock threw all their spar and rubbish into a brook which washed away the dirt and left the heavy material behind. 'I long to be Fossiling amongst them,' Wedgwood told Bentley. 'Pray tell me when you will see me there.'[10]

'Fossilling' still carried its old portmanteau meaning – from the Latin *fodere*, to dig – of anything a spade could turn up, earth, metals, minerals and bones. But the canals had also unearthed fossils in the word's more specific modern sense. In July 1767, Wedgwood was reading a letter from Darwin, who was grieved, he said, that he could not come over to Burslem that night to see Wedgwood's curiosities.[11] Wedgwood had sent him some fossils, and Darwin had been trying to identify them. They were curious indeed. One was a cylindrical stone 'set round with stellations' which took its form from the roots of the water-flag, a species of iris. The next was a bone, perhaps 'the third vertebra of the back of a camel', Darwin suggested jokingly, utterly at a loss. And a horn, that was 'larger than any modern horn I have measured, and must have been that of a Patagonian ox I believe'.[12] He could hazard no more guesses until he knew the precise strata in which they had been found – another joke.

These treasures had come from the great Harecastle Tunnel, being dug for the Trent and Mersey Canal. At its southern end, the tunnellers uncovered huge, unidentified bones swept down in the drift of glaciers:

Darwin's horn may have been a tusk from some long-extinct elephant. To the north, they unearthed a wealth of fossils from far earlier eras, lying beneath a bed of coal in strata that we now know to be 300 million years old. Darwin's curiosity was piqued: he joked about the camel and the Patagonian ox, because a doctor, of all people, should be able to name a bone when he saw one. (The famous doctor William Hunter was also puzzling over giant bones, this time from America, in 1768.[13]) Wedgwood was sceptical of Darwin's guesses – he pasted them among 'Humorous Letters' in his commonplace book. But jokes were in Darwin's nature anyway. Below his signature he added a postscript: 'I beg my compliments to Mrs W– and the fine little Wedgwoodikins.'

However light the tone, the Harecastle discoveries stirred their imagination. When Darwin asked for information about the precise strata, he was making the vital leap from the random – the particular, accidental find – to methodical scientific enquiry. But he had already become intoxicated with things underground: at the end of his first letter to James Watt about steering and carts, he noted, 'I have got another and another new Hobby Horse since I saw you.'[14] And when he asked Wedgwood for more details of the fossils, he added:

> I will in return send you some mineral observations of exactly the same value (weighed nicely) for I must inform you I mean to gain *full as much* knowledge, from you who can spare it so well, as I return you in exchange.
>
> I have lately travel'd two days journey into the bowels of the earth, with three most able philosophers, and have seen the Goddess of Minerals naked, as she lay in her inmost bowers, and have made such drawings and measurements of her Divinity-ship, as would much *amuse*, I had like to have said *inform*, you.[15]

Darwin's mineral observations had come from a trip into a Derbyshire cave and his tumbling style inadvertently demonstrates the fusion, and confusion, of his excitement. The conventional, quasi-biblical cliché, 'the bowels of the earth', has echoes of the traditional naming of the caves, such as 'the Devil's Arse'; but this swiftly gives way to a semi-mystical classicism, the naked Goddess in her 'inmost bowers' before returning briskly to the philosophical, the 'drawings and measurements' that will inform as well as entertain.

Darwin's Derbyshire underworld is alive, sensuous, intensely physical. And the trip was no fantasy, despite the mythic language. In wig and breeches, sweating under his flapping topcoat, Darwin had climbed the hill and then clambered down the narrow fissures into the earth. His companions were John Whitehurst and the two Tissington brothers, 'subterranean Genii!', as Darwin called them.[16] (The older brother, Anthony Tissington, now in his sixties, was mine agent for the Duke of Devonshire, and himself owned coal and copper mines in Derbyshire, Yorkshire, Durham and Scotland. A friend of Franklin, and an authority on minerals, he was elected to the Royal Society this same year.)

The caves they visited lay in Treak Cliff, near Castleton. Even then, this was a famous tourist spot. The cottages and inns of the village, at the head of the verdant Hope Valley, sheltered below the medieval Peveril Castle from beneath which a stream gushed from the largest natural cave-opening in Britain, where the rope-makers worked, exploiting the damp atmosphere. Across the valley soared the heights of Mam Tor, scarred by a vast landslip that exposed its great shelves of grit and sandstone, and between the Tor and the rocky defile of Winnat's Pass lay Treak Cliff. Miners had worked here since Roman times, following the veins of lead into the cliffs. In the 1760s they were working in two great cave systems, where underwater streams had gouged swirling, tube-like tunnels: in the lower system (now known as Treak Cliff Cavern), the caves are a fantasy of stalactites and stalagmites; above, entered through a narrow miners' passage at the top of the hill, Blue John Cavern winds its way three hundred feet into the depths. Descending here is indeed like entering the bowels of some great beast, the narrow walls lined with glistening mineral streams of barytes and calcite, milky white and yellow and green, like the inside of an alien ribcage. Darwin and his companions descended ever deeper, squeezing through the narrow passages, hearing only their own rasping breath and the constant drip of moisture and rush of water through hidden channels. Their guide was a man with a lamp, its flickering candle pale against the dusky heights and dark abysses: in its feeble light gleamed glittering, iridescent crystals of dark grey and dusky blue, bronzy gold and milky green. Thrilled by his subterranean adventure, Darwin wrote not only to Wedgwood but to Boulton, with the same flurry of images:

> I want to see you and Dr Small much if you will fix a Day . . . I have been into the Bowels of old Mother Earth, and seen Wonders and learnt much curious

knowledge in the Regions of Darkness – . . . And am going to make innumerable Experiments on aquaeous, sulphureous, metallic and saline Vapours. Food for Fire-Engines![17]

It sounds as though as soon as he stubbed his toe on any interesting-looking lump of quartz or spar, Darwin would chip off a fragment, rush home and heat it, hoping in his exuberant manner to 'fuse' it into glass, or make astounding discoveries of new 'airs' or gases, as Black had done.

Over the next few years, the identification of minerals and their properties would become one of the most fruitful areas for the work of the whole group, including Watt. His canal work and pottery interests had already heightened his awareness, and through Joseph Black he became a good friend of the 'father of geology', James Hutton. On his second visit to Birmingham in the showery summer of 1768, he stayed for a fortnight at Soho with Keir. They had a mutual connection in James Lind, Keir's cousin and Watt's friend, and at the end of the stay, when they 'decamped together', Small chafed at their silence, imagining they had fallen into 'some bottomless pit in Derbyshire, or had been dipped into a cistern of air compounded with the inflammable principle'.[18]

Much later Keir undertook his own mineralogical and geological survey of Staffordshire. The work of the whole group stimulated the growth of geology as a subject: they entertained visiting experts – Hutton, J. A. de Luc, Faujas de St Fond – and corresponded with many others in Britain and abroad. And Darwin never lost his early enthusiasm. As soon as his son Charles was old enough they went out together collecting fossils and minerals and 'descended the mines, and climbed the precipices of Derbyshire, and of some other counties, with uncommon pleasure and observation'.[19]

For Boulton, however, Treak Cliff promised profit as well as curious knowledge, and soon after Darwin's excited letter he began to exploit this potential. The cliff had unique properties: it was – and is – the only place in the world that one could find the beautiful banded purple fluorspar 'Blue John'. This is the result of a potent geological collision, for here the limestone of the southern 'White Peak' meets the 'Dark Peak' of peat and bogs and strange-shaped stones, the granite and sandstone and millstone grit of the Pennines, formed by vast washes of mud and sand flowing south from the Caledonian Highlands, three hundred million years ago. Mineralizing fluids, seeping from the rocks of Mam Tor into the limestone crevices of the ridge opposite, fused in the chemical reactions induced by heat, pressure and inversion into this ravishing fluorspar.[20]

Although it had been known since the beginning of the century, and some objects had been made over the previous thirty years, mining of Blue John really began only around 1765. With great skill and care, local stone-workers dried the soft, brittle crystals, sawed them into rough shapes and then turned and polished them to a high sheen of glowing blue and purple, cut through with bands of yellow and white. Ornaments were soon on sale in Castleton, and Richard Brown, a local sculptor with interest in a mine, used Blue John as an inlay in a fine marble fireplace at nearby Kedleston Hall. Wedgwood added it to his list of materials, as 'No. 52. *radix amthyst*'; and Boulton seized on it as the perfect body for a new line of luxury products.

Here the narrative of the geological past brushes against a different history, the excavation of the classical past, crucial to Boulton and Wedgwood, since it created the phenomenon of 'vase mania'. A whole set of influences, building over the past two decades, came into play. One was the work of the architects, especially James 'Athenian' Stuart (co-author with Nicholas Revett of *Antiquities of Athens*, published in 1762) and the Scottish brothers Robert and James Adam. Another leader in the introduction of the antique taste was the more conservative William Chambers, architect to the King, followed by Boulton's up-and-coming friends Samuel and James Wyatt. In the late 1750s Stuart designed interiors for Kedleston Hall, only to have his work decried by his younger rival, Robert Adam, as 'pityfulissimo'. Adam, who had returned from Rome determined to 'blind the world by dazzling their eyesight with pomp',[21] now took over as designer with a bravura array of colourful, neo-classical designs. The architects' commissions extended from ceilings to candlesticks: at one point, Adam Smith reckoned three thousand people in different trades were dependent on the Adam brothers for work. In 1765 an order for gilt brass door furniture for Kedleston came to Soho via Samuel Wyatt, and Lord Shelburne ordered girandoles from Boulton for Adam's remodelled rooms at Bowood in Wiltshire.[22]

The antique craze was fuelled by news of the finds at Herculaneum and Pompeii and elsewhere, and great collections of engravings from classical sites were published.[23] Craftsmen and designers ransacked these for ornament – frames and medallions, swags of laurel, wreaths of acanthus leaves, friezes, figures and scenes. In 1764 Boulton bought Bernard de Montfaucon's classic, *Antiquitée Expliquée* (1719); the next year he snapped up Robert Adam's *Ruins of the Palace of Diocletian at Spalatro*; eventually he acquired five volumes of the lavish *Antichita de Ercolano.* The catalyst for Wedgwood was the first collection of 'Etruscan' vases put together by Sir William Hamilton, Ambassador to Naples.[24] As soon as he arrived, in 1764, he collected vases unearthed in nearby tombs, buying them from dealers or other collectors and eyen opening tombs himself. Even more important, Hamilton employed the brilliant, eccentric Baron d'Hancarville to publish the collection, with over four hundred and fifty engravings. A tangled series of events – including d'Harcanville's flight from Naples pursued by debt – meant that the complete set of *Antiquités, Etrusques, Grecques et Romaines* was not published until 1776, but the first volume appeared in late 1767, the year of Darwin's Derbyshire adventure. Wedgwood's patron, Lord Cathcart, who was married to Hamilton's sister Jane, lent him prints, and soon the complete volume was given to him by Sir Watkins Williams Wynn. He wrote to Bentley:

> The colours of the Earthen vases, the paintings, the substances used by the Ancient Potters, with their method of working, burning &c . . . Who knows what you may hit upon, or what we may strike out betwixt us; you may depend on an ample share of the profits arising from any such discoverys.[25]

Hamilton had wanted exactly this response. His writing had a complex note of 'patriotism', a determination to raise British taste and improve design. In his preface he declared:

> We think also, that we make an agreeable present to our manufacturers of earthen ware and china, and to those who make vases in silver, copper, glass, marble etc. Having employed much more time in working than in reflexion, and being in besides in great want of models, they will be very glad to find here more than two hundred forms, the greatest part of which are absolutely new to them; there, as in a plentiful stream, they may draw ideas which their ability and taste will know how to improve to their advantage, and to that of the public.[26]

A Tomb at Nola, engraved by A. Cléner after Christoph Heinrich Kniep, showing Sir William Hamilton and Emma on the right. This was the frontispiece to Hamilton's second vase collection, 1790.

This could hardly have been more direct. In 1767 P. J. Wendler, travelling in Italy for Boulton, subscribed to Hamilton's *Antiquities* on his behalf, telling Boulton that he thought this would be a good source of designs. The idea alone inspired Boulton, as it did Wedgwood.[27] The year before, Wedgwood had developed a richer form of the old Staffordshire 'Egyptian Black', called 'black basalt', perfect for classical-style vases. For the red 'Etruscan' painting he developed a matt 'encaustic' enamel, the only process for which he ever took out a patent.[28] Soon he would take the name 'Etruria' (probably suggested by Darwin) for his new pottery.

This summer Boulton decided that the market to target, with regard to vases, was that for ormolu ornament. Two years earlier, in Paris, he had noted the fashion for mounting vases in gilt metal and had since been busy finding out about gilding.[29] He swiftly added Blue John, and other Derbyshire marbles to his potential vase-making list. Soon he was hounding Whitehurst to find out about mine leases and adding, 'I beg you will be quite secret as to my intentions and never let M. Boulton & John Blue be named in the same sentence.'[30] As a friend of both Boulton and the local sculptor and mine lessee Richard Brown, Whitehurst was awkwardly placed: he ended by supporting Brown. Nothing daunted, in early 1769 Boulton was in Derbyshire, contacting mine agents, buying vases and stones from local lapidaries and purchasing an amazing 14 tons of Blue John.[31] And even though he never won a monopoly, he managed to make it look as though he had. A few years later Wedgwood found some Soho vases in 'a very rich shop' in Bath. The owner told a long story of Boulton paying £1,000 to engage 'the only mine in the World of Radix Amethyst, and that nobody else could have any of that material':

> I heard him patiently, but afterward took the opportunity of advising him when we were alone in the corner of his shop not to tell that story too often, as many Gentlemen who came to Bath had been in Derbyshire, seen the mine, & knew it to be free & open to all the world, on paying a certain known mine rent to the Land owner. The Gentleman star'd, & assur'd me upon his honor that he had not said a word more that Mr Bolton had assur'd him was true. 'Well done Bolton', says I *inwardly*.[32]

There was another link, however, between Italy and Derbyshire, Hamilton and the Lunar men – hinted at in Wedgwood's naming of 'black basalt'. As well as collecting antiquities, Hamilton had become a keen observer of Vesuvius, currently in a very active phase with frequent, dramatic explosions. There was one this year, 1767, which Hamilton described vividly in letters to the Royal Society, reported in the *Transactions*.[33] But a debate was already arising concerning volcanic forces – did they spring from a molten layer deep within the earth, or was the seat of volcanic fire simply at the core of each individual mountain?

These arguments tied in with an increasing interest in the forces active below the earth's crust, which had been shown horrifically in the great Lisbon earthquake of 1750. It was becoming accepted, especially after Buffon's first *Natural History* of 1749, that the earth was a 'theatre of active processes', constantly destroying and rebuilding itself – its tumultuous, dangerous energy held in a benign overall balance.[34] Moreover Buffon estimated that the cooling of the earth must have taken at least 75,000 years. Soon men were talking in terms not of thousands of years, but millions. By the 1780s men such as Hutton would argue that natural forces had changed the face of the globe in an endlessly repeated sequence over aeons, reshaping continents, throwing up mountains, carving rivers, building deltas – without any divine intervention.[35]

Rocks and minerals now became something more than objects of interest in themselves, or materials to be exploited commercially or aesthetically. They were evidence in a fundamental argument about the way the earth was formed. Theorists and fieldworkers focused on strata, realizing that rock formations were not local but stretched across vast tracts, as could be seen in exposed quarries, cliffs, caves and mines. The Lunar men were particularly aware of this through their connection with John Michell, who gave the first clear explanation of stratification and whose work on earthquakes was immensely influential, being the first to suggest how shocks could move in waves.[36] But almost more important was the work of Whitehurst, who for years had been collecting the data on Derbyshire strata that would eventually form the Appendix to his *Inquiry into the Original State and Formation of the Earth*, in 1778.

Whitehurst's findings seemed to back up the idea that subterraneous fire had shaped the earth, a theory already suggested by Nicholas Demarest, who claimed that the basalt found in the Auvergne was of volcanic origin, and of Rudolph Erich Raspe, who had studied basalts in

Hesse in Germany.[37] Whitehurst established that basaltic lava and volcanic ash known as 'toadstones' could be found in his own Peak District and he illustrated this with sections of interleaved strata and toadstone at Matlock, showing (wrongly) how they slid into a 'vent' at a central fault. The bed of the river Derwent, he proposed, was 'a great fissure or chasm filled up with the fragments of the upper and adjacent strata', which looked as though they had been 'burst, dislocated and thrown into confusion, by some violent convulsion of Nature'.[38]

Whitehurst was proposing a new 'system of Subterraneous Geography', a terrain of deep chasms, channels and fiery forces. Writing from Naples in 1774, Joseph Wright asked his brother:

> when you see Whitehurst, tell him I wished for his company, when on Mount Vesuvius, his thoughts would have centr'd in the bowels of the mountain, mine skimmed over the surface only; there was a very considerable eruption at the time, of which I am going to make a picture.[39]

Wright painted his picture. He made many of them over the years, thirty or more, flames against darkness, tinting the belly of the clouds, lava sizzling down the slopes, a burning glow dousing a pale moon reflected in the sea. 'The most wonderful sight in Nature,' he told Richard. 'The grandest effect I ever painted,' he boasted to his sister Nancy.[40] In Naples, Wright also made careful studies of rock formations and caverns on the seashore, the archives of Nature.[41] Later he worked these into luminous pictures, peopled with *banditti*, deliberately contrasting the rough cave to the calm sea and the serenity of a classical past.

A few years later Wright painted Whitehurst himself, showing the old man drawing the strata of Derbyshire's Matlock High-Tor, with a smoking volcano in the background. Yet, unbeknown to his friends, Whitehurst faced a terrible dilemma. As a careful observer, he should have been content to make deductions from the 'facts' he had collected, but these collided with his deeply held religious beliefs, which had taught him allegiance to the Mosaic story of creation and the Deluge as the great shaping catastrophe. The theological and the scientific fought each other within him.

Vases and volcanoes – the Neapolitan duo – thus came together in Lunar thought in Derbyshire in 1767. But discoveries in the realm of nature, just as much as ancient urns, also called forth aesthetic responses, and these too were changing. The landscape of the Peak, once thought so violent and ugly, was now being viewed as 'picturesque' and 'sublime', just as Wright thought Vesuvius 'the grandest effect' he ever painted. Wedgwood expressed some of this wonder when he told Bentley about the Harecastle fossils. He marvelled at the delicate imprints of vetches and crowfoot, hawthorn and withy. But how could these be found *beneath* the coal? They seemed to be part of a strata that dipped a hundred feet deep, that seemed to have been liquid, '& to have travel'd along what was then the surface of the Earth, something like the Lava from Mount Vesuvius'. They twisted and turned like serpentine rivers:

> and we have one great Hill, *Mole Cop*, which seems to have been formed entirely by them, as the mines are all turned by it, some to the East, & others to the West. But I have done. I am got beyond my depth. These wonderful works of Nature are too vast for my narrow, microscopic comprehension. I must bid adieu to you for the present & attend to what better suits my small capacity, the forming of a Jug or Teapot.[42]

As so often, Wedgwood found a perfect image for the relation of the minute to the general, the trivial to the universal.

That leap was also made by Darwin. When he stumbled on the slippery floor of the Castleton caves, and clutched at the rock wall for support, his fingers traced more marvels, veritable sheets of fossils – screw-like trilobites, bivalve shells, and corals from the reef that had once enclosed a vast tropical lagoon.[43] Not long after this he shocked Lichfield by adding a motto to his family crest of three scallop shells: '*E conchis omnia*' – 'Everything from shells'. All species, he felt increasingly, had descended from one common microscopic ancestor, a single filament. From this idea, slowly, over the years, Darwin would construct the first coherent theory of evolution, of competition and survival.

Here too science met the classics. Darwin's motto echoes the Ovidian myths he loved, Venus, Castor and Pollux hatching out of their shells. But its deeper implication was clear: when he painted it on his carriage door in 1770 Canon Seward was outraged. Darwin, he sputtered, was a follower of Epicurus, who claimed that the world was created by accident and not by God:

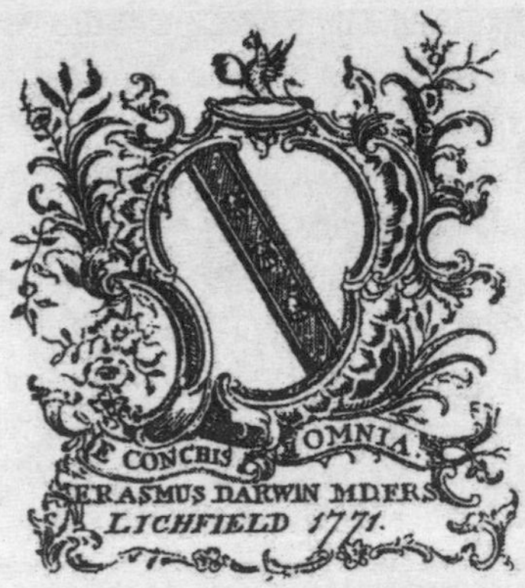

Darwin's bookplate

He too renounces his Creator
And forms all sense from senseless matter
Great wizard he! By magic spells
Can all things raise from cockle shells . . .
O Doctor, change thy foolish motto
Or keep it for some lady's grotto.[44]

Fearing for his practice, Darwin caved in and painted out his blasphemous Latin. (By a nice coincidence, only two years later Piranesi published an intensely evocative plate suggesting that vases, too, had evolved their shape from shells.[45]) Darwin often had to defend himself to his clerical friends, and in a passionate letter to the Reverend Richard Gifford, he swore that he did not mean to attack Christianity, 'and yet I do not mean to go up to the first Cause of anything, but endeavour to trace it one step higher, than others have done'.[46] The final imaginative narrative of this Derbyshire adventure – not of the formation of vases, or minerals, or strata, but of the long evolutionary struggle of organic life – would have to wait until the great works of his last years, *The Botanic Garden*, *Zoonomia* and *The Temple of Nature*.

14: CHEMICAL REACTIONS

The group now comprised eight men: Darwin, Wedgwood, Boulton, Small, Keir, Whitehurst, Edgeworth and his friend Thomas Day. When they could not meet, they wrote or sent messages and shipped over samples and objects of interest, from bones and fossils to vases and urns. The post-chaise, trundling down the turnpike in its cloud of dust, carried laden baskets and boxes. When Boulton was considering making porcelain, for example, Watt despatched a typical crammed bundle:

> I shall also send petuntse tobacco pipes kaolin to stop them & a bit of China made of Cornish Soap rock (which can be had at £10 pr tun) & of petuntse but Quartz will do if pure. I send also a bit of pure pipe crumble converted into porcelaine by melting in window glass, Bone ashes, minimume & nitre, but which trick I cannot play again. Let Mr B. keep in his age a China Factory as a good trade.[1]

Their enquiries ranged over the whole spectrum, from astronomy and optics to fossils and ferns. One person's passion – be it carriages, steam, minerals, chemistry, clocks – fired all the others. There was no neat separation of subjects. Letters between Small and Watt were a kaleidoscope of invention and ideas, touching on steam-engines and cylinders; cobalt as a semi-metal; how to boil down copal, the resin of tropical trees, for varnish; lenses and clocks and colours for enamels; alkali and canals; acids and vapours – as well as the boil on Watt's nose. Anything could set off a train of enquiry: 'You must often have observed the apparent dancing of dust

Detail from 'Chemistry' in the *Encyclopédie* (1755)

and visible objects in the spring in windy weather,' wrote Small, before turning to problems of optics and refraction.[2]

Derbyshire may have prompted a new interest in minerals and spars, but geology and chemistry went together – the discovery of new substances demanded their analysis. In 1769, on Franklin's recommendation, the mineralogist Johan Jacob Ferber visited Whitehurst, who reported his coming humorously to Boulton:

> Mr Ferber was a pupil of Linnaeus, hence bottony is an Object. I dont See the Utility of bottony, and therfore cant say a word about that pursuit. Tho' cannot help wishing it did not interfere with the Study of the Fossil Kingdom.[3]

Ferber met Boulton and the two doctors and showed them his drawings of the Derbyshire strata, very similar to Whitehurst's. When he wrote of his visit some time later, he remembered Small ('a man of much understanding in chemistry') explaining Derbyshire porcelain-making to him, and Whitehurst showing him how to make the gold-coloured brass called 'tombac' from calamine, iron and copper.[4]

The group's chemical interests were given new impetus by Keir. After leaving Edinburgh he had some adventurous years, including a narrow escape from yellow fever in the West Indies. From his porthole in the hospital ship he watched sharks tear the bodies of fellow soldiers who had died of the disease until 'the sea was tinged with blood, and their mangled limbs were seen floating on its surface'. He gradually weakened, unable to speak or move, until the army surgeon declared 'he has gone too', but just as a man arrived to throw him overboard he managed to scrawl a note asking for antimony, an unusual medicine, although long ago recommended by Paracelsus. This was granted 'only from the persuasion that his case was hopeless', and to the surprise of all he recovered.[5]

Keir was a strong man physically and mentally: it was like him to keep his wits *in extremis*. Yet he was no typical army man. Early each morning he re-read the classics and translated the military historian Polybius (the manuscript was 'unfortunately burnt' at the publishers) and continued his study of geology and chemistry.[6] He was clever and funny, a patient listener and straight talker, tolerant of others' eccentricities. In religion he was a broad churchman, in philosophy a devotee of the Stoics and Marcus Aurelius, in politics a solid Whig and a believer, like Priestley, in the democratic power of knowledge. Edgeworth found him blessedly down to earth, having 'turned the energy of his powerful mind to science, with a view to make some discovery, by which he might increase his fortune, and in the pursuit of which he might find interesting occupation'.[7]

Keir stayed with Edgeworth for some months while he worked on his annotated translation of a famous standard work, the first dictionary of chemistry, the *Dictionnaire de Chymie* of the French Académicien Pierre Joseph Macquer, published anonymously in 1766. Although it was hard to make a living as a writer, Keir's *Dictionary* was in part a money-making project, designed – like Priestley's writings – to appeal to the ever increasing readership among students, laymen and manufacturers. It was also a way to teach himself, as he admitted in his Preface: 'I thought I could not take a better method of fixing in my mind a knowledge of chemistry, than by employing the leisure which I then possessed, of making a complete translation, and giving it to the public.'[8]

Macquer's provisional, open approach was in tune with the strong preference of the Lunar group for facts over abstract systems and with their questing spirit – like Whitehurst mapping his strata, he gave them the feeling of being on an uncharted, shifting terrain. He made no claim to be definitive. Indeed he covered all aspects of the subject in about five hundred substantial articles, choosing an alphabetical arrangement precisely because he wanted to avoid the false connections of a linked, discursive treatment. Macquer was a keen Newtonian, fascinated by chemical affinities and by the 'reciprocal gravitation' of atoms, as he called it. But his emphasis was on discovery, not theory. In his preliminary note he declared that the chemistry now known 'is scarcely entitled to the name of science, so defined. It is at present little more than a collection of facts, the causes of which, and their relations to each other, are so imperfectly understood, that it is not yet very capable of the synthetic and analytic modes of explanation.'[9]

Macquer's articles summed up what was known – say, the different kinds of acid – and what was believed, such as theories of what went on during combustion or putrefaction. Above all – and this too would appeal to the Lunar men – he was immensely practical. The chemistry of the laboratory met that of the trades, the household and the body: 'Bones, Borax, Brains, Brandy, Brass, Bread, Bronze'. Keir went ahead with his translation without contacting Macquer, but when it was published in 1771 the Frenchman was grateful rather than insulted. He incorporated many of the notes in his second French edition and sent this for translation to Keir (who was so efficient that his second English edition of 1777 actually appeared before the French one).

Keir's great contribution to writing on chemistry in Britain was his determined attempt to bring the subject up to date – his dictionary included a long section on 'fixable air' with quotes from Black, and his notes gave news of recent experiments and findings, ranging from chemical analyses of plants and minerals 'so far as they are known', to colouring for enamels, glass and dyes. (He and Macquer were two of the first chemists to be interested in the chemical aspects of dyeing and calico printing.[10]) His comments were backed by references to European authorities, to contemporary British work and to his own experience. The notes are like windows into the workrooms of chemists across Europe, catching the German chemist Marggraf, for example, patiently working on the acid of ants (not yet named 'formic acid'), which he claimed could be perceived by smell 'on turning up an ant-hill in spring or summer', and obtained by distilling '24 oz of ants' in a warm bath, until the acid formed crystals.

There was still something magical about such science. Indeed Keir's book was one of the sources used by Joseph Wright when he painted *The Alchymist, in Search of the Philosopher's Stone, Discovers Phosphorus, and prays for the successful Conclusion of his operation, as was the custom of the Ancient Chymical Astrologers.*[11] The long title says it all. Macquer had described the production of phosphorus, the white vapours and strange jet of blue light (which illuminate the alchemist's figure in Wright's picture), presenting the discovery as a positive, accidental outcome of the pursuit of gold. The real story of this discovery, made by around 1670 by Hennig Brand, 'variously called the last alchemist or the first chemist', puts Marggraf's buckets of ants in the shade. It involved boiling down fifty buckets of human urine until it became a wormy, putrefying paste, leaving it in the cellar for months and then distilling until it glowed in the dark.[12] But however bizarre such goings-on appeared, they did bring results: Boyle and Newton had been interested in alchemy, and Priestley and Darwin were among those who argued, perceptively, that alchemy should be respected as a genuine search for wisdom, in contrast to the self-interested, obscurantist priestcraft of their own day.[13]

Keir himself, however, thought alchemy misguided and his bias was generally extremely down to earth. In late 1771 Wedgwood noted, 'I have bought a Chemical Dictionary translated by Mr (late Captn.) Keir with which I am vastly pleased . . . I wod. not be without it at my elbow on any account. It is a Chemical Library! I think you should buy yourself another & charge them to the Company.'[14] Keir had also been working on Wedgwood's current obsession, fusible spars, noting how he had managed, 'by a very considerable heat', to fuse some of the white, opaque, *petuntse*-like fluorspars.[15]

Inspired by Keir's arrival in the Midlands, Darwin was soon busily brushing up his chemistry and getting out his retorts and crucibles. 'I rejoice exceedingly to see you study chemistry so eagerly,' wrote Keir ironically, packing up a crate of materials for him to play with:

> I shall send you some bismuth, and some zaffre, from which you may extract regulus of cobalt, by fusing it with alkaline salt and a little charcoal dust, by which you will get a blue glass, similar to the smalt of the shops, a regulus of cobalt, and probably a regulus of bismuth adhering to the upper or under surface (I forget which).

Darwin also borrowed the works of Vogel and William Brownrigg, the Cumberland chemist who had produced influential papers on air, and an 'Essai sur la Putréfaction' – about the airs, or gases, arising from decaying matter. And Keir provided an alarming reading list, beginning with 'All Stahl's Works', which he said would give the foundation of the modern theory of chemistry and were full of genius. ('They are large,' he warned laconically.[16]) Then he listed thirteen more books, many of several volumes, including ones he had not yet got round to reading himself. No one knows how far Darwin got with this pile, but presumably several tomes went into the wicker basket in his carriage, solace for stops on the road.

Keir also listed three works on metallurgy, pointing to studies of the method of assaying ores, tables of affinities, experiments on the density of metallic alloys, and accounts of smelting and forging 'very valuable to commercial metallurgists'.[17] This, of course, was of considerable interest to Boulton, not only for his ornamental business, but because of his interest in steam-engines, whose cylinders would need metals that could be cast and worked easily but would also resist corrosion and heavy wear.

This search, like Wedgwood's hunt for clay, had stimulated Boulton's interest in unusual metallic ores and fossil materials, and he too asked his friends and foreign agents to send samples back to him. But much remained to be understood. In June 1768, Darwin had told Wedgwood, 'Mr Boulton has got a new metal which rivals Silver both in Lustre, Whiteness, and endures the Air with as little tarnish. Capt. Keir is endeavouring to unravel this metal.'[18] In 1769 Small groaned to Watt, 'I study Metallurgy, but make small progress' – the books were superficial and full of errors, the metallic substances were impure, while 'undiscovered substances produce vast effects'.[19]

That summer, Boulton, Small and Keir were all working on cobalt, nickel and manganese and were borrowing books from the London metallurgist Peter Woulfe, who also offered to do chemical trials.[20] They focused particularly on new metals. One was manganese, the strange metal that dissolves in water, only recently discovered by the Swedish chemists Johann Scheele and Johann Gahn. Another was the malleable, gleaming *platina del Pinto* – platinum – discovered in the 1730s and now being used for jewellery and for chemical apparatus since it proved to be as immune to corruption as gold. Nickel had long been known as an ore that acquired a green tinge when bathed in acid (which is why German miners called it 'Old Nick's ore'), but it was only in 1751 that another Swede, Axel Cronstedt, isolated this silvery, magnet-attracting ore. Experts were still arguing that it was a compound, and not an element like iron, and it was much in demand: in his letter accompanying the booklist, Keir told Darwin, 'Nickel is not to be got for love or money.'

Keir's chemistry was not confined to the study. He had been hunting around for some kind of opening where he could use his chemical knowledge to make money. In industry, the greatest need was for sulphuric acid – already being made by Garbett and Roebuck at Prestonpans and for the two main alkalis, soda ash (sodium carbonate) and caustic soda (sodium hydroxide) and for their potassium equivalents (potash and caustic potash).[21] These had been used for centuries in the glass industry and in glazing pottery, and were now also wanted for soap-making and in bleaching processes for the textile industry, especially in Scotland. Natural alkali was found in the ashes from burning different forms of vegetable material and in the past the main source had been 'pearl ash', from charcoal, but as Scotland's forests gradually diminished, this was replaced by expensive imports of ashes from Spanish *barilla* (the plant *salsola soda*) and potash from northern Europe, and most of all by pearl ash from the American colonies (Franklin had sat on a committee to promote the trade). Another source was seaweed or 'kelp', and from the 1750s a huge kelp industry grew up along the west and east coasts of Scotland. We are so used to thinking of the Western Isles as being depopulated in the nineteenth century when crofters were driven off their land and forced to emigrate, that it is startling to realize that most of this 'native population' was deliberately imported from the mainland to comb the rocks at each turn of the stormy tide, and to burn the seaweed. The isles were an industrial as much as rural area: the wind was heavy with black salt smoke; the land lay uncared for, given over to heather and gorse, and the landlords made vast annual profits while their workers could scarcely pay the rent.

The race was now on to find a new, chemical substance to replace kelp and other natural sources. Since 1770 Keir had been telling Small his plans to make alkali. It seemed a winning idea. But here he unknowingly faced a potential clash with Watt, who was doing research in the same area with Joseph Black, backed by Roebuck.[22] Both Keir and Watt had the same thought: to turn common salt (sodium chloride) into the alkaline soda (sodium carbonate), probably by using lime. The belief was that salt was a mixture of hydrochloric acid and alkali: according to the contemporary theory, if you added a substance that had a greater 'elective affinity' to hydrochloric acid you could extract it, almost seduce it out of the salt, leaving the alkali behind.

As the confidant of both Keir and Watt, Small was in an awkward position, telling Watt to get a patent for his alkali process quickly, while knowing that Keir – his 'particular friend' – was on the same track. Keir's trials had been fairly successful and Small warned Watt that he might hit on the technique first, '& as you had told me your secret, my situation in such an event would not have been pleasing'.[23] The rivalry brought out Watt's ingrained suspiciousness: had he said anything in front of Keir when he was in Birmingham? Small wrote soothingly, explaining that Keir had been conducting trials for years. Indeed he himself had helped in some of these, but only, he added hastily, with the aim of persuading him not to waste time on an 'impossibility'. Very cleverly, he took the simplest route out of his dilemma by putting his two friends in touch with one another, sure that collaboration would be better than competition. In 1770 Roebuck invited Keir to Kinneil, and from then on they thought of themselves as a team. Keir wrote openly and plainly to Watt, telling him of his methods and his difficulties. The quest took over his life: he kept his mass of salt in his cellar, 'but in other trials I found any part of the house from the uninhabited Garrets to the cellars fit for the purpose'.[24]

This is an interesting example of how the cohesion of the group was threatened by competing interests yet saved by diplomacy and compromise. As it happened, Keir soon put his alkali plans on one side, daunted by the slowness of the process and the poor financial prospects since the amount of soda they could make was too small, while the tax on salt was too high. In 1771 he entered a Caveat at the Patent Office (a formal marker to ensure that no patent was obtained without his knowledge), but he assured Watt that this would not prevent the Scottish team from taking out a patent if they wished.[25]

Soon Watt and Black also turned to other things. But four years later, during the American Wars of Independence, imports of alkali from the colonies and Spain dried up, and it became desperately scarce. And when a petition to reduce the salt tax was put forward by other projectors, the uncle–nephew duo of Alexander and George Fordyce, the Lunar men leapt into action. Inevitably, the networks overlapped: Keir was friendly with George Fordyce, who was Professor of Chemistry at St Thomas's Hospital, while Watt – who was always neurotically convinced any new idea had been stolen from him – loathed Alexander, abusing him as 'a lyar, an Irishman, a bankrupt and a blockhead'.[26] To protect their corner, Watt and the Scottish group each sent their own petitions for exemption, with Whitehurst and Boulton as witnesses.[27] But nothing came of this: the salt tax stayed unchanged and the search for a chemical alkali process continued for the next twenty years. Not until the early nineteenth century was it produced commercially on a large scale – incidentally driving thousands of Scottish kelp-workers into poverty.

When Keir left alkali aside, he turned his attention to glass-making, long one of the major industries of the West Midlands. In 1770 he bought a lease on a glasshouse making white flint glass at Holloway, near Amblecote, from Thomas Rogers, a leading glass-maker, and former Sheriff of Worcester. Keir moved into a house behind the works which also provided a useful space for him to continue his experiments.[28] The following year, he married the dashing Susanna Harvey in St Philip's Church, Birmingham, in October. 'Mr Keir has turned glassmaker at Stourbridge and has married a beauty,' Small reported.[29] He now entered into a full partnership with a local manufacturer of sulphuric and nitric acid, Samuel Skeys, and with 'old John Taylor', perhaps the famous and very wealthy button-manufacturer.[30] Both these men later became bankers, and at Amblecote they provided most of the cash while Keir was the manager, having 'a sellary and a share of the profits' beside the share that came in proportion to his own investment.

Urbane and financially astute, Keir was successful at everything he turned his hand to and making glass was no exception. He became a dynamic local figure, one of the original shareholders in the Stourbridge Canal Company formed in 1775 to link Amblecote with the Staffordshire and Worcestershire Canal. He worked closely with Boulton and Fothergill, sending over vases, glass bottles, sugar castors and blue-glass liners for Soho silverware – dozens at a time. But their chemical collaboration continued as well. In 1772, sending Boulton a bill, he added that he was doing some 'chemical operations' in an old glasshouse, including producing nitric acid, 'and I hope to undersell you in Aqua fortis'. The only thing he couldn't, or wouldn't, do was 'one part of your Scheme, viz. the rectification of your old stuff', recovering the acid by distilling, although he thought this 'a very advisable bit of Economy'.[31]

Keir was constantly called upon to analyse samples for his friends, and to make equipment for them, promising Boulton that his order for 'chemical glasses' would be 'executed as speedily as possible, especially those for your own Experiments as I well know the Impatience of my fellow-schemers, and I should also be sorry to check by delay your present hobby-horsicality for chemistry'.[32] His industrial and experimental interests ran side by side, and would come together – with an echo of that fascination with mineral structure in the Derbyshire caves – in a paper for the *Philosophical Transactions* four years later, 'On the crystallizations observed on glass'.[33]

A couple of years later, Keir sold his share in the Amblecote glassworks

Glassworks, from Croker's *Complete Dictionary of the Arts and Sciences*, 1764–66

and moved to help Boulton out at Soho. But his real skill was in chemistry, and he never lost his own industrial ambitions. Eventually he set up a chemical works of his own at Tipton, a few miles from Birmingham, with his friend Alexander Blair, another ex-army officer. They made nitric and hydrochloric acids, litharge and red and white lead, and began making alkali (sodium carbonate) to sell to the local glass-makers and manufacturers such as Boulton. Here, it seemed, Keir's long experimental interest in crystallization proved immensely useful.

From the description given by his daughter, and the experiments by his own great-grandson (a leading research chemist at ICI) Keir based his new process on the reaction of the sulphates of potash and soda with lime.

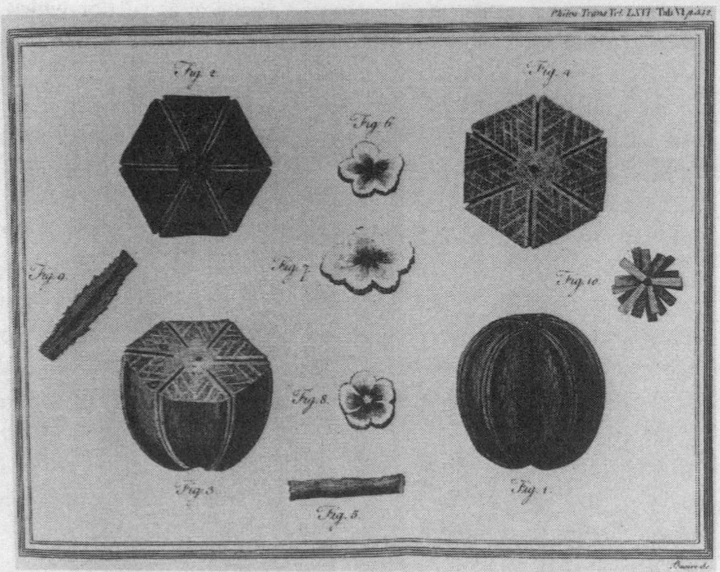

Illustration of Keir's crystallizations on glass, from
the *Philosophical Transactions*, 1776

Preferring experience to theory, he threw out the accepted wisdom of the doctrine of elective affinities, which ruled that sulphuric acid had a 'greater affinity' for the alkalis than for the lime (so it should have bonded with them), and realized the opposite, that the acid would unite with the lime. If the sulphates were mixed in a weak solution and forced slowly through a thick sludge of lime, they would gradually decompose: the sulphuric acid would mix with the lime, leaving the alkalis free so that they 'had then only to be brought into a concentrated form for sale'.[34]

To deal with the very slow nature of the chemical reaction Keir used the new technique of percolation. And as he heated and cooled the mixture in huge open-air vats, he saw that the extra, unwanted sodium sulphate formed crystals, which would be left behind while the clear liquid he wanted was run off. Science and personality both played a part. He succeeded where others had failed, not only because he was a brilliant empirical researcher but also because he was immensely patient. As his daughter put it, he called in 'the aid of a chemical agent (for which he always expressed the highest respect, and the functions of which in natural operations were, he thought, greatly underrated) *Time*'.

The kind of alkali that Keir made was more suitable for soap than for glass-making, and it was from this that Tipton became famous and Keir became rich. A simple process, old as the hills – boiling oil or fat with alkali, then adding salt to make it 'curd' – now turned into a large-scale operation.[35] Soap pans could be huge, fourteen feet deep and fourteen feet wide, full of boiling gunge, ranged alongside dozens of enormous cast-iron vats, and more pans for evaporating waste. The Tipton soapworks covered twenty acres, with two water-wheels and two steam-engines and a work-force specially trained in Keir's new techniques. (He even invented his own steam-pump, a rather inefficient, coal-guzzling version of a Savery engine – which avoided him having to trespass on Watt's patent.[36])

By the end of the century Tipton was producing a million pounds in weight of soap a year. In 1788, running through the group's news, Darwin told Edgeworth:

> Mr Keir amuses his vacant hours by mixing oil and alcaline salts together, to preserve his Majesty's subjects clean and sweet – and pays 1000 Guineas every six weeks to an animal call'd an Exciseman.[37]

Darwin was one of Keir's best customers. The next day he told his son Robert that the soap was packed in boxes of 100, and he had just ordered '400 for ourselves, 400 for Mr Strutt and 200 for Mr French'.[38] The fall in the mortality rate towards the end of the century has often been put down partly to increased hygiene; it could be that Keir's soap did as much for public health as Darwin's medicine. His crates travelled by canal to Hull, Liverpool, Bristol and London, reaching all corners of the kingdom.[39] Tipton became a showpiece, rivalling Etruria and Soho, attracting visitors from across Europe. Through Keir, and through Roebuck in Scotland, the Lunar group and their friends were pioneers of the future chemical industry, just as they were of the great age of steam.

15: TRIALS OF LIFE

The intense activity of the Lunar men in the 1760s took place against a picture of increasing instability. It was true that the Peace of Paris garnered great imperial conquests, and many prospered from the boom in trade. The arts and sciences flourished: the Society of Artists, established in 1765, was outshone by the new Royal Academy in 1768; in the same year James Cook was commissioned to view the transit of Venus from the South Pacific, and set sail on his first voyage round the world. In contrast to such glittering endeavours, however, these years also brought problems. At Westminster there were eight administrations between 1757 and 1770. The manufacturers had to tread carefully and Wedgwood's letters show his keen interest in who was in or out – Grenville, Townshend, Rockingham, Chatham, Grafton. Boulton, too, ensconced in his fine Soho House, had to keep a keen eye on national events.

The greatest crisis looming concerned America, although no one realized yet the great rifts that were to come. The spark that lit the fire seemed small and insignificant. Between 1763 and 1766, desperate to pay for the war, the Government extended a whole series of legislative measures to America. Hoping that it would hardly be noticed, in 1765 it proposed a Stamp Tax, a levy on paper used for official documents, legal transactions and newspapers: a trifling thing, were it not that it affected every liberty-minded lawyer, writer and thinker on the eastern seaboard. What infuriated this articulate, dangerously influential group was the

Soho House

assumption that Westminster had a right to impose direct taxes on colonists, even though they were not represented in Parliament. Tax without consent created uproar, first in Virginia, then Boston, then across the thirteen colonies to the West Indies: short of force it was impossible to collect the dues. As usual, Wedgwood was percipient: the Government, he later remarked to Bentley, 'seem determin'd to *Conquer England in America* . . . I tell them the Americans will then make Laws for themselves & if we continue our Policy – for us too in a very short time.'[1] When the Americans threatened to boycott British goods, British merchants and manufacturers joined the outcry, flooding the Government with petitions: Garbett was among the organizers from Birmingham.[2] Alarmed, the new Rockingham ministry repealed the Act, and the cheers in Parliament were echoed in every trading town. The threat of lost business hit hard in the Midlands, where men and women were flocking to the towns in search of work. This was the era of enclosures, of Goldsmith's 'Deserted Village', and in 1767 'Mechanicus' of the *Coventry Mercury* devised a new song to the tune of the patriotic 'Hearts of Oak':

> Inclosing of fields and of commons I'm sure
> Can be called nothing better than robbing the poor
> He that stops honest labour his country must wound,
> Though in Charity's deeds he may seem to abound . . .[3]

The whole nation – apart from the very rich – was also suffering from soaring post-war prices. The arrival of French silks provoked riots among the Spitalfields silk-weavers and the unrest spread to other trades while the countryside suffered terrible food shortages. The winters were unusually severe: men froze to death in country lanes and newly planted woods were blasted in a single night of frost. Floods, gales and hail followed; the corn grew tall but the grain failed to ripen. In the autumn of 1766 troops dealt with riots against farmers and corn merchants from the Wash to the West Country. In Birmingham the soldiers stayed all winter, and emergency supplies were brought from London. Making swift notes in the margins of a pamphlet, which accepted the colonists' claims but blasted their 'riotous and seditious ways of asserting them', Franklin wrote:

Do you Englishmen then pretend to censure the Colonies for Riots? Look at home!!! I have seen within a Year, Riots in the Country about Corn, Riots about

Elections, Riots about Workhouses, Riots of Colliers, Riots of Weavers, Riots of Coalheavers, Riots of Sawyers, Riots of Sailors, Riots of Wilkites, Riots of Government Chairmen, Riots of Smugglers . . . &c &c. In America if one Mob breaks a few windows, or tars and feathers a single rascally Informer, it is called REBELLION: Troops and Fleets must be sent, and military execution talk'd of as the decentest Thing in the World. here indeed one would think Riots part of the Mode of Government.[4]

The issues were thrashed out in clubs and taverns, such as the Birmingham poet John Freeth's Free Debating Society in his Leicester Arms Coffee-House. And Birmingham men were ready to take sides when a new political crisis loomed in 1768, following the election of the outlawed John Wilkes as MP for Middlesex. The Government fanned the flames by harassing Wilkes, banning him and imprisoning him and suppressing gatherings in his support. In May 1768, thousands poured into the open space outside the King's Bench prison where he was held, and in the 'Goodman's Fields Massacre' several demonstrators were shot. Yet this was not a protest of the poor: the pressure came from wealthy Londoners and progressively minded gentry and merchants.

As employers, Boulton and Wedgwood were unlikely to approve of unrest but to begin with Boulton backed the Americans to protect his trade, while Wedgwood, as a Whig and a Dissenter, supported them throughout on grounds of principle. He spoke out for Wilkes too, noting wryly how the duo 'Wilkes and Liberty' were 'tagged together like Hobgoblins & darkness'.[5] In these tumultuous years the Midlands was strong in its support for Wilkes, raising petitions and sending presents (including forty-five gross of clay pipes). But patriotism and protest often ran side by side as if the left hand ignored what the right was doing, or as if, indeed, no contradiction existed at all. The year 1769 saw David Garrick's Shakespeare Jubilee at Stratford – an epoch-making event, even though not a line of the bard's words was spoken and the triumphal procession was drowned in the rain. That year, John Baskerville's famous type was used for a special edition of Shakespeare; soon it would serve for a local edition of Wilkes's life and writings.

If you were committed to questioning authority, it was impossible to separate national struggles from personal principles. As a minister in

Leeds, Joseph Priestley courted controversy: his religious position became more clearly defined, adopting the Socinian heresy that Christ was not divine, but a man like other men. In 1769 he founded the *Theological Repository*, the first journal for free religious enquiry, and three years later, with liberal Anglicans such as Theophilus Lindsey and John Jebb, he helped draft the 'Feathers Tavern Petition' to relax subscription to the Thirty-Nine Articles. The petition was howled down by the Church and the universities, and swiftly stamped on in the Commons. Supported by Priestley and Richard Price, and powerful patrons such as the Earl of Shelburne and Duke of Grafton, Lindsey became minister of the first formal Unitarian chapel, in Essex Street, off the Strand.

Unitarianism was the most rational and open of all sects, proclaiming only belief in God and arguing for toleration for all, including Catholics. Many of the old Presbyterian congregations came under its banner and Priestley was one of its leading spokesmen.[6] But he could not win: scientific friends ridiculed his faith and Christians damned his materialism. On the secular front, Priestley had already ruffled the Establishment with the call to arms in his *Essay on the First Principles of Government*, in 1768:

> Let all the friends of liberty and human nature join to free the minds of men from the shackles on narrow and impolitic laws. Let us be free ourselves, and leave the blessings of freedom to our posterity.[7]

Priestley's *First Principles* took the important step of distinguishing between political and civil liberty. True political liberty would mean everyone having an equal opportunity of reaching high office – a state so far unknown. Society should be for the benefit of all, not for its rulers, who should be viewed as the servants of the people:

> It must necessarily be understood, therefore, that all people live in society for their mutual advantage; so that the good and happiness of the members, that is the majority of the members of any state, is the great standard by which every thing relating to that state must finally be determined.[8]

From this perspective, violent revolution could certainly be justified under certain circumstances. Civil liberty, by contrast, was a sphere in which government had no place: in matters of religion men and women should follow their own conscience, in education they should be free to experiment.[9] In his writings Priestley was helped by his Dissenting friends, and sometimes by Franklin and Wilkes. And in February 1769, he dashed a few lines on the end of a letter to Franklin, wishing him well with his 'laudable endeavours in the cause of science, truth, justice, peace, and, which comprehends them all, and everything valuable in human life, LIBERTY'.[10]

While opponents were roundly attacking him on political or religious grounds, Priestley won the lasting affection of his Leeds flock, 'a liberal, friendly, and harmonious congregation'.[11] He continued with his teaching, his science and his educational writing, and won backing to found a public-subscription library. In their own communities most of the other Lunar men were also seen as enlightened improvers abounding in 'Charity's deeds'. Darwin, for example, gained a reputation as a doctor of uncanny skill, with a generous determination to treat the poor without charge. Charity patients lined up at his door in the morning and his annexe was an informal town hospital. He subsidized this by riding far across neighbouring counties collecting fees from richer patients (who took it for granted that the doctor would visit them at home). He made calls almost every day, was a Commissioner for the local turnpike trust, and through the Wychnor iron mill could see himself too as an entrepreneur, benefiting the local economy.

Boulton's star was also in the ascendant. Although he was still listed as 'Mr', rather than 'Esq.' like the older, grander manufacturers Garbett and Taylor, his name appeared on all the key projects from the Canal Company to the New Hospital, the subject of a long campaign led by Dr John Ash and William Small. This was part of a general movement, spreading outwards from London: twelve provincial hospitals were founded in the early 1760s, all funded by subscriptions, their lists bringing together the local aristocracy and gentry and the prosperous of the towns – a way of welding a community. After the first public meeting in December 1765 land was bought and building began but the funds soon ran out and despite a much lauded music festival, including the inevitable *Messiah*, the project was abandoned for over ten years. Now and in the future the Lunar men were closely connected with the hospital, first through Ash and Small, then through William Withering. Small, like Darwin, was said to have spent two hours, every day of his life, caring free for the poor of the city. In 1769, with Baskerville and Garbett, he was one of the Commissioners supervising the new Act for widening and improving Birmingham's streets. Here, too, the provinces were staking their claim to rival London, to become modern cities, pulling down the medieval alleys and replacing them with neat squares and open highways, proper sewerage and lighting. So great was provincial zeal that a satirical wit suggested that there would soon be lamps every thirty yards along highways to the capital, with carpet laid down for 'genteel travelling'.[12]

Boulton stood back a bit from this – his real platform was not the town but Soho. In 1768 he was forty. Wedgwood was thirty-eight and Darwin thirty-seven. These were strong, stubborn men. They really did believe they were living in an age of miracles, as Wedgwood put it, in which anything could be achieved. But they also thought you had to make things happen. Darwin's son Robert remembered his father declaring, 'The world was not governed by the clever men, but by the active and energetic.'[13] As employers, Boulton and Wedgwood were unashamedly paternalistic: Boulton described himself openly and proudly as a father controlling a potentially wayward, anarchic family.

As a doctor, Darwin too showed a streak of forcefulness. As Keir noted affectionately:

> He despised the monkish abstinences and the hypocritical pretensions which so often impose on the world. The communication of happiness and the relief of misery were by him held as the only standard of moral merit.[14]

When he chose, Darwin had a gift for instinctive sympathy: Keir saw how keenly he felt for others and 'could enter into their feelings and appreciate their different constitutions and characters'. Anna Seward, no uncritical observer, greatly admired his diagnostic skill and his ability to read between the lines. 'Extreme was his scepticism to human truth,' she wrote, remembering how he preferred rather to glean information from patients 'by cross-examining them, than from their voluntary testimony. That distrust and that habit were probably favourable to his skill in discovering the origin of disease, and thence to his preeminent success in effecting their cure.'[15] He still had to be wary of his name, however. Seward also told the story of Lady Northesk, who stopped at a Lichfield inn some years later and told the landlady she was dying. All the doctors thought her case hopeless. 'I wish, Madam,' said the landlady, 'that you would send for *our* Doctor, he is so famous.'[16] After she stayed with Darwin for a fortnight – thin, pale, huge-eyed, breathing painfully – he contemplated a blood transfusion through a syringe, a daring challenge to his skill. But after a night's sleep he thought better of it, deciding he did not 'choose to stake my reputation upon the risk'. Instead he suggested a diet of fruit and vegetables: Lady Northesk was cured: she died a year later, not from consumption, but from setting fire to her clothes.

Darwin inspired confidence simply by arriving at the bedside, but his reputation depended less on his cures – which tended to be standard remedies or placebos, laced with generous helpings of opium – than on observation and understanding. Edinburgh had given him a lasting interest in the nervous system, which made him alert to personal and social factors. In his *Zoonomia*, written towards the end of his life but based on years of case notes, the afflictions he listed included not only scarlet fever and scabies, measles and mumps, but anger, boredom, credulity, ambition, sentimental love. In 1767 he wrote a humane and far-sighted letter about infanticide, often provoked, he thought, by the public shame of illegitimacy. The women who committed such an act should be the objects of pity, he said, not blame: 'What struggles must there be in their Minds, what agonies!'[17]

Yet he could be dictatorial. When he was approached by a man who feared for the sanity of his young second wife (she had slashed his first wife's portrait in a fit of jealousy), one might think this would evoke all his famed sympathy. Instead he 'told her in the plainest manner many unpleasant truths, among others that the former wife was infinitely her superior in every respect, including beauty'.[18] Not surprisingly, the mere threat of sending for him again calmed her down – hailed as a miracle cure.

So there was something of the bully beneath the benevolence, despite Darwin's back-slapping humour and broad smile. Many suffered from his 'jocose but wounding irony', directed at vanity and affectation. Robert remembered him declaring that common sense would improve,

> when men left off wearing as much flour on their heads as would make a pudding; when women left off wearing rings in their ears, like savages wear nose rings; and when fire-grates were no longer made of polished steel.[19]

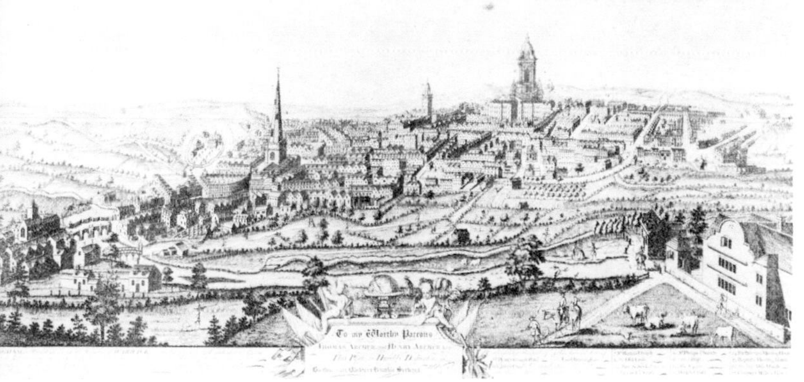

1. *Erasmus Darwin*, by Joseph Wright of Derby, 1770
2. *Matthew Boulton*, by Lemuel Francis Abbot
3. Westley's *Prospect of Birmingham from the East*, 1732

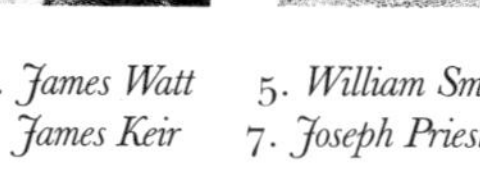

4. *James Watt* 5. *William Small*
6. *James Keir* 7. *Joseph Priestley*

8. *Richard Lovell Edgeworth*, miniature on ivory by Horace Hone, c. 1765
9. *Honora Edgeworth*, jasper medallion modelled by John Flaxman, 1780
10. *Thomas Day*, by Joseph Wright of Derby, 1770

11. *Josiah Wedgwood* and *Sarah Wedgwood*, by George Stubbs, 1780
12. *Thomas Bentley*, after John Francis Rigaud, 1778

13. Oval platter from the Frog Service showing Etruria Hall
14. Glacier from the Frog Service
15. *Sir William Hamilton*, black basalt portrait plaque, 1772

16. *Vesuvius in Eruption, with a view over the Islands in the Bay of Naples*, by Joseph Wright, c. 1766–80
17. *Joseph Wright*, self portrait, c. 1782–5
18. *John Whitehurst*, by Joseph Wright, 1783

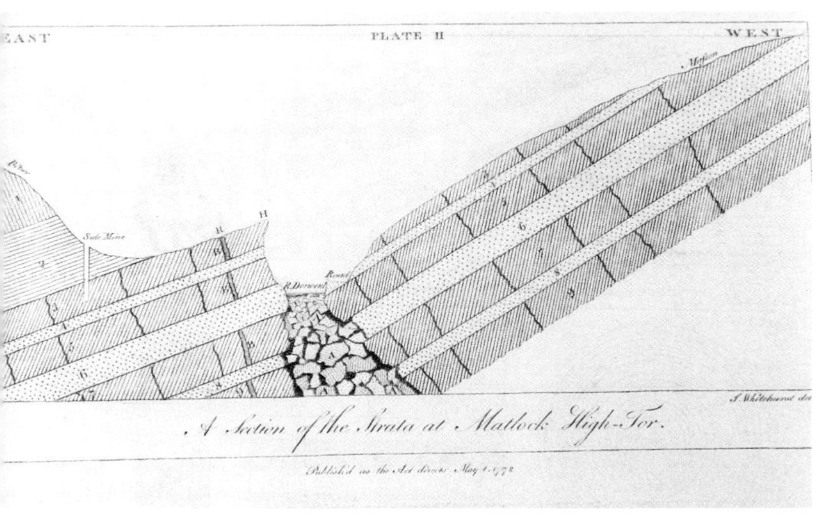

19. *Matlock Tor by Daylight*, by Joseph Wright, mid-1780s
20. 'Strata at Matlock Tor', from John Whitehurst's *Inquiry into the Formation of the Earth*, 1778

21. *A Philosopher giving that Lecture on the Orrery, in which a lamp is put in place of the Sun* by Joseph Wright, 1766

22. *An Experiment with a Bird in an Air Pump*, by Joseph Wright, 1768

Perhaps he was slightly arrogant about his local status: in 1765, in a punchy quatrain, he managed simultaneously to insult Canon Seward and the city's other famous son, Samuel Johnson, both of whom had produced editions of British dramatists:

> From Lichfield famed two giant critics come,
> Tremble ye Poets! hear them! 'Fe, fo, fum!'
> By Seward's arm the mangled Beaumont bled,
> And Johnson grinds poor Shakespeare's bones for bread.[20]

But despite icy glances in the Close, Darwin still shone at the town's literary dinner tables. He even made his mark on the Cathedral itself, where a lightning conductor was installed in September 1766 on 'the advice of some philosophic gentlemen'.[21]

None of the Lunar men suffered fools gladly but they knew they had to avoid making enemies. Boulton and Wedgwood were adept at the emollient application of charm, and Darwin told his son firmly not to react violently to gossip, and to take backbiting coolly. He cited the example of Small, who 'always went and drank tea with those who he heard had spoken against him; and it is best to show a little attention at public assemblies to those who dislike one; and it generally conciliates them'.[22] Their public roles and social lives show how crucial this kind of negotiation was: to achieve their goals they had to operate skilfully and learn to manipulate opinion.

However confident they seemed as they manœuvred through the knots of enemies and rivals and friends, in private they were still vulnerable. Small was dogged by ill-health, and, although he thought whimsically from time to time about marriage, his men friends were really more important to him. Keir was embarking on his new life with Susanna. Priestley worried about his wife Mary's health, and the need to support his growing family – in addition to little Sally, two sons, Joseph and William, were born during his time in Leeds. Several of them now had children to care for. By 1768, Charles and Erasmus Darwin junior were nine and eight; a baby girl, Elizabeth, had died at four months in 1764 but the Darwins' next child, Robert, was now a sturdy two-year-old. Yet clouds were gathering: Polly was increasingly ill, suffering spasms of pain, probably from a biliary tract infection or from gallstones. Far from strong, in July 1767 she gave birth to a fifth baby, William Alvey, who died at three weeks, his tiny corpse buried in Lichfield Cathedral.

Darwin hoped for much from his sons. He noted delightedly how Charles observed natural objects, played with tools, collected fossils and tinkered with chemistry. The only problem was that he took after his father in another way as well – he had a terrible stammer. The Darwins decided on radical measures. In October 1766, a month after Charles's eighth birthday, they sent him to France for a year with Darwin's old Cambridge friend, the kindly Reverend Dickenson, stipulating that he should speak only French. As soon as they reached London, Polly wrote anxiously, apologizing for her style – 'however if my pen mistakes, and puts in, and leaves out letters, it generally conveys my ideas and is a faithfull index to my heart'.[23] In January she enclosed a special note for Charles, telling him that Bob had got two teeth and that 'Rassy has done nothing but throw snow these last two or three days. Your papa is very well . . . I hope you will write a little letter.'[24] A month later, from Paris, the letter came. The people were very civil, Charles wrote solemnly, and 'tho' the Frosts have been very severe I have not had a Cold since I left England'.[25] Late in the year, having toured France (and spent about half Darwin's annual income), tutor and pupil returned. Anna Seward reported that from now on Charles never stammered, although his speech was 'somewhat thick and hurried'.[26]

As the leaves turned, falling on the still Lichfield ponds, Polly took her boys to stay with the Boultons, later thanking Anne for 'the happy days I passed lately at Soho'.[27] By then Anne herself was pregnant and early in 1768 she and Boulton had their first child, a daughter, also called Anne: a couple of years later, Small could announce joyously to Watt, 'Mr Boulton has a son' – christened Matthew Robinson.[28]

Boulton could not have been prouder of his children: in 1772 he wrote to Anne from London: 'The queen showed me her last child which is a beauty but none of 'em are equal to the General of Soho or the Fair Maid of the Mill. God bless them both and kiss them for me.'[29] But at the age of one, little Anne found difficulty in learning to walk, perhaps suffering from a congenital dislocated hip, which Small found hard to diagnose. They consulted William Hunter and appealed to Darwin, who wrote a touchingly detailed reply, joking Boulton out of his panic: 'WOMAN, says Aristotle, is a two-leg'd animal *without feathers* – he certainly meant very young Women!'[30] He explained the correct alignment of the bones and possible risks of malformation of the spine, illustrating this with his fast, jerky sketches, and asked Small to check for problems with the calf muscles or shortening of the thigh bone due to a fall. If Boulton watched the baby playing, sprawling naked on the bed as well as tottering on her feet, he said, then he would see the defect. If not, they must all come over to Lichfield, 'that we may have a general Consultation, Ladies and Gentlemen. For if it be remediable, it must be NOW. And for this purpose the Defect must be absolutely ascertain'd – Delay is damnable.'

The Wedgwoods also sought modern medicine for their children. In the spring of 1767 Wedgwood and Sally had two-year-old Sukey and one-year-old John inoculated against the smallpox that had affected Josiah so terribly as a child. Although this revolutionary technique had been introduced long ago from Turkey by Lady Wortley Montagu, this was the first year that inoculating infants was recommended. The procedure was still risky and Wedgwood held his breath. Both children 'had Convulsions at the first appearance of the eruption' and,

> have had a pretty smart pox as our Docter terms it. I believe they have had no dangerous symptoms, but have been so very ill that I confess I repented what we had done, and I much question wether we should have courage to repeat the experiment, if we had any more subjects for it.[31]

In early July they had a second son named Richard. He too was inoculated and on New Year's Eve Wedgwood told Bentley that he had recovered 'and a Charming Boy he is, his mother tells me. You must try your hand at manufacturing something of the kind when you become an inhabitant of this *Prolific* country.'[32]

At the time of John's and Sukey's inoculations Wedgwood had been jubilantly planning his new pottery, vowing to 'work away like fury next year'. In May he was hunting for a London showroom, but by 13 June he was home, full of plans. The next morning all this was wiped out. The same post that took his letter brought the news that his brother John was dead, drowned in the Thames, slipping from the riverbank after seeing the fireworks at Ranelagh. Wedgwood was more distressed than he could have imagined. For weeks a kind of listlessness hung about him, a miasma of depression, cutting off his energy at source, stopping him doing the simplest things. That summer his grief turned to illness and this in turn 'sunk my spirits & disheartened me greatly in the prosecution of my schemes'.[33] He recovered, but slowly. By late summer he was on track again, his pen skimming the page:

> Why you never knew so busy a Mortal as I am, – Highways – surveying Ridge House Estate – experiments for Porcelain, or at least – a new Earthenware, fill up every moment almost of my time & would take a good deal more if I had it – besides all the hands in the Country are not hired but are still coming to me – '*to know when they must begin*'.

Soon he found the perfect London showroom in Great Newport Street, where he hoped to install a colony of enamellers, modellers and carvers. Nothing, it seemed, could slow him down – except his own body.

Wedgwood's old pain in his leg, intensified by illness and injury, had never left him. He suffered constantly from osteomyelitis, and the gradual thickening of the bone left him almost disabled, unable to walk at all. The 'rational' answer was to get rid of the offending limb altogether. Darwin spoke to the Newcastle surgeon James Bent, and an amputation was arranged. In an age before anaesthetics the operation was both agonizing and dangerous: if he survived the shock, the threats of haemorrhages and gangrene still lay ahead. Yet on 31 May 1768, with Bentley and Sally to support him, Wedgwood went ahead. Insisting on watching everything that happened, sedated by laudanum, he sat upright in his chair and never uttered a groan as two surgeons tightened their tourniquet around his thigh and briskly sawed off his leg above the knee. Matching his stoicism, the devoted Peter Swift attached a note to an invoice he was sending to Cox in London:

> Sir, Your favour of the 26th is just come to hand, but can make no reply to the contents. Mr Wedgwood has this day had his leg taken of, & is as well as can be expected after such an execution. The rev. Mr Horne's Goods are packed, and one Crate for the warehouse . . . Mr Chester's Goods will be delivered on Thursday next. I am &c . . . Peter Swift[34]

It was business as usual. Five days later the bandages came off: the operation was successful. But on the same day the Wedgwoods' baby son

Richard died. For Sally and Josiah, the June sun was darkened. Yet with his amazing, characteristic resilience, by the end of the month Wedgwood had revived: 'Have been at the Workhouse, & had two Airings in a chaise – have left off my laudanum & do better without it. The skin on the upper part of the wound is healed, & got down to the bone, which I tell you to confute all those who deny the present to be an Age of Miracles.'[35] He turned off the surgeon, dressed his wound himself and proudly measured it with his compasses (two inches by one and a half). Soon his new wooden leg could be heard clacking and tapping around the works and he became famous for wielding his stick to smash substandard pots.

There were other scares: in late 1769 he feared he was going blind after seeing floating spots before his eyes, until Whitehurst (who had suffered the same thing) briskly told Sally 'he would insure my eyes for 6d', passing on Darwin's advice that this was common and he should just ignore it and wait until it passed. During this time, as after his amputation, Sally wrote his business letters for him. They were very close and Wedgwood was altogether the most family-minded of this group and his business letters were always sprinkled with home news. He wrote exuberantly to Cox when Sally gave birth to another 'fine Boy', Josiah, in 1769; two years later he was waiting eagerly for another baby – 'There is no being from home at such a time on any account,' he told Bentley – and he was furious when 'little Tom' was born while he was out. He was distraught when Sally became ill and miscarried in the autumn of 1772: 'You never saw such a changeling, nothing but skin and bone, pale as her cap, and does not seem to have a drop of blood in her.' Darwin had left him to 'act as Physician' and he took his job seriously, prescribing what none of the nurses 'durst think of':

When nothing would stay upon her stomach I gave her fruit, ripe plums &c as often as she would eat them and she has never vomited since. For the wind, I have given her Cyder that blows the cork up to the Ceiling. She relishes it and it does her good.[36]

Sally coped with the cider. But during the autumn she became seriously ill with rheumatism, and Wedgwood despaired again, constantly calling in Darwin, sending her to stay at Lichfield and taking her to Buxton (as he had to Bath the year before). He was exhausted, 'not so stout at heart' as he used to be, but despite his wooden leg he found the best prescription for stress was 'an hour or two of real labour in the fields every day that time or the weather will permit'.[37] Miserably Wedgwood pulled out his waistcoat to show Darwin how much weight he had lost, and Darwin sensibly but sternly told him that nothing was seriously wrong, and 'to live pretty well, to take moderate exercise & to keep free from care & anxiety'.[38] He also assured him that Sally would survive – as indeed she did, soldiering on for another forty-two years and outlasting them all.

Darwin, Sarah Wedgwood's 'favourite Esculapius', was now a close friend. He often rode over to Burslem in the weeks after the amputation, and Boulton too called in on his way home from Buxton. When he could not visit, Darwin wrote, telling Wedgwood of Boulton's new metal and Keir's analysis and Edgeworth's grand inventions, ending cheerily: 'This is all the news I can think of to amuse you with in the philosophical Arts.'[39] But ironically, just as Wedgwood recovered, Darwin himself was lamed.

His latest carriage design was an extremely high-wheeled light phaeton, with a 'garden chair, fix'd upon the axle tree'.[40] By a clever device, the horse was free to swing from one side of the road to the other, but the driver had to concentrate every second and Darwin was thrown several times. In July 1768, after seeing a wealthy country patient whom he had been inoculating against smallpox, he gave the Reverend Robinson of Lichfield a lift to Burton. When the horse shied at a barrow of gravel thrown down by a labourer working on the Grand Trunk Canal, the carriage tipped violently and ran one of the wheels on top of the hedge. The next day the jarred axle-tree snapped, flinging the doctor on to the road. He fractured his right knee, 'an accident irretrievable to the human frame' as Anna Seward put it, and limped from now on.[41] He never used the carriage again.

Darwin was not used to being laid up and did not take it well. The usual invalid was Polly, whose illness gradually grew worse. 'Mrs Darwin has been ill for some time,' Small told Watt in February 1769.[42] For much of this year and the beginning of the next she was in a haze. She had attacks of violent pain in her head and on her right side, near her liver, followed by convulsions. Darwin prescribed opium, as he did for many patients, however minor the complaint – for some people he even advocated half a grain a day, 'as a habit'.[43] But the more Polly took the less effect it had and she began supplementing it with wine, then with brandy and spirits. Often she was drunk or delirious. 'I well remember', Darwin told his son Robert much later, 'when your mother fainted away in these hysteric fits (which she often did), that she told me, you, who was not 2 or 2½ years old [would] run into the kitchen to call the maid-servant to her assistance.'[44]

In early 1770 Anna Seward often sat with her. Polly had nothing but praise for Darwin, whom she said had given her great happiness and tried desperately to fight her disease; but in the years of their marriage he had so often been away, leaving her alone in the house with the children. Now he sat for nights by her bed, and wept when on his rounds, thinking of their shady garden walk, their arbour and the wild primroses she had planted. Polly hid her drinking from him and by the time he discovered it, it was too late. She was dying of cirrhosis of the liver. 'A few days before her death,' he noted, 'she bled at the mouth or whenever she had a scratch, as some hepatic patients do.'[45] She was frightened, crying to Darwin that it was 'hard so early in life to leave her Children and her Husband she loved so much – pray take care of yourself and them'. Fear was followed by terror and delusions: she saw someone striking at her husband with a dagger, hurled herself forward to protect him and shrieked that her children be spared: 'Don't kill them all, leave me one, pray leave me one!' Next day she was calmer but still delirious, babbling half-sentences about household affairs and candlesticks. She died on 30 June 1770.

Two days later, stricken with remorse and grief – like other literary husbands, honouring their wives once they have gone – Darwin sent a harrowing account of her death to one of her friends. The act of writing, even more than her funeral in the Lady Choir of the Cathedral, put Polly into the past. By the end of the long document, his tone had settled into convention, much as Boulton's had done in his tribute to his first wife. If Polly had lived on with her disease, he consoled himself, life would have been 'twisted with many threads of black'. Perhaps it was better thus. He must live for the children: 'A few months will soften these Ideas and smooth the remembrance of her down to Pleasure, and turn my Tears to Rapture.'[46] He was genuinely wretched, for a time. He kept the brief notes of her funeral expenses. He wrote few letters, did few experiments. Slowly, things returned to normal. He picked up his medical practice and his travels to patients. His sister Susannah moved in as housekeeper. Charles and Erasmus seem to have gone briefly to school in London. And a month after Polly's death a seventeen-year-old girl, Mary Parker, came over from his old home village of Elston to look after four-year-old Robert.

In August 1770 Wright painted his first portrait of Erasmus, a broad-shouldered man in solid brown broadcloth leaning on a table, his muslin shirt-cuffs emphasizing his delicate hands: an imposing, intriguing personality. He looks attentive, but his keen eyes are sad.[47] The melancholy did not last long. In December Edgeworth arrived in town, in a bizarre new equipage that revived Darwin's interest in invention and soon he was writing to Wedgwood in his old cheerful mode about his windmill and his Wychnor ironworks. When Franklin visited in May, they plunged into detailed discussions of phonetics and chemistry and went happily dabbling in ponds in the evening.

By this time Darwin's eye had begun to roam. Different liaisons have been hinted at, including one with a married Derbyshire woman, Lucy Swift, but one thing is sure: by the end of 1771, Robert's young governess, Mary Parker, had become his mistress (however liberal his views, class seemingly ruled out the possibility of marriage).[48] The first of their two daughters, Susan, was born the following May 1772. She and her sister Mary, born two years later, were part of the family and were brought up in the Darwin home with Susannah as housekeeper. None of his friends turned a hair. Boulton the Anglican, Wedgwood the Dissenter, the open-minded Small, the sensible Keir, and even Watt, with his Calvinist background, all accepted Darwin's household without demur. A morality where the greater happiness was the goal could be wondrously flexible – and convenient.

16: ROUSSEAU & ROMANCE

Whether dealing with their families or forging their careers, these successful men in their late thirties were far more worldly wise than the new recruits, Edgeworth and Day, who were still in their early twenties. The gap between them also reflects a notable turn in the guiding philosophy of the time. The younger men had an altogether more personal, radical view of society, freedom and 'Nature'. Into the sober, reformist thinking of the group they brought a breeze of fashionable Rousseauism.

Since the late 1750s a new respect for 'sentiment' had entered British culture, the idea that the mark of civilized women and men was their capacity for feeling, as well as reason. Within no time, to be 'natural' and informal was all the rage. Edmund Burke's *Philosophical Enquiry into the Origin of our Ideas on the Sublime and the Beautiful* of 1756 stressed the power of the passions in aesthetic response, embracing awe and terror as well as harmony and delight. To many, the primacy of feeling was the message of Laurence Sterne's *Tristram Shandy*, the surprise hit of 1759 (and a lasting favourite of Wedgwood), and sentiment was given a still firmer push by the translation of Jean-Jacques Rousseau's *Julie, ou La Nouvelle Héloïse* and *Emile*.

Darwin met Rousseau when he was in England in 1766. Having been driven out of different countries and states across Europe, Rousseau was befriended by Hume, who found shelter for him and his mistress Thérèse le Vasseur as guests of Richard Davenport at Wootton Hall in Staffordshire. He walked the moors to Dovedale, he read, he played with

Honora Sneyd, engraving by Hopwood after a painting by George Romney

Davenport's grandchildren and began to write his *Confessions*. The story went that Rousseau used to hide in a cave on the terrace, 'in melancholy contemplation', and as he hated being interrupted, 'Dr Darwin, who was then a stranger to him, sauntered by the cave, and minutely examined a flower growing in front of it.'[1] Rousseau, who was interested in botany, was winkled out – they talked, and allegedly corresponded for several years.

Darwin was interested but not wholly won over by Rousseau's ideas. The younger Lichfield men, by contrast, were captivated – Brooke Boothby, who called on Rousseau in Paris ten years later and was charged with the editing of his posthumous *Premier Dialogue* in 1780, was painted by Wright reclining in a wood, his hand lying on a volume of Rousseau's work. Edgeworth was another fan. Rousseau's *Emile*, which argues for education through kindness and freedom, had made an indelible impact on him as a student. It had 'all the power of novelty', he remembered, 'as well as all the charms of eloquence'. Inspired, in 1765 Edgeworth had decided to apply its ideas to the upbringing of his one-year-old son. Freedom and fresh air and practical understanding were in: discipline and rote-learning were out. He let Richard go barefoot, in loose jacket and trousers, determined to free him from the fetters of book learning: 'He had all the virtues of a child bred in the hut of a savage, and all the knowledge of *things*, which could well be acquired by a boy bred in civilized society' (including, of course, mechanics). By the age of seven, he was 'bold, free, fearless, generous'. But to Edgeworth's innocent bafflement, somehow 'he was not disposed to *obey*' and had 'an invincible dislike to control'.[2]

Edgeworth was not entirely eccentric or unique in his experiment. Even Darwin, Anna Seward said, 'made it a rule never to contradict his children, but to leave them to be their own master' (although Robert remembered him being rather hard on them).[3] Wright let his children play ball on the gravel walk, and hide-and-seek in the house, and painted fashionable portraits of children alone in the countryside in loose, free clothes, or in relaxed, affectionate groups with their parents. Dr Johnson's crony, Bennet Langton, had ten children, whom he used to spoil dreadfully, exasperating all the friends who visited him.

At Hare Hatch Edgeworth had found a fellow devotee of Rousseau in the nineteen-year-old Thomas Day, a student at Edgeworth's old college,

Corpus Christi, now staying near by with his mother and obstreperous stepfather. He inherited a substantial fortune when he was a year old: as he grew up, unconcerned with money, he was free to express himself. At school at Charterhouse he gave his pocket money to the poor, was ostentatiously kind to animals and learned to box but refused to continue the fight if he was winning. He decided against being a vegetarian, however, because rearing animals for food meant that at least there were more of them, and they could indulge their appetites in ignorance of the slaughterhouse to come.

Rousseau appealed to Day instantly. If all the books in the world should be destroyed, he declared to Edgeworth:

> except scientific books (which I except, not to affront you) the second book I should wish to save, after the Bible, would be Rousseau's Emilius. It is indeed a most extraordinary work – the more I read, the more I admire – Rousseau alone, with a perspicuity more than mortal, has been able at once to look through the human heart, and discover the secret sources and combinations of the passions. Every page is big with important truth.[4]

In many ways Day was Edgeworth's opposite – depressive rather than excitable, misogynist rather than womanizing, politically idealistic and totally lacking in any social graces. He was tall and stooping and dishevelled. 'Mr Day's exterior was not at that time prepossessing,' admitted Edgeworth. 'He seldom combed his raven locks, though he was remarkably fond of washing in the stream.'[5] He also gave vent to 'long and dismal catalogues' of the evils of women and swept Edgeworth off for hours to discuss metaphysics. For some reason, which Edgeworth never fathomed, Anna Maria took a strong dislike to him.

From the time they met the two men were inseparable. Then in 1768, while taking Edgeworth's son to stay with his grandparents in Ireland, they bumped into Darwin and Whitehurst at a Staffordshire inn – much to Edgeworth's embarrassment, since he and Day were both playing roles. Day was posing as 'a very *odd*' misanthropic gentleman, travelling to get over the death of his wife and devoted to his young son, while Edgeworth acted his snooty servant, ordering everything delicate or costly in the pub. Once Darwin had exposed their charade and introduced the astonished Whitehurst ('one of the most simple, unassuming philosophers I have ever known', remembered Edgeworth) they settled down to talk. At first Darwin wrote Day off as an idiot because of his ignorance of mechanics but when the topic changed and Day stopped sulking and waxed eloquent, Darwin was charmed, and immediately invited him to Lichfield.[6]

The personal dramas of these two young Rousseauians would be a source of constant amazement and entertainment to their Lunar friends for years to come. When Edgeworth wrote about his first encounter with Darwin in his memoirs, he noted, 'How much of my future life has depended upon this visit to Lichfield!' And from now on Day too was embraced by the whole circle. Small, Boulton and Darwin made allowances for his youth, seeing his genuine philanthropic zeal and his real concern for freedom. He became touchingly devoted to Small, who sensibly argued him out of studying medicine, persuading him he might not be the most effective of doctors. Instead Day poured his idealism into education, anti-slavery campaigns and political causes, and into writing his famous children's book *Sandford and Merton.* For all his oddities, he became, in many ways, a remarkable figure and the older Lunar men rightly took him seriously. (But one can't help feeling that they might also have valued Day's money: the first thing Anna Seward mentioned was his inheritance from his father, his 'possession of a clear estate, about twelve hundred pounds per annum'.[7])

After that first meeting in the inn, Edgeworth and Day travelled on to Ireland. The trip was a revelation to Day – and to Edgeworth's family. On one hand, the Irish rural squalor shook Day's Rousseauian belief in the noble savage. On the other, his table manners shook Mr Edgeworth's view of a gentleman, rousing a violent antipathy. Day merely 'smiled with philosophic indifference at these prejudices in favour of politeness', and fell in love with his friend's witty, clever sister Margaret. Wisely, she gave him a year to think about it, and told him to smarten up in the meantime. Edgeworth senior caved in (possibly thinking of Day's cash and his own impoverished estate) and accepted the idea of a proposal. Day went off to London to become a lawyer, and Edgeworth returned home, to work on his inventions. He made frequent trips to Lichfield and Birmingham, 'to visit my friends Dr Small, Mr Keir and Dr Darwin', leaving gaping crowds behind him as he drove across country in his specially designed, low-slung, one-wheeled chaise, with leather sides which folded up when he went through water – like a sort of high-speed black banana.

Before the year was out, Day and Edgeworth's sister had realized what everyone else had seen from the start, that they were eminently unsuited. Day brooded miserably for a while, before deciding on a radical plan. Although he was only twenty-one, he was determined to find a wife (out of 'duty', thought Anna Seward rather kindly). His demands were modest:

> He resolved, if possible, that his wife should have a taste for literature and science, for moral and patriotic philosophy. So might she be his companion in that retirement, to which he had destined himself; and assist him forming the minds of his children to stubborn virtue and high exertion. He resolved also, that she should be simple as a mountain girl, in her dress, her diet and her manners, fearless and intrepid as the Spartan wives and Roman heroines.[8]

As Anna admitted, 'There was no finding such a creature ready made.' Since no woman fitted his ideals – and those whom he had stooped to fancy had jilted him – he would have to create the wife he wanted, all by himself. The plan was to adopt two girls and bring them up according to the best Rousseauian scheme. And as they grew up, Day 'might be able to decide, which of them would be agreeable to himself for a wife'.

In some ways, this was a version of the 'we can do anything' approach of the older Lunar men, a stunningly confident attempt to control nature, 'to mould them as he pleases', as Darwin had said of Edgeworth's magnetic conjuring. The shift from mechanical to social engineering was also related to Day's political ideals, and could be seen as creating a fitting mother for a 'manly' race. (The rhetoric of uncorrupted simplicity and Spartan hardiness link it to Rousseau's *Discours on the Arts and Sciences* of 1750, and to a strong tradition within British political thought which contrasted classical virtue to modern luxury.[9]) It was alarmingly easy to procure guinea-pigs for this experiment. With an old schoolfriend John Bicknell, Day went first to the orphanage in Shrewsbury, picked out a girl of 'remarkably promising appearance' and named her Sabrina Sidney (after the river Severn, and his hero, Algernon Sidney).[10] The next stop was the Foundling Hospital in Coram Fields in London, where he chose a second girl, 'Lucretia'. They were eleven and twelve respectively; pre-pubertal dolls. Anna described Sabrina as 'a clear, auburn brunette, with dark eyes, glowing bloom and chestnut tresses', and Lucretia as her balanced opposite, 'fair, with flaxen locks and light eyes'.[11]

All Day had to do was to promise, in writing, that within a year he would apprentice one girl to a trade, or set her up in business and give her £400 on her marriage; the other he intended to marry, and if he did not, he would place her in a good family and give her £500. He also 'solemnly engaged not to violate her innocence'. One other stipulation, with regard to Sabrina, was that she had to be apprenticed to a married man, and so (without telling him), Day simply named Edgeworth. He began by lodging the girls in London, but then took them to France, in the belief that as they knew no French, corrupting influences would be shut out and their minds open only to him. The trio settled in Avignon, causing a minor stir. Slowly he taught the girls to read and lectured them 'to imbue them with a deep hatred for dress, and luxury, and fine people, and fashion, and titles'.[12] The twenty-one-year-old Day saw himself as an intrepid traveller, engaged in a bold social experiment. 'Had I staid at home,' he wrote cheerfully to Edgeworth,

> perhaps at this moment I might be in a warm comfortable room, calculating the vibrations of your wooden horse's legs; but should I, my friend, should I have been what I now am, – the traveller, the polite scholar, and the fine gentleman?[13]

Unfortunately, apart from 'Excellent Rousseau! best of humankind!', Day soon decided that he despised the French. The whole population displayed 'perfect vacuity', and the women were imbecilic, obsessed by fashion and unnaturally dominant. Even French roads were terrible. He longed for news from home: 'Have you got a house yet? – have you got a patent? – a title? – a fortune? – a child? – a medal? – a new chaise?'[14] He wrote reams, was pleased with his pupils, and sent a dictated letter from Sabrina:

> Dear Mr Edgeworth, I am glad to hear you are well, and your little boy – I love Mr. Day dearly, and Lucretia – I am learning to write – I do not like France so well as England – the people are very brown, they dress very oddly – the climate is very good here. I hope I shall have more sense against I come to England – I know how to make a circle and an equilateral triangle – I know the cause of night and day, winter and summer. I love Mr Day best in the world, Mr Bicknell next, and you next.[15]

Then things began to go awry. Bored, unable to speak a word of the language, the girls squabbled and pestered Day constantly for attention. They caught smallpox, demanding that he sit by their beds night after night. If he stole out of the room, they woke up and wailed. No one else was allowed near them and Day had to look after their every need. The smallpox passed, leaving no scars, but Day was exhausted. He struggled on, dealing with multiple disasters: a capsized boat on the Rhône; a French officer in Lyons who spoke to them too freely for Day's liking and found himself challenged to a duel. After eight months, the odd trio came back to England. Having decided that Lucretia was either invincibly stupid, or impossibly stubborn, Day apprenticed her to a milliner on Ludgate Hill. Soon she married a linen-draper and received her promised dowry.

Sabrina, he kept. He sent her for a few months to stay in the country with Bicknell's mother while he looked for a house and then early in 1770, largely to be near Darwin, he rented Stowe House, gazing from its green park across Stowe lake to the Cathedral. Lichfield was agog. At thirteen, Sabrina was a sweet-voiced beauty, with long eyelashes and auburn ringlets, a marked contrast to fashionable girls with hair plastered with powder and pomatum. Edgeworth, on his way back from his father's deathbed in Ireland, was curious, he remembered, to see how his friend's 'philosophic romance' would end.

Nobody regarded Day's act as one of personal gratification: it was seen as noble, principled, philanthropic – like adopting an orphan from a war-torn country. He himself caused a stir in the Close. At twenty-two, he absolutely looked the philosopher, thought Anna, meditative and melancholy, awkward yet dignified:

> Powder and fine clothes were, at that time, the appendages of gentlemen. Mr Day wore neither . . . There was a sort of weight upon the lids of his large hazel eyes; yet when he declaimed,
>
> ——Of good and evil,
> Passion, and apathy, and glory, and shame,
>
> very expressive were the energies gleaming from them beneath the shade of sable hair, which, Adam-like, curled about his brows.[16]

Wright painted Day's portrait this August, posed against a tempestuous sky, his poetic figure draped in a swirl of russet silk, a flash of lightning playing on his sable locks, a book in his hand, almost certainly *Emile*.[17]

Day now embarked on another stage in the training of the model wife. If she was going to be able to teach her children fortitude and endurance,

Sabrina must learn Stoicism, and be able to withstand both immediate pain and fear of coming danger. Unfortunately, 'his experiments had not the success he wished and expected'.[18] When he dropped hot sealing-wax on her neck and arms Sabrina forgot she was a Spartan maid and screamed; when he fired pistols at her petticoats, she leapt aside and shrieked. When he tested her loyalty by telling her he was in grave danger, which would increase if she told anyone, he found that the servants and all her friends soon knew all about it. She detested books and scorned science, 'which gave little promise of ability, that should, one day, be responsible for the education of youths, who were to emulate the Gracchi'.[19]

Day gave up, deciding he could never mould her into a worthy mother. Early in 1771 she was packed off to boarding school in Sutton Coldfield. He had failed to train her, which proved, of course, that she was hopeless material to start with, like a water-spaniel that always gets distracted instead of picking up the fallen bird it is supposed to fetch.[20]

And what of Sabrina's future? When she left school, Day gave her an allowance and she lived first near Birmingham and then in Shropshire but often came back to stay in Lichfield, particularly with the Darwins. She made friends everywhere she went, and when she was twenty-six she married John Bicknell, Day's barrister friend who had helped pick her out at the orphanage in Shrewsbury. Day gave her a dowry of £500 and a slightly grudging blessing: 'I do not refute my *consent* to your marrying Mr Bicknell; but remember you have not asked my *advice*.'[21] Within a decade she was left a widow with two small sons, dependent on an annual £30 from Day, and a generous £800 raised by Bicknell's fellow barristers. She then became housekeeper to Charles Burney, son of Dr Burney, music teacher, writer, friend of Samuel Johnson, and brother of the author Fanny Burney. Here she stayed, loved and respected, until her old age.

When the Sabrina experiment ended, Day changed course. 'His trust in the power of education faltered,' wrote Anna Seward, adding enigmatically, 'His aversion to modern elegance subsided.'[22] This latter change of heart had nothing to do with Sabrina, but everything to do with Anna's close friend, Honora Sneyd.[23] The girls had grown up together in the Bishop's Palace since Honora's mother died. Honora was beautiful and clever and had been briefly romantically involved with a Derbyshire merchant, the Swiss-born John André (he joined the army in 1771 and in 1780 was hanged as a spy in America, his death causing outrage in Britain and occasioning a famous poem from Anna[24]). Over Christmas 1770 Edgeworth and Day spent much of their time at the Sewards', and although Edgeworth claimed that the attraction was the literary talents of Anna, the real draw was Honora. Small, Keir and Darwin all agreed she was delightful. Edgeworth fell in love: among other charms, Honora was genuinely interested in science, and a blissful change from Anna Maria who (fairly understandably) was not good company: 'My wife was prudent, domestic and affectionate; but she was not of a cheerful temper. She lamented about trifles; and the lamenting of a female, with whom we live, does not render her delightful.'[25]

At first Day was less smitten: Honora danced too well, she was too fashionable, her arms were insufficiently round and white. Possibly influenced by Edgeworth's raptures, he slowly changed his mind. After Edgeworth returned to Hare Hatch, Day sent him a vehement letter, telling him that he was wrong to cherish a hopeless passion, but that Day would never marry if it would affect their friendship. Edgeworth, obviously rattled, adopted the familiar language of experiment, replying that the only way he could see if he could overcome his feelings was to test them, so he would come back to Lichfield – but would bring his wife and three children as a safeguard.

He watched stoically as Day and Honora became closer and even took a heavy packet to her containing Day's 'plan of life' for her to approve. This was rashly based on Day's conviction that if she once decided to live a secluded life, she would never want to return to the gaiety of polite society. Honora's reply 'contained an excellent answer to his arguments in favour of the rights of men, and a clear dispassionate view of the rights of women'.[26] Astonishingly, she would not admit 'the unqualified control of a husband over all her actions'; she did not feel that virtue and happiness depended on becoming a recluse. Day was literally thrown into a fever. Darwin bled him and talked to him sternly. Luckily, just as he was recovering, Honora's father settled in Lichfield with four more daughters, including Honora's younger sister Elizabeth, 'very pretty, very sprightly, very artless and very engaging'. It was high summer and Edgeworth organized an archery party, with fencing and vaulting, music and dancing. Honora and Elizabeth were there, and so was Day. Like Mr Collins in

Pride and Prejudice, Day immediately switched sisters. For one thing Elizabeth seemed far less polished and threatening:

> her dancing but indifferently, and with no symptom of delight, pleased Mr Day's fancy; her conversation was playful, and never disputatious, so that Mr Day had liberty and room enough, to descant at large and at length upon whatever became the subject of conversation.[27]

Although stunned – as who would not be – Elizabeth was not silenced. Like Margaret Edgeworth before her, she told Day that perhaps she could love him, if he had more normal manners. Day now put all his failures down to this, and told Elizabeth he would go to Paris for a year, learn to fence, dance, pull back his shoulders and point his feet, and practise 'the military gait, the fashionable bow, minuets and cotillions'. On her part, eyes twinkling, Elizabeth promised not to go to Bath or London or other frivolous places and to undertake a heavy course of improving reading.

Edgeworth, too, left for France, telling everyone that when he returned he would go straight to Ireland. Obsessed by Honora, but also thinking of his family, he wrote, 'I knew that there is but one certain method of escaping such dangers – *flight*.'[28] He took seven-year-old Richard with him, with a tutor, hoping to teach him to speak French without an accent. Even in old age, he wondered if he had been right. In Paris, they met Rousseau, who pronounced Richard a boy of abilities, but was disconcerted by his praise of everything English, from carriages to shoe-buckles. Day sent rapturous letters to Anna Seward, mixing incoherent flights of philosophy with accounts of their travels and requests for home news: 'I suppose Mr E & I are clean forgotten by this time . . . Addio caro mia. I have forgotten nobody that I esteem'd, or lik'd, or lov'd at Lichfield.'[29] Eventually they settled in Lyons, where Day went through the prescribed tortures. He fenced, he rode, he danced and for hours on end, with a book in his hand, he was screwed into a contrivance of tight, narrow boards, to bend his stubborn knock-knees outwards. In December, he told Anna,

> I am a lac'd coat, a bag, a sword, and nothing else. I am become a Type, a parable, a Symbol. Eyes have I which see nothing but Absurdity, ears which hear nothing but Nonsense, a mind which thinks not, etc etc etc. But in return I speak french very prettily, I bully, I vapour, la la, cut capers and am what a gentleman should be.[30]

Edgeworth's designs for a machine for loading and unloading boats

Edgeworth, meanwhile, learned French, concentrating on the technical language of mechanics, writing a treatise on watermills and becoming involved in the city architect's ambitious schemes to enlarge the town, hemmed in by cliffs and rivers. In order to get more land it was proposed to divert the Rhône to meet the Saône a mile further south; the empty riverbed would be filled and new suburbs built. Edgeworth was in charge of building an embankment to dam the river. He designed special high platforms to pour gravel and stones into the roped ferry boats carrying material across the violent current, and an aerial trestle-bridge which was supposed to take a stream of wheelbarrows across without labourers pushing them. This was practical yet imaginative work: using his mechanics in the service of civic reform, Edgeworth felt understandably proud.

Day returned to England in early 1772, his sufferings over, determined to claim Elizabeth. In Lichfield, however, his French polish merely looked ludicrous. 'The studied bow on entrance, the suddenly recollected *assumption* of attitude, prompted the risible instead of the admiring sensation,' noted Anna drily.[31] Elizabeth decided she preferred the old Day to the new, but neither would do as a husband. Mortified, Day returned briefly to Paris then travelled around England before taking lodgings in the Middle Temple, which he shared with the brilliant young Orientalist, William Jones. The following March, he told Anna, 'I do not believe I was ever much in love, and I scarcely believe I shall ever be again.'[32]

This time, their positions were reversed. Day was now the confidant advising philosophic resignation and Anna the subject of romantic scandal, due to her long friendship with John Saville, Vicar's Choral of the Cathedral. In late 1771 rumours grew so strong that Mrs Saville refused to let Anna into her house and early the following year Saville walked out on his family and moved into the house next door. Canon and Mrs Seward responded by virtually keeping Anna, now aged twenty-nine, under house arrest, calling down her understandable rage against their 'deaf and inexorable cruelty'. In May she wrote fiercely:

> If they attempt further persecution the consequences will be more desperate. I lov'd Saville for his virtues. He is entangled in a connection with the vilest of women and the most brutally despicable. He cannot be my husband but no law of earth or heaven forbids that he should be my friend or debars us the liberty of conversing together while that conversation is innocent.[33]

The self-justification and passionate outpourings of this circle of friends in their twenties are all couched in abstractions: 'sensibility', 'liberty', 'sublimity'. This was the rhetoric of Rousseau, but also the language of the poetry that Seward and her circle admired: her fight too, belongs to the Lunar story – both as an assertion of the rights of the individual woman, and of the pride in a provincial poetic, at odds with metropolitan taste.

In 1772, while Day and Anna Seward were suffering, Edgeworth was enjoying himself in France. That summer, his wife and her sister joined him in Lyons but Anna Maria did not take to French society: she was pregnant, and as winter set in the women made their way back to England. But Edgeworth made himself at home here. He mixed with the expatriate community and with aristocrats on their Grand Tour. He dined with dowagers, drank with officers, and became entranced by crystals and minerals after a visit to the Grotto de Baume. The only dampener was his son Richard. His tutor could not control him and Edgeworth's final resort was to send him to the Jesuit school – the opposite pole to his earlier Rousseauian experiments. 'I dwell on the painful subject', he wrote in his memoirs, 'to warn other parents against the errors which I committed.'[34]

Another anxiety was his Rhône embankment. He had been warned of floods from autumn rain in the Alps, but the company would not lay out more money to complete the dam in time. One day at dawn he was woken by a great roar; crowds were rushing towards the swollen river; the work of weeks had been swept away; timber, barrows, tools, machines were hurled downstream and lay broken on the banks. The onset of winter now stopped the work, and Edgeworth turned to designing flour-mills for the reclaimed land. They were never built. In March 1773 he heard that he had a daughter, Anna. A few days later, a second letter brought news that his wife had died. With Richard he travelled slowly north through Burgundy to Paris, and returned to England. But Edgeworth shed few tears for Anna Maria: his heart (the true guide) was still set on Honora Sneyd. The year before there had been alarming rumours that she might lose her sight but now Day told him she was well, and though 'surrounded by lovers, still her own mistress'. At once, Edgeworth headed straight for Lichfield.[35] Arriving at Darwin's house he found the doctor out, but Susannah Darwin received him kindly. She was on her way to tea with the Sneyds – would he go with her?

On 17 July 1773, Edgeworth and Honora were married by special licence in the Cathedral. Canon Seward officiated and his wife showered them with smiles. Only Anna seemed morose. The following month, Day (who was staying with Dr Small and driving over to dine with Darwin) sent his blessing and declared that he himself was obviously marked out by fate for a bachelor, destined to buy hobby-horses for Edgeworth's grandchildren.[36] For both these brave followers of Rousseau, their first romantic experiments had ended.

17: VASES, ORMOLU, SILVER & FROGS

At the end of the 1760s, while Lichfield was feeling the impact of Rousseau and the country was riven by riots, Wedgwood and Boulton looked to the polite world where spending still soared, seemingly immune to disorder. Riding the wave of vase mania, the two manufacturers launched themselves on the élite London market. Boulton's main line was his ormolu. Defined strictly, this was the gilding of metal with ground gold, '*or moulu*', mixed with mercury; it was chiefly associated with France and in Britain the French bronziers had led the field. Now Boulton seized the chance to challenge them. 'To understand the character of Mr Boulton's mind', as Keir put it, 'it is necessary to recollect that whatever he did or attempted, his successes and failures were all on a grand scale.'[1] The use of vases extended beyond mere ornament: with metal branches added they could become candelabra and candle-holders; with a clock or watch fitted they were lavish timepieces; with perforated lids they turned into 'cassoulets' or perfume-burners. (Inviting Boulton to her London house in 1771, his patron Elizabeth Montagu added '& then you will be sensible how agreeable the aromatick gales are from these Cassolettes when they drive away the vapour of soup and all the fulsome savour of Dinner'.[2]) And Boulton soon looked to other products, planning ormolu tripods and girandoles, ice-pails and clocks.

Wedgwood's vases inevitably caught his eye as suitable for mounting, and in the spring of 1768 he broached the subject of collaboration. After

Canalside scene at Etruria, engraving, 1769

inspecting Darwin's horizontal windmill at Lichfield, Wedgwood spent a weekend at Soho: 'We settled many important matters', wrote Wedgwood, '& laid the foundation for improving our manufacture, & extending the sale of it to every corner of Europe.'[3] The main idea was that Boulton would set Wedgwood's vases in metal but he also showed him intriguing specimens of gilding and printing on ceramics and stirred his competitive spirit by telling him that French artists were coming over to England and picking up 'all the whimsical ugly old things they could meet with', taking them back to Paris, ornamenting them and then selling them back as great rarities to 'Millords d'Anglise'. The prospect of expanding in this direction was mouth-watering: 'This alone (the combination of Clay & Metals) is a field, to the farther end of which we shall never be able to travel.'

Later that year, when Wedgwood had recovered from his amputation, he and Boulton met in London to go 'curiosity-hunting'. It was a festive party. Sally Wedgwood and Catherine Willet came too, 'pushing me off the stool', wrote William Cox plaintively from the showroom, and grabbing Cox's paper to pen comic joint accounts to Bentley of Wilkesite riots, of their visits to the House of Commons and Drury Lane, of hearing the fiery Methodist preacher George Whitefield.[4] One can see why Wedgwood responded teasingly to one of Sally's high-spirited letters the following year, that 'she did not conclude it so meekly as one might wish, but with a Toss of her head crys, I shall not trouble my head – Oh Fye Sally Fye, wilt thou never mend?'[5] But Sally's only complaint during this trip was that,

> my good man is upon the ramble continually and I am almost affraid he will lay out the price of his estate in Vases he makes nothing of giving 5 or 6 guineas for . . . if we do but lay out half the money in ribband or lace there is such an uproar as you never heard.[6]

Wedgwood was certainly vase-hunting. He and Boulton visited Lord Shelburne together and raided Harrache's exclusive shop, with their wives in tow. 'Mr Boulton is picking up vases', Wedgwood reported, 'and going to make them in bronze.' (He also sent for an artist from Soho to draw the 'pretty things' he saw.) But should he work with him or not? If he refused, Boulton might set up as a potter himself or get vases made by Garbett, who had a pottery up in Prestonpans. Boulton was aiming for the top, as Wedgwood saw, ignoring the court of St James and 'scheming to be sent for by his Majesty! I wish him success. He has a fine spirit, and I think by going hand in hand we may in many respects be useful to each other.'[7]

By the end of the year Boulton had ordered Wedgwood vases in blue, green, or other simple colours.[8] But the scheme petered out, with Boulton complaining that orders had not been filled and hinting that he was thinking of a deal with the Chelsea factory, which was briefly under the management of the astonishing goldsmith and showman James Cox. Earlier he had been approached about a possible partnership by the Comte de Lauraguais, who had discovered kaolin and petuntse near Alençon, and had taken out a patent for making porcelain vases in England. Faced by the threat of competition Wedgwood was anxious but determined, uttering defiance like Horatio keeping the bridge. If Etruria must give way to Soho:

> It doubles my courage to have the first Manufacturor in England to encounter with – *The match likes me well.* I like the Man, I like his spirit. He will not be a mere snivelling copyist like the antagonists I have hitherto had, but will venture to step out of the lines upon occasion, and afford us some diversion in the combat.[9]

In fact, china was far too fragile to bear the heavy ornaments, especially for big pieces such as candle-holders. Instead Boulton turned to marble and Blue John – it was now that he wrote to Whitehurst about trying to lease a mine – and explored the possibility of glass and gilt, lacquered and japanned metal.

Wedgwood had nothing to fear from Soho. He was already running to keep up with demand, even getting the woodcarver John Coward to doctor discarded seconds, adding bases, lids and snake-shaped handles in black wood.[10] His new works at Etruria were finally opened on 13 June 1769. On the slopes behind the factory a great feast was laid out on trestle tables in the shade of the trees. The factory was nearly finished, covering seven acres of land, bounded by walls except on the canal. Land was marked out for Wedgwood's own house – Etruria Hall – and a village was being built for the workmen with houses for two dozen (and eventually two hundred) families. In the sunshine, surrounded by Wedgwood relations and old friends such as the Whieldons and the Brindleys, Wedgwood put on his 'slops', the old potter's smock. Sitting down at the wheel he threw six perfect copies of a black Etruscan vase, while Bentley turned the crank. The shape was the ancient '*lebes gamikos*', a ceremonial bowl to hold water at marriage rites. After the vases were fired they were sent to London to be hand-painted with the encaustic enamel in Wedgwood's decorating studio at Chelsea. On one side were three figures from what was considered the most beautiful vase in Hamilton's *Antiquities*, the 'Meidias Hydra'. Below was an inscription: *Artes Etruriae Renascuntur*.

The 'First Day's Vase'. black basalt with red encaustic enamels. Six of these were thrown by Wedgwood on 13 June 1769, and four survive today.

It was a time of great celebrations, but it was also at this point that the partners decided that it made better sense for Bentley to work for the firm in London rather than in Staffordshire. In September he took a house in Chelsea, to be the headquarters for David Rhodes and his painters, and Wedgwood went down to help prepare the enamelling works there. Meanwhile, until Etruria Hall was finished, he and Sally and the children moved into the house planned for Bentley, 'Little Etruria'. Life was not easy. The winter was harsh and when Sally had to go and care for her ill father in Cheshire Wedgwood nearly collapsed with insomnia, headaches and the trouble with his eyesight. But as spring and summer came and Sally returned home, his mood lifted. In July 1770, when the first dinner service ordered for the Russian court went on show in London, he rushed around the capital so briskly that he damaged his wooden leg, and although he always carried a spare, he now ordered a 'veritable wardrobe of peg-legs'.

The previous September Bentley had inspired him with the notion that they could, with effort, outshine their rivals abroad as well as at home. 'And do you really think that we may make a complete conquest of France?' a delighted Wedgwood asked:

> Conquer France in Burslem? My blood moves quicker, I feel my strength increase for the contest. Assist me, my friends, and the victorie is our own. We will make them (now I must say Potts, and how vulgar it sounds), I won't though, I say we will fashion our Porcelain after their own hearts, and captivate them with the Elegance and simplicitie of the Ancients. But do they love simplicitie?[11]

French and 'frippery', not simplicity, went together in his mind. Boulton too vowed to make his ormolu simpler than the ornate French ware. He too saw his trade as a patriotic battle and so did his patrons, such as the wealthy colliery-owner Elizabeth Montagu, a distant relation of his wife, 'Queen of the Bluestockings' and friend of Samuel Johnson. She took greater pleasure in victories over the French in arts than in arms, she told him: 'Go on then Sir, to triumph over the French in taste & to embellish your country with useful inventions & elegant productions.'[12]

In 1769 Boulton began to get orders for his ormolu from Thomas Anson, the Earl of Shelburne, Henry Hoare the banker, and many others. He made no pretence to originality. Seeing that 'Fashion hath much to do with these things & that the present age has distinguished itself by adopting the most Elegant ornaments of the most refined Grecian artists', he declared, 'I am satisfyd in conforming thereto, & humbly copying their style, & makeing new combinations of old ornaments without presuming to invent new ones.'[13] But once he and Wedgwood had established a demand, both manufacturers had to think fast about increasing the range of their designs and raising production. Sometimes a note of panic sounds through their letters and – particularly on Boulton's part – a string of ingenious excuses (Mrs Montagu had to wait about three years for a 'plaited tea-vase'). And admiration did not always bring orders, as he found on a visit to the Palace in 1770: 'Never no man was so much complimented as I have been but I find compliments don't fill the pockett nor make me fat.'[14] He was cheered by the prospect of home and a round of dinners – with Lord Shelburne's brother, David Garrick, Franklin, Mrs Montagu and others, 'nowhere on Wednesday, & sup with my own dear Wench on Thursday'.

In late 1770, one particular order brought several Lunar men together. Just before Christmas Wedgwood spent two days at Soho, where Boulton was making 'an immense large tripod' for the MP Thomas Anson of Shugborough Hall, 'to finish the top of Demosthenes Lanthorn building', designed by James Stuart.[15] 'The Legs were cast', wrote Wedgwood,

> & weigh'd about 5Ct wt. but they (the Workmen) stagger'd at the bowl & did not know which way to set about it. A Council of the Workmen was called and every method of performing this wonderful work canvassed over; They concluded by shaking their heads, and ended where they had begun. I then could hold no longer but told them very gravely they were all wrong . . . they must call in some able Potter to their assistance, and the work would be completed. Would you think it? They took me at my word & I have got a fine job upon my hands in consequence of a little harmless boasting.[16]

Boulton, Darwin and Wedgwood arranged to dine with Anson on New Year's Day to talk the matter over. Darwin wrote happily, 'I shall be glad to see you Birmingham Philosophers-and-Navigators', but the meeting never happened, as Anson was ill and Wedgwood's lameness made him fear the frosty weather.[17] Although Boulton and Keir did visit that month, the project was ended by Anson's death soon afterwards. Regretting the lost meeting, Wedgwood warmly invited Boulton and Anne, and 'the Infantry' to stay with them in the summer.[18]

All along, the two men had been facing the same challenges. The first problem was design. They borrowed works of art from rich clients and from friends and rivals, shopkeepers and sculptors. (The surgeon John Hunter sent Boulton a ghoulish figure of 'Death'.) They wangled introductions to the British Museum, and Wedgwood managed to charm his way into seeing collections at the Oxford colleges and at Blenheim. Woodcarvers such as John Coward and sculptors like John Bacon made models for them and they bought plastercasts from John Flaxman in Covent Garden, father of the more famous sculptor, whose precocious talent Wedgwood was quick to spot.[19] They also begged models from architects: lions' heads and tritons and gryffins, winged figures and sphinxes, caryatids and Persian slaves. This last – used by both manufacturers – was based on a design by the King's architect, Sir William Chambers. Boulton breakfasted with him in 1770, and he gave him, he told Anne, 'a present of some valuable, usefull and acceptable models'.[20]

Although Wedgwood turned moulds from the models himself he still needed more craftsmen.[21] His letters are full of troubles with modellers (particularly the talented Voyez, a wayward man and ultimately a plagiarist who copied his designs). Hunting for skilled painters for Chelsea, where the pottery was finished and painted under Bentley's supervision,[22] he looked for men and women from the porcelain works in Derby, Worcester and Liverpool, from among the fan painters with their talent for miniatures, and from the coach and fresco painters.[23] Boulton faced similar problems. For some time, on his trips to London he had been noting down the names of both French and English ormolu-workers in his diary. But most were not to be poached, so in 1770 he told Robert Adam proudly that he was training up 'plain Country Lads, all of which that betray any genius are taught to draw, from whom I derive many advantages that are not to be found in any manufacture that is or can be establishd in a great & Debauchd Capital'.[24] That regional note is significant. Wedgwood divided the work between Etruria and Chelsea, but Boulton did everything at Soho, apart from making the stone and pottery vase bodies and the movements and dials for his clocks.

It was in the production that Boulton's and Wedgwood's scientific knowledge counted. Unlike Wedgwood, Boulton did not do all the experiments himself but had them carried out by trusted workmen or by friends such as Small and Keir. He knew how different alloys produced different colour metal: from the yellowy gold of standard brass (70 per cent copper to 30 per cent zinc), to the 'mock-gold' that comes with higher copper content, like 'Bath metal' or 'pinchbeck', and the paler tint as more zinc is added, fading to silver-white and grey. He made numerous notes on 'fire gilding', the highly toxic process that involved the blending of leaf or powdered gold with mercury at an intense heat. The cooled solution, squeezed to get rid of the excess mercury, left a butter-like wax which was smeared over the metal, already prepared with nitrate of mercury. When heated, the mercury evaporated, leaving the gold alloyed to the underlying metal surface.[25]

Boulton was frustrated at first by the poor quality of Soho gilding and searched for colouring processes that would give a lasting lustre and create dramatic contrasts between matt and polished surfaces. This could involve dipping in solutions of turmeric, sulphur or even urine,[26] but the Soho workmen tended to use a gilding wax, and Boulton's notes are full of recipes, from a simple mixture for a matt surface, of salt, 'sal petre' and alum and salt (sodium chloride and potassium nitrate) to pale colours using 'verdigris' or 'blue vitriol', and elaborate combinations including 'copperas' and red chalk for a 'red middle yellow colour'.[27] Not above a bit of industrial espionage himself, he was still trying to spy out the secrets of fine Parisian gilding years later.[28]

By 1772, however, the ormolu craze had peaked, as had the vogue for vases as a whole and Boulton was left with cartloads of beautiful, unsold stock. Wedgwood had faced a similar crisis with his creamware the year before: 'The General trade seems to be going to ruin on the gallop,' he groaned, 'large stocks on hand both in London and the Country, and little demand.'[29] He had been saved then by a huge order from Russia, and in 1772 Boulton followed suit, shipping tons of ormolu to an admiring Catherine the Great. Wedgwood, meanwhile, saw that the loss of élite trade could be offset by addressing a different market.

> The Great People have had these Vases in their Palaces long enough for them to be seen and admired by the *Middling Class* of People, which Class we know are vastly, I had almost said, infinitely superior, in number to the great, and though a *great price* was, I believe, at first necessary to make the vases esteemed *Ornament for Palaces*, that reason no longer exists. Their character is established, and the middling People would probably buy quantitys of them at a reduced price.[30]

Indeed they did.

Another response to the glut was to diversify. Increasingly in 1772 and 1773 Wedgwood and Bentley worked on a different line, the production of seals in *cameo* and *intaglio* form, using black basalts and white biscuit clay. This fitted in with another current craze, the collecting of carved antique gems with classical or mythological motifs. These were already being copied in a newly invented glass paste by James Tassie, who now began supplying Wedgwood with casts. Many of these small, delicate objects were sent from Etruria to Birmingham, where they were set in metal by Boulton & Fothergill and others, to be used in brooches and buckles or on snuff-boxes or tortoiseshell boxes. And soon Wedgwood was supplying a taste even more prevalent than that for urns or gems: the love of portraits. His portrait medallions sold brilliantly, displaying heads of popes and kings and queens, and especially the 'Heads of Illustrious Moderns from Chaucer to the Present Time'. With such a success on his books, he began to hunt even harder for a new ceramic body that would suit them still better, dense and translucent, moving slowly towards his great development of 'jasper'.

Among the first of his 'Etruscan Portraits', made in April 1771, was one of Sir William Hamilton. Two months earlier Hamilton had sold his collection to the British Museum for £8,400, but Wedgwood later calculated that Etruria had made three times that sum in profits from the vases Hamilton had inspired – as a boon for the nation, of course.[31]

Boulton's answer to the slipping sales was also to try something different. In 1773, Hamilton had urged him to try making glassware in the antique style, sending specimens via his nephew, Charles Greville.[32] But Boulton was more realistic: he now intensified his attention to silver on the basis that, unlike ormolu, silver was never out of fashion. It had long been used in his button- and buckle-making and toy trades.[33] He already had many fine craftsmen and now began adapting all his new techniques and classical moulds to make expensive ornaments for the gentry, and cheaper versions in plate for the 'middling classes'.[34]

Working in silver, Boulton had to compete with well-known London craftsmen. He also had to have his work stamped by an assay office to guarantee its quality. If he sent it to London, he feared that his designs would be copied and his prices undercut, so instead it went to the nearest provincial assay office, at Chester, seventy miles away. His fine goods were often bashed and broken on the rough northern roads and in 1771, furious at the near ruin of two pairs of candlesticks, he told Lord Shelburne that although he was 'desirous of becoming *a great silversmith*' he would never take it up wholeheartedly unless Birmingham had an assay office of its own.[35] In late 1772 he decided to bring a petition to Parliament.

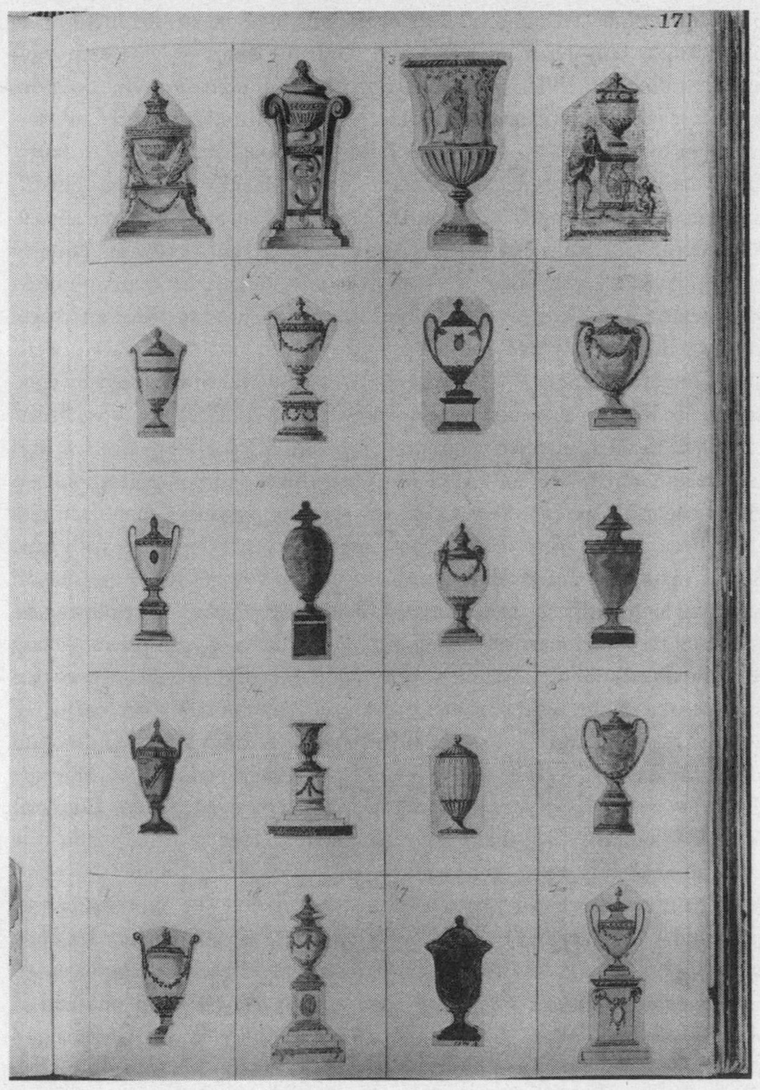

Urns and perfume burners, from the Boulton & Fothergill catalogues

This soon became another war of the provinces versus the metropolis. As soon as they heard, the Sheffield Cutlers Company (who sent their goods to be assayed in Newcastle) suggested a joint petition. But Boulton, working with his parliamentary lobbier – none other than Samuel Garbett – wanted to run his own campaign. Indeed, although he claimed to represent the whole Birmingham toy trade, he presented the petition himself, unsigned by other makers. Meanwhile, Sheffield prepared its own appeal. All Boulton's influential patrons were lobbied relentlessly. In January 1773, he jotted in his diary a list of over forty dukes and earls and lords he planned to call on, as well as three bishops.[36] 'I hope the King and royal family, the Nobility and the Ministry and your other friends are well,' Small wrote teasingly.[37] In February Parliament was bombarded by rival petitions. Two separate committees were set up, with Boulton's allies firmly in place: the chairmen were Thomas Skipwith, MP for Warwickshire, and Thomas Gilbert, MP for Lichfield. Newspapers and pamphlets took up the fight, with the Londoners claiming that the provincial manufacturers had been plating ware made of base metal and marking them with false silver marks. Part of Boulton's aim was precisely to get rid of the slur on Birmingham goods as cheap, shoddy, 'Brummagen' fakes. In retaliation he accused the Londoners of poor design, low value and high prices. Acting on Boulton's tip-off, Gilbert's committee secretly bought twenty-two pieces of London-assayed plate and found that the silver content of all except one had been diluted well below standard. With this coup, Boulton won the day. He took over the Sheffield campaign, and on 28 May the Royal Assent was given. A weary Boulton, 'damned sick of London', crawled into the post-chaise and returned to Handsworth, where the church bells rang out to greet him.

The new provincial manufacturers had routed the monopolists of London. On 31 August 1773 the Birmingham Assay Office opened in New Street. Boulton was the first manufacturer among the town's forty licensed silversmiths to submit his wares – and the first to have them smashed as below standard. Week by week the load from Soho grew: buckle-rims and sauce boats, tankards and tea-vases, church plate and fish trowels, sword-hilts and salvers. Silver production reached a peak around 1776–77 and then slipped, both because it proved unprofitable and because Boulton's energies shifted to his engine business, but silver and Sheffield plate remained a staple of the Soho trade. In 1780 a staggering near half-million objects were made here, ranging from thousands of buttons, salt-cellars and candle-snuffers to new-fangled coffee machines (a central urn dispensing hot water to teapot and coffee-pot); they even made silver cheese-toasters with hinged lids, glamorous great-great-grandparents of today's sandwich-toasters.[38]

With their vases and ormolu and silver, Wedgwood and Boulton established themselves as men who could not just follow the fashion but set it, entrepreneurs of taste. Perhaps the most bizarre aspect is that they borrowed models from patrons – vases, china, candlesticks – and sold them back again to those same people, not as replicas of something old, but transformed into something new: products of a modern, scientific manufacture. They now had a name for making fabulously desirable, expensive articles. But they wanted even more prestige – to present their products as objects of art, not just as good 'manufactures'. Yet how could this be, when they were produced by mechanized techniques from standard moulds, rather than by a single craftsman? Could one apply the same aesthetic standards? This remains a fascinating problem. And in the 1770s two contrasting achievements challenged received notions: Boulton's 'philosophical' clocks and Wedgwood's great 'Frog' dinner service.

Boulton made several elaborate allegorical clockcases in ormolu but in 1772 he showed two particularly remarkable ones.[39] The first was a 'geographical' clock, surmounted by a revolving globe which showed the position of the sun at the appropriate date. The movements (influenced by James Ferguson's astronomical clocks) were made by John Whitehurst. The globe, engraved by the London instrument-maker Nathaniel Hill, was enclosed in glass and revolved once every twenty-four hours, so that 'you see at every instant what part of the world is enlightened and what is in darkness'. It was supported by what Boulton called 'a group of Hercules and Atlas' which he had bought from Flaxman.[40]

The second was a sidereal clock: this is a 'star clock', rather than a solar mean-time clock; it measures time between successive appearances of the same star on the observer's meridian (apparently astronauts use sidereal time, because it is more accurate). Boulton's clock had a twenty-four-hour dial which showed the motion of the sun against the fixed stars. The three-foot-high case was decorated with a bas-relief of Science explaining the laws of Nature with the motto '*Felix rerum cognoscere causas*', and its face showed the signs of the zodiac, with gold stars riveted on to a sky of blue enamel. On the top was a figure of Urania, crowned with stars and resting on a celestial globe. A special feature was the addition 'of some of the inventions of Mr Harrison's longitudinal timekeeper which are essential to the accurate performance of the clock'.[41] Boulton took a minute interest in every detail. 'I am determined', he told Whitehurst, perhaps remembering Wright's recent success with the *Orrery* and the *Air Pump*, 'to make such like Sciences fashionable among fine folk.'[42]

The making of parts by so many people in so many different places made Boulton very anxious. He was also still dealing with odd individual requests – such as the green glass ear-drops that Joseph Banks hoped to take as trading goods to Tahiti on Cook's next expedition, planned for April 1772. The haste and muddle show in a rushed letter to his manager John Scale, dashed off from Stratford, *en route* for London.

Mr Boulton begs Mr Scale would not forgett to bring the *Drum* of the Syderial Clock He ought & intended to have brought it but forgott. To procure ye Glass & make ye lines upon it will require time.

Send also per Coach the Modell of ye annihilation of Friction. Charles Wyatt says it is in Jukes's ware room. The Roads to Stratford are exceeding bad being 5 Hours & ½ coming here.

. . . I beg the Gilders will be particularly careful about the Titus's faces & not Scratch off the matt. Grenvills dog. Stuarts Clock must be cleaned & Frenchifyd. bring a few pairs of the Otehite Ear rings as I may shew them to Banks & also one of their queer Gods. dont forgett the Pendulum Rod nor Ball.[43]

Finally, despite the panics, the clocks were finished. Both were expensive, the geographical clock priced at £180, and the sidereal clock at £275, and to Boulton's intense disappointment neither of them found purchasers. In an April blighted with unseasonable snow, he wrote bitterly to Anne, comparing his failure to the success of the automata produced by men such as James Cox:[44]

I find Philosophy at a very low ebb in London & therefore I have bought back my two fine Clocks which I will send to a markett where common sense is not out of fashion. If I had made the Clocks play Jiggs upon bells and a dancing Bear keeping time, or if I had made a horse race upon the faces I believe they would have had better bidders. I shall therefore bring them back to Soho and some time this summer send them to the Empress of Russia . . .[45]

He might have been right about the need for a dancing bear. The geographical clock stayed in England, but its partner was eventually sent to Russia in 1776. Despite much admiration, Catherine did not buy the clock, nor did any of her courtiers: 'They all praised it . . . but it did not strike the hours, nor play any tune.'[46] The market there was only for 'gewgaws and French baubles'. In the end it returned to England.

Wedgwood was luckier. His commission for a dinner service for fifty people came directly from the Empress (who had already ordered a smaller dinner service in 1770), through the Russian Consul in London. It was designed for the palace of La Grenouillère in St Petersburg – a site known originally, in wonderfully onomatopoeic Finnish, as Kekerekekskinsk, 'Frog Marsh' – and each plate was to bear the emblem of a frog and to be decorated with a view of the British Isles. This became known as 'The Green Frog Service'.[47] Ironically the basis of this magnificent work came not from his ornamental but from his useful line: his creamware. Although it was his very best – technically immaculate, of an even colour with every shape perfect, plates that stacked neatly, lids that fitted, handles that were easy to hold – the basic service itself, 952 pieces, was cheap and easy to supply. The views were another matter; Wedgwood hunted down every available print and hired artists to make new ones. Once again he leaned heavily on his wealthy patrons, seeking their permission to show their estates – and found that one major difficulty was selecting the prints without upsetting anyone: it needed all his tact to explain why one person's stately home was squashed on to a saucer while another was spread lavishly across a huge dish.

As the larger pieces had two scenes, 1,244 views were needed, ranging from landscape gardens to wild, romantic views of Fingal's Cave and the Derbyshire crags. It was an aesthetic summing up of the moment, picturesque meeting Gothic, polite nodding to sublime. As Bentley put it, it represented 'all ages and styles, from the most ancient to our present day; from rural cottages and farms, to the most superb palaces; and from huts of the Hebrides to the masterpieces of the best known English Architects'.[48] It was also a set of definitions of 'Britishness': Celtic dolmens and Roman remains; landscape gardens and London streets; Coalbrookdale ironworks and the Bridgewater Canal. In March 1773 the painter Edward Stringer took to the road to sketch new views, carrying his camera obscura, a pyramid of rods covered with cloth, with a lens at the top and a mirror reflecting the scene on to the paper inside the little tent. Among his scenes was Etruria Hall, sitting proud on its hill with a barge passing by on the canal below.

It took over a year to put the whole thing together. Two artists in particular, James Bakewell and Ralph Unwin, worked on the delicate scenes in dark-purple enamel. The dinner service had a border of an oak wreath and the dishes had handles of oak saplings with acorn knobs; the dessert service was bordered with an ivy wreath. The cost was immense: the china cost only £51, but the expense of the decorated whole reached over £2,500 – only a fraction below the price paid by Catherine. As a prestige advertisement, however, it was priceless. When the Frog Service was displayed in Wedgwood's showroom the nobility flocked to see it, the Queen herself paid a visit, and it was the talking point of London for weeks.[49]

In the early 1770s both men were responsible for objects of lasting beauty. Years later, in *The Botanic Garden*, Darwin would say that Wedgwood's vases were works of 'uncopied' merit, while the poetic convention manages to make it sound as if Wedgwood and his factory were one:

> Etruria! next beneath thy magic hands
> Glides the quick wheel, the plastic clay expands,
> Nerved with fine touch, thy fingers (as it turns)
> Mark the nice bounds of vases, ewers and urns;
> Round each fair form in lines immortal trace
> Uncopied Beauty, and ideal Grace.[50]

The entrepreneur, the master, makes all his workmen's skills his own.

18: RUNNING THE SHOW

'I left London with a full resolution to *simplify*', Wedgwood wrote in September 1769, 'and you shall soon be convinced I am in *earnest*.'[1] He meant styles of design, but while he was busy shipping off his old cauliflower ware and rejecting 'frippery' French rococo, he was also streamlining his methods of work.

During the vase craze, he and Boulton both reduced costs by having an interchangeable range of ornaments and designs, varying them to provide customized articles. They sent drawings to clients who could supposedly choose what they wanted, but if a commission meant new designs or special tools Boulton often managed to 'delay' to the point of non-delivery, while Wedgwood told Bentley to avoid orders for 'any *particular kind* of Vases . . . at least till we are got into a more methodicall way of *making the same sorts over again*'.[2] Sighing over one such order, he added, 'It is this sort of *time loseing* with *Uniques* which keeps ingenious Artists who are connected with Great men of taste poor.'[3] Good business worked against originality.

True, Boulton and Wedgwood were copying antique designs, playing on the fashion for the past. Soho, they joked, could be the new Corinth, the classical city of metal, just as Burslem was the new Etruria. In 1771, Wedgwood – Potter to the Queen – told Boulton of a witty article by 'Mr Antspuffado', who declared he should not wonder if the 'Genius of Birmingham should be tempted to make *Roman medals*, & *tenpenny nails*, or *Corinthian Knives* & *Daggers*, & stile himself Roman medal & Etruscan

The original Flint Crushing Room at Etruria

tenpenny nail maker to the Empress of Abyssyinia'.[4] But as they sharpened up their businesses, from design and manufacture to marketing, they were tackling practical problems in ways that mirrored contemporary ideas: the application of 'scientific' method; the ideal of neo-classical order (already reflected in the façades of their new factories), a broad triangle of labour culminating in an apex – the master; the model of the body, with capital circulating like blood, marketing reacting like the nerves, discipline keeping the whole system in equilibrium.

Yet another dominating image, of course, was the machine. When they sold their ormolu, creamware and vases, far from downplaying their scientific advances and new industrial processes, both Wedgwood and Boulton were proud of them. Their 'machines' were part of their cachet. 'It was always in Mr Boulton's mind', noted Keir (in connection with Boulton's clock-making activities), 'to convert such trades as were usually carried on by individuals into great manufactures by the help of machinery, which might enable the articles to be made with greater precision and cheaper than those commonly sold.'[5] Etruria had a water-wheel and a windmill for grinding flint and enamels; Soho was powered by water and, increasingly, steam. Both factories had the latest machines. When Watt visited Soho in 1768, he was astounded, to put it mildly, at the variety and modernity. He saw the watermill 'employed in laminating metal for the buttons, plated goods etc., and to turn laps for grinding and polishing steel work'. He heard that Boulton was the first to use a mill this way, and saw several other ingenious improvements: a lap turned by a hand-wheel for cutting and polishing the steel studs for ornamenting buttons, chains and sword-hilts; a shaking box for scouring metal ('also a thought of Mr Boulton's'); a steelhouse for converting iron into steel, recycling all the metal cuttings and scraps.[6] There was nothing that could not be put to use, or improved upon. Etruria needed fewer machines, but Wedgwood made great strides by developing an improved turning lathe (first seen at Soho), which allowed patterns to be cut on a fired clay vase, or cup, or even coffee-pot. Later he acquired his new rosette lathe, and added 'rose-engine-turning' to cut curving patterns. He gave the turning room pride of place and the best possible lighting in his plans for his factory.[7]

Yet it is easy to overstate the role of the machines. It was skilled hands that made the difference in the toy trades and in pottery. But machines

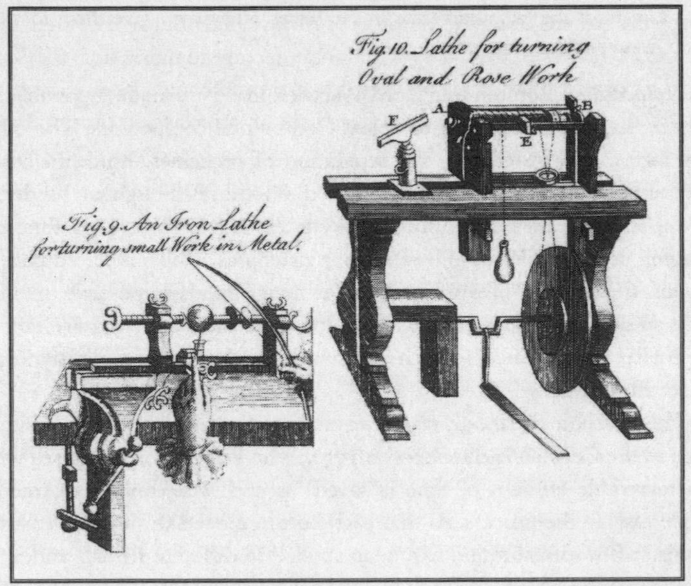

Lathes, from Croker's *Complete Dictionary of the Arts and Sciences*, 1764–66

were new and intriguing and helped make their factories into showpieces. Carriages full of gentry rolled up to Soho and Etruria, often without warning. They were like patrons visiting an artist's studio – Wedgwood knew they must be humoured but he often resented the time spent on them, and their patronizing comments.[8] Boulton, by contrast, revelled in showing off his factory. 'I had lords and ladies to wait on yesterday,' he boasted. 'I have Spaniards today; and tomorrow I shall have Germans, Russians and Norwegians.'[9] Soho could be claimed as a triumph for modern mechanics, for the fusion of science and art, for Birmingham and for Britain – and for him personally. 'I must again thank you for the most agreeable day I pass'd at the Soho Manufactory,' wrote Mrs Montagu:

> The pleasure I received there, was not of the idle & transient kind which arises from merely seeing beautiful objects. Nobler Tastes are gratified in seeing Mr Bolton & all his admirable inventions. To behold the secrets of Chymistry, & the mechanick powers, so employ'd, & exerted, is very delightful. I consider the Machines you have at work as so many useful subjects to Great Britain of your

own Creation: the exquisite Taste in the forms which you give them to work upon, is another national advantage.[10]

Mechanization, Boulton told Lord Warwick in 1773, made it possible for Birmingham manufacturers to defeat Continental competitors. The other key factor, he insisted, was the separation of processes. Lord Shelburne had anticipated him when he reported on the Birmingham hardware trades seven years before, putting its success down to three factors: the shaping of malleable metal by stamping machines, which replaced human labour, the division of work between as many hands as possible, making tasks so simple that even a child could do it (and often did), and the 'infinity of smaller improvements which each workman has and sedulously keeps secret from the rest'.[11]

The division of labour was always remarked on by foreign visitors, such as the German Lichtenberg in 1775, who was astonished to see what 'an incredible amount of time is saved' as each workman performed a single task in the process, so that each button passed through at least ten hands.[12] But specialization had been applied in different British trades for some time; the added efficiency of employing a workman for one particular operation was common knowledge.[13] The large-scale textile manufacturers had long issued precise specifications to their out-workers, producing a standardized range of goods that could be ordered from samples and pattern books; in the metal trades, buckles made by a brass-worker were finished with iron spikes from the chape-maker, while fine brass harnesses went elsewhere to be inlaid with silver or tin. Even in the potteries, pioneers such as Whieldon had allocated different tasks to different men.

Although Boulton now had all the different processes in one place, each workshop still operated independently, under a separate foreman: the only rough move towards a production line was in trying (not always successfully) to orchestrate the making so that it proceeded stage by stage. His ormolu products were a good example.[14] The vases, in Blue John or other materials, were bought in or commissioned. At Soho, in the first stage the sheet metal or ingot was cut into long strips and rolled in water-driven flatting mills for linings for vases and for decorative bands. The rolled metal was then sent to the stamping shop and hammered over special dies, or shaped in new presses. Meanwhile, the brass-founders used moulds in sand or clay, or cast the ornaments by the lost wax, *cire-perdue* method. The cast ornaments and the stamped bands were then sent on to different workshops, to be rubbed and gilded and chased, the detail enhanced by fine chiselling (there were thirty-five chasers here in 1770, noted Wedgwood with awe[15]). Finally the gilt metal was coloured and polished and burnished, and attached to the vase bodies with nuts and pins. But when demand for ormolu faded, the men had to turn their hands to other things; flexibility was just as vital as specialization.

Wedgwood, notoriously, wanted perfection and saw it in mechanical terms. 'I have been turning models', he told Bentley in October 1769, '& preparing to make such *Machines* of the *Men* as cannot err.'[16] He had looked carefully at the Soho organization in 1767, and also knew of the Worcester factory, where the workroom spaces were laid out in an order that reflected the steps in the manufacturing process.[17] At Etruria, although it would be quite wrong to talk of 'assembly lines' or 'mass production', there was certainly a flow of work from start to finish. And whereas Boulton had a mass of workshops and small partnerships, Wedgwood's organization was clarity itself. First, he ruthlessly divided his manufacture in two, dividing it between his 'useful' and 'ornamental' ware.[18] These operated under separate partnerships, one with his cousin Thomas, the other with Bentley: workshops, workforce, ovens, decoration, sales, even costs and accounting were all kept rigidly apart. When Bentley rashly suggested uniting them Wedgwood firmly laid out the lines of the division, the need to balance the solid everyday ware, always in demand, with the more risky, imaginative, artistic trade.[19]

By 1769, he had therefore two specialized factories and a hundred and fifty employees, thirty working specifically on vases. In 1772, he finally moved the 'useful works' from Burslem to Etruria. Both companies were now working beneath the same roof. And, within them, he made a clear distinction between unskilled and skilled labour: men (and to a lesser extent women) were trained for particular tasks – some to make black basalt ornaments, say, others to make creamware sauce boats. But although this meant that high-quality goods could be made by less-skilled workmen, new practices still brought problems. Jealousies and fears arose, particularly among the painters, anxious that they would be threatened by the new transfer printing. Workmen were baffled by new demands, and on 20 July 1772, Wedgwood faced a full-scale demonstration over piece-work: all the men of the ornamental works gathered outside the gates to meet him at half-past six in the morning, to protest about his plan to reduce their rates.

Accounting, 'calculating' – the weighing and measuring to judge results accurately that proved so central to the new science – were now applied to industry. Wedgwood became puzzled that although he was selling so much and his turnover was high, he was short of money. 'How do you think, my dear Friend,' he asked Bentley in 1771, 'it happens that I am so very *poor*, or at least, so very *needy* as I am at the present time, when it appears by my accounts that I clear money enough by the business to do allmost anything with.'[20] He discovered that part of the problem was the great amount of cash tied up in expenses and unsold stock, but also that the prices of his ornamental work were completely unrelated to the cost of manufacture. In response, he made a pioneering analysis of his costs of production.[21] He drew up his 'price book of workmanship', costing every stage carefully, from raw material to the shop counter. Then he set about increasing production runs and lowering manufacturing costs and therefore prices.

This was entirely novel to the workmen who met him outside the factory gates. Wedgwood dealt with them sharply, but in negotiations over the next month he explained that if overheads could be spread over more products by '*making the greatest quantity possible in a given time*', the cost of each individual article, and consequently the selling price, could be reduced.[22] This was the time of his new resolution to sell to 'the middling People': lower prices meant more sales. So although the pay per piece might be lower, theoretically the higher productivity meant that overall wages would rise. Pragmatic capitalism had arrived in the potteries. In April 1773 the Potters' General Assembly agreed to lower prices by 20 per cent.

In the same month John Scale, Boulton's manager at Soho, put forward a revealing set of proposals to transform Soho work practices. He too recommended paying skilled craftsmen by the piece, but apart from the die-sinkers, who were traditionally paid this way, Boulton & Fothergill kept to the standard eighteenth-century practice of paying by the week, and continued to supply the men with free tools and work space.[23] Boulton was always impulsive, never costing products properly, balancing debts, swapping money around, and sacrificing profits to prestige so chaotically that he faced disaster more than once. Yet he was canny: although terms were clearly agreed, none of his partnerships (there were thirteen by the time he died) had a formal contract, which meant that if his partner went bankrupt he stayed in the clear. He looked slapdash, but his spreading of interests reduced costs; he thrived on the adrenalin surge of risk. Wedgwood was an entirely different personality. No huge loans or astronomic interest payments blotted his books. To begin with he was fairly casual about accounts, shoving aside the mountain of paper arriving monthly from London because he had no time and saying that he found the previous week's figures boring, 'like reading an old newspaper'.[24] But after a nasty scare in 1772, when he discovered that Benjamin Mather, his clerk at Newport Street, had been embezzling considerable sums, he was scrupulous in checking the figures weekly.

John Scale not only felt that Boulton should change his mode of paying, but that he overloaded Soho by never turning orders down and wilfully refusing to recognize his workers' capacity and the strain they were under: people were assets too, and had to be carefully treated. In fact Boulton knew this well, and had his own clear views on labour and training. Rich fathers offered him big payments to take their sons as apprentices but he felt this encouraged set ways of working and that well-off boys would be less part of the team, or do what they were told. He let his managers accept one or two apprentices, but preferred, he declared roundly in 1768, to take only 'fatherless children, parish apprentices, and hospital boys, which are put to the most slavish part of our business'.[25] This common practice, in which boys were handed over by the Overseers of the Poor Law in return for a lump sum, could indeed amount to slavery. But at Soho the boys received wages, lived in a special prentice house and worked under individual managers. Several were taught to draw and design by top chasers and engravers. No scandals erupted and lasting loyalties were built.

Boulton also improved working conditions generally by whitewashing the factory and having proper ventilation and both he and Wedgwood contributed to schools and health schemes and provided houses for key workers – the Soho cash books show that several of their workshop managers rented houses for between £5 and £8 a year. His men and women earned good wages, on a par with top craftsmen elsewhere, ranging in the 1770s from 2s 6d for boys and 5s a week for a labourer to 10s a week for a journeyman and 15s or £1 for a skilled man – and far more for the silversmiths who ran individual workshops. The hours were long, from 6 a.m. to 7 p.m. in summer, and 7 a.m. to 8 p.m. in winter, broken by half an hour for breakfast and an hour at midday. Yet the work was fairly secure and the Soho insurance club, which seems to have been started in the early 1770s, was one of the first schemes created by manufacturers to provide benefits in case of accident, sickness and death.[26] (Wedgwood hired three apothecaries and a male midwife to look after his workers.[27])

At Soho the boys and men paid contributions according to earnings, from ½d a week for those earning 2s 6d to 4d for those who earned £1; but if they fell sick the lower earners got bigger benefits in proportion to their contribution – about four-fifths of a full week's wages. Fines were levied for malingering and if an illness was due to 'drunkenness, debauchery, quarrelling or fighting', the first ten days' payment was cut. The scheme had a sound financial rationale, preventing sick workers applying for poor relief, and so keeping Soho's contribution to the poor rate down.[28] It also had one or two telling clauses: a worker who left could not take out the contributions he had paid (a disincentive to moving) and each foreman had to give an account of the men who worked under him (a way of checking standards in different workshops). And the workshop committees, which dealt with bad language or carelessness, look like a move towards shop-floor democracy, but were also a method of control. But still, it was a generous model.

Generosity was part of Boulton's make-up: he would keep drunken men on, joke and swear with the best, hand out presents when things went well. Boswell, on a Midlands tour, watched a workman complaining that his landlord had distrained his goods for arrears of rent:

> Your landlord is in the right, Smith (said Bolton), But I'll tell you what; find you a friend who will lay down one half of your rent, and I'll lay down the other half: and you shall have your goods again.[29]

That tone of voice sounds just like Boulton, who made a point of knowing all his workforce by name. But still, he could get exasperated, especially in future years when the engine business was under way. 'Our forging shop wants a total reformation,' he wrote testily. 'It is worse than ever. Peploe has been drunk ever since his wife's death almost. Jim Taylor has been drunk for nine days past.'[30] It was hard to manage temperamental craftsmen, and accusations often flew – of drink, debts, neglect, theft. Poor John Scale had to cope with most of this but Boulton's stomping presence was very much felt: it was said that he could feel when something was amiss simply by the change in the sound of the machines.

Wedgwood too was a striking presence on the factory floor, a hands-on manager, fierce on quality and prone to smash bad ware with gusto, as potters traditionally did. But he was even keener on rules and fines, penalties and rewards.[31] At Etruria – as at Arkwright's mill in Crompton and at Soho and in many other factories – there were fines for every misdemeanour: for being drunk, writing graffiti (obscene or otherwise), playing fives against the wall. There were rules of cleanliness, partly to protect workers against lead poisoning, but also to protect the clay itself. The bell rang at six each morning, and anyone turning up after quarter past was locked out until the breakfast bell at eight-thirty, losing two hours' work, and the pay that went with it. In later years, Wedgwood introduced a checking-in system, where workers left name cards at the gate, and even a primitive time-clock, based on a design by Whitehurst.

The combination of machines, cost accounting and paternalist discipline was something Boulton and Wedgwood shared with other leading manufacturers, a model for the century ahead. Factories run on these lines improved living standards for many, but they also restricted their movements and deprived them of independence. Workers were left open to exploitation by unscrupulous masters, while repetitive work, as Adam Smith foretold, ran the danger of reducing minds to 'hands'. In later years, James Watt would declare that compared to the inventors, most workers should be considered 'as mere acting mechanical powers . . . it is scarcely necessary that they should use their reason'.[32] As William Blake lamented, the factory was both a product of, and a powerful image of, the bleak rationality of his age, 'Washed by the Water-wheels of Newton':

> . . . cruel Works
> Of many Wheels I view, wheel within wheel with cogs tyrannic

Moving by compulsion each other, not as those in Eden, which
Wheel within Wheel, in freedom revolve in harmony and peace.[33]

Beyond the factory, Wedgwood and Boulton focused on marketing.[34] Patronage was still essential, as they well knew. Boulton's correspondence with the wealthy Mrs Montagu was flirtatious as well as flattering but he might have felt piqued if he knew how some critics mocked her salons, where she sat enthroned in the middle of a circle,

> . . . like a statue of the Athenian Minerva, incensed with the breath of philosophers, poets, painters, orators, and every votarist of art, science or fine-speaking . . . She can make a mathematician quote Pindar, a master in Chancery write novels, or a Birmingham hardware-man stamp rhimes as fast as buttons.[35]

Oblivious to any sneers, Boulton's happy letters home to Anne from London in 1770 showed how hard he worked to get patronage. One Sunday he spent an hour with Lord Shelburne in his library, and hearing that Lady Shelburne had a 'putrid sore throat' and 'wished that she could have a few of my pretty things to amuse her', he dashed off and brought back a load, 'and sat with her Ladyship two hours explaining and hearing her criticisms'.[36] Next day came his breakfast with William Chambers and then an audience with the Duke and Duchess of Northumberland: 'I was very politely received and drank chocolate with them. The Duke himself shewed me his picture gallery and many of his curiositys. He made me sit down with him till 3 o'clock and talked about various arts.' He visited the Earl of Dartmouth and the Dowager Princess of Wales. Then came the ultimate: a visit to the palace. The Queen, he decided, was 'extremely sensible, very affable, and a great patroness of English manufactorys'. After she and the King talked to him for nearly three hours, 'the Queen sent for me into her bedchamber, shewed me her chymney piece and asked my opinion how many vases it would take to furnish it, for says she all that china shall be taken away'. Scale must send a set of blue-john vases within the fortnight: there is still a Blue-John candelabra at Windsor.

As well as selling privately, he and Wedgwood also organized exclusive public sales. Wedgwood had always been concerned to find a showroom that would suit his élite customers, 'for you well know they will not mix with the rest of the World any farther than their amusements, or conveniencys make it necessary to do', he told Bentley in 1767. It had to be chic and private but also 'Large', as vases would decorate the walls and a full display of at least six or eight complete dinner services was essential 'in order *to do the needful* with the ladys in the neatest, genteelest & best method'.[37] In August 1768 the showroom moved from Charles Street to Great Newport Street, and six years later would move again, to the grand Portland House in Greek Street, Soho. Bentley lived next door with his second wife Mary, the daughter of a Derby merchant friend, whom he had married two years earlier.

Shopping was now a fashionable diversion, and manufacturers were alert to the increasing influence of women, as Wedgwood's remarks on 'the ladys' and Boulton's wooing of Mrs Montagu make clear. (When Wedgwood heard of the new craze for women bleaching their hands with arsenic in 1772, he promoted his sale of black basalt teapots to make a good contrast at the table.[38]) And their sales of vases and ormolu were deliberately planned as spectacles, rivalling the shows at the Royal Academy or the Society of Artists. In Wedgwood's showroom the display changed constantly so that smart visitors always had something new to see. The black basalts were shown off by yellow backgrounds, the Queensware by blue or green, the rarest vases kept enticingly in a locked room for private view.

Boulton had no London base of his own. He nearly formed an alliance with the Adam brothers, on Lord Shelburne's recommendation, while Wedgwood thought of taking showrooms in the Adelphi, their ambitious development off the Strand.[39] But both men liked to run their own show. Between 1770 and 1772, Boulton held grand week-long sales at Christie's and Ansell's showrooms in Pall Mall. In 1771 there were three days of viewing, the first for the nobility alone, so that they could place orders in comfort. The following year, Keir corrected his draft of an advertisement, dryly pruning Boulton's wilder claims:

> I have omitted mentioning that you are content to work without profit for the advantage of your Countrymen, because in these days, such an instance of disinterestedness would not be credited, and if it was, it would rather excite admiration of your generosity than of your understanding.[40]

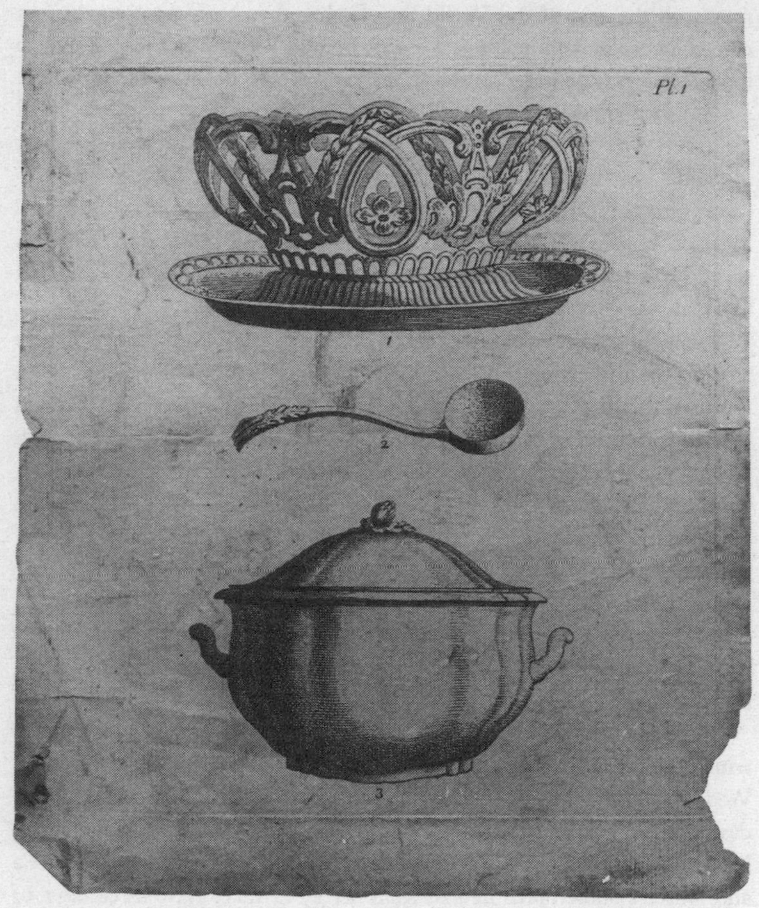

Plate I, from first Useful Ware Catalogue issued by the Etruria Factory, 1774

Nervous and excited, Boulton paced the coffee-houses waiting to see the notice in the newspapers. He need not have worried: Christie's room glittered with Blue John and gilt, 265 lots, with over four hundred pieces, ranging from candlesticks at four guineas to a Persian candelabra at £200, a staggering price. His show was the talk of the town, but in this world of spectacle, just as in manufacture, there was competition. Somehow,

Boulton had to divert the flow of people to his exhibition, rather than to others'. In the same year William Duesbury was displaying his Derby china, and James Cox was displaying fabulously expensive bejewelled automata in what he called his 'Museum'. Wedgwood rightly foresaw that Boulton might be outshone, or even eclipsed:

for what with the fine things in Gold, Silver & Steel from Soho, the almost miraculous magnificence of Mr Coxes Exhibition, & the Glare of the Derby & other China shews – What heads or Eyes could stand all this dazzling profusion of riches & ornament if something was not provided for their relief.[41]

That much needed relief, he thought, might just be provided by his own black, Etruscan and Grecian vases.

Auctions were perfect for the metropolitan crowd. Advertisements, 'puffing' articles and salesmen travelling the provinces (as Tom Byerley did for Wedgwood) could reach still more people, while agents and contacts with ambassadors could help exports. In 1768 Wedgwood met Lord Cathcart, now Ambassador to St Petersburg, '& we are to do great things for each other'.[42] He quickly offered to include Soho in his export drive. Increasingly, another major tool was the catalogue. And after Boulton showed Wedgwood how the lists could be illustrated with engraved plates (much as botanists used plates of plants), Wedgwood and Bentley issued their first catalogue of ornamental ware in 1773, followed by a useful ware catalogue the following year.[43] With their global ambitions Boulton and Wedgwood transformed the crafts they grew up in – and also put an old skill on a new scientific basis, the ancient art, as Wedgwood called it, of '*money-getting*'.

19 : EDDIES

In an idle moment in the ormolu fever of 1771 Boulton opens his notebook. The first part is crammed with Soho price lists: buckles, candlesticks, instrument cases and steel chains – over six hundred patterns. On a few blank pages in the middle he daydreams. He makes notes on a possible round building for a church, calculating how many it will hold. And for himself:

> A round building for my Study, Library, Museum or Hobby Horsery to hold 6 handsome Book Cases with drawers in the lower parts to hold things which relate to subjects of books which are in the upper parts. e:g: a Book Case containing Chymical Books should have drawers under whch contain Metals Minerals and Fossils – between each Book Case should be a Sophi & under the Space between the upper parts of the Cases should be fixed such instruments as Barometer, Thermomotor, Pyromotor, Quadrants, all sorts of Optical, Mathematical, Mechanical, Pneumatical & Philosophical instruments also Clocks of Sundry kinds both the Geographical & Syderial, Lunar & Solar System & one good regulator of time.
>
> A table in the middle of the room & a skylight in the middle of the doomical roof which roof may be covered either with Sail Cloth or brown paper. out of this round room should open a private door into a passage in which passage should open doors into sundry convent. rooms such as Cold & Warm Bath, a Labritory a dressing & powdering room & an observitory for a transit instrument &c.[1]

Lounging on the sofa between his cabinets he will be a man of leisure, of money, of science. The round room is never built.

*

John Baskerville

The following year, James Brindley is dying, aged fifty-five. Wedgwood visits him daily. In the middle of the night of 27 September 1772, Brindley wakes from a sound sleep, asks for a glass of water, drinks it and says, ''Tis enough – I shall need no more.' He turns over, goes back to sleep and dies nine hours later without waking again. Darwin has been looking after him and asks Wedgwood for reminiscences, 'and I will some time digest them into an Eulogium. These men should not die, this Nature denys, but their Memories are above her Malice – Enough!'[2] Years later he writes in *The Economy of Vegetation*:

> So with strong arm immortal BRINDLEY leads
> His long canals, and parts the velvet meads;
> Winding in lucid lines, the watery mass
> Mines the firm rock, or loads the deep morass,
> With rising locks a thousand hills alarms,
> Flings o'er a thousand streams its silver arms,
> Feeds the long vale, the nodding woodland laves,
> And Plenty, Arts, and Commerce freight the waves.[3]

Two months after Brindley's death, the Scottish winter closes in and frost, sleet and lack of cash turn Watt's work on the Monkland canal into a nightmare.

> I am now, in spite of a most inclement season, from five to six hours in the fields every day, and ride about ten miles . . . This is the one side. On the other, I am extremely indolent, cannot force workmen to do their duty, have been cheated by undertakers and clerks, and am unlucky enough to know it. The work done is slovenly, our workmen are bad, and I am not sufficiently strict . . . I would rather face a loaded cannon than settle an account or make a bargain. In short I find myself out of my sphere when I have anything to do with mankind. It is enough for an engineer to force Nature, and to bear the vexation of her getting the better of him.[4]

By the summer of 1773 Sukey Wedgwood is eight. She is very like her father, out-going, quick-thinking, stubborn; Wedgwood will always love her most among his many children. Although Stringer and the artists are working full tilt on the Frog service, Josiah is off to the sea with Sukey and her cousins.

14 June. I am going to Liverpool . . . with poor Sukey, who after sitting & sewing at school for 12 months is so full of pouks, & boils, & humors that the salt water is absolutely necessary for her.

18 June. She was in high spirits, playing her pranks upon a high Horse Block, miss'd her footing, & pitch'd with her head upon a stone, which was sharp enough to make a wound, but I hope no other harm will insue.

21 June. We set our faces for Runcorn Gap, on the North side of the River, & as it blew pretty fresh, & was a terrible, slippery, dirty way to the Boat, I left my Girls at the Public House, & ferried over by myself to the other side, to behold the wonderfull works of his grace of Bridgewater, & truly wonderfull they are indeed . . . In short to behold ten of these Locks, all at one view, with their Gates, aqueducts, Cisterns, sluices, bridges &c, &c, the whole seems to be the work of the Titans, rather than a production of our Pygmy race of beings.

26 June. I left the Lasses very well at Liverpool, poor Suke with her broken head, but I hope her skull is safe – it certainly is rather of the *thick*, than *paper* species . . .

For all his strength of purpose, he cannot persuade the girls to bathe, 'but they promise fair'.

By now Edgeworth is settled on his Irish estate with Honora, who will soon have two children of her own to care for as well as four from his first marriage. The Edgeworths read and talk together every evening and Honora becomes 'an excellent theoretick mechanic'. Day, meanwhile, is still looking for a wife and William Small thinks he has found one for him, a Yorkshire heiress called Esther Milnes:

'But has she white and large arms?' said Mr Day.

'She has,' replied Dr Small.

'Does she wear long petticoats?'

'Uncommonly long.'

'I hope she is tall, and strong, and healthy.'[5]

'Remarkably little, and not robust,' replies an exasperated Small. Can Day really imagine that a charming, benevolent woman of twenty-three – the only woman in the world who thinks on the same lines as him – should be formed exactly to fit his fantasy? If he is not satisfied, he should give up any idea of marrying at all.

'My dear Doctor', replied Mr Day, 'the only serious objection which I have to Miss

Milnes, is her large fortune. It was always my wish, to give to any woman whom I married the most unequivocal proof of my attachment to her self, by despising her fortune.'

'Well, my friend,' said the Doctor, 'what prevents you from despising the fortune, and taking the lady?'

Day thinks this over.

In May 1773, Mary Priestley, of the sharp gaze and sudden smile, is getting the family ready to move from Leeds to Wiltshire. Mary packs the bags and Joseph offers to rope them up. She leaves him to it but when she undoes the trunks after the long trip south she finds that under every lid lies a mass of flasks and chemicals and minerals, tenderly packed in the linen. She is not to worry if the clothes are 'a little injured', Priestley says, as his equipment has survived the journey beautifully.[6]

Baskerville's body. The journeys of men who profoundly dissent do not end with death. In January 1775, John Baskerville will die. He leaves instructions that he is to be buried in a lead coffin in a vault under 'a conical building, heretofore used as a mill, which I have lately raised higher and painted'.[7] This is done. After the house is sold in 1788, the small conical mausoleum is removed and so, the owner thinks, is the body. Not so. Workmen stub their spades on lead and hastily cover it up. Here Baskerville stays until 1821, when a new canal wharf is built right across his resting place. The coffin is disinterred and propped up in 'Messrs Gibson and Son's warehouse', where a few curious people come and see it. (Gibson charges 6d a head.) Baskerville looks good, in a pure white shroud with a branch of laurel on his chest – but he smells terrible, 'strongly resembling decayed cheese'. After eight years at Gibson's he is moved to the shop of John Marston, a plumber. An artist comes to sketch him. A surgeon comes to peer, pinches a piece of the shroud, and is dead within days.

As more and more people fall ill, by 1829 John Marston is keen to be rid of him. He has rented a vault for himself at St Philip's and suggests that Baskerville move there, but his request is refused on the grounds of the printer's 'atheism'. A bookseller, Mr Knott, offers his own vault at Christchurch, and poor Mr Marston asks the churchwarden, Mr Barker, 'a solicitor of eminence, well known to florists as a collector of orchids'. He says no, but his eyes twinkle and he says the keys are always on the hall table. Off Baskerville goes, 'on a hand barrow covered with a green baize cloth'. But people still doubt he is there, until 1892 when the records show that there are 136 vaults, and only 135 recorded burials. Is the great printer in the unmarked odd lot? Crowds follow as the last disinterrment begins, and here he is. On the head of the lead coffin, in printing types, soldered on, is the name: 'John Baskerville'. Finally, in 1898, he moves to Warstone Lane Cemetery in the middle of Birmingham's jewellery quarter. He is still there now. I think. And his beautiful type lives on, and you are reading a version of it in this book.

THIRD QUARTER

For QUICK-SILVER is spiritual and so is the AIR to all intents and purposes.
For the AIR-PUMP weakens & dispirits but cannot wholly exhaust.
For SUCTION is the withdrawing of the life, but life will follow as fast as it can.
For there is infinite provision to keep up the life in all the parts of Creation.

Christopher Smart, *Rejoice in the Lamb*

20: EXPERIMENTS ON AIR

On a hot, sunny afternoon at the end of August 1774, a couple of weeks after Wedgwood's exhibition of the great Russian dinner service came to a close, Priestley was trying out his latest toy, a new burning glass. This was a convex lens 'of twelve inches diameter, and twenty inches focal distance' which concentrated the sun's rays with tremendous power.[1] The story went that this particular glass had belonged to the Grand Duke Cosimo III of Tuscany, who entertained himself by burning his subjects' diamonds. Priestley was thrilled with it and 'proceeded with great alacrity to examine, by the help of it, what kind of air a great variety of substances, natural and factitious, would yield'.[2] He put a series of metals and ashes in glass vessels, inverted in a mercury-filled basin. One of his substances was the brick-red ash left after heating mercury itself. What followed, as he told his readers with his characteristic authorial stance of innocent amazement, furnished a 'very striking illustration' of his tenet, 'that more is owing to what we call *chance*, that is, philosophically speaking, to the observation of *events arising from unknown causes*, than to any proper *design*, or preconceived theory in this business'.[3]

When he heated the mercury ash, he saw globules of quicksilver forming, and a colourless air rising from it. 'But what surprised me more than I can well express was that a candle burned in this air with a remarkably vigorous flame.'[4] It seemed like ordinary air, yet as well as the bright flame a bit of red-hot wood crackled and burned, and threw out sparks like

Priestley's pneumatic trough, from *Experiments and Observations on Different Kinds of Air*, 1774 white-hot iron. Over the next months he kept up his trials with the mercury, until one day in March 1775:

> On the 8th of this month, I procured a mouse, and put it into a glass vessel, containing two ounce-measures of the air from mercuric calcinations. Had it been common air, a full-grown mouse, as this was, would have lived in it about a quarter of an hour. In this air, however, my mouse lived a full half-hour; and though it was taken out seemingly dead, it appeared to have been only exceedingly chilled; for, upon being held to the fire, it presently revived, and appeared not to have received any harm from the experiment.[5]

Priestley kept a great many mice, all pretty short-lived (one lasted less than two seconds in a batch of foul air, he told Franklin cheerfully). But this particular one is remembered for its part in a key experiment, the isolation of oxygen, or, as Priestley called it, 'dephlogisticated air'.

He was right to ascribe the discovery to chance, but ingenuous to pretend he had no preconceived theory. Like nearly all his contemporaries, when he was examining anything to do with combustion, he held fast to the idea that everything that would burn or was changed by heat contained a special inflammable substance, 'phlogiston'. When something burned, or when an animal breathed out, phlogiston was released into the air until it became saturated or 'phlogisticated' and could absorb no more. At this point combustion ceased and the air could no more support life.

This theory had developed from the Arabic and medieval idea of the three principles, sulphur (the 'stone which burns', the combustible principle), mercury (the volatile) and salt (which gave solidity). Anything that contained sulphur, it was thought, would burn. At the end of the seventeenth century the theory was redefined by Johann Becher, and then by the German chemist Georg Stahl, who named the mysterious inflammable element 'phlogiston', from the Greek verb to inflame.[6] The richer a substance was in phlogiston, the faster and more fiercely it would burn. A counter-theory was put forward in the seventeenth century (supported by Robert Boyle), that life in the atmosphere depended on the 'nitro-aerial spirit'. When Priestley tried restoring the air breathed out by animals, or contaminated by putrefaction, he went back to this idea. Would a candle stay alight in the 'air' gained from burning saltpetre itself? It did – and here, too, Priestley had stumbled on oxygen, without knowing it.

The phlogiston theory appealed because it could be applied to so many experimental findings: it seemed an underlying cycle of nature. True, observers noted that air was always essential to burning, but whereas we now see the process as 'adding' oxygen (and the ash as an oxide) they saw air as the medium required for phlogiston to escape. There were some awkward facts to negotiate – no one could see, or hold, phlogiston, for instance. And why was it that a metal ash actually weighed *more* after burning than the original metal? Surely, if it had parted with its phlogiston, it should weigh less? This was variously explained by giving it a 'negative' weight, or by saying that it was not exactly an element but a force – like light, or heat, or magnetism.

Priestley's puzzle in August 1774 was that in his eagerness to use his glass, he had seized on the mercury calx, which in other moods he might not have done since it was already an ash and should therefore have no phlogiston left in it at all. How then, could it burn again, and release a different kind of air? At first he doubted the purity of the calx, but from another specimen he got even more air.[7] Over the next few months, as each experiment confirmed the air's extraordinary properties, he struggled to bend the findings to his theory. The only conclusion he could come to was that the phlogiston must have been taken from the air itself – this is why he called his new, pure air 'dephlogisticated'. The discovery was exhilarating, literally. When he breathed his new air his chest felt 'particularly light and easy for some time afterwards. Who can tell', he asked, 'but that, in time, this pure air may become a fashionable article in luxury. Hitherto only two mice and myself have had the privelege of breathing it.'[8]

Priestley had been working on different 'airs' for several years, with remarkable success. Some chemists (including Keir) used the word 'gas', following the earlier physician Jan Baptista Van Helmont, who likened the different vapours he found to unformed matter, or pre-matter, and called them 'chaos' – which sounded like 'gas' in his heavy Flemish accent. But the common term was still 'airs' since most people still regarded them as variants of the common atmosphere. Priestley was indeed setting out to map an unexplored territory hitherto taken for granted.

He was hardly a lone explorer. As with his work on electricity, he followed enthusiastically in the wake of others, testing and adapting their ideas. It had been known since ancient times that air, one of the four elements, could be 'good' or 'bad'. Miners had found bad air, 'choke-damp', lurking at the bottom of shafts, or wells. Since the start of the century a line of investigators had worked in this field. In the 1720s Stephen Hales – whose *Vegetable Staticks* Priestley quoted often – had examined the air given off by distilled vegetable, animal and mineral substances, inverting a vessel full of water over a separate trough of water and bubbling the air up through it – the first time that a gas had been easily collected. A little later, William Brownrigg, a Leyden-trained doctor working in Cumberland, managed to extract the air from spa water, developing a special apparatus, a 'pneumatic trough', with bottles and tubes in a tub filled with water.

Brownrigg expanded the notion of several different kinds of air, and identified the inflammable 'fire-damp' (methane) floating at the top of enclosed shafts in the local mines. Crucially, once Joseph Black's work in the 1750s identified the air given off by chalk, limestone and magnesia as a specific *kind*, with particular properties, the way was open to identify others. Black's fixed air – carbon dioxide – turned out to be the same as choke-damp and the airs from fermentation and in mineral water; it was created when charcoal was burned and was in the air we breathed out. The incurably shy, eccentric aristocrat Henry Cavendish then experimented with acids and metals, producing yet another 'inflammable air', hydrogen.[9] Nothing seemed to put these men off: Hales distilled anything at hand, including hog's blood, deer's horn, peas, dry tobacco, oil of cloves, beeswax, bones and even stones from a human gall-bladder. Cavendish went one better: 'From 7640 grains of putrefying broth . . . one grain of inflammable air was produced,' reported the *Transactions* in 1766.[10]

Priestley claimed to have made his first notable discovery again by accident, although this may be deliberately ingenuous. On first moving to Leeds he lived near a brewery and noticed that above the liquor fermenting in the vats there hung a permanent supply of 'mephitic air', Joseph Black's fixed air. Enlisting the workmen as helpers, he experimented with candles, burning wood splints and heated pokers and found that the smoke caught in this air swirled and fell over the sides of the vat: it was obviously heavier than common air. It certainly was not good to breathe but it wasn't poisonous: butterflies and insects became torpid but revived

Henry Cavendish, aquatint by C. Rosenburg

in fresh air; a mouse had convulsions and 'a large strong frog was much swelled, and seemed to be nearly dead, after being held about six minutes over the fermenting liquour', but it too perked up outside.[11] There was one fatality: 'a snail treated in the same manner died presently'. However when he put water in dishes suspended in the fixed air, it absorbed it, gaining a pleasantly acid taste, like mineral water. Priestley had discovered soda water.

Another man – Boulton or Keir – would have pounced on this as a money-spinner. Spas and mineral waters were immensely popular and had prompted a mass of medical and self-help literature. But Priestley initially set it aside as merely interesting. His findings, however, were of great interest to the medical community, currently experimenting with means of combating 'putrid diseases' such as scurvy or gaol fever. Because doctors linked the causes of these to putrefaction, it was thought that material based on fermentation – such as malt, or vegetables, which released fixed air, might be an antidote or cure.[12] Four years after Priestley's discovery, at a dinner in London, the Duke of Northumberland showed him some distilled water, which had been proposed as useful on sea voyages. Priestley immediately suggested that the Navy should try making some of his own water, which tasted better. Furthermore, since fixed air was thought to have curative properties, it might even, he thought, help against scurvy. Back in his lodgings he borrowed bottles and bowls from the kitchen and rapidly set up some makeshift apparatus; next day, with the help of another of his patrons, the MP Sir George Saville, a proposal was put to the Admiralty. Although he himself might have missed the chance of sailing with Captain Cook, his fizzy water circumnavigated the globe in the *Resolution* and the *Endeavour*.

His pamphlet, 'Directions for impregnating Water with Fixed Air', was snapped up across Europe. He wanted credit for the discovery, not profit. Taking the opposite stand to most inventors, he described his apparatus and his method clearly, with diagrams to show how chalk and acid were mixed in a bottle, and the fixed air led by a tube through a bladder (which could be used to control the flow) so that it bubbled up through water in another inverted bottle. Immediately, other experimenters suggested complicated refinements, much to his annoyance. He was particularly piqued by the comments of Dr John Nooth, the manufacturer of an enduringly successful version, who said that Priestley's use of a simple bladder gave the water a 'urinous quality' – which Priestley fiercely refuted.[13] Once the apparatus was on the market, it was a towering success. Lecturers bubbled up brews for excited audiences. Doctors recommended machines to their patients. Entrepreneurs sold them across the globe and many middle-class homes featured a 'gasogene' apparatus into the middle of the next century. Within five years, a thousand models of the modified Nooth machine had been sold as far as the East Indies; within twenty years artificial waters were being made under pressure by a successful London firm, a certain J. J. Schweppe. (Boulton seems to have drunk gallons of fizzy water, illustrated by many bills from Schweppe, including a letter assuring him that 'Your hamper of 12 dozen ½ pints soda water was sent off on Monday by Deykins wagon.'[14])

Priestley regarded soda water as one of the happiest of his finds, because it so clearly brought benefits to so many. It was an early high point at the start of 'pneumatic medicine', the movement to apply the chemistry of gases to health, which would develop throughout the century.

Good and bad air came to have almost moral connotations: in Tobias Smollett's satirical novel, *Humphry Clinker*, Matthew Bramble tells his doctor about country dances in Bath. 'Imagine to yourself', he says:

> a high exalted essence of mingled odours, arising from putrid gums, imposthumated lungs, sour flatulencies, rank armpits, sweating feet, running sores and issues, plasters, ointments and embrocations, hungary-water, spirit of lavender, assafoetida drops, musk, hartshorn and sal-volatile; besides a thousand frowzy steams, which I could not analyse. O Dick! is this the fragrant aether we breathe in the polite assemblies of Bath?[15]

Jokes apart, Priestley and his peers took measuring the purity of air very seriously, and another of his discoveries that looked promising was the 'nitrous air test'. He had heard of a gas found by Stephen Hales, which formed cloudy red fumes when mixed with ordinary air and in 1771, at the suggestion of Cavendish, he made a series of experiments adding spirit of nitre to metals. With all of them, he found the new gas, and when he shook it together with ordinary air over water, he found that the brown fumes were gradually absorbed in the water: the air remaining was then measured and the loss used as an indicator of the quality of the original air. Once again, imitators leapt in. A fashion developed for 'eudiometrical tours', testing air in towns, by the sea, around marshes in Britain and across Europe.

In these years Priestley made a staggering number of discoveries. But the sciences, like the arts, had their fashions, and if England buzzed with activity, the Continent matched it. In Germany, in France, in Sweden, many chemists were now studying airs. The Swedish apothecary's assistant, Karl Scheele, for example, discovered the choking, greenish gas chlorine in 1773. Three years before Priestley found dephlogisticated air, Scheele had also produced oxygen, which he called 'fire air', by heating several compounds including mercury oxide (as Priestley had done), silver carbonate, magnesium nitrate and saltpetre. He too found that a candle burned far more brightly in this new, colourless, tasteless, odourless air. Due to his modesty and his publisher's muddles, these experiments were not published until 1777, by which time Priestley had scooped the laurels. (Like Priestley, Scheele was another inveterate tester and taster: it was said that what he did not breathe was hardly worth mentioning, and what he did breathe probably killed him. He would die in his forties, in 1786.)

*

Fame came from taking the given – whether it be the shapes of antique vases or the latest knowledge about air – and pushing the boundaries forward to find *new* forms, new answers. But Priestley had an additional driving motive. All his work was subsumed within the over-arching framework of his political and religious beliefs: with each discovery he felt that he was adding to a shared body of knowledge that would lead to a better future. In his time in Leeds, in addition to his books on electricity and optics and his work on air, he had published twenty-eight non-scientific works, including two controversial books, *The Institutes of Natural and Revealed Religion* and his *First Principles of Government*, as well as historical charts, texts on education, a work on perspective, and an influential attack on the remarks on Dissenters in Blackstone's *Commentaries on the Laws of England*. It was vital, he felt, to turn often from research to religious writing and duties:

> We must make frequent intervals and interruptions; else the study of science, without a view to God and our duty, and from a vain desire for applause, will get possession of our hearts.[16]

To a large extent though, it did possess his heart. Once air became his focus, Priestley read everything he could. He fixed up his apparatus, using anything to hand, whether it be a tub for washing linen in or a beer mug, and tried catching his gases over water, or, following one of Cavendish's ideas, over mercury, so that he could capture gases that were soluble in water (such as ammonia or hydrogen chloride). As he had done with electricity, he tested earlier experiments, making fixed air by adding sulphuric acid to chalk, as Black had done, and Cavendish's inflammable air. Both, he found, quickly dealt a death blow to his mice – and to frogs and snails and flies. He established that common air did not have to be wholly used up before a candle flame burned out; it was diminished only by about a fifth. Trying to create fixed air by heating chalk in a long gun barrel he found he had also made yet another kind of inflammable air, which burned with a blue flame, quite different from Cavendish's (carbon monoxide). Was this a result of the gun barrel's iron reacting to the acid in the chalk?

Sometimes, as here, he had no answer to the questions his work threw up. The reclusive Cavendish was a staunch ally. He did not bolt from the room in fright, as he usually did, when Priestley entered and, for his part,

Priestley was not put off by Cavendish's squeaky voice, or by his clothes, assembled from the handed-down garments of a previous generation. Both men stuttered, but their conversations could not have been more fruitful. Inspired by Cavendish's openly admitted failures and frustrations as much as by his dazzling successes, Priestley worked on. He explored the properties of nitrous air and the fumes of charcoal; he created 'marine acid air' (hydrogen chloride) and brewed viciously poisonous airs from lead and tin. In this last case he incidentally isolated nitrogen, which was also described around this time by the Scottish chemist Daniel Rutherford.

His researches were hardly programmatic. Without apparent method – employing what Watt later called 'his usual way of Groping about' – he followed up odd leads and curious phenomena in no particular order, just wondering what they might turn up.[17] One of his most famous 'accidental' discoveries concerned plants. People knew that no animal could live in a confined space if the air was not changed and it was simply assumed that the same went for plants, especially if the air was already foul. This was what Priestley expected in 1771 when he put a sprig of mint into a glass jar containing air in which two mice had died.[18] He stood the glass in water, left it and forgot it. A week later he saw that the mint was still growing. A candle flame burned clear in the jar; a mouse breathed happily within it. And mint was not the magic ingredient – the same trick worked with groundsel, and best of all with spinach.

His experiments attracted attention, and he was always keen to share what he had found. In May 1771 Benjamin Franklin set out an a tour with some scientific-minded friends – a mixture of a holiday and an industrial fact-finding mission and called on Priestley, 'who made some very pretty Electrical Experiments and some on the different properties of different kinds of Air'.[19] The following June Franklin returned with Sir John Pringle, President of the Royal Society, and saw Priestley's experiments on restoring putrid air. It was absolutely clear that the restorative power lay in the plant. Franklin was startled, wondering later if the 'air is mended by taking something from it, and not by adding to it'.[20] In fact both occur: Priestley was on his way to discovering photosynthesis, in which sunlight enables a plant to absorb carbon dioxide from the air, to make organic matter, and release oxygen. Percipiently, Franklin added, 'I hope this will give some check to the rage to destroying trees that grow near houses.'[21]

Priestley's first papers had been read at the Royal Society in March 1772 and the men of Lichfield and Birmingham were fascinated. When Franklin visited Darwin, he found him keen to collect some airs of his own, from a pond near by. Several tests, later described by Keir, had shown that decaying matter released a combination of two gases, one flammable and the other not. Almost ten years before, a friend had shown Franklin how to collect 'marsh gas' in the wetlands of New Jersey. You had to choose a muddy place, where the bottom could be reached by a walking-stick, stir up the mud and, when bubbles began to rise, touch them with your candle: 'The flame was so sudden and so strong that it catched his ruffle and spoiled it, as I saw.'[22] Now Franklin set out to try this with Darwin, but with no luck. After he left Darwin tried again.

The apparatus you constructed with the Bladder and Funnel I took into my Pond the next day, whilst I was bathing, and fill'd the Bladder well with unmix'd Air, that rose from the muddy Bottom, and tying it up, brought it Home, and then pricking the Bladder with a Pin, I apply'd the Flame of a Candle to it at all distances, but it shew'd no Tendency to catch Fire. I did not try if it was calcareous fixable air.[23]

Franklin had also helped Boulton in the same vain quest, and felt he had paid badly for it. He had been ill for the last fortnight, he groaned to Darwin, blaming his gout and fever and headache on 'the Amount of Dabbling in and over your Ponds and Ditches and those of Mr Bolton, after Sunset, and Snuffling up too much of their effluvia'.[24] Ludicrous as they sound, Darwin's dabblings in the mud explained something of the profound fascination with this research: through chemistry one might find the secrets of life in death itself, in putrefaction, in mud, in slime.

At the end of 1772, Priestley published his first account.[25] Soon his work was common knowledge and a source of continuing excitement. The following spring, Small's correspondence with Watt was full of chemistry – of making ether, of producing phosphorus from bones, of the new 'acid of Tartar', of a London chemist who had found that powdered tin exploded when added to copper nitrate. Most important of all,

Dr Priestley has found the Vapor discharged from all metals excepting Zinc dissolving in Nitrous Acid to be the most powerful antiseptic known and an excellent test of the purity of air . . . He has also found, as you must have heard, that the vapors which exhale from growing vegetables mixed with mephitic air render it respirable.[26]

In November, in recognition of his many discoveries, Priestley was awarded the Copley Medal by the Royal Society. The President, Sir John Pringle, had a keen interest in this area since he himself had written on putrefaction as a cause of diseases such as scurvy. He had already supported medical research on fixed air, and in an unusual address he placed Priestley's work as the culmination of a long quest, gradually revealing the workings of nature – all in the service of man:

From these discoveries we are assured, that no vegetable grows in vain, but that from the oak of the forest to the grass of the field, every individual plant is serviceable to mankind; if not always distinguished by some private virtue, yet making a part of the whole which cleanses and purifies our atmosphere. In this the fragrant rose and deadly nightshade cooperate.[27]

Even the most remote plains and woods played their part, for the winds carried the used air of men to them 'for their relief, and our nourishment'. And even if the wind rose to gales and hurricanes, pronounced Pringle, we should still trace the ways of a 'beneficent Being', who 'thus shakes the waters and winds together, to bury in the deep those putrid and pestilential effluvia', which the vegetables had been unable to consume.

When Pringle uttered these pieties, Priestley was no longer in Leeds. In 1772, through Richard Price, he had been offered a post with Lord Shelburne, to act as his librarian, oversee the education of his two sons and help him collect information on parliamentary issues. Priestley thought hard. He liked his work in Leeds, but the salary Shelburne offered was more than double, and he was tempted by the magnificent library. When Bentley told Wedgwood, he said he was glad to hear of the appointment, 'taking it for granted that he is to go on writing & publishing *with the same freedom* he now does, otherwise I had much rather he still remain'd in Yorkshire'.[28] This was Priestley's view too, but he was won over when the Earl called in August 1772 and offered a London house, where he could meet his scientific friends, as well as a house on the Wiltshire estate and funds for experiments. In December Priestley told his congregation, and next May he preached his final sermon. Soon afterwards the family headed south.

It took frustratingly long to get back to his experiments – his books were all over the place and the family was stuck in one room while Mary energetically papered the rest. But he soon established a rhythm, setting time aside for his work in the library and with the Shelburne children, for his research and demonstrations for Shelburne's guests, and for his writings on politics, education, religion and science. In the country he took up vegetable gardening and played three games of chess with Mary before dinner. He became central to the Bowood circle, with its linked interests in natural philosophy.[29] In a controversy with the Scottish 'Common Sense' philosophers led by Thomas Reid he argued that appeals to 'common sense' could be used to justify politicians' repressive attitudes), and in a new edition of Hartley in 1775, he turned the notion that morals were formed by association and habit into a sharp argument for political toleration.[30]

In the autumn of 1774 Priestley published the first volume of *Experiments and Observations on Different Kinds of Air*: two more volumes would appear in 1775 and 1777, published by the radical Joseph Johnson, who issued all his works. As before, he described all his tests and his equipment, and even told his readers how to keep their mice. (In Leeds he had kept his 'on a shelf over the kitchen fireplace where, as it is usual in Yorkshire, the fire never goes out'.[31]) He had a host of imitators, which was exactly what he wanted, part of his plan to make everyone a scientist. Many public lecturers adapted his experiments, to the bafflement of Dr Johnson, who detested his politics and religion:

> In the course of the experiments frequent mention being made of Dr. Priestley, Dr. Johnson knit his brows, and in a stern manner inquired, 'Why do we hear so much of Dr. Priestley?' he was very properly answered, 'Sir, because we are indebted to him for these important discoveries.' On this, Dr. Johnson appeared well content; and replied, 'Well, well, I believe we are; and let every man have the honour he has merited.'[32]

Priestley cherished the honour. And his findings were valued for their usefulness as much as their extension of knowledge. Wedgwood, for instance, made notes on each volume as it appeared, seeing how Priestley's work on combustion and heat, acids and metals could help in preventing air bubbling and blistering his clay or in preparing glazes. When he found the phrase 'Fluor acid dissolves glass very freely', he immediately saw that this could be a valuable discovery in perfecting enamel colours, which were currently ground, or 'levigated', to a fine powder, 'but every chemist will perceive the difference there is – one may almost say "infinite" between the degree of fineness produced by levigation and the chemical solution of any body'.[33] Priestley was delighted by such practical applications, and even more delighted by new discoveries. It was at Bowood, in 1774, that he eagerly got out his brand new burning glass and found his astonishing dephlogisticated air. That autumn he accompanied Shelburne on a tour through Flanders and Holland and Germany. He had already been corresponding with foreign electricians and chemists like Alessandro Volta and Tobern Bergman, but he was touched and excited to find that he was held in such respect abroad. Before returning home they stayed in Paris for a month, where Priestley met leading scientists and carried out experiments in their laboratories. He was still puzzled by his experiment with the mercury calx, and in Paris he obtained another specimen and tried again, 'and at the same time I frequently mentioned my surprize at this kind of air which I had got from this preparation to Mr. Lavoisier, Mr. le Roy, and several other philosophers'.[34]

Antoine Lavoisier, the son of a wealthy Paris lawyer, was then twenty-nine. By profession he was an agent of the hated 'Ferme Générale' which collected taxes for the government, but he was also a brilliant natural philosopher who had been elected to the Académie des Sciences in 1769 at twenty-three, having worked on astronomy, geology, agriculture, minerals and chemistry.[35] Two years later he married the thirteen-year-old Marie-Anne Paulze, who became his secretary, learning English so that she could translate papers from the Royal Society. Priestley noticed that Lavoisier listened intently when they met at dinner, but at the time he had simply no idea, he said, 'to what these remarkable feats would lead'. What they led to was Lavoisier's new theory of combustion, which trounced the old phlogiston theories, and at last the identification of oxygen.

'Knowledge is important,' declared Keir, 'but whether the discovery is made by one man or another is not deserving of consideration.'[36] You get a vivid sense of this from Keir's own *Treatise on Gases*, appended to his third edition of Macquer's *Dictionary* in 1777. As Keir ranges through the findings, it is like listening to an international debate, with voices from the present and the past: 'Mr Scheele affirms . . .', 'Mr Cavendish observes . . .', 'Dr Black is of the opinion . . .', Mr Macquer very ingeniously conjectures

. . .', 'Stahl maintains . . .', 'Mr Bayer remarks . . .' His text is punctuated by new communications from Priestley, spilling out his latest thoughts. It was natural to Priestley to speak so openly to his French colleagues: his whole aim was the sharing and dissemination of knowledge. Yet this generosity of exchange was undercut by keen rivalry. Like Wedgwood with his urns, Boulton with his ormolu and Watt with his steam-engine, Priestley was catching the moment and taking things a step further. And like them he was intensely competitive: he was not interested in originality *per se* but he would hate to be thought a 'slavish imitator', as Wedgwood put it, and as with his apparatus for soda water, he certainly wanted what 'is strictly my due, *the sole merit of the discovery*'.[37] He could never quite accept the way Lavoisier picked up his findings and ran with them. He was proud as well as modest. 'It may be my fate', he wrote,

> to be a kind of comet, or flaming meteor in science, in the regions of which (like enough to a meteor) I made my appearance very lately, and very unexpectedly; and therefore, like a meteor, it may be my destiny to move very swiftly, burn away with great heat and violence, and become as suddenly extinct.[38]

21: 'WHAT ALL THE WORLD DESIRES'

In 1774, six years after he first met Boulton, James Watt finally moved to Soho, the final step in what at times seemed an endless saga of correspondence and negotiations.

Watt's patent 'for a new method of lessening the consumption of steam and fuel in fire engines' had been granted on 5 January 1769. The specifications had to be enrolled within the next four months, but although Watt made precise drawings, Small persuaded him to 'give neither drawings nor descriptions of any particular machinery, but specify in the clearest manner that you have discovered some principles'.[1] This outrageously wide remit effectively allowed Watt to claim property in all new steam developments – even those involving expansive steam, which he never intended to use. It was to lead to many battles in future.

At the end of January Joseph Black urged Watt to meet Roebuck at Kinneil, despite the long ride in the bitter cold. 'I am persuaded the ride will do you good . . . I am getting a pair of boots of sufficient capacity to contain two legs and six stockings.'[2] When they did meet, Roebuck gave Watt the go-ahead to build a full-scale engine in the outbuildings behind Kinneil House. The real work could begin. He set aside his experimental engines with inverted cylinders and tried applying his separate condenser to an old-fashioned beam-engine. Through the spring and summer, living in the grey-stone cottage beneath the trees at Kinneil, Watt struggled with his great machine. He was again dogged by crude workmanship and by

'Beelzebub', Watt's early lap engine, 1775

the problem of sealing his piston so that it stayed air-tight – when he tried an oil seal, the oil went thick and ropy and white and clogged the pump. 'Of all things in life,' he lamented, 'there is nothing more foolish than inventing.'[3]

Watt was also taking on more and more surveying work, for a canal through Strathmore, for the improvement of the Clyde and the docks at Fort Glasgow. Then in 1770 came his appointment to supervise the construction of the Monkland Canal, and for the next four years one surveying contract followed another. He was constantly away from home, and Peggy's letters followed him – to Kinneil, or to distant posts near his surveys. Peggy ran the instrument business briskly while he was away and kept him going through his depressions and frustration over his engine: '*If it will not do, something else will,*' she had written in August 1768; '*Never despair.*'[4] Now she kept his spirits up by writing of business and pleasure, bathing in the sea, and family news: 'Meg is very well but allready forgot you'; 'Little James has got a tooth without any trouble . . .' She begged him to take care of himself and told him how much they missed him: 'Your daughter does nothing but look throw the window to see you Come back again . . . I will be glad to hear the engin is going on and when will you be home.'[5]

Small almost gave up thinking that Watt might return full time to engine work. 'Nothing of late years', he wrote, 'has vexed me so much as the peculiar circumstances that have retarded your fire-engine . . . you have as much genius and as much integrity, or more, than any man I know.'[6] Watt had always been impatient with the pressure that Small put on him. 'You talk to me about coming to England just as if I were an Indian that had nothing to remove but my person. Why the devil do we encumber ourselves with anything else.'[7] But he was encumbered. Although he constantly urged Roebuck to talk to Boulton about a partnership, he felt a genuine loyalty to him: 'As to the Doctor,' he wrote, 'he has been to me a most sincere, generous friend, and is a truly worthy man.'[8] And he loved his family, and felt Peggy was settled among her friends. 'Is her douceur still as touchante as it was?' teased Small.[9] If so, he might consider getting married himself, if only to relieve his boredom.

In 1773 circumstances finally conspired to uproot Watt. His life had never looked bleaker. A depression hit in 1772, following bad harvests and the calling in of credit. Work on the Monkland Canal stopped for lack of funds. In the same year came the collapse of the Scottish banking house Neale, James, Fordyce and Down. It was a terrible jolt. Alexander Fordyce, brother of a noted London doctor often mentioned in Lunar group letters, had been a star of the boom that followed the Peace of Paris in 1763, making a fortune by speculating in Indian stock, marrying the daughter of a Scottish earl and building a handsome seat in Roehampton. A swashbuckling character, on the day of the crash he allegedly came home in wild high spirits, vowing that he had always told the wary ones, 'and the wise ones, with heads of a chicken and claws of a corbie, that I would be a man or a mouse; and this night, this very night, the die is cast, and I am . . . am . . . A man! Bring Champaign; and Butler, Burgundy below! let tonight live for ever! . . . Alexander is a man.'[10]

Instead of vituperation, Fordyce and his partners won universal sympathy. Yet his collapse caused the failure of almost every private bank in Scotland. 'I have no doubt that the faces of the People in Glasgow have been for some time past screwed up with an unusual degree of Care and Anxiety,' wrote Black in October.[11] Matthew Boulton had deposited a large sum with the Fordyce bank: he was out of pocket on his ormolu and was already being chased by a panicking Fothergill to sort out the tangled debts in their Bill Account. On the brink of disaster, he was rescued by a loan of £3,000 from the unworldly Thomas Day, against security of ten shares in the Birmingham Canal Company. Yet the Fordyce disaster suddenly offered an opportunity.[12] Boulton knew that Roebuck had been virtually ruined by the crash. In effect he had lost everything, having sunk all his private fortune, and his wife's, into the mines at Bowness. He had no cash for engine experiments and Watt ended by paying for these himself. Watt also blamed himself bitterly for having involved his friend so deeply, financially, in the engine scheme, and appealed to Small to get Boulton to help Roebuck out by taking a bigger share. But Boulton bided his time. In March 1773 Roebuck was declared bankrupt: Garbett took over some of his Carron shares, and the Steelhouse Lane vitriol plant in Birmingham – he was never to rise high in commercial life again.

Boulton was one of Roebuck's creditors, being owed £1,200: at the end of March, he wrote to Watt, making an offer for Roebuck's share in the engine. For two months, Roebuck held out, and then, on 17 May, wrote accepting Boulton's terms. On the same day, an anguished Watt wrote a formal discharge of all the money Roebuck owed him under their agreement, 'in consideration of the mutual friendship between Dr Roebuck and myself and because I think the thousand pounds he has paid more than the value of the property of the two thirds of the inventions'.[13] In return, he took the Kinneil engine, dismantled it and sent it to Soho. The trustees of Roebuck's estate saw absolutely no value in the creaking, unfinished engine. 'None of his creditors', Watt told Small, 'value the engine at a farthing.'[14] For the money Roebuck had owed him, Boulton had at last got the two-thirds share in the invention he had wanted all along. And it was all his own. Fothergill would have nothing to do with it and happily, if foolishly, took his share of the £1,200 instead.

But although Boulton now had the engine, he still did not have the man. Ideas, models and plans were nothing without Watt's expertise. And in August 1773 Watt set off once again to survey the wild country between Inverness and Fort William for a proposed canal through the Great Glen.[15] Peggy was expecting another baby, and at first an untroubled Watt sent her the usual instructions for the shop and messages about instruments. But as the days passed without news he became increasingly anxious. On 15 September he wrote from Inverness that he was very uneasy at not having heard from her. Soon she told him why: 'I have been very ill this last 4 days but thank God I am now able to sit up.' She was very grateful to the doctor:

> I will not die if he can help it for he said and did everything he could to keep up my spirits which I think were never so low . . . the Complaint I was so afraid of has left me quiet. I hope I will be strong before I bear the Child and that you will be at home.[16]

Within a week, he received another letter, noted in his diary:

> 1773 Sept. 26th Sunday about half-past ten received letter from Mr. Muirhead advising that Mrs Watt was dangerously ill and her life despaired of & desiring me to come home with all speed. I immediately set out, with a very sad heart & came to Fort William that night having a heavy rain the whole way.[17]

The next day he reached Tyndrum, and on the next, Dumbarton. Racked by premonitions, he ordered a chaise for the morning:

> Sept 29th. About ten o'clock I saw the chaise arrive and Mr Hamilton in it; by his black coat and his countenance I saw I had nothing to hope.[18]

Peggy had died on the 24th, two days before Muirhead's warning letter even reached him.

He could not bear to go home, fearing 'to come where I had lost my kind welcomer', and instead stayed at the house of his friend, Gilbert Hamilton, the Glasgow agent for the Carron company. As he wrote to Small, the journey back seemed a mirror of his wretchedness, struggle and tears.

I know that grief has its period; but I have much to suffer first. I grieve for myself, not for my friend . . . I am left to mourn . . . I had a miserable journey home, through the wildest country I ever saw, and the worst conducted roads: an incessant rain kept me for three days as wet as water could make me. I could hardly preserve my journal-book. Adieu, God bless you.[19]

'Come to me as soon as you can,' wrote Small.[20]

In his next letter, Small urged him to lose himself in work, the best solace for grief. It was indeed the only distraction, and in November Watt reported that he had been surveying on the Firth of Forth in the bitter cold. But Watt's mind was as numb as his chilled fingers. He had 'lost much of my attachment to the world, even to my own devices . . . I long much to see you – to hear your nonsense and communicate my own; but so many things are in the way, and I am so poor.'[21] His debts were huge, his earnings were pitiful and all he could think of was coming to England or getting a lucrative post abroad. His emotional ties were now cut as surely as his agreement with Roebuck. 'I am heartsick of this cursed country,' he sighed.[22] Over the winter he settled his affairs. In the spring of 1774, leaving Margaret and James with relations in Glasgow, he packed his bags and headed south. On 31 May he arrived in Birmingham.

Ironically, over the past year, while Boulton was trying to lure Watt south, plans were afoot to find Small a job in Edinburgh, to fill the professorial chair in the medical school left vacant by the death of Dr John Gregory. The suggestion came from some of the professors via Roebuck, who (like all who knew him) had a soft spot for Small, signing his letters, 'your very affectionate friend'.[23] There was hot competition from Cullen's favoured candidate, and although Small demurred, in January 1774 Roebuck was still scheming to call for support on 'Dr Franklin, Dr Priestley and others who are to be the most conspicuous Puppets in this Drama moved by us who in our turn are to be acted upon by invisible springs from your Self who are constantly to be concealed behind the Curtain'.[24]

Small stayed, as he so often did, behind the screen. No Edinburgh job was forthcoming and instead he concentrated on manipulating his dour, anxious, talented Glaswegian puppet, James Watt. He heard from Watt in April that his friend James Hutton, 'the famous fossil hunter', would be coming to Birmingham with him.[25] Hutton was descending on England for a geological tour, soon whipping Watt off to look at salt-mines and doing elaborate experiments with Darwin and Edgeworth involving air-guns and thermometers. Boulton set Watt up in his old home in Newhall Walk and over the summer Watt fiddled with his current distraction, the doomed 'wheel-engine'. But Boulton also made sure that his senior mechanic at Soho, John Harrison, unpacked the rusting Kinneil engine and set it up between two wings of the works, where it shuddered and groaned, pumping water back to one of the water-wheels. This both helped the summer power problem for the factory and meant that Watt could make advances with his separate condenser.

In late October Watt reported cheerfully to Roebuck that all was going well, receiving a touchingly warm reply from the ruined man, cheering Watt for having finally justified the principles he was working on and assuring him that 'the generous and spirited gentleman you are connected with will never suffer it to fail for want of exertion to carry it into execution'.[26] A month later, Watt told his father in tones of surprise that his work in Birmingham had been 'rather successful'. His fire-engine was going and was working better than any made so far, 'and I expect the invention will be very beneficial to me'.[27] Small told fifteen-year-old Erasmus Darwin, to whom he wrote like a sort of adopted uncle, advising him affectionately on books and microscopes, that he must get his father to come over to see Watt's engine, 'at last in a condition that satisfies even me'.[28]

His mood matched Boulton's, who certainly showed no lack of exertion. Many deep mines were in desperate need of more efficient engines: news had spread, and since 1771 companies from Yorkshire to Cornwall had delayed replacing old engines in the hope that Watt's would be ready soon. Although no formal partnership agreement was drawn up, for his two-thirds share, Boulton agreed to bear all the costs of getting patents, carrying out experiments, providing stock, managing the workmen and making deals. But now that they were working together, he could see that they needed to invest time as well as capital: Watt's patent was good for only eight more years and that might not be long enough. If their investment was to pay off they needed a good spell free from competition to get the engineering right. Going straight for his goal, as always, Boulton set about extending the period of protection. Instead of applying for a new patent he used his well-honed lobbying skills to push a Private Bill through Parliament to extend the current patent for twenty-five years. On tenterhooks, Watt went to London on 22 February, warned by friends such as James Hutton of all the manœuvering that lay ahead: 'The honestest endeavour must to succeed put on the face of roguery but what signifies the dress of a rogue unless you have the address of a wise man; come and lick some great mans arse and be damned to you.'[29]

Then came an unexpected blow. At the start of the year, William Small, whose letters had supported Watt during six long years, became increasingly ill. He had often complained of lassitude and a weakness that made him unable to concentrate or do any heavy work. 'The *ennui mortel*', he joked bitterly, 'has totally ruined me for an experimental philosopher . . . I flatter myself that I shall soon be *pulvis et umbra* and fold my arms to sleep. Who will call me projector now?'[30] His health had been poor ever since his years in Virginia and he seems to have suffered from a form of malaria, recurring at intervals and weakening him more and more. When he heard that Watt was finally thinking of heading south, he was delighted, saying it would preserve him 'one year longer at least from this lethargy'.[31] This spring his familiar 'ague' returned. Young Erasmus Darwin told a friend that in early February, travelling to see a patient at Tamworth, ten miles away, Small vomited the whole way in the coach and collapsed on his arrival. Back in Birmingham he was feverish and delirious. Boulton told Watt anxiously in February that the poor doctor was much worse, and sent an express for Darwin. On 25 February 1775, just after Watt had reached London, Small died. He was conscious to the end, pathetically convinced he would recover. In his last days he was visited constantly by Keir, who 'loved him with the tenderest affection', and by Darwin and Boulton, who were at his bedside when he died.

That evening Boulton wrote to Watt.

> The last scene is just closed. The curtain is fallen, and I have (this even) bid adieu to our once good and virtuous friend, for ever and for ever . . .
>
> If there were not a few objects yet remaining for me to settle my affections upon, I should wish also to take up my Lodgings in the Mansions of the Dead.[32]

Three days later he added, 'My loss is as inexpressable as it is irreparable. I am ready to burst.' He added a curt postscript, 'Acquaint Dr Roebuck, I can't.'[33]

The funeral was delayed for three weeks, in the hope that Small's brother could come down from Scotland. Thomas Day, whom Small had gently persuaded not to take up medicine, and for whom he had even tried to find that impossibly suitable wife, rushed back from Brussels. Day was one of several people who composed tributes, but it was hard to catch the elusive spirit of this clever, kindly man. 'Messrs Keir, Darwin, Day and self have never yet agreed about a monument for the church,' Boulton later told Small's brother. Instead he erected a monument in his own garden, in a 'sepulchred grove', where he had built a summer house.[34] From here, he could look across at St Philip's where Small was buried. The monument was inscribed with the doctor's name and dates and with an ode by Darwin, ending:

> Cold Contemplation leans her aching head,
> On human woe her steady eye she turns,
> Waves her meek hand, and sighs for Science dead,
> For Science, Virtue and for Small she mourns.[35]

Small had left a blank space. His goods were sold, his books and papers dispersed and lost, his letters to Watt among the few to survive. Yet he had been the living filament that bound them together. All these men were shocked. As they rushed through their lives, focusing on pistons and steam, alkali and airs, china and carriages, they stopped, stunned for a moment, moved by the realization of how much they meant to each other as friends as well as colleagues. He was the first of them to go. In their forties, with everything ahead, they were reminded that they were mortal.

They picked up the threads of life again, some more quickly than others. Darwin was brutally matter of fact, writing the same day to a possible applicant for Small's job. Surprisingly, Watt was almost equally brisk. He consoled Boulton by repeating Small's own advice to him when Peggy died: the dead man was at peace; grief was pointless, unprofitable, even selfish. Better to plunge back into work:

> Come, my dear Sir, and immerse yourself in this sea of business as soon as possible, and do not add to the griefs of your friends by giving way to the tide of sorrow. I again repeat that it is your duty to cheer up your mind and to pay a proper respect to your friend by obeying his precepts. I wait for you with impatience, and assure yourself no endeavour of mine shall be wanting to render life agreeable to you.[36]

At this critical juncture Watt had second thoughts. Since 1771 he had had offers of work in Russia,[37] and now a serious offer came from the Imperial Government, with a salary of £1,000 – prompted, ironically, by Boulton boasting of Watt's abilities to the Russian Ambassador. Mentally kicking himself, Boulton employed all his charm, playing on Watt's hypochondria and anxiety. 'Your going to Russia staggers me,' he wrote. 'The precariousness of your health, the dangers of so long a journey or voyage, and my own deprivation of consolation, render me a little uncomfortable; but I wish to assist and advise you for the best, without regard to self.'[38] There was dismay elsewhere in the Midlands. 'Lord, how frighten'd I was', wrote Darwin jovially:

> when I heard a Russian Bear had laid hold of you with his great Paw, and was dragging you to Russia. – Pray don't go, if you can help it: Russia is like the Den of Cacus, you see the Footsteps of many Beasts going there but of few returning. I hope your Fire-Engines will keep you here.[39]

Watt's doubts were fostered by genuine fear over the Private Bill: all parliamentary business was held up because of the American crisis. After the furore over the Stamp Tax, most of the taxes so high-handedly imposed on the colonies had been removed. But the banking failure in 1772 had threatened even the East India Company, and to protect its trade the Government kept one crucial American import duty – on tea. When the ships arrived in Boston in 1773, enraged citizens gathered. A boarding party dressed as axe-wielding Mohawks stormed the ships, ripped open the chests and poured three hundred cases of tea into the harbour. The Boston Tea Party was cheered as a gesture towards 'Liberty', in defiance of British high-handedness and tyranny. On both sides, temperatures rose. Coercive Acts were passed against Massachusetts and the first Congress of the Colonies took place in Philadelphia in September 1773. The following year, the Protestant colonists were alarmed when the Quebec Act gave Roman Catholics in Canada full rights. By 1775 Massachusetts was declared in rebellion and Burke's conciliation plan was roundly defeated. On 19 April, a group of British redcoats were sent to take a munitions store in Lexington: shots were fired and local militiamen killed. It was the start of the American War.

It was against this background that the patent battle took place. But in this small sphere passions were also roused. Boulton and Watt faced howls of protest from competing engineers at the patent's deliberately vague, all-embracing terms. In the House of Commons, their opponents – including the eloquent Burke – levelled bitter charges of monopoly. The partners and friends such as Darwin scurried to find votes, and for three months Watt waited, 'which is a very long time to be kept in suspense'.[40] Finally, on 22 May 1775, the new Act received the Royal Assent and its coverage was extended to Scotland, a vital concession.

Having won the breathing space they needed, Boulton and Watt tackled the technical problems. The engine sent down from Scotland had a block-tin cylinder with a copper bottom, enclosed in an outer steam-case of wood. The piston had to pass through this outer case, through a sort of gland, and this was where all Watt's old problems of keeping it air-tight were caused. Again he tried every possible sort of packing – horse and cow dung, papier mâché, felted cloth used by the hatters – even adding weights to fix it. Eventually, with huge reluctance, he fell back on the materials used by Newcomen, hemp and oakum, well oiled with tallow.

In the course of all these tests, the inner tin cylinder itself finally gave way. This was the cue for the entrance of an extraordinary man, Joseph Priestley's brother-in-law, John 'Iron Mad' Wilkinson. The Wilkinsons rivalled the Darbys of Coalbrookdale. John's father Isaac had been a Cumbrian farmer and ironworker before taking over the Bersham furnace near Wrexham. John now managed this, and other nearby foundries. He won the soubriquet 'Iron Mad' because he firmly believed that anything could be made of his metal – he launched an iron boat on the Severn, confounding his friends who vowed it would sink. He had an iron desk and an iron bed and kept an iron coffin propped up in his office to persuade others to place orders. (Luckily he couldn't see the future: when he died, the coffin was too small and he was buried in a wooden one while a new one was cast; it was too large for the hole blasted in the rock, so he was reburied in his garden, with an iron obelisk overhead; a new owner, disliking this macabre spike, had him moved yet again; in all, he was buried four times, allegedly by the same gravedigger.) Stories stuck to Wilkinson like iron filings to a magnet. He had a will of iron, and a temper as hot as his furnaces. But he was the man who finally made the engine viable.

Part of the difficulty in getting the large-scale engine to work efficiently had always been the uneven, bumpy nature of the inside of the cylinder itself. The first one, cast at Carron, had quickly been abandoned. Wilkinson, like Roebuck at Carron, planned to make cannon, and knew that the trajectory of a cannonball would be much surer if the inner cylinder was bored smoothly. Working on this, he patented a new boring machine, a fifteen-foot bar with a revolving cutting head, which could slide along it: the rigid bar acted as a kind of ruler, keeping the cutter straight and even. Boulton knew of this, and suggested they order a cylinder for their engine, to replace the broken tin one, and it was delivered in April 1775. At Soho Watt's slowness and obstinacy and tendency to see the worst were always offset by Boulton's energy and pace. When Watt went to London to deal with the Patent Bill, Boulton took over the engine tests and made his own suggestions, including fitting a gauge to measure the vacuum: 'I think some sort of meter should be annexed to it, by which one may see the rate of vacuum, for without an outward & visible sign it is impossible to judge the inward and spiritual grace.'

His humour reveals his real delight in his new toy. Where Watt would have preferred to work a step at a time, Boulton wanted working engines and he wanted them now. In the summer of 1775 he cajoled Watt into agreeing to make two that were not only far bigger but designed for immediate use – escaping from Watt's diligent care, like unready adolescents taking on the world. One was an engine with a 38-inch cylinder, to provide blast for Wilkinson's furnaces at New Willey in Shropshire. The other, even larger, with a 50-inch cylinder, was for a pumping engine for the Bloomfield colliery near Tipton. This was the test – could Boulton and Watt beat the competition from the old Newcomen engines? Excited at the prospect, Boulton reported that if Wilkinson's engine worked, his neighbouring iron-masters were all planning to have theirs adapted 'and that work alone will be sufficient for our lives'.[41]

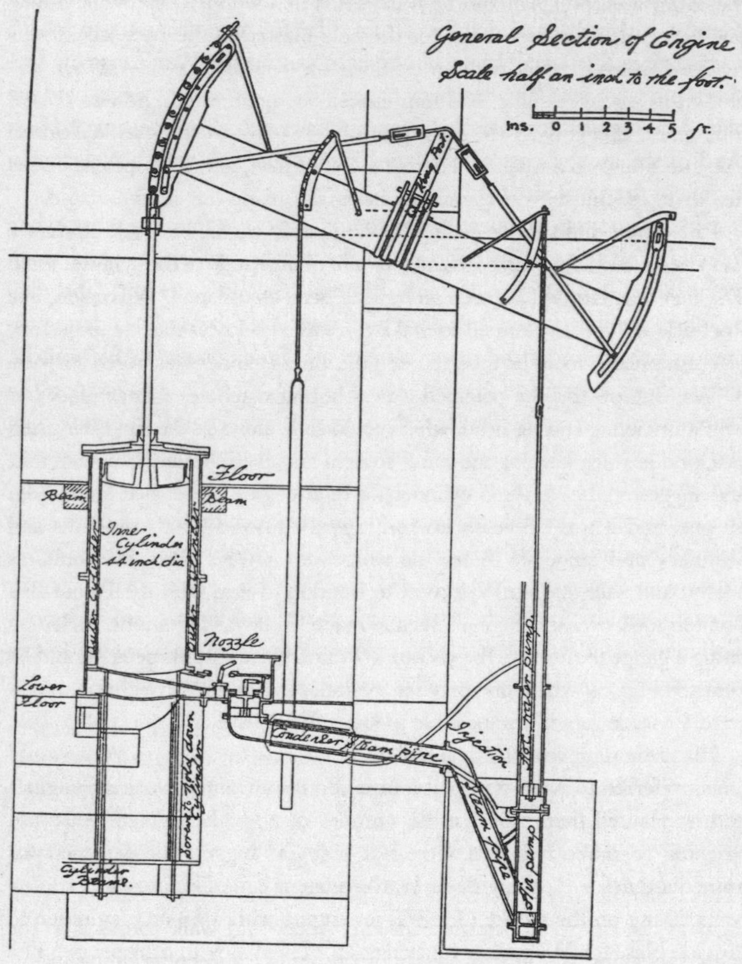

General section of Watt's early steam-engine, a drawing of 1776

The two big cylinders were cast and bored by Wilkinson, who virtually took over the construction of his own engine. Boulton and Watt, however, were wholly responsible for installing the engine at the Bloomfield colliery.

These hulking engines were never supplied whole but were more like a complex kit, with parts delivered from different manufacturers to Watt's specifications. Instead of laying out capital, Boulton and Watt chose the suppliers and gave them the designs, but the purchasers of the engine paid the bills. The owners got their cut from a royalty, which was estimated at a third of the annual cost of coal saved in comparison with a Newcomen engine of the same size.

For the first two engines, Wilkinson supplied the foundry work while John Harrison's team at Soho made the valves and pistons and small metal parts. Now and in the future (often to Watt's despair), the vital building, carpentry and final erection was done by independent contractors on site under the supervision of a Soho engineer. At Bloomfield, after various hiccups, the engine was started up at a special ceremony in March 1776. From the first thrust the great beam-engine, standing high and proud in the March winds, managed about fifteen strokes a minute and drained the pit of nearly sixty feet of water in less than an hour. Success assured, the 'Gentlemen' and the workmen then went off – separately – to dinner, and presumably to drink: 'After which, according to Custom, a name was given to the machine, viz., THE PARLIAMENT ENGINE, amidst the Acclamations of a number of joyous and ingenious workmen.' *Aris's Birmingham Gazette* carried a good puff, mentioning Watt's years of experiments, and pointing out that the engines were made

> under his and Mr Boulton's Directions at Boulton and Fothergill's manufactory near this Town; where they have nearly finished four of them, and have established a Fabrick for them upon so extensive a Plan as to render them applicable to almost all Purposes where mechanical Power is required, whether great or small, or where the Motion wanted is either rotatory or reciprocating.[42]

(This venerable engine is still in Birmingham. After closing time on 'steam days' at the city's old Science Museum, you could walk round it in the calm of evening and hear it quietly breathing like a sleeping dinosaur.)

Boulton's round marketing phrases ring through the *Gazette*'s 'impartial' report. Both engines needed adjusting, but both eventually fulfilled their promise, bringing their owners more power for less fuel. Slowly the orders came in, from a small pumping engine for a London distillery in Bow to the largest one yet at Hawkesbury colliery in Bedworth. Best of all, orders were expected from Cornwall, where the real money was to be made: 'Our plot begins to thicken apace, and if Mr Wilkinson don't bustle a little as well as ourselves, we shall not gather our orders before sunset.' The potential market was dizzying. And even at the start they were targets for spies. In the summer of 1776 a group of Cornish managers, or 'mine captains', visited Soho. When they left, Watt found that a key drawing of the general design had disappeared. Enraged, Boulton fired off a letter, lashing Thomas Ennis of Redruth who had led the group: 'We do not keep a school to teach fire-engine making, but profess the making of them ourselves.' The drawing reappeared. It had been taken 'under a misapprehension' by Richard Trevithick, captain of three mines and father of a five-year-old boy, another Richard, later the most dashing engineer of all.

With happy visions of world domination, Boulton urged Watt on with thunder-cracking rhetoric:

> If we had a hundred wheels ready made, and a hundred small engines like Bow engine, and twenty large ones executed, we could readily dispose of them all. Therefore, let us make hay while the sun shines, and gather our barns full before the dark cloud of age lowers upon us and before any more Tubal Cains, Watts, Dr Fausts or Gainsboroughs arise with serpents like Moses's that devour all the others.[43]

But they still had many details to sort out. To begin with Wilkinson made the outer casings, instead of the wooden steam-case used in earlier models, as well as the working cylinders with the piston. But he insisted on demanding the same price for the simpler work, and a furious Watt promptly redesigned the engine so that the outer cylinder, instead of containing steam to keep the inner one hot, was merely a casing which they could make cheaply from cast-iron segments. This also cut out a steam valve between the two cylinders, always a nightmare to fix.

Other improvements followed: in 1776 Watt designed an engine for a Scottish mine with separate valves for admitting the steam and letting it out. A less successful idea was a small 'Topsey Turvey Engine' for John Wilkinson, with the piston inverted and the piston rod linked directly to the pump rod. Boulton even persuaded him to try using expansive steam, building a wayward, noisy engine quickly called 'Beelzebub' that jolted on at Soho for seventy years: in time, familiarity feminized the old engine and its name was fondly shortened to 'Old Bess'.

The journey to successful engines had taken its toll. Watt had brought his children down to Birmingham but felt strongly that he needed someone to care for them. He also wanted to keep his close links with home and in 1776 he went back to Glasgow and returned with a new wife, Ann Macgregor, daughter of a prosperous linen manufacturer. This finally pushed Boulton into a proper agreement when an embarrassed letter arrived from Watt saying her father wanted to see the contract of partnership. Watt had hedged, 'lest he should have called my prudence into question I have been obliged to allow him to suppose such a letter did exist'.[44] To get him out of this hole, Boulton now added another lie, saying his lawyer was in London and no one could find the deed, but outlining the terms in a detailed, practical letter.[45]

Both men were now caught up in something unstoppable. Steam was so visibly powerful – demonic in its strength, a hissing, heaving animal of fire and water and air. There is something about this moment, the start of the engine partnership in 1775, that throws Lunar science into a new phase. In place of the slow, careful improvements, the new inventions, the gradual process of mechanization, came a vision not of industrial evolution, but of revolution. Boulton was a large-souled man as well as an ambitious one. When James Boswell came to Soho in 1776, he 'contemplated him as an *iron chieftain*, and he seemed to be a father to his tribe'. As Boulton swung his arm in a huge expansive gesture across his empire of machines, and his seven hundred workers, Boswell wished Johnson had been with him:

> for it was a scene which I should have been glad to contemplate by his light. The vastness and the contrivance of some of the machinery would have 'matched his mighty mind'. I shall never forget Mr Bolton's expression to me. 'I sell here, sir, what all the world desires to have – POWER.'[46]

22: 'BANDY'D LIKE A SHUTTLECOCK'

There was a poignant coda to William Small's death. Months after his funeral, a letter from his old pupil Thomas Jefferson made its slow way across the Atlantic, telling him that three dozen bottles of madeira were on their way, 'half of the present which I had laid by for you'. Jefferson wrote passionately about the violent clashes between British and American troops, which had cut off all hope of reconciliation and threatened to unleash a fury of revenge. 'I express my constant wishes for your happiness,' he ended. 'This however seems assured by your philosophy & peaceful vocation. I shall still hope amidst public dissension private friendship may be preserved inviolate.'[1]

This was the hope of many. Small's death was not the only disaster that threatened to fracture the cohesion of the Lunar group in 1775. In the arguments over the American War they found themselves on different sides; Boulton with his Tory allies became a staunch supporter of the ministry against the rebels; Keir thought the Government incompetent and untrustworthy; Wedgwood, Whitehurst, Darwin and Priestley sided openly with the Colonists' cause. One of the most troubled reactions was that of Thomas Day, whose attitude was complicated by his views on American slave-owners. In 1772, he had been profoundly moved when Bicknell showed him a newspaper item about a slave, working for a captain on the West India route, who fell in love with a white servant and ran away to have himself christened. After being

The Colonies Reduced, cartoon by Benjamin Franklin, 1765

recaptured and locked up on a boat on the Thames, he shot himself in the head. Bicknell and Day felt this tragedy could dramatize the evils of slavery, and together wrote a long poem, *The Dying Negro*, published in 1773. Coloured throughout by the Rousseauian romance of the noble savage, the poem tells the slave's story in the first person, from the point where European traders drug him in Gambia and fling him into the hold of the slave ship:

When bursting from the treach'rous bands of sleep,
Rouz'd by the murmurs of the dashing deep
I woke to bondage and ignoble pains
And all the horrors of a life in chains.
Ye Gods of Afric! in that dreadful hour
Where were your thunders and avenging pow'r!
Did not my pray'rs, my groans, my tears invoke
Your slumb'ring justice to direct the stroke?[2]

The Dying Negro was one of the earliest propagandist texts of the abolition movement in Britain, and it found a receptive readership. Attitudes towards slavery had long been shifting. It was no longer seen as part of the natural order, and few intellectuals defended it whatever their politics: Johnson opposed it as did Rousseau; Burke attacked it vehemently and so did Adam Smith.[3] In England and America, the Quakers in particular worked to stop slave-holding and in the late 1760s Evangelical campaigners such as Granville Sharp repeatedly brought cases to the courts, culminating in the case of James Somersset in 1772 when Lord Chief Justice Mansfield ruled that a master had no right forcibly to remove his slave from the country – effectively the end of slavery in England.

Impassioned and indignant, *The Dying Negro* was an immediate best-seller. But by the time Day was preparing a third edition in 1775, when Britain was stumbling into war with America, he found himself flummoxed. He wanted to support the Americans on grounds of principle, but how could he, if they were also slave-holders? In his thundering dedication he blasted America's hypocritical clamours for liberty: 'Let her aim a dagger at the breast of her milder parent, if she can advance a step without trampling on the dead and dying carcases of her slaves.'[4] When he received a hurt response from Americans he retorted, 'If there be an object truly ridiculous in nature, it is an American patriot signing resolutions of independence with one hand, and with the other brandishing a whip over his affrighted slaves.'[5]

But by early 1776, even before the publication of the Declaration of Independence on 4 July, Day had accepted the justice of the rebels' cause. He now poured out more verse, this time in support of the Americans, denouncing the British Government and painting lurid pictures of English atrocities.[6] But even Day was shaken by the impact of the war on British trade, and began thinking nervously of his own pocket and his generous loan to Boulton: 'America is unconquer'd, the King is —, England will be — and then what will become of button making?' he groaned.[7]

This kind of swerving between high-flown principle, nationalism and self-interest was typical. Boulton's response, however, was unashamedly self-interested. To begin with he had supported the Colonists, alarmed that their boycott would damage trade. Now this seemed less worrying than the threat of American competition, so he swung back to arguing for restrictive laws to prevent American production, organizing a Birmingham petition.[8] From this time on, there were no more Soho visits from Benjamin Franklin. Wedgwood, by contrast, was wholeheartedly pro-American, while taking care not to alienate his pro-Government patrons. He produced intaglio medals for rebel sympathizers, showing a coiled rattlesnake, with the motto 'DON'T TREAD ON ME' (quietly limiting these to 'the *Private Trade*'), and contributed to an appeal to help rebel prisoners in British gaols (carefully insisting his donation be anonymous).[9] His disgust with the waste of 'blood and treasure' in a 'wicked and preposterous war' constantly broke into his letters. 'How could the nation bear such an insult,' he stormed when he heard Lord North's speech in February 1778 introducing the Bill of Conciliation. It was not the contents but the tone that astounded him: 'After spending 30 millions & sacrificing 20 thousand lives, to tell the house the object was a trifle – a something, or nothing worth the trouble of collecting!'[10] Even the French support of America could not shake him. As so often, in a spontaneous, laughing juxtaposition of images, he set Boulton-style narrowness against the broader horizon of freedom. 'How could you frighten me so in your last?' he accused Bentley in March.

It was very naughty of you. I thought nothing less than some shelves, or perhaps a whole floor of cases and crocks had given way and you were sinking down with them, 'till reading a little farther I found it was only the nation was likely to founder in a french war, and having been fully perswaded of this even for some time past, I recovered from my shock and blessed my stars and Lord North that *America was free.* I rejoice most sincerely that it is so, and the pleasing ideas of a refuge being provided for those who chuse to flee rather than submit to the iron hand of a tyranny has raised so much hilarity in my mind that I do not at present feel for our own situation, as I may do the next rainy day. We must have war, and perhaps continue to be beat. To what degree is in the womb of time. If our drubbing keeps pace with our deserts, the Lord have mercy upon us . . .[11]

Darwin was equally vehement. He burst out to his friend Charles Greville in December this year, 'I am sorry to see by the papers that we are likely to have a violent French war – I hate war!'[12] If Chatham's line had been followed years before, with America providing the mother country with raw materials and Britain sending back its manufactures, what an empire this might have been: 'But the Lord and King George thought otherwise! perhaps it is for [the] better – Britain would have enslaved mankind, as ancient Room [Rome], and have been at length enslaved themselves, adieu.' Perhaps taken aback by his own forcefulness, he added, 'I never talked so much politicks in my life before.'

Everyone was talking politics: it could not be avoided. But in a close group of friends, the rifts had somehow to be smoothed over. In late 1775 Boulton sent Priestley a box of Derbyshire spar that he needed for his experiments. Priestley was delighted, and hoped they would meet again soon:

It will be a great pleasure to me to see your improvements on *fire engines* and all your other valuable improvements in *mechanics* and the *arts* . . . I shall not quarrel with you on account of our different sentiments in Politiks – When I tell you what is fact, that the Americans have constructed a canon on a new principle, by which they can hit a mark at a distance of a mile, you will say their *ingenuity* has come in aid of their *cowardice*. I would tell you the principle of it, but that I am afraid you would set your superior ingenuity at work to improve upon it, for the use of their enemies.[13]

From this time on, political divisions rippled beneath the surface, with the Tory Boulton a supreme pragmatist, while his friends increasingly challenged the *status quo* and the power of authority. Several of them, for example, sponsored the newly founded David Williams Chapel in London, a venture grounded in Franklin's proposal that morality could be taught without mentioning faith. This was backed by a group with strong Deist leanings, 'the Philosophical Club', founded by Franklin, Bentley and Colonel Dawson, Lieutenant-Governor of the Isle of Man, with members including Wedgworth, Whitehurst, Day, the architect James Stuart and the naturalist Daniel Solander (and their colourful new friend Raspe, geologist, gem expert, probable spy and anonymous author of *The Adventures of Baron Munchausen*[14]). One evening in Lichfield in May 1776 Wedgwood reported that Darwin was caustically planning a publicity drive for the chapel:

> He advises us by all means to hire some Parsons to abuse it in the Papers – To call upon the Government for immediate help, & advise the burning of them – Parson & Congregation – altogether – To lay the disturbances in America, & any other public Disasters which may happen at their Door – & he offers his service if you should be at a loss for an abuser of this new Sect.[15]

Sure enough, a correspondence in the *Morning Post* did link Williams with the 'deluded rebels'. That summer, in Paris, Bentley showed the pro-American Rousseau a prospectus and won his support (and gave him a copy of Day's *Dying Negro*).[16]

If the national obsession of 1775 was the American War, during that year the Lunar men all had their own concerns, and these too tended to keep them apart. Boulton and Watt were concentrating on their engine; Keir was busy with his glass; Whitehurst had a new job in London as Keeper of Stamps and Weights, a post created after an Act of 1773 regulating the weights for gold and silver coins. Whitehurst was in charge of the authorized weights and had to stamp any new sets brought to him as valid; he still lived in Derby but spent much time in his London office, in Adam Ferguson's quarters off Fleet Street.

Wedgwood had also been busy. By January 1775 he had developed the first form of the longed for 'jasper', a hard ceramic body that could be coloured and polished on a lapidary's wheel, and had employed Flaxman as a designer. 'What do you think of vases of our fine blue body, with white festoons, medallions etc?' he asked hopefully.[17] In addition, he had been fighting a patent battle of his own: in May, when Watt won his restrictive patent, Wedgwood and the Staffordshire potters managed to defeat an equally restrictive one, the extension of the patent for the use of Cornish china clay, applied for by Richard Champion (who had acquired the rights from its discoverer, his fellow Quaker William Cookworthy). Champion's rights were limited to production of porcelain only, not earthenware. This was seen as a great local victory and at the end of May, as soon as the Act was passed, Wedgwood headed west to look for clay, accompanied by Thomas Griffiths, John Turner (another Staffordshire potter), and a Plymouth chemist, the eighty-seven-year-old Henry Tolcher, remarkably fit, 'and nothing flatters him so much as telling him how young he looks, & how many years he may yet expect to live'.[18] In Cornwall Wedgwood visited Thomas Pitt, later Lord Camelford, on whose estate Cookworthy had found china clay and stone – but his trip was also an adventure: he scrambled on his peg-leg across the rocks at Land's End and gazed 'with a kind of silent awe, veneration, & astonishment' at the vast Atlantic.[19]

Edgeworth was in Ireland, Day in Holland and Brussels. Darwin, it seemed, was the only one who did not travel. He was growing in reputation and in girth. He had always been burly, and after his carriage accident he put on weight, his linen shirts ballooning like a tent at a country fair. But he had kept up his experiments, and in December 1774, sounding rather embarrassed, had sent Franklin a report for the *Transactions* on experiments by the lecturers Young and Warltire and others, on 'Animal Fluids in the Exhausted Receiver' – to see if gases could be absorbed by the blood and fluids of pigs and sheep. He was still lording it in Lichfield, and was distinctly put out when a rival passed through in the shape of Samuel Johnson. They met once or twice, reported Anna, 'but never afterwards sought each other'. The problem, in her view, was that 'Johnson liked only *worshippers*', and Darwin had no intention of becoming one.[20] 'The surly dictator felt the mortification, and revenged it, by *affecting* to avow his disdain of powers too distinguished to be an object of *genuine* scorn.'[21]

Darwin, in his turn, always spoke of Johnson dismissively, perhaps showing some resentment on the part of the large fish in the small provincial pond towards the London leviathan. He was a great networker, and it was he who leapt into action after Small's death in recruiting a replacement. Acting on a tip (probably from Dr Ash or Boulton), he at once wrote to William Withering, a young doctor at the Stafford infirmary, suggesting he apply for the post. Darwin had come to know him when he was a medical student and spent summers in Lichfield with his uncle, Dr Brooke Hector. Knowing Withering was keen on chemistry and botany, Darwin dangled the carrot of patronage before him, underlining his words: '*your philosophical Taste would gain you the Friendship of Mr Boulton, which would operate all that for you which it did for Dr Small*'.[22] Going through Small's papers, he worked out that Small had earned a decent £500–£600 a year, and had managed to live a fairly reclusive, studious life by keeping his practice chiefly in the town – adding with a slight note of envy, '*without the Expense and Fatigue of Travelling and Horsekeeping*'. Withering took the bait. The whole thing was kept firmly under wraps as the Darwin–Boulton cabal manœuvred through delicate local politics and saw off opposing candidates (including John Roebuck's son). Eventually Withering got the job, and in May he moved to Birmingham with his wife and three-month-old baby Helena.

The death of Small, and the arrival of a new, 'philosophically inclined' young doctor, as well as the realization that their work and their politics threatened to make them drift apart, all prompted them to decide to weld the group together. New Year's Eve 1775 saw the start of a more formal 'Lunar Society'. The idea was to meet each month for a few hours, beginning with dinner at two and carrying on at least until eight in the evening. For the first five years they chose the Sunday nearest the full moon, a common date for gatherings at a time when social life hung on the phases of the moon in a way that seems incomprehensible today, with a very different relationship to the natural rhythms of light and darkness. In Kent in the 1780s, for example, concerts and assemblies were all squeezed into the second and third quarters of the moon: the musician John Marsh and his wife were 'always rather glad than otherwise when the dark nights came that we might have a few evenings to ourselves'.[23]

From the start it was hard for everyone to be there, but they were determined to keep the plan going. 'Pray where were you last full moon?' Boulton asked Keir. 'I saw Darwin yesterday at Lichfield. He desires to know if you will come to Soho on Sunday 3rd March, in which case he will not fail to meet you, although he says he has inoculated some children which will probably be ill about that time.'[24] At this meeting Boulton told both Keir and Watt that he was planning to 'make many motions to the members', concerning laws and regulations 'such as will tend to prevent the decline of a Society which I hope will be lasting'. For the next meeting Boulton ordered an electrical machine, asking John Wyatt junior to go to a particular supplier whom Priestley recommended: 'as this is for a philosophical society beg everything be most accurately fitted'.[25] Sometimes they brought guests: young Charles Darwin came along, and Whitehurst asked if a friend's name could be added 'to the assemblage of Conjurers at Soho'.[26] At such gatherings, they were really operating as a small, high-powered research group. In the summer of 1776, the friends collected to test Buffon's theories that heat affected weight, setting up experiments using Boulton's fine balances to see if there was any difference between a cold ball of iron and a heated one.[27] They carefully consulted Black's notes on heat, and Wedgwood copied a number of experiments in his commonplace book, testing the way bodies of different densities responded to and retained heat.

They all looked forward to their meetings, but their correspondence is scattered with excuses and new dates – and at times over the next couple of years, when Boulton and Watt were in Cornwall on engine business, Wedgwood on his travels, Whitehurst in London, or Darwin stuck with patients, it seemed that it would be impossible to keep going. Some interests fell by the wayside: in 1778 Boulton sold his telescope, which had lain outside, suffering from rain and neglect. And in the same year Darwin wrote to him, apologizing for missing a meeting:

> I am sorry the infernal Divinities, who visit mankind with diseases, and are therefore at perpetual war with Doctors, should have prevented my seeing all you great Men at Soho today – Lord! what inventions, what wit, what rhetoric, metaphysical, mechanical and pyrotecnical, will be on the wing, bandy'd like a shuttlecock from one to another of your troop of philosophers! while poor I, I by myself I, imprizon'd in a post chaise, am jogged, and jostled, and bump'd, and bruised along the King's high road, to make war upon a pox or a fever![28]

But when they did meet, all political divisions were forgotten. By the light of the moon, warmed by wine and friendship, their heads full of air pumps and elements and electrical machines, their ears ringing with talk, the whirring of wheels and the hiss of gas, the Lunar men would clamber on their horses or into their carriages and unsteadily head for home.

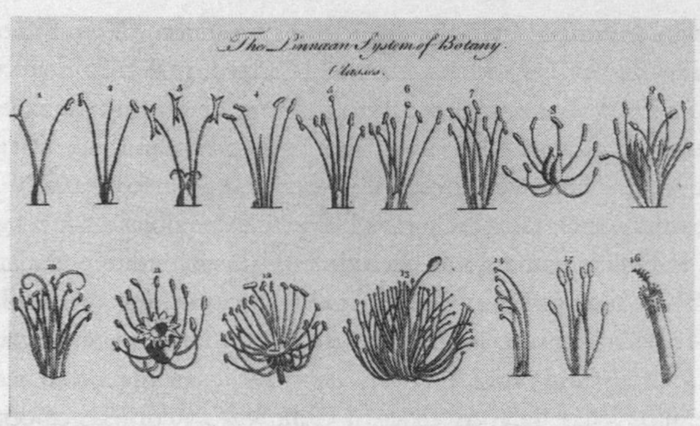

23: PLANTS & PASSIONS

Ordering, tracing lineages, classifying and naming – taking control of all realms of knowledge – were key eighteenth-century projects. The impulse was found everywhere, from the great encyclopaedias of the French *philosophes*, to Johnson's *Dictionary* and *Lives of the Poets*, from Burney's *History of Music* to Keir's *Dictionary* of *Chemical* or Priestley's *History of Electricity* and the naming of gases.

The pressure to classify plants had been intensified by the discovery of thousands of new species in remote parts of the world, brought home by colonists, traders and explorers. Newton's contemporary John Ray, an experimental biologist of genius, had long ago drawn up a 'natural' scheme for classification based on the likeness of the main parts – root, flower, calyx, seed and seed-vessel. Other botanists favoured a simpler 'artificial' system, founded on the minimum number of characteristics that would define a group: the French botanist Tournefort had named nearly seven hundred genera in this way, principally on the form of the corolla.[1]

Both systems ran alongside until Carl Linnaeus cut through the muddle with confident, startling clarity. Argumentative, conservative, ill dressed and uncultured, shunning the 'new science' and keeping his daughters almost illiterate so that they avoided the pollution of 'French fashions' – in every way the type of a stubborn Swedish provincial – Linnaeus revolutionized taxonomy.[2] He took two guiding principles: the idea of species put forward by Ray, and the crucial notion of a hierarchy of organic

The Linnean Orders, from Croker, 1764

life (a vital foundation to later theories of evolution). Most important of all, he saw the animal and plant world as linked by shared characteristics. In the 1730s he spent three years in Holland, helped by the now elderly Boerhaave, from where he visited France and England: on his return he became a professor at Uppsala. His ideas were diffused across Europe in several works, especially the *Systema Naturae*, first published in 1735 when he was in Holland, and going through edition after edition, expanding to a solid four volumes.[3] His system of naming was stunningly simple. In the 1690s it had been shown that plants, like animals, were sexual, with 'male' stamens and 'female' pistils. The flower, stamen and pistil then became seen as the keys to structure (followed by curiosity about flowerless plants such as fungi). Linnaeus divided plants into classes according to the number of stamens (Monandria, one stamen; Diandria, two stamens) and then into orders.[4] His second great breakthrough was to replace the long descriptive names of the past by two names only: the genus, plus a 'trivial' name to identify the species.

Linnaeus's easy, economical binomial system offered a map of nature to suit everyone from scholars to schoolgirls. He provided handbooks and instructions and lists of equipment from pocket knives to microscopes; he named plants after discoverers, farmers, women friends. His students sent specimens back to him from across the world, and he tried to cultivate these in his own Botanic Garden at Uppsala. It is easy to see why his botanical system spread rapidly across Europe. But it had fierce opponents, including the influential Georges-Louis Leclerc, Comte de Buffon, Director of the Jardin du Roi in Paris. Buffon's great *Histoire Naturelle* read almost like a novel, challenging the fictions of Genesis and re-telling the story of the earth through seven epochs – the animals appearing in the fifth and man in the last.[5] While Linnaeus saw species as static and the state of nature as fixed by God, to Buffon the essence of the natural world was constant change, dynamic force and movement. Imposing a crude and rigid grid killed the very spirit of life. He damned the Linnaean system as 'vile and inferior' and 'further exceedingly forced, slippery and fallacious'.[6]

It was inconceivable that the Lunar group would be left out of this new argument. Botany was immensely fashionable in Britain, especially after Cook's first voyage, which yielded a rich harvest, including 1,300 previously unknown species of plants. The Birmingham men cherished their links to Cook's explorations: through Keir's cousin, the naval surgeon James Lind (who was invited on Cook's second voyage, but went to Iceland with Joseph Banks instead); through Priestley and his soda-water; and especially through the high-flying young naturalist Joseph Banks. A further appeal was the idea that Cook's botanical discoveries might be of practical use: seeds for new colonial crops, plants for new medicines.

Botany also acquired status in Britain because it was a court pursuit. It had been patronized by George III's mother the Dowager Princess Augusta, whose gardens at Kew Palace were transformed in the 1760s under the eagle eye of Lord Bute, and when Augusta died, in 1772, Banks took over as adviser at Kew. This too was a patriotic project, developed in conscious rivalry with Buffon's magnificent Jardin du Roi. Natural history societies sprang up across Britain and a flurry of books appeared.[7] Linnaean primers, such as James Lee's *Introduction to Botany*, and Philip Miller's *Short Introduction*, both published in 1760, went through numerous editions. By early 1775 all the plants at the famous Apothecaries' Garden in Chelsea were being rearranged 'according to the system of Linnaeus' and the gardeners were quailing at the way their work had been increased,

> by the numerous Collection of new Plants raised and cultivated in the Garden; and as many of them are of foreign Production, of tender natures and especially such as are raised from seeds, in Hot Bedds; and require frequent Shifting and changing of situation and constant watchfulness, attention and care to preserve them . . .[8]

Doctors and apothecaries had often taken the lead in work on plant physiology and classification because of the use of plants in *materia medica*, and both the Lunar doctors, Darwin and the new recruit William Withering, now made this their field. Withering's current project in 1775 was an application of Linnaeus to British flora: he was therefore a highly desirable addition to the Lunar group. But if Darwin thought that he was going to be a grateful protégé, like Edgeworth, ready to share all his knowledge in a jovial, exuberant exchange, he was sorely wrong. Withering was stubborn, ambitious, reserved and wary. He bristled at interference, held jealously to his own line, and was unforthcoming in company. In the Lunar circle only the broad-shouldered charm and generosity of Boulton really managed to break down his defences. He was thirty-four when he came to Birmingham, ten years younger than Darwin, the son of a prosperous, property-owning apothecary from Wellington in Shropshire. He had studied at Edinburgh under Cullen and John Hope, a pioneer in lecturing on Linnaeus in Britain. Ironically, when Hope offered a gold medal for the best student in botany, Withering wrote home stuffily that although such a prize might produce 'emulation in young minds' it would hardly have 'charm enough to banish the disagreeable ideas I have formed of the study of botany'.[9]

After a brief stay in Paris – where he went to Nollet's lectures, visited the Hôtel Dieu and gave heartfelt thanks that he lived in a democracy rather than this absolutist state – Withering set up practice in Stafford at the new General Infirmary. Having fallen for Helena Cook, his future wife, who was keen on sketching plants, he began collecting specimens for her; in 1772, the year of his marriage, he embarked on his Linnaean classification of native plants. In the same year, the *Philosophical Transactions* published his paper on Staffordshire marle.[10] He had gathered different samples of this rich, fertilizing soil (a mixture of clay and calcium carbonate). Then he had added nitric acid to remove the carbon dioxide; burned the clay and calcined the ash, and added water to see if it would produce lime-water. He might well have been following Priestley's experiments on 'fixed air' and plants, but his chief bent was solidly practical, to see if burning marle to make quicklime would help agriculture.[11]

Withering was an obsessive note-taker, with a painfully neat mind and a compulsion to record and order. Classification was absolutely in his line, whether of soil, plants, medical case studies or minerals. By early 1775 his book was almost ready for the press, and Darwin leapt in eagerly – too eagerly – to help. Just as Whitehurst's interest had fired a passion for geology, and Keir's arrival made him an instant chemist, now botany became his new obsession. In March, when the manœuvring for Withering's appointment was in full swing, he wrote to Watt in London, interrupting his patent fight with a stream of requests for books to fill up the gaps in his already considerable botanical library. He already had twenty-seven volumes of Buffon's *Histoire Naturelle*, and wanted the most recent ones; he had seventeen volumes of another massive reference work, the 'Leipzig commentaries', but needed two missing volumes and anything published since 1771; he had the first part of Linnaeus's *Systema Naturae*, and wanted the rest.

Not content with getting briskly up to date, Darwin waded in to slap Withering's book into shape. In May, he told him heartily:

The title of your book should be easily remember'd, and easily distinguish'd from Lee etc. as 'The scientific Herbal', 'Linnean Herbal', 'English Botany', 'Botanologia anglica in which the science of Botany is reduced to English etc'. But we'll settle all this at Mr Boulton's with the assistance of Mr Keir and Mr Watt.[12]

Spotting a potential takeover, Withering would have none of it. His two-volume book appeared in 1776 with a title that took up a whole page.

Withering's volume was the best national flora to date. He illustrated it with twelve plates, annotated it with references to different authors and periodicals, from the *Transactions* to the popular *Gentleman's Magazine*, and gave a sound practical introduction to the finding and preserving of specimens. But as far as Darwin was concerned classification was not enough. Once he applied his mind to families of plants, and particularly when he went back to Buffon's *Histoire Naturelle*, he became more and more interested in plant development. Just as with the fossils in the Derbyshire caves, it was clear – although Linnaeus dogmatically clung to the contrary view – that some species had vanished over time, new ones had emerged, and older types had mutated. How had the 'families' that formed the basis of classification come about? Buffon had made a radical suggestion. If families exist, 'they can only have been formed by crossing, successive variation, and degeneration of the original species'. And once it was admitted there were families among plants and animals,

then one can equally well say that the ape is of the family of man, that it is a generative man, or that man and ape have had a common ancestor like the horse and the ass, that each family, whether of animals or plants, came from a single stock . . .[13]

The more Darwin thought upon this, the more convinced he became.

Flowers, plants and their families, however, were different from rocks, gases, engines. They had always been part of the artistic culture as much as the agricultural and the medical, and – as many Linnaean names acknowledged – they were connected with myth, poetry, painting, embroidery and love. They were decorative, and they were feminine.

Botany was often considered the one scientific pursuit suitable for women, yet Linnaeus seemed rather shocking for a female audience. When he organized plants into class, order, genus, species and varieties he chose sexuality as the key, classifying flowering plants by the stamens, the male 'genitals'. He grouped them into twenty-three classes, which were then divided into orders by the structure of the stigmas, the female 'genitals', while the supporting structure, the calyx, became the 'nuptial bed'. This meant, of course, that some flowers had far more than a single male sharing a bed with the female – and the sexual naming went further, with some structures compared to *labia minora* and *majora*, let alone a whole class of flowers named *Clitoria.* There was no escaping the link between Linnaean botany and sex. Even the fungi, mosses and ferns, which have no flowering parts, were labelled the *Cryptogamia*, 'plants that marry secretly'. In Britain, several botanists reacted with horror to the sexual emphasis and William Withering was certainly embarrassed. In his *Botanical Arrangement*, rather bizarrely, he changed 'stamen' to 'chive' and 'pistil' to 'pointal' for fear of offending the ladies. Darwin had little time for such linguistic niceties: for the translation of 'pedunculis axillaxilus' (taken from *axillus*, armpit), he offered 'Flower-Stems in the angles', commenting impatiently, 'Surely in the angles will be as intelligible as armpit, or Groin, or any other Delicate idea.'[14] But although they might not have acknowledged it, the botanical studies of both Withering and Darwin were linked to physical desire. Withering began by collecting flowers for Helena, and Darwin was now increasingly involved with a woman who was not only a passionate gardener, but whom he addressed in a deluge of intensely emotional pastoral verse.

His return to poetry was partly prompted by his second son, Erasmus. Suddenly he seemed to realize that his boys were growing up. At the start of 1775, Charles was sixteen, and Erasmus a year younger. They were like the two sides of Darwin's personality: Charles brilliant, scientific-minded, outgoing and ambitious; Erasmus literary and sensitive. In March 1775 Charles went to Christ Church, Oxford, leaving after a year (to Darwin's delight) and transferring to Edinburgh to study medicine. It was said that Darwin found his quiet second son Erasmus effeminate and irritating, but it was partly with him in mind that he founded a small literary circle in Lichfield. This included, of course, Anna Seward, as well as

Edgeworth and Day when they were there, the poetry-writing squire Francis Mundy and the Rousseauian Brooke Boothby, leisured, indolent and frequently drunk. Revelling in his role as mentor, Darwin began versifying again. In 1776, four poems appeared as tailpieces to Mundy's poem 'Needwood Forest'. They were signed 'E.D.', 'A.S.', 'B.B.', and 'Eras. Jun.', but Anna claimed that Darwin wrote three of them himself, putting Erasmus's name to the best and hers to the worst. When she tackled him, 'He laught it off in a manner peculiar to himself.'[15]

There was another, deeper spur to poetry. Darwin's relationship with Mary Parker had faded after the birth of their second daughter, Mary, in 1774. The following year, in return for a moralizing book of 'Village Memoirs' sent by an acquaintance, he enclosed a poem provoked by his intercession with 'a Derbyshire lady to desist from lopping a grove of trees'. Darwin's poetic trees appeal directly to 'the sweet Belle', to stay her 'dread commands' and 'unpitying hands'. If she reprieves them, they promise that her name will be carved on every trunk; for her the blossoms will scent the wind, and,

> The love-struck swain, when summer's heat invades,
> Or winter's blasts perplex the billowy sky,
> With folded arms shall walk beneath our shades,
> And think on bright Eliza – think, and sigh.[16]

It takes some effort to see Darwin in his mid-forties – pock-marked, overweight, limping and stuttering – as a love-struck swain, but struck he certainly was. And this was the ultimate romance, for the object of his desire, Elizabeth Pole, was not only beautiful but married, and unobtainable. His 'bright Eliza' was in her late twenties, dark-haired, witty and impetuous, with alluring aristocratic connections. Anna Seward summed her up: 'Agreeable features; the glow of health; a fascinating smile; a fine form, tall and graceful; playful sprightliness of manners; a benevolent heart, and maternal affection.'[17] The illegitimate daughter of the Earl of Portmore, she was half-sister to Caroline Colyear, wife to Sir Nathaniel Curzon of Kedleston Hall. Here she met Colonel Pole from neighbouring Radburn Hall, four miles away. He was almost thirty years her senior, a hero of the Continental wars who had been left for dead in at least three battles (at Minden a bullet smashed into his left eye and came out at the back of his head) and an undeniably dashing figure. Elizabeth was seven months pregnant when she married Pole in 1769. By 1775 they had three children: Sacheverel aged six, Elizabeth five, and one-year-old Milly.

Four years before, probably on the recommendation of Joseph Wright, who painted several portraits and groups at Radburn, the whole family had become Darwin's patients. Untroubled by thoughts about medical ethics, he laid siege to Elizabeth. Since she was the inviolable possession of a hoary warrior, there was far less risk involved than in flirting, say, with the unmarried Anna Seward. He could play troubadour, feel young again, and – at least partly – sublimate his sexual longing. Darwin thoroughly approved of sex. He had none of the worries that would plague Victorian England; masturbation was fine; an erection on waking was a sign of health.[18] Flagrantly heterosexual himself, he had homosexual friends and made no adverse comments.

Verses to Elizabeth poured from his pen over the next few years: he mourned when he had to leave Radburn after a consultation ('Dear distant Towers! whose ample roof protects/All that my beating bosom holds so dear').[19] Then in late 1777, when three-year-old Milly Pole became apathetic and listless for no apparent reason, Elizabeth brought her to stay in the Darwin household for several weeks. Drugged up to the eyebrows, Milly probably slept her illness through: 'Nov. 16,' Darwin noted diligently. 'She has now taken a grain of opium thrice a day for a month and seems to have perfectly recovered.'[20] Soon after Milly and her mother went back to Radburn, Darwin sent a poetic order to Boulton for the finest tea-urn Soho could produce:

> With orient pearl, in letters white,
> Around it, 'To the Fairest', write;
> And where proud Radburn's turrets rise,
> To bright Eliza send the prize . . .
>
> Perch'd on the rising lid above
> O place a lovelorn turtle-dove,
> With hanging wing, and ruffled plume,
> And gasping beak, and eye of gloom.[21]

Given Darwin's poetic excesses and Anna Seward's infatuation with John Saville, the atmosphere in the sedate Cathedral Close sounds intensely emotional. (When Richard and Honora Edgeworth visited in 1776 and decided to return to England for a few years, their arrival reduced Anna to hysterical weeping.) But if Darwin's romantic yearning was unassuaged, at least his botanical passion could take physical shape. He bought a few acres of land a mile to the west of his house and began to create his own botanic garden. The plot followed 'a little, wild umbrageous valley', as Seward put it, cut into one of the few rocky hills on the rich Lichfield plain.[22] The valley was swampy and tangled with undergrowth, and in its midst stood a huge stone bath, built over a spring for the citizens of the town a century before by the zealous Dr Floyer, a proselytizer for cold baths.

Darwin dammed the streams to make small lakes and forced the brook into the prescribed 'serpentine' course. On each bank he planted trees and shrubs, 'and various classes of plants, uniting the Linnaean science with the charm of landscape'. His favourite feature was a rock which dripped water continually, about three drops a minute, with water plants hanging from the rocky crevices. In this magical spot, unaffected by drought or flood or frost, Darwin had a poem inscribed, spoken by the Naiad of the Fountain, warning away all those with thoughts of avarice and greed and the busy world. He would not let Anna Seward see the valley until it was finished, about three years later. She went alone, as he was called away to a patient, and 'seated on a flower-bank, in the midst of that luxuriant retreat . . . while the sun was gilding the glen, and while birds, of every plume, poured their song from the boughs', she at once penned about fifty lines dedicated to the Genius of the Place. Darwin – the Genius – was properly flattered. It was at this point, she said, that he declared the Linnaean system to be 'unexplored poetic ground', as rich in metamorphic possibilities as Ovid. Ovid had turned people into plants: they would turn plants into men and women. Even the project itself was gendered: she would write the verse and he the notes, 'which must be scientific'. As in the Linnaean names themselves, science met the classics.

When Anna urged Darwin to write the poem himself, he demurred, on the grounds that his reputation as a doctor might suffer if he was known to be dabbling in poetry. His excuse echoes one made earlier to a friend, when he confessed that he had written many medical tracts but dared not publish them: 'Faults may be found or invented; or at least ridicule may cast blots on a book were it written with a pen from the wings of the angel

Gabriel.'[23] But in private, since the early 1770s, he had been writing a long prose work on diseases, which eventually became *Zoonomia*. And now he began a long poem, which would grow over time into *The Loves of the Plants*. He certainly sent Seward's verses to the *Gentleman's Magazine* in her name, but eleven years later her lines appeared again without acknowledgement, much to her fury, as the opening to *The Botanic Garden*.

In social situations, wit and poetry could take the place of accurate data. When Darwin later talked to Anna about the newly introduced Kalmia – which he had never seen – and she asked him what colour it was, he blithely said, 'Madam, the Kalmia has precisely the colours of a seraph's wing.'[24] In medicine, however, accuracy was all important: many plants listed in old herbals were poisonous if misused. Withering had deliberately left medical usage out of his *Botanical Arrangement*, dismissing traditional lore as a mix of ignorance and superstition.

Curiously, the work that made him famous was provoked by the very lore he so despised. When he first came to Birmingham he used to drive back every week to see patients at the Stafford Infirmary. The horses had to be changed on the thirty-mile journey, and one day during this stop he was asked to look at an old woman with dropsy. His prognosis was bad and he was later astonished to find that she had recovered, helped by a mysterious herb tea, made according to an old family recipe, 'kept secret by an old woman in Shropshire, who had sometimes made cures after the more regular practitioners had failed'.[25] It led to vomiting and purging, and Withering saw that it contained twenty or more different herbs, 'but it was not very difficult for one conversant in these subjects to perceive that the active herb could be none other than the Foxglove'.

The foxglove, however, is deadly poisonous. When he wrote up his scrupulously recorded cases ten years later in his *Account of the Foxglove*, Withering included a chilling note on possible side-effects:

> The Foxglove, when given in very large and quickly repeated doses, occasions sickness, vomiting, purging, giddiness, confused vision, objects appearing green or yellow, increased secretion of urine, with frequent motions to part with it, and sometimes inability to retain it, slow pulse, even as slow as 35 in a minute, cold sweats, convulsions, syncope, death.[26]

He had learned all this partly through the 'accurate and well considered experiments' he considered so vital. His subjects – as disposable as Priestley's mice – were the poor of Birmingham. William Small had always set aside two hours a day to care for the poor of the town for free: Withering kept this up, estimating that he saw two to three thousand patients a year. It was on them that he first tried digitalis.

Withering's belief was boosted by hearing of other cures; but instead of using the root, which seemed too variable, he tried the leaves, drying these in the sun or by the fire, then grinding them into a powder. His 'discovery' translated folk remedy into accepted experimental medicine. He could see that digitalis was a good diuretic and worked on the heart, although he did not know exactly how – in fact it slows down the heart muscle so that the heart fills and empties properly, providing a better flow of blood to make the kidneys more efficient. But the more Withering studied the plant, so he heard still more stories. A Stourbridge apothecary reported digitalis as good for tuberculosis and epilepsy. A Yorkshire salesman told him that his wife had stewed a great handful of leaves into a drink to cure his asthma: he was violently sick and his pulse raced, but he recovered. 'This good woman knew the medicine of her country but not the dose of it,' decided Withering.[27]

One of Darwin's cases in the summer of 1776 was a feverish middle-aged woman, with a pain in her side and a perpetual cough, spitting up buckets of phlegm. After six weeks, she was practically suffocating, with a clammy, leaden look and a raging thirst. She was distended with fluid yet could not pass water, her urine reduced to mere spoonfuls. Darwin called in Withering for advice. The only thing Withering could think of was digitalis but at first he hesitated, since the case looked so bad, and her death might 'discredit a medicine which promised to be of great benefit to mankind, and I might be censured for a prescription which could not be countenanced by any experienced practitioner'.[28] However, when he proposed it, Darwin agreed and Withering made up the draught. The results were dramatic: the poor woman was very sick, but 'within the first twenty four hours she made upwards of eight quarts of water'. Over the next fortnight, both doctors prescribed a cupboardful of other medicines. Despite, rather than because of these, she recovered. Medically, this shared case was a success; but in five years' time it was to be the cause of an almighty row.

'The Foxglove', frontispiece to *An Account of the Foxglove* by William Withering, 1785

Perhaps inspired by his methodical young colleague, in 1776 Darwin too began keeping notes on all his cases. Never one to do things on a small scale, he acquired a huge folio commonplace book. But somehow his notes would not stay in order – they escaped, flew sideways, turned into poems, or sketches of inventions, or grand speculative theories. He began earnestly enough, with diligent pages of case notes: from worms and dropsy and drunkenness to scarlet fever and smallpox. But even here the commonplace book makes alarming reading: for melancholia and 'causeless fears' following childbirth (which we would judge by hindsight as post-natal depression) he orders the extraction of three double teeth (the operator made a botched job and had to come back next day to take the roots out), followed by 'a giant purge'.[29] He was clearly impressed by digitalis and doubtless knew it as a folk remedy. Now that Withering had worked out 'scientific' dosages Darwin used it at once:

> Mr Harrison took four draughts of foxglove, vomited a little and then purged twenty times with great debility; had next day but one a violent inflammation of the liver with much pain.

Another is worse – blunt and horrifying:

> Mrs —, was asthmatic and dropsical, but did not appear near her end. She took four draughts of the decoction of Foxglove. She vomited two or three times, and then purged twice and died upon the close stool.[30]

Withering always held to his belief that digitalis was 'one of the mildest & safest medicines we have, as well as one of the most efficacious'.[31] Ten years later, hearing that the Quaker doctor John Lettsom had used digitalis in eight cases that proved fatal, he merely suggested that he might have got the dose wrong. Experimental medicine could kill, as well as cure.

In 1778, Darwin experienced the tragedy of sudden death himself. This too was a result of a medical accident and the victim was no nameless patient, but his own son Charles. In his second year as a medical student Charles had been helping his teacher, Andrew Duncan, in treating the poor, had been awarded a gold medal for an essay on the criterion for distinguishing between pus and mucus, and was writing a thesis on the lymphatic vessels. He was easy-going, brilliant, and much loved by his father's friends, corresponding on the crystallization of glass with Keir and standing as godfather to his baby son Francis: 'I wish you might infuse into him some of your thirst for knowledge,' wrote Keir happily.[32] But the hopes of both fathers were dashed. Little Francis died in infancy, and in April 1778, performing a dissection on a child, Charles Darwin cut his finger. It was a scratch, a mere nothing. Yet within hours he was felled by severe headaches, followed next day by delirium and haemorrhage.

Cullen, Black and Duncan rushed to treat him. Darwin drove north as fast as he could, believing at first that Charles would recover and then admitting, in despair, that he was sinking fast. On 12 May he told Erasmus, 'I fear in two or three days he will cease to live.'[33] On 15 May, Charles died; four days later he would have been twenty. Darwin stayed in Edinburgh with James Hutton until Charles was buried. He wrote an inscription for the tomb and very probably had a hand in a long, anonymous elegy, published later that year, mourning Charles as a lost lover of Science, the goddess whose light would one day sweep ignorance and terror away.[34]

He told Hutton that he planned to publish Charles's thesis and two years later he did so, silently embellishing it with cases of his own.[35] *Pus and Mucus*, as it is disconcertingly known, included five accounts of cases treated with digitalis, the first publication of such a cure. There is no evidence that Charles had ever used digitalis: indeed one case was clearly that of the woman whom Darwin had called Withering in to see in 1776. At no point did Darwin even mention the younger doctor's name. Withering, who had been patiently piling up his notes, had been scooped, and was enraged. He had had to deal with intense grief himself when his baby daughter died at the age of one. But by the late 1770s the two doctors had become increasingly competitive – and both of them now began to publish.[36] Darwin's editing of Charles's work might have been an attempt to hold the field – or an ill-considered act of paternal love, stealing from a rival to give the prize to the son he had lost. Whatever the reason behind it, after this row, he and Withering hardly spoke again.

24: CONQUERING CORNWALL

The Chacewater Mine in Cornwall, where James Watt found himself in the summer of 1777, was a harsh contrast to Darwin's fertile dell. On the moors there was hardly a single tree. The earth was dotted with engine houses and sheds, heaped with great piles of rubble and pitted with shafts and reeking vents and fissures, like the cracks through which ancient heroes entered the Underworld.

The mine captains ruled, and their workers fell into two classes, peculiar to Cornwall. The less-skilled 'tut-workers' were paid by the fathom for sinking shafts and driving levels. The 'tributers' who brought up the ore were speculators, paid by the ton according to its richness. In theory, they were profit-sharers; in practice, since the mine-owners charged them for their ropes and candles and paid them nothing if nothing was coming in, they were the most economical of labour forces, constantly competing with each other. Every day, lines of men strode towards the shafts, their drill trousers caked with red dust. They carried pickaxes and ropes slung over their shoulders and wore hardened felt hats with candles stuck to their brims by lumps of sticky clay; more candles hung from their belts. In some places temperatures below ground reached 100°F; in others they had to wade waist-deep through icy water; sometimes the air was so foul that no candle would light at all. A six-hour shift was all they could manage before hauling themselves back up the ladders, hand over hand, six hundred or even twelve hundred feet.

Dolcoath Mine, Cornwall

The great baskets or 'kibbles' of ore were pulled to the surface by 'whims', circular wooden winches driven by mules and horses, plodding in an endless circle, poked and whipped by small boys. Men crushed and sorted the ore, breaking the rock into lumps to find the patches of yellow chalcopyrite and grey chalcosine, the rich source of copper. Near by worked the balmaidens in their brown gowns and winged Breton bonnets, helped by an army of children, tipping the smaller pieces on to the mill, an iron plate where more women whacked the ore with flat-headed bucking-hammers into a coarse powder. Whole families worked here and the life of the community hung on the copper mine. Tin-mining was less labour intensive since machines and water could do some of the work. Huge stamping mills made of iron-bound logs pulverized the ore, and then grinding stones turned it to fine sand which was washed in shallow pans – 'buddles' and 'slimes' – where the black tin settled as the lighter impurities were washed away.

Every now and then, amid the thumping of the pumps and mills, came a hollow boom from gunpowder blasting deep below. Accidents were common: rockfalls, burns, blinding from explosives. Most miners were broken by their mid-thirties, if they lived that long. And as the mines drove deeper so the risk of flooding grew: adits were tunnelled out to the cliffs or valley sides to reduce the cost of raising water. They had used Newcomen engines since 1715 and by the late 1770s sixty of them humped and puffed across the county, but still the mines were often forced to close down during wet seasons. Worse still, the Newcomen engines devoured costly fuel. In the Midlands, with coalfields near by, this did not matter greatly, but here coal had to be imported from South Wales and carried from the harbours by packhorse and mule.

The mines were known by their ancient Celtic names of 'wheals'. Tin had been mined here since the Romans and had been a great industry in Tudor times; copper lay deeper and large-scale mining had really begun only at the start of the century. From the 1720s, especially with the rise of the brass trades, hundreds of tons of ore were sent for smelting in Bristol and Swansea and Hayle on the north Cornish coast. At auction times the streets of Redruth were filled with huge parcels of ore and a tight ring of smelters bid by a complicated process of 'ticketing'. Meanwhile, dreaming of profit, the Cornish adventurers flung their family fortunes down the mines, subscribing for shares and receiving endless calls for more money to pay for sinking and draining new shafts. In 1768 a new blow fell. Run by the formidable Thomas Williams, the great Parys mine in Anglesey was a single hill of ore so easy to work that it caused an earth-shattering slump in prices. Since there was no way the Cornish mines could cut down on labour costs, the only possible economy was in fuel and local engineers worked feverishly to improve their boilers so that they could get as much steam as possible from every shovelful of coal. John Smeaton was called upon, and by 1775 he had managed to double the efficiency of the Newcomen engine. But, even so, over forty of the county's engines were often idle, while even the best could not cope with the water.

As Small had foretold, all this made the Cornishmen leap on the news of Watt's fuel-saving engine. As well as the separate condenser, Watt's engine used steam to work the power stroke of the piston, making it more powerful than any Newcomen design. But the Cornishmen were proud and suspicious, even among themselves. The captains managed the mines and the engineers kept the machinery going for an annual retainer, often overseeing a number of different mines divided between two rival areas, eastern and western. The mineworkers supported their respective landlords Lord Falmouth and Sir Francis Basset – bitter political opponents – like feudal armies.

The deputation who came up to Soho in the summer of 1776 included the stormy, impatient Richard Trevithick, and obstinate John Budge, familiarly known as 'Old Bouge', a noted engineer, both of whom worked for the western mines, the most hostile to 'incomers'. Not surprisingly, Boulton & Watt's first orders in November 1776 were from the rival eastern group: a 52-inch cylinder engine for the Tingtang mine near Redruth, and a smaller 30-inch engine for the Wheal Busy mine at Chacewater. The Tingtang engineer, the sixty-year-old Jonathan Hornblower, was the son of a Staffordshire man who had come here in the 1720s to erect engines for his friend Thomas Newcomen; all seven of his sons went into the business. Hornblower rivalled Budge as the duchy's leading engineer and all through late 1776 he sent Watt detailed accounts of mines, engines and engine-houses, informing him that there were 'some very good Engine smiths in Cornwall & some bad ones, (all of them love drinking too much)'.[1]

The Cornish business got off to a slow start; when the Tingtang cylinder arrived at Chester for shipment it was too big to go through the hatch, and although Watt offered to pay to make the hatch bigger (a mark of desperation in a thrifty man) it lay forlornly on the quay as the ship sailed. Wheal Busy – often called Wheal Spirit – was therefore the first to receive its new engine. Watt set off for Cornwall with Annie to ensure the success of this vital commission. The journey was a marathon. They left as dawn broke and jolted to Bristol, arriving at nine that night. Next day, even at six o'clock, all the coaches were booked, and the parsimonious Watt was forced to hire a post-chaise, racing the Exeter coach so they could leap into its cheaper seats when people got off at Bridgwater. Another chaise was needed from Exeter to the Tamar ferry across to Saltash. At last, exhausted by four days' travelling, they reached Truro. Nothing Watt saw pleased him. The people had 'the most ungracious manners' of any he had met. The engines in general were 'clumsy & nasty' while the engine-houses were all 'crackt & everything dripping with water from their house cisterns'.[2]

The Watts stayed with Thomas Wilson, another northern speculator making a fortune in the west. He owned partnerships in ships, farms, copper mines and the Neath Abbey Iron Works; he had a brewery and a candle-making business and was chief agent for a company that ran a Swansea smelting works. Most significantly he had major investments in Wheal Virgin and in the Chacewater Copper Company, a local smelting venture and soon he would become Boulton & Watt's financial agent, running their affairs while they were away. He and his wife and large family were the soul of kindness, although the Cornish servants, Ann Watt told Anne Boulton, were the laziest wretches on earth and Mrs Wilson, a Yorkshire woman, was 'teazed to death by them'. She herself hardly knew how to describe the country: 'The spot we are at is the most disagreeable in the whole county. The face of the earth is broken up in ten thousand heaps of rubbish and there is scarce a tree to be seen.' But all Cornwall was not like this and indeed some places were beautiful – life was an exciting adventure in a foreign land:

The sea-coast to me is charming, but not easy to be got at. In some places my poor Husband has been obliged to mount me behind him . . . I assure you I was not a little perplexed at first to be set on a great tall horse with a high pillion. At one of our jaunts we were only charged twopence a piece for our Dinner, you may guess what our fare would be from the cost of it, I assure you I never ate a dinner with more relish in my life nor was I ever happier at a feast, than I was that day at Portreath. I hope to see you enjoy the manner of living here with as much pleasure as I have done.[3]

Cornwall, however, was having an odd effect on her husband. 'One thing I *must* tell you', she added, 'is to take care Mr Bolton's principals are well set before you trust him here poor Mr Watt is turned AnaBaptist, and duly attends their meeting, he is indeed and goes to chapel most devoutly.'

Watt needed to pray. Ann was a devoted help, but instead of reassuring and sweetening him as Peggy had done, her support took the form of backing up his resentments, of pursed lips, dark comments, money-pinching schemes. As soon as he arrived Watt found life difficult. Every word he said was misrepresented in the mining community, so he hung back – in itself a mistake. He found himself accused of making several speeches at the huge, drowned-out Wheal Virgin, when he had only 'talked about eating, drinking, and the weather'. The adventurers there, he decided, were 'a mean dirty pack, preying upon one another'.[4]

To make things worse, for several years it was very hard to find skilled men who could 'put engines together according to plan as clockmakers do clocks', as Watt complained to Smeaton: so far engineers had been improvising on site, and were unused to precise instructions and prejudiced 'against anything new'.[5] 'Engineer' itself was a new word, coined by Smeaton, according to Watt, only a few years before. But even with a shortage of men they made progress with the engine at Wheal Busy. 'All the world is agape to see what it can do,' wrote Watt.[6] Many in the neighbourhood had been convinced it would never 'fork water' and it was crucial to persuade the sceptical engineers, including old Hornblower. By September 1777, it was up and running, supervised by the Soho-trained Thomas Dudley, and most of the spectators, Watt thought, were highly impressed:

The velocity, violence, magnitude, and horrible noise of the engine, give universal satisfaction to all beholders, believers or not. I have once or twice trimmed the engine to end its stroke gently, and to make less noise; but Mr Wilson cannot sleep unless it seems quite furious, so I have left it to the engine men; and, by the by, the noise serves to convey great ideas of ye power to ye ignorant, who seem to be no more taken with modest merit in an engine than in a man.[7]

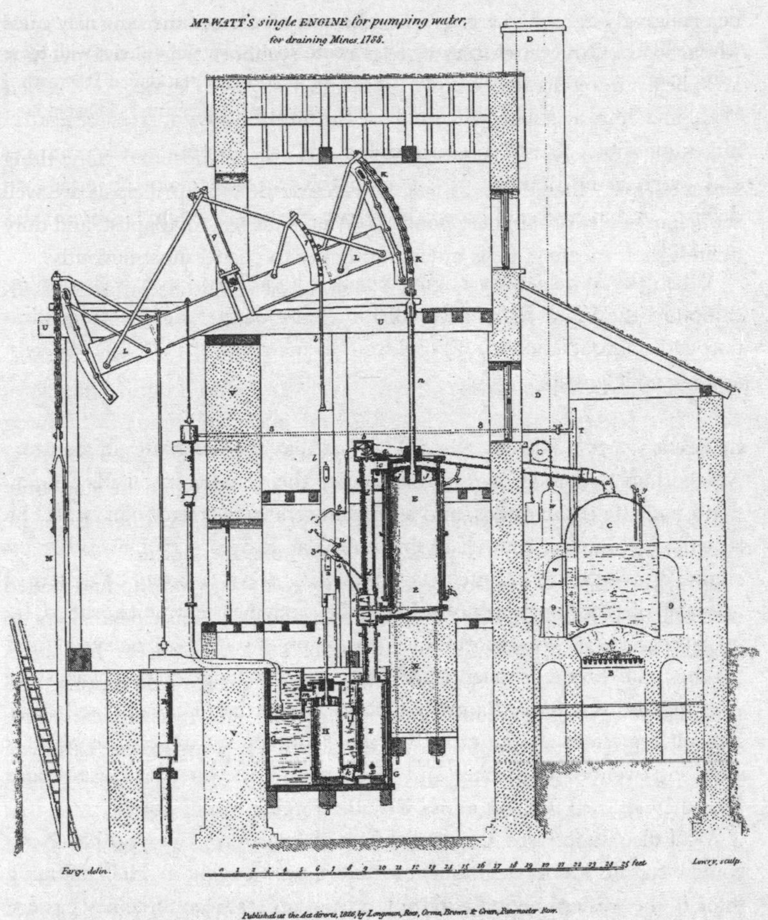

Watt's single-acting engine of 1788, engraved in 1826

Before the summer was over Dudley took on a new order for a huge engine, with a 63-inch cylinder, for Tregurtha Downs in Marazion, in the western area: the regular engineer was 'Old Bouge', who still refused to touch Watt's engines, declaring that they were rubbish and that his own improved atmospheric engine was just as good. At an interview with Watt he grudgingly agreed to recant when he was convinced the new machines would work. Other engineers were equally stubborn. Richard Trevithick at Wheal Union scoffed openly at the fuel savings claimed, and when Watt met him at a later meeting of the mine's adventurers he found him impossible: 'I was so confounded with the impudence, ignorance, and overbearing manner of the man that I could make no adequate defence, and indeed could scarcely keep my temper, which, however I did to a fault.'[8]

When the Wheal Busy engine reached its full load and was running smoothly, the Watts returned to Birmingham for the winter – to civilization and Soho, to friends and children, to meetings of the Lunar Society and scientific gossip.

In January 1778 Hornblower told Watt that the Tingtang adventurers wanted him to come down himself to see their engine installed. 'I fancy that I must be cut in pieces, and a portion sent to every tribe in Israel,' he lamented.[9] Shutting himself in his study at Harper's Hill, working on engine drawings and only reluctantly coming down to Soho, he managed to stave off the journey until May. On arriving he was mortified by mistakes made by Soho workmen – wrong pipes sent, parts delayed – and by the state of the copper trade. But the Tingtang engine was soon working and was judged a great success. Tregurtha Downs proved more difficult, especially as the crowd-pleasing Dudley set the engine off at a storming twenty-four strokes and 'soon got all his water boiling hot and then they seemed to be at a loss why the Engine would not go'.[10]

Watt also supervised the adaptation of Smeaton's two-year-old Newcomen engine for Chacewater, erected a new engine at Hallamanning mine in the western region and took crucial orders from three key eastern mines; Poldice, Wheal Chance and United.[11] The critical parts, especially nozzles and valves, had to be sent from Soho; the cylinders from Wilkinson or Coalbrookdale; the pumps and rods and beams were made locally. The finished structures were both intricate and vast, with engines standing as high as a three-storey house.

In the summer Boulton was urgently pressing Watt to get payment for the engines, most of them not even erected.[12] The truth – well hidden from prospective backers – was that Soho was plunging into crisis. A severe slump in trade had brought troubles in the Potteries and riots in Nottingham and Lancashire: Arkwright's factory at Birkacre was burned. Soho's agent Matthews sent bad news from London and Fothergill bombarded Boulton with woeful letters about their Bill Account, even recommending bankruptcy: 'Better stop payment at once, call our creditors together, and face the worst, than go on in this neck and neck race with ruin.'[13] At Soho, a fire destroyed the roof and several workshops and damaged the engine, old Beelzebub. Boulton was forced to lay people off, reducing his workforce from six or seven hundred to a hundred and fifty. Yet through all this he was unabashed. Far from paying off his creditors, he borrowed more, on a terrifying scale, including a substantial loan of £7,000 from the Dutch banker Wiss. Altogether he raised almost £17,000 on annuities and mortgages against the engine patent. Even the fire was an opportunity: '*Now*,' he said, metaphorically rubbing his hands, 'I shall be able at last to have the engine house built as it should be.'[14]

In Cornwall, Watt felt helplessly cut off. Everything seemed to be happening without him. Whitehurst had published his book and Withering his account of scarlet fever; Keir was writing his 'Treatise on Gases', Priestley completing volume after volume of his *Experiments and Observations*, and Wedgwood was causing a stir with his medallions of British greats, from Shakespeare to Garrick. At last, in October, Boulton swept into Redruth and toured the mines cancelling out Watt's sourness by a brave display of back-slapping and bonhomie.

Boulton's main task was to settle terms. As in the Midlands the engines were erected at the mine-owners' expense, and Boulton & Watt's cut came from a premium, a third of the annual cost of coal saved, payable for twenty-five years. (The amount used was computed by a special counter attached to the mine engines, invented by Whitehurst and improved by Watt.) Soon, however, it was agreed that the Cornish adventurers should not pay the full premium until their engines were working at full load and that a fixed sum should replace the variable and hard-to-work-out royalty.[15] Watt was bitter, but Boulton was firm: a hard bargain, he said, was a bad bargain. Image was a bankable asset: it paid to appear generous. Watt simmered, but at least the deal guaranteed some definite income. And before Boulton went home in early December, he scooped a further £2,000 loan from the normally cautious Truro bankers, Praed and Elliot, with the engine royalties as security.[16] The work was so intense that Watt stayed on alone, while Annie went home, pregnant with their son Gregory, born in 1779. But as Christmas approached Watt was almost cheerful. They had seen no profits yet, he told Joseph Black, and 'the struggles we have had with natural difficulties and with the ignorance prejudices and villanies of mankind have been very great but I hope are nearly come to an end or vanquished.'[17] Even Smeaton had begun recommending customers.

By now their fame had spread to France. Several different approaches were made, some direct, some covert.[18] In 1775, J. C. Perier, backed by the powerful House of Orleans, had the idea of providing a steam-powered water supply to Paris, and by 1778 had been authorized to order two engines. Meanwhile, Joseph Alcock (son of the Birmingham foundry-owner at la Charité) visited Birmingham and passed on news of Watt's engine to the French Government: the *Académiciens* Macquer and Marigny then recommended that Boulton & Watt be offered a monopoly *privilège* to erect engines in France for fifteen years. Negotiations dragged on, others became involved, and the Perier plans seemed shelved. It was a time of immense frustration, since their licence, an *arret de conseil*, had no force until an engine had been built and tested. Eventually an order came from Joseph Jary, owner of a coalmine near Nantes: Jary paid not a penny, but the engine was there to trumpet their name.

Cornwall was clearly going to be part of their lives for the next few years and, to make life easier, Boulton hunted for a house. He found one in the village of Cosgarne, in the sheltered Gwennap valley just to the south of the Redruth mining area. At the foot of this warm fold in the moors, trees overhung a clear stream, flashing with dragonflies. Cosgarne House stood at right-angles to the lane on a south-facing slope, a long low house with worn flagged floors and burnished panelling. Even Watt saw its charm:

> a most delightful place, a neat roomy house with sash windows double breadth, the front to the south covered with vines loaded with young grapes. A walled garden with excellent peaches and plums – plenty of currants – two orchards, a lawn before the door.[19]

By mid-1779 things were going better. Orders were coming in; Dudley had apologized for messing up the Tingtang engine; to Watt's delight, two wayward engineers had got the sack (the normally easy-going Boulton exploded that 'Sam Evans and young Perrins at Bedworth are two drunken, idle, stupid careless conceited rascals').[20] Briefly, in June, even their money worries seemed over when the arbitrators ruled in their favour in a long dispute with the Hawkesbury Colliery Company. Writing to Boulton, Watt positively capered:

> Dear Sir
> Hallelujah! Hallelujee!
> We have concluded with Hawkesbury – £217 – pr annum – from Lady Day last, £275.5 for time past, £117 our account.
> We make them a present of £100 guineas.
> Peace and good fellowship on earth –
> Perrins and Evans to be dismissed –
> 3 more Engines wanted in Cornwall –
> Dawdle repentant and amendant –
> Yours rejoicing
> James Watt[21]

Abashed, he then scribbled, 'Please burn this nonsense.'

His joy did not last. Although the income from the engines now seemed secure, the huge debts still made Watt tremble. He wrote agonized letters; he collapsed with headaches; he begged Boulton not to accept any more orders before Christmas: 'It is, in fact, impossible, – at least on my part; I am quite crushed.'[22] Stalled in his wrestling with the adventurers he begged Boulton to come down – and to bring waxed linen coats for the dreadful Cornish rain.

Boulton's greatest achievement was to find a good assistant for Watt. He was hopelessly unable to get on with the local engineers and erectors: they resented 'foreigners'; they were tough, hard bargainers; and in Watt's view they were brash, careless and frequently drunk. Repeatedly, he asked for engineers to be sacked and repeatedly Boulton calmed him – 'Have pity on them, bear with them, give them another trial.'[23] There was no point work grinding to a halt because 'perfect men were not to be had'.

They needed to find at least one perfect man. Thomas Wilson was a good agent but no engineer and was busy with his own affairs. They did have an engineering manager, a young ex-officer of the marines called Logan Henderson, who had learned about drainage and machines during a brief spell as a West Indies sugar-planter. But Henderson's easy-going ways alienated Watt, while Annie wrote an indignant letter about his reception of them (no maids, no saucepans, house full of people).[24] Instead of providing intellectual company for Watt, Henderson set up house with his mistress, the notorious 'Miss Peggy', and Watt morosely warned Boulton not to advance him cash, 'as if he continues to keep Miss P. he can scarcely avoid running into our debt'.[25]

William Murdoch

Neither Wilson nor Henderson could do the job Watt wanted. But in 1779 the problem was solved by the arrival of William Murdoch. Another Scot, he was the son of an Ayrshire millwright who worked on James Boswell's home estate of Auchinleck and was well known locally as the inventor of cast-iron gear for mills, cast for him by the Carron ironworks. Two years earlier, aged twenty-three, aware of Soho's fame, Murdoch had come south in search of work and his meeting with Boulton is enshrined in engineering mythology, courtesy of Samuel Smiles. As Boulton was questioning him, so Smiles's story goes, his eye was caught by the hat that Murdoch was twirling. 'It was not a felt hat, nor a cloth hat, nor a glazed hat; but it seemed to be painted and composed of some unusual material.' What was it made of:

'Timmer, sir' said Murdoch, modestly. 'Timmer? Do you mean to say that it is made of wood?' 'Deed it is, sir.' 'And pray how was it made?' 'I made it mysel, sir, on a bit laithey of my contrivin'.' 'Indeed!'[26]

A man who could turn an oval hat to size on a lathe of his own making was a rare talent and Boulton snapped him up. Murdoch worked first in the Soho pattern shop, then in casting and machining parts for the steam-engine, and supervising engine work at Bedworth and in Scotland. He became the only man Watt really trusted and in the autumn of 1779 he was sent to Cornwall, where he would stay for twenty-one years. Tall, strong, and a fine wrestler – capable of thrashing any hecklers – he worked until he dropped, dashing from mine to mine, and became immensely popular, even winning the friendship of the fiery Trevithicks.

With Murdoch in place Watt's burden was eased, and Boulton now set about solving their money problems. The state of affairs at Soho can be judged by the reaction of Keir. Early in 1777, needing someone on the spot while he was in Cornwall, Boulton had suggested that Keir take on the management of Boulton & Fothergill for a quarter of the profits.[27] Keir gave up his partnership in the Stourbridge glassworks, moved to Winson Green and began to examine the Soho books, only to find, as he put it with dry understatement, that 'considerable amendments in the economical part were requisite'.[28] He wrote forceful letters about cuts and improvements. But the more he learned, the worse the losses looked, and the idea of a partnership was dropped, although 'Mr B. continued to consult me about his affairs which he had the most sanguine hopes of re-establishing by the profits he expected from the fire-engine business.'[29]

Keir managed the Boulton & Fothergill business efficiently, undertaking special tasks such as making silk-reels for the East India Company (helped on the design by Whitehurst). Then in 1779 he took up a personal project in collaboration with Boulton, a new metallic alloy, containing copper, iron and zinc.[30] Boulton was immensely enthusiastic, informing the Earl of Sandwich, First Lord of the Admiralty, that since the alloy did not corrode in water it would be far better than copper for sheathing ships and for bolts and nails. He persuaded the Navy to undertake trials. Over the next three years it did so, before finally rejecting Keir's claims. Although he patented his alloy, instead of sheathing the nation's fleet his 'Eldorado' metal was used merely for window sashes – some remain at Soho House today.

After waiting two years Keir gave up hope of any partnership. He had already lost badly by giving up his Stourbridge glassworks without compensation and he needed a more lucrative job: it was now that he and Alexander Blair set up their chemical factory at Tipton. When this partnership began, Keir severed all formal links with Soho. It was a tense time. But here, as so often, Boulton's extraordinary charm saved their friendship, as Keir humorously acknowledged:

> I can make no doubt that Mr B. wished he had had it in his power to have made a different result, but I believe his difficulties & restraints were insurmountable. I confess that I was vexed that he was not sufficiently explicit to tell me but that soon passed, and a very sincere & hearty affection for him continued to the last.[31]

Boulton had already cut out his japanning line and dropped the ingenious 'mechanical painting' process invented by the painter Francis Egginton.[32] Everything now depended on the engine business but for this he needed Watt, and Watt was becoming almost incapable. He complained of headaches, rheumatism and stomach pains, and dosed himself with 'Dr James's Powders', stuffed with phosphate of lime and antimony, a known depressant.[33] The effect was disastrous. Like many depressives he slept for hours or sat motionless and miserable. 'I am plagued with the blues,' he wrote, 'my head is too much confused to do any brain work'; another time, bracing himself to go to a concert with Annie, he sighed, 'I am quite eat up with the mulligrubs.'[34] With patient desperation Boulton jollied him on, suggesting drily that he pray (like other Scotsmen) and sit back and wait to tot up his royalties in a year or so.

Sometimes even Boulton felt Cornwall was an exile. On his birthday in September 1780 he wrote to Anne, his 'dear Joan',

> So sure as there are 1728 Inches in a cubic Foot so sure was I born in that year & so sure as there are 52 weeks in the year, or, what you will remember, so sure as there are 52 cards in a pack, so sure am I 52 years old this very day and yet I fear you think so little of me that you will neither have a plumb pudding for your dinner nor drink my good health . . . I find a great deal of the old Derby about me, at least that part of his Character that I am ever uneasy asunder from my Joan.
>
> . . . Let me hear from you often & tell me all you know of my dear children who with yourself are objects of my constant prayer. I long to meet you; till then, farewell. Remember me who am yours & only yours M.B.[35]

Middle age made little difference to Boulton. His hair was touched with grey and his feet stabbed by gout, but he seemed astonished to be growing older. When he rode across Cornish moors in November winds and floods he admitted it was 'very disagreeable that one cannot stay out until dark upon the most emergent business without risking one's life'.[36] But by the fire at Cosgarne, the world righted itself: 'The greatest comfort I find here is being shut out from the world and the world from me. At the same time I have quite as much visiting as I could wish for.' He tinkered in his laboratory, collected fossils, and told Matt of his fruit trees and his mineral experiments, or his visit to Pendennis Castle with its view of the moors and the glittering sea, white with sails. He was amused by the Cornish people, briskly dividing them into demons and friends. The Quaker manager of Chacewater, Mr Phillips, was a great favourite and in years to come Boulton would go with the family to the huge Friends meeting at Truro, where Catherine Phillips preached, rising from her sick bed to do so. They built a great tent, he told his daughter, to hold many thousand people for this 'grand Fete a la Tremblant or rather what the junior Quakers consider it, Le Fete d'Amour as there is much Brotherly love going forward at these meetings & many matches are made'.[37]

Women other than grey-clad Quakers also caught his eye. Boulton, like Darwin, was fond of women. He even applied his charm to Annie Watt, sending her 'an extra kiss' in his letters, and was particularly fond of the beautiful Mrs Keir. Away from home, he may have strayed. Around 1782, in Cornwall, John Phillp was born, said by some to be Boulton's natural son. At fourteen Phillp travelled up from Falmouth to train as an artist and die-sinker at Soho. His letters to Boulton are affectionate and respectful, and his later drawings of the manufactory and estate would vividly evoke Soho in the 1790s.

But if Cornwall could please, it could also infuriate. Boulton's letters were rarely free from energetic, aggressive language. '*Grace au Dieu*! I neither want health, nor spirits, nor even flesh, for I grow fat,' he told Watt roundly, soon adding that somebody must be down in Cornwall all next summer: 'I shall be here myself for the greater part of it, for there will want more kicking than you can do . . .'[38] By now he and Watt were adventurers themselves. Partly to keep their market going, in 1780 Boulton had made them partners in at least four mines, bringing in

Wedgwood and Wilkinson as fellow investors.[39] Both these old allies now had their own interests in Cornwall: Wilkinson as a supplier of parts for the engines, Wedgwood as co-owner of leases on clay mines near St Austell, spied out for him by Watt, who received some of the profits. Boulton put backs up by trying to tighten the rough and ready Cornish accounting and by standing out against the harsh truck system, and by suggesting that meetings might go more smoothly if held before, rather than after, the drunken dinners. But the mine captains liked his style and a shaky truce was won.

It did not last long. The adventurers, who now saw their engines as their own property, resented the annual dues and threatened to petition Parliament to repeal the patent. Faced with accusations of monopoly, Watt was apoplectic:

> They say it is inconvenient for the mining industry to be burdened with the payment of engine dues; just as it is inconvenient for the person who wishes to get at my purse that I should keep my breeches pocket buttoned. It is doubtless also inconvenient for the man who wishes to get a slice of the squire's land that there should be a law tying it up by entail. Yet the squire's land has not been so much of his own making as the condensing engine has been of mine. He has only passively inherited his property, while the invention has been the product of my own active labour and of God knows how much anguish of mind and body.[40]

In time, this storm passed too, and in a year or so the profits began to trickle in. Boulton & Watt would erect twenty-one engines in the duchy in the first six years, and a further eighteen in the next four. The contrast between the partners could not have been sharper. Watt was pale, round-shouldered and anxious, thrifty and full of fears; Boulton robust and ruddy and loud, extravagant and incorrigibly hopeful. Yet the odd alliance worked: together, despite the battles, they conquered Cornwall.

25: DULL EARTH & SHINING STONES

One thing that kept Watt going in his Cornish exile was a new flow of letters from Joseph Black, who asked if he might lay out a few guineas on mineral specimens for him. As soon as he could, Watt sent a box by sea to Glasgow, full of the best tin ores, 'pickit out of many tons' and a variety of copper ores and other stones.[1] He also told Black of seeing cast iron softened by the greeny-blue 'vitriolic water' in the mines, 'so soft that I cut it with my knife as easy as cheese'.[2] It was still metallic, resembling black lead, 'very soft and unctuous to the touch and full of very bright white metallic particles' and when it dried it was said to become so hot that it could burn paper.

These letters show a different side to Watt, a mood of relaxed curiosity ruffled only when Black makes the mistake of describing new inventions – like a blowing engine for furnaces – in which case Watt inevitably retorts that he had invented it first. With minerals they are on safe ground. Black was a close friend of James Hutton (now developing his own theory of the formation of the earth) and he himself was now working on a variety of minerals and their reaction to sulphuric acid. He asked Watt especially for fluorite, since Cornish samples were more phosphorescent than those from Derbyshire, and for a curious grey copper ore that had small veins of a deep transparent red, like cinnabar. Britain had its own exotica. A couple of years later Black wrote, 'I am told there are found in Cornwall some Coarse Rubys Saphirs Hyacinths and Garnett in their native Crystallized State . . .'[3]

James Hutton, from Kay, 1836

Minerals were a universal passion among anyone with an interest in natural philosophy – and with money, since building up a collection could swallow fortunes, as Darwin's friend Charles Greville later found. (Greville is probably best known today for handing on his teenage mistress, Emma Hart – Nelson's later love – to his uncle Sir William Hamilton, in return for payment of his debts, but his collection of some fifteen thousand minerals became the core of the British Museum's collection.)

The Lunar group's own knowledge had developed and widened since their caving adventures ten years before. Their Derbyshire jaunts still continued: in October 1780 Darwin and Edgeworth would work together on the origin of the warm springs at Matlock, which Darwin was convinced were heated by deep volcanic fires. But their study of mineralogy was now far more systematic and methodical. It was becoming increasingly clear that the history of the earth could best be explored not through sweeping argument but through minute analysis. The first drive, once again, was towards classifying and naming. Here, as in the plant world, Linnaeus attempted a system of artificial classification, but this soon gave way to a more 'natural' system of classifying rocks on their placing within the earth's crust. Torbern Bergman, Professor of Chemistry at Uppsala, whose *Physical Description of the Earth* was translated into English in 1772, was the leading light in this field, developing a theory of the sequence in which strata were deposited: the lowest strata were often crystalline, lacking in fossils; those above had fewer crystals and tended to be sedimentary, with fossils in abundance.

Watt and Black knew Bergman's work well. And Watt's own geological skills had grown since his early surveying days. In 1777 he supplied notes and sketches to his fellow countryman John Williams, an engineer, antiquary and geologist, who had asked about the curious construction of the fort at Craig-patrick near Inverness, which Watt had studied during his canal surveys.[4] The local rock, Watt told Williams, was a granite mixed with quartz and the fort's walls had been made by burning this, melting the granite into a glassy cement around the hard quartz. In such inquiries the history of his native country, the ingenuity of his Scottish forebears and the formation of the land itself came together in a single image.

Wedgwood too remained fascinated by minerals. His book of 'Notes and Experiments' was dotted with questions: on the artificial crystallization of alabaster; on steatites or soaprock; on the formation of flints such as the beautiful 'Egyptian pebbles' with their irregular concentric rings, 'Nature's formation of some of her most beautiful products'. He jotted down tables of different earths in various stages of crystallization – siliceous, calcareous and gypseous – and followed up inquiries in dissolving and precipitating cobalt and nickel.[5] He paid vivid, keen attention to odd things that puzzled him. In one small notebook, with a picture of an Indian bee-eater on the front, he recorded his attempt to grow the greenish, translucent, moon-glinting crystals of selenite.[6]

Darwin was another devotee: the notes of his *Botanic Garden* are full of information on granite and moorstone, on natural crystals of nitre, 'in beautiful leafy and hairy forms'.[7] He was intrigued by Hutton's theories and came to believe in the great age of the earth, formed 'millions' of years ago. But he was also interested in Wedgwood's practical trials and was immensely cheered when one of his old favourite schemes, the horizontal windmill for grinding flint and enamels, at last saw the light of day. In April 1779, Whitehurst proposed some improvements, while 'Mr Watt and the philosophers of Birmingham' came over 'to criticise all the parts of the model'.[8] That summer Edgeworth helped him to build a proper scale model: 'I write this at the joiner's shop,' Darwin told Wedgwood happily, 'with the machine whirling before me.'[9]

Wedgwood himself was also in a practical mood. He read Priestley's *Experiments on Air* with great care, noting points in his commonplace book. One thing he fixed on was the 'great uncertainty' in some experimental results, due to their being made in a gun barrel ('which furnishes phlogiston as a result'), or in glass tubes, so liable to melt. 'These considerations', he wrote, 'have induced me to attempt the making of them in a kind of Crucible Composition.'[10] He now began to make ceramic equipment: retorts and crucibles, filters and tubes. His stoneware mortar was invaluable, stopping the danger of contamination (and even poisoning), which came with the old metal mortars, and avoiding the acid and oil damage associated with marble. After six months of tests, he demonstrated a model at the Apothecaries' Hall in London in July 1779. To begin with he gave this equipment out free to friends and fellow researchers such as Watt, who sent very precise requirements. But after Priestley described the new equipment in his writings innumerable requests arrived, and this too became a successful commercial line.

Wedgwood's greatest effort during the last seven years, however, had been devoted to his jasper. Since 1772 he had identified the materials he needed: the feldspar, moorstone and the elusive 'spath fusible', carbonate of barium. He finally discovered great masses in the lead mines near Matlock on an expedition with his father-in-law Richard in 1774 and with this he achieved a fine white body. But specimen after specimen cracked or blistered in the kiln: there was still a long way to go. 'Fate I suppose has decreed that we must go on,' he wrote wearily in May 1776. 'We must have our Hobby Horse, & mount him, & mount him again if he throws us ten times a day.'[11] It was not until late 1777, after five thousand experiments, that he could genuinely say, 'I am now ABSOLUTE in this precious article.'[12]

Geological knowledge and chemistry were crucial to the development of jasper, as they were in another set of experiments in 1776, in which Wedgwood worked with Keir. Wedgwood often used ground glass in his glazes and Keir suggested he use the raw materials of flint glass, giving him the recipe, rather than grinding the finished product. He also spent hours patiently teaching Wedgwood about annealing – the slow cooling of vitreous material.[13] Then it was Wedgwood's turn to try to help Keir, who was puzzled by the streaky veins or 'cords' that often appeared in his glass, making certain batches unfit for fine products such as achromatic lenses. Wedgwood set his mind to this, undertaking experiments at Etruria and at glass-makers in Liverpool and London. Keir had wondered if the streaks were caused by heavier particles moving through the molten glass, leaving a kind of wake.[14] Wedgwood disagreed. 'The waviness of flint glass', he wrote in his notebook, 'resembles that which arises when water and spirit of wine are first put together before they become perfectly united.'[15] He mixed fluids of different densities and decided that the cause – as he showed by having samples specially ground for him and testing them with light beams – was the different specific gravity of glass at different heights in the melting pot. Constant stirring should make it homogenous. This was a real discovery. Although the heavy duties on glass stopped him testing his ideas on a big scale he was twenty years ahead of his time in this conclusion.[16]

Keir's own paper on glass of 1776 was also a pioneering study on mineral crystallization. He had noticed that crystals formed when glass was cooled slowly and wondered if this might happen with all bodies slowly

'Fingal's Cave in Staffa', from Pennant's *Tour in Scotland*, 1776

turning from fluid to solid.[17] Noting the greater density of the crystals he asked whether the 'great native crystals of *basaltes*' in formations such as the Giant's Causeway (a site of perpetual fascination in the later eighteenth century), might have been created by the crystallization of vitreous lava. Keir's paper was cited as an early inspiration by Sir James Hall twenty years later, when he proved that basalts could indeed be melted to a glassy fluid and cooled back to a crystalline form.

Behind all these researches lay wide-ranging ideas about the structure of the earth from which the vital minerals came. Keir had linked his paper on crystallization to the work of Nicholas Demarest on volcanoes and basalt, while a note against one of Wedgwood's experiments reads, 'The granulating of this body in the fire may lead to some curious facts tending to confirm the idea of porphyry and granite being the production of volcanic fires.'[18]

Both men were taking a stand on a growing dispute. The debate about the role of volcanoes, which had begun to stir in the 1760s and would later develop into an intense argument between the 'Neptunists' (exemplified by the German chemist A. G. Werner), who believed that the earth once had a solid core, surrounded by water dense with chemicals that settled to form rocks, and the 'Plutonists' (led by Hutton), who judged from the way that horizontal layers ran over vertical ones that cycles of sedimentation had often been interrupted by violent periods of uplift caused by subterranean heat and pressure. In Hutton's view the earth was continually, slowly, being destroyed and re-formed, a process wheeling through the millennia, with 'no vestige of a beginning, – no prospect of an end' – a shocking, Bible-defying extension of time.[19]

Although Lunar opinions varied, most of the group were early Plutonists – and they believed that Whitehurst, the principal geologist among them, felt the same way. They had all looked forward eagerly to the publication of Whitehurst's *Inquiry into the Original State and Formation of the Earth*, which had been slowly gestating over the past fifteen years. Wedgwood, Boulton, Darwin, Day, Keir and Priestley all put their names down for copies when the book was announced in 1776, and the subscription list included Shelburne, James Adam, Banks, Henry Cavendish and John Wilkinson. But when at last it was published in February 1778, most of the author's friends felt a certain degree of consternation.

On the whole, British vulcanists such as Sir William Hamilton were cosmopolitan men with a secular, European mind-set. They wanted to clear away the mumbo-jumbo that decreed earthquakes and volcanoes to be instruments of God's wrath and to see them as purely natural phenomena, and to throw out old religious theories of creation. But John Whitehurst, the middle-class clockmaker made a Fellow of the Royal Society in 1779, was deeply religious. For years this had tormented him: how could he reconcile his scientific observations with his theology?

What he did was to split his book in two: an Argument and an Appendix. The main part of the book was a convoluted wrenching of new ideas on the history of the world into a pattern that could just about fit the old Mosaic account. Readers such as Wedgwood were dumbfounded. He liked the Appendix, with its careful studies of Derbyshire, but was, he said,

> fully perswaded his manuscript has undergone as many alterations since its first formation by the *fine philosopher* of Derby as his world has suffer'd by earthquakes,

& inundations . . . I own myself astonish'd beyond measure at the labour'd & repeated efforts to bring in & justify the mosaic account beyond all rhime or reason.[20]

It almost read like a parody. 'I should like to tumble a little of his world about his ears,' Wedgwood sighed in another letter, 'but I shall forebear, for I love the man.'[21] Besides, he knew it was easier to find fault with a system than to make one: he would read it through a few more times, before he spoke out.

It certainly had been tough for Whitehurst, who desperately tried to demonstrate that the operations of Nature revealed the same truths as Scripture. The earth was originally chaos, pulled into the shape of a spheroid, as Newton had proved, by the laws of gravity. The fusion of particles released air from the mass, forming a dense atmosphere. Water came next, completely surrounding the earth; then the pull of the moon and tides created islands on which plants and animals developed, in a primitive Edenic state. Meanwhile, sedimentary particles had formed concentric shells around the earth's core and had gradually solidified into rock. But at the same time great heat and fire was being generated by pressure *inside* the inner shell, its expansive force pushing out the strata. Cracks formed, water seeped into the core and the resulting steam exploded the ocean bed. Now came the Great Flood. As the molten lava met the sea it created a titanic explosion 'which tore the globe into millions of fragments'. The islands became the sea-bed, while the explosions created an infinity of gaping caverns into which the waters surged, 'and left the mountains and continents naked and exposed which had no existence prior to that era'.[22]

Whitehurst was old-fashioned in his emphasis on the Flood, on God as the original creator. Yet he was very modern in his refusal to admit any active cause of change other than the processes of Nature. Above all he challenged the old catastrophe theorists in insisting on the slow, progressive formation of strata, and in his stress on the role of subterraneous fire – his Deluge was a consequence of, not a primal cause of, violent change.[23] The most startling aspect was the contrast between his main argument and his Appendix. This had an entirely different tone, detailed and practical, emphasizing the value of detailed knowledge of strata, particularly to miners. Its content, too, seemed to undercut his own Argument. His carefully recorded observations included a study of fossils in various strata – some containing marine fossils, others remaining blank, others containing land fossils – that conflicted with his own theory of one major flood and inversion of land and sea. Moreover, it was clear that fossilized creatures found in different levels could not have existed at the same time, while his own findings in Derbyshire supported the view that volcanoes had played a major role in forming the European land-mass.

Whitehurst's drawings were the first published illustrations of the arrangement of strata. Most importantly, he produced a reliable theoretical guide to work out what rocks might lie beneath the younger ones on the surface, showing a regular succession. His work was revolutionary in juxtaposing grand theory and precise observation – but the fusion was distinctly uncomfortable. When he added more data to his second edition, including accounts of strata in North Wales and of the Giant's Causeway, it was all too obvious that the 'Plutonist' Appendix represented his true views.[24]

Boulton in particular saw mineralogy as a ripe field for Lunar science: a source of useful knowledge for metallurgy and pottery and the chemistry of gases. Putting off one of Day's many requests for repayment of his loan, in late 1780 he explained cheerily that he was just leaving for a couple of months in Cornwall, taking young Erasmus Darwin with him; perhaps he could send Day some geological specimens as a basis for a natural history museum? Day replied testily that he was too busy in his own unrewarding pursuit, 'the study of man', to turn to minerals. But, he added pointedly (having not got his money), he was very happy upon Erasmus's account, as knowledge was always useful:

> and besides the science of mineralogy, there are two others for which he could not possibly find an abler master than yourself, the knowledge of the world, and the art of saying handsome things; I hope by some expressions in your letter, you have by this time become an adept in a third, le moyen d'y parvenir, vulgarly called the art of getting a fortune.[25]

The fortune was still some way off, but the mineral passion was present and lasting.

Collecting, identifying and testing became a leading project. In October 1781, Wedgwood was in Cornwall 'hunting clays & soap rocks, Cobalts, &c' with Watt, who lost a day and a half's work by his visit but wrote drily, 'nevertheless I dont grudge that, as I am glad to see a Christian'.[26] By now

Boulton had told Logan Henderson eagerly that he had got Bergman's last volume translated, and he had his own collection of minerals, many garnered on 'fossiling' hunts with Wedgwood. One of these, a glassy, milky-white mineral with pyramidal crystals, found either in Cumberland or Cheshire, caught Withering's attention. Peering at it closely, he was sure he had found a 'terra ponderosa aerata' (barium carbonate), which Bergman had said was not found in its native state, but only as barytes (barium sulphate, or 'heavy spar'). Withering's experiments proved him wrong. Two years on, he published his findings and much later Werner listed this as 'witherite'.[27]

Withering had already done experiments on mineral waters and wells and on local toadstone and ragstone,[28] and his translation of Bergman's mineral classification appeared in 1783 as *An Outline of Mineralogy*. The friends' investigations often overlapped. In 1783, for example, Withering, Priestley and Wedgwood were all considering the puzzle of the strange mineral 'Black Wadd', or black woad, a powerfully dangerous form of manganese dioxide, often blended with lamp-black and mixed with linseed oil to make paint, which sometimes inexplicably flared into flame. Gradually, almost all the group won a reputation in this field.[29] The shining stones and dull rocks in their cabinets, with their carefully written labels, were all part of the pattern they sought, all clues to the mysterious workings beneath the crust of the natural world.

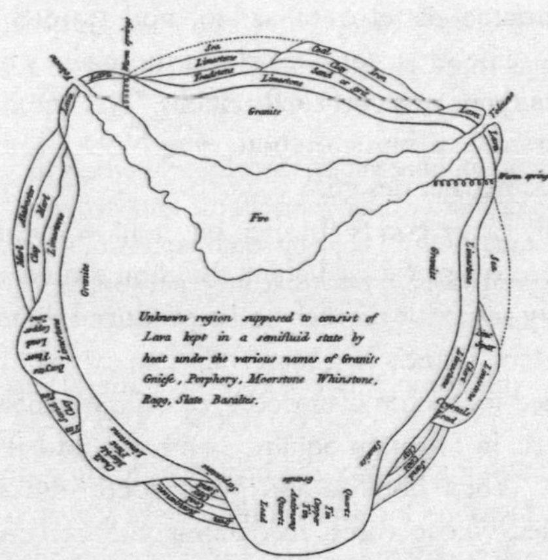

Darwin's 'Section of the Earth', *The Economy of Vegetation*, 1791

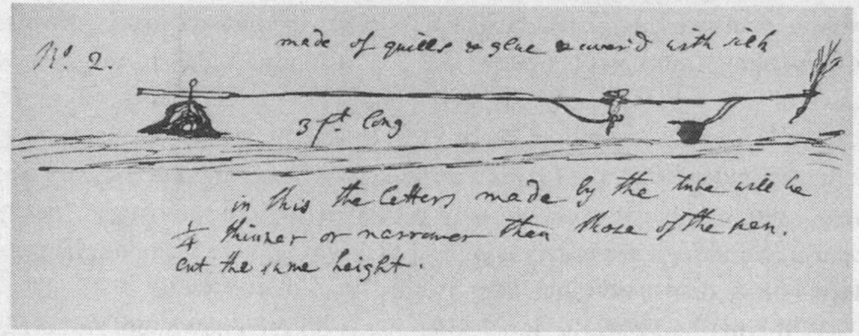

Darwin's early sketch for a bigrapher, Commonplace Book, 1777

26: CREATIVE COPYING

'Man', Darwin wrote in *Zoonomia*, 'is termed by Aristotle an imitative animal; this propensity to imitation not only appears in the actions of children, but in all the customs and fashions of the world.'[1] Copying, in its innumerable forms, displays the very mind-set of this age. Experiments were judged by the principle that they should be repeatable; botanists and zoologists studied replication and variation; teachers saw the urge to 'imitation' as the drive to learning; manufacturers valued their machines for their predictable patterns of action, and trained their workers to resemble them. Fine artists argued about the virtues of imitation but in the applied arts there were no such qualms. The standardizing of styles, whether it be transfer printing on pottery or block printing on calico, saw the start of mass-market design.

This not only made goods cheaper, but it allowed the middling classes to imitate the lifestyles of their betters. Boulton's silversmiths made some designs in heavy silver for the rich and reproduced them in Sheffield plate for the less affluent. Francis Egginton's mechanical paintings were a similar wheeze, designed for niches over doors, or ceiling panels – as in Elizabeth Montagu's house in Portman Square – with the added cachet of novelty and cheapness. When Boulton and Wedgwood addressed the powerful men and women whose vases, medallions and ornaments they copied, they claimed only to imitate the taste of the past in order to raise the standards of the present, but when Wedgwood advertised his portrait medallions, he suggested that orders from wealthy patrons would not only ensure that their image was 'copied' for posterity, but encourage the arts themselves:

> If the nobility and gentry should please to encourage this design, they will not only procure to themselves *everlasting portraits*, but have the pleasure of giving life and vigour to the arts of modelling and engraving. The art of making durable copies, at a small expense, will thus promote the art of making originals . . .
>
> Nothing can contribute more effectually to diffuse a good taste through the arts than the power of multiplying copies of fine things, in materials fit to be applied for ornaments; by which means the public eye is instructed; good and bad works are nicely distinguished, and all the arts receive improvement.[2]

Mechanical inventions were no good unless they could be copied, and even the most wayward and original among the Lunar men, such as Darwin, became obsessed with this idea.

The late 1770s were emotional years for Darwin, with the grim shock of Charles's death and his burgeoning love for Elizabeth Pole. In the autumn of 1778, Elizabeth became dangerously ill and Darwin dashed to Radburn to treat her, spending the night in the grounds, watching the lights in her window. In his panic he wrote more poems, one a dark Petrarchan lyric describing his 'dusky dream' of Eliza's death, another a cheerful ode on her recovery, addressed to the river Derwent, tumbling down from Matlock towards her, carrying the burden of his bleeding heart.

As a counter to grief and middle-aged passion, Darwin had his commonplace book, a resource and a retreat. This heavy notebook was stuffed with ideas, including notes on the 'Wheel of Offyrent', a puzzling perpetual-motion machine. There was a sudden craze for such things at the end of the 1770s, and even the sober Watt tinkered with the idea. ('I am persuaded it is some deception or mistake,' wrote Black anxiously – he thought it more likely to be something like the electrical jack that Franklin had invented for roasting fowls.[3]) Between medical case notes Darwin pondered on everything from the 5-inch worms in cats to the function of the placenta or an amazing dried cucumber, which 'burnt by a spark like tinder'. In between, he worked out a host of often brilliant mechanical inventions, from telescopic candlesticks to new stocking-frames, from locks and lamps to an artificial bird with wings that would flap briskly up and down – like a bat, or a lady's fan. At first he thought this might be powered by a watch spring, set off by tiny gunpowder charges, but decided a simpler way would be to use a compressed-air bottle, 'a copper globe fill'd with condensed air' – a kind of primitive guided missile.[4] Some of his designs over the next few years were pure fun, like the 'factitious spider' with metal legs, which could be drawn across a silver plate by a hidden magnet, to make friends laugh and children squeal. Some came to nothing, like his many-bladed steam-wheel. Others were prompted by his medical work, including a widely acclaimed generator, an 'electrical doubler'.[5]

These inventions were not copies but utter Darwinian originals of real excellence and promise, and all these were for interest, rather than for profit, but in 1777 he began flirting with the idea of a mechanical copying machine, or 'bigrapher', which he thought *might* make some money. In the earliest version, the quill of the writer is linked by a long arm to another pen, held on a swivel, which repeats the action on a separate piece of paper. The really tricky aspect was designing a balance to raise the pen at the end of each word and bring it down smoothly at the next. The writing looks rather clumsy, spidery and narrow – but the machine does work once you get the knack.[6] Then he developed a far better design, which reproduced the writing so exactly that one could hardly tell which was the copy and which the original. Since a patent was expensive, Darwin entered a caveat to prevent anyone else taking one out for a similar scheme, but his friend Charles Greville, soon destined to become Lord of Trade, urged him to develop it. As a reward he received the world's first duplicated letter, in December 1778 (full of Darwin's fulminations about the looming war with France).

The following May Darwin felt his machine was as ready as it ever would be. In June he dismantled it and sent it to London, hoping Greville would find someone willing to market it commercially. Greville might have followed the nightmarishly complex instructions for re-assembly – but probably not, as nothing was heard of it again. Meanwhile, however, it had been an obvious thing to show off at Lunar meetings, perhaps to lure Boulton to invest. Boulton kept quiet, but this spring, Watt borrowed Darwin's machine and promptly worked on one of his own. Part of his frustration in Cornwall came from having to redo innumerable sets of instructions for every machine erected; but he never liked to follow another inventor. He had to take a different tack, as he did here. The original letter had to be written in a special ink, thickened with sugar or gum (Keir worked hard on finding the right formula for this), and a sheet of dampened, unsized tissue paper was then pressed on with rollers. This produced a mirror image, but the paper was so thin that the text could be read – and it can be to this day – through the back of the sheet. Darwin's machine produced far clearer copies, and used ordinary paper. Yet in early May, just as Darwin was telling Greville his machine was ready, Watt wrote to him triumphantly:

> I have fallen on a way of copying writing *chemically* which beats your bigrapher hollow. I can copy a whole-sheet letter in 5 minutes. I send a copy of the other page inclosed for your conviction, and I tell you farther that I can do still better than that copy.[7]

In June, when Darwin was waving his packing cases off to London, Watt was already reporting to Boulton: 'I send you enclosed some of Mr Nobody's draughts with authentic copies of them.'[8]

Watt designed two kinds of presses, one worked by rollers and the other by screwing down the plate. Boulton stumped up for the patent and with Boulton and Keir as partners, 'James Watt & Co.' was set up in 1780. Boulton took the press to London, showed it to businessmen and MPs and demonstrated it to a crowded audience at the Royal Society. Everyone was enthusiastic, apart from nervous bankers imagining forgery: Boulton's agent told him that 'the bankers mob him for having anything to do with it; they say that it ought to be suppressed'.[9] His best moment came at a coffee-house with Smeaton and other engineers, 'when a gentleman (not knowing me) exclaimed against the copying machine and wished *the inventor was hanged*, and the machines all burnt, which brought on a laugh, as I was known to most present'.[10] By the end of the year, 150 presses had been sold. Keir ran the business well and versions of Watt's copier were used in offices in Britain and abroad until the advent of carbon paper swept them away.

It was Watt, at the end of his life, who took the obsession with copying to a surreal conclusion. In his late seventies, with Murdoch's help, he built two machines for copying sculpture, one that would make reduced-scale copies and the other equal-sized. He had no artistic skill at all – but after all, this was the man who in his youth had made a perfect organ although he was tone-deaf. In February 1811, he tackled 'Sappho', cementing the bust to moveable plates, and roughening the stone 'with the tearing-drills to within the thickness of a halfpenny of the truth':

			Hour
Saturday	2	Doing the face with the 1-8th drill to the truth, from the outer corner of one eye to do. of the other (went too slow)	5
	3	Doing her breast with do.	1
Monday	4	Do. one side of the head	4

Altogether, this masterpiece took 39 hours. Watt might have lacked confidence, but never self-esteem – one of his most successful efforts, copied from a cast by Chantrey, was apparently the bust of himself.

27: SONS & DAUGHTERS

If learning was 'copying', then what pattern, the Lunar men asked, should their own children follow? In 1779, as well as working on the horizontal windmill, Wedgwood and Darwin also had this question very much on their minds (though Wedgwood felt he was still educating himself, confessing that Sally told him that 'I must buy no more books 'till I build another house and advises me to first read some of those I already have'[1]).

In July this year the Wedgwood family day was full:

We sally forth, half a dozen of us, by 6 or 7 o'clock in a morning, and return to breakfast with appetites scarcely to be satisfied. Then we are very busy in our hay, and we have just made a new garden. Sometimes we try experiments, then read, and draw a little, that altogether we are very busy folks . . . Poor Sukey is quite out of patience with her old Spinet and often asks me when the new one will come.[2]

By August the weather had changed. Floods swamped the Midlands. Wedgwood, stacking his 'poor daggled hay', thought it hardly worth the effort. Rain was followed by blistering heat, '*dog days hot*' and in the midst of the heatwave the Wedgwoods' last child, Mary Anne, was born.[3] The family were bowling and Sally made them tea and sent for the midwife without telling anyone – 'Slipt upstairs just before supper, and we had not risen from the table before the joyfull tidings of a safe delivery, and all well was brought down to us.'[4] When the baby was brought down to the dining room Wedgwood stoutly declared that Sally was staying upstairs

Silhouette of Elizabeth Pole with her dog

only as part of a feminine conspiracy to impose on men, 'and make us believe what sufferings they underwent for us and our bantlings'; he was trying to persuade her that 'this farce will no longer pass upon us in this enlightened age'.[5] She too came down the next day and soon Wedgwood and the older children were arranging their new shell collection, which taught them all 'something of arrangement in general, & induces us to read a good deal of french without the formality of saying lessons'.[6]

The Wedgwood children were like a ladder of the years. Sukey was now fourteen, John thirteen, Josiah ten, Tom eight, Catherine five and Sarah four. The only cloud was the plight of Mary Anne. She suffered convulsions when teething which left her partly paralysed, and Darwin was called, as he was in every Wedgwood crisis, and launched all his armoury on the poor baby. Two or three times a day, for several weeks, she was given electric shocks on her paralysed side; then her gums were lanced; then the drugs were administered – up to seven drops of laudanum twice a day. The convulsions simply became worse, and Darwin more adamant: ' I am clear of the opinion', he told her father, 'that, whenever these fits recur, the exhibition of opium, so as to induce intoxication, is the only method, which can prevent (if any can) the sensibility of her mind from being impaired.'[7] In fact Mary Anne did suffer some sort of brain damage. When she was about seven her older sister Sarah reported touchingly that she liked bathing, and 'cryed out distinctly mamamama & smiled but she said gna gna when I asked her if she would send her love to you & when I asked her if she wanted to see you'.[8] She died in 1786, aged eight.

In 1777 John (known as Jack), Jos and Tom were at boarding school in Bolton, while Sukey was sent south to London. But in early 1779 the lecturer John Warltire, Priestley's friend, was in the Potteries and Wedgwood arranged for him to give some private demonstrations. Jack came back from Bolton to hear him, and Robert Darwin came to stay. He was now nearly fourteen. He had always adored his elder brother Charles, who had sent curious pebbles from Edinburgh, asked after his fossil cabinet and despatched a 'specimen of the Cabbage Tree Bark for the Medecin Cabinet'.[9] When Charles died, Darwin was adamant in his own mind that Robert would take up his mantle. The problem was that Robert simply wasn't interested in medicine or mechanics: he never wanted to be a doctor and was laid low with headaches and depression. The Lunar enthusiasms could be too much for him, as Susanna Keir recognized when she invited him to stay, kindly promising, 'Mr Bob, if he will come to Winson Green, that he shall have the pleasure of Soho, without being lectured.'[10]

Luckily Robert had always hero-worshipped Wedgwood and breathed easily in the easy-going Etruria household. Continuing his informal teaching, Wedgwood talked on clays, copied down by Darwin in his commonplace book as 'spoken extempore by Mr W. and taken down from his mouth', and developed a neat system of colour coding for simple and complex elements and chemical changes. There was hardly time enough in the day: 'The boys drink in knowledge like water, with great avidity . . . Jack is very deep in chemical affinities, and I have no fear of his making a tolerable progress in the science, for it is much pleasanter to him than grammar.'[11]

In the autumn Jos was ill and when Wedgwood fetched him from Bolton he rode through violent mobs in the cotton towns. He then decided to take all the boys out of school, where 'They have more business, confinement, and phlogisticated air than their machinary can dispense with.'[12] He was prompted both by concern for their health and by the need to provide an education more suited to their future lives. A knowledge of classics, he had come to feel, was largely ornamental, useful only for sprinkling conversation with mottos and quotations. Darwin agreed, declaring it was 'a very idle waste of time' for boys destined for trade to learn Latin; they would be far better off doing French and accounts. Wedgwood had his boys' futures all mapped out: Jack would be an 'improving' gentleman farmer, leaving Jos and Tom to be potters: 'Tom to be the traveler & negociator, & Joss the manufacturer.'[13] Boulton and Watt were similarly single-minded, determined that Matt Boulton would take over Soho, Jimmy and Gregory Watt would be engineers. For the last two years Matt had been a day boy at a small academy near Birmingham, but in 1779 his teacher advised Boulton to make him a boarder so that he would not pick up the 'vicious pronunciation & vulgar dialect' of the local day-boys.[14] The following year he went to school in Twickenham, to polish him still further. Jimmy Watt, by contrast, left a local school at fourteen and went straight to Wilkinson's foundry at Bersham to learn machine-drawing and carpentry.

While the Soho men sent their sons away, Wedgwood kept his at home. In the autumn of 1779 his three girls were also at Etruria, Sukey sadly needing cheering after the death from tuberculosis of her cousin Kitty Willet. Wedgwood now included his daughters in his schemes and by November had worked out a regular schoolday. His accountant, Peter Swift, could be one teacher (Sukey joined in doing accounts) and Tom Byerley and a local Unitarian minister, Mr Lomas, could teach some basic Latin. In December, the Wedgwood boys went to Lichfield, where a prisoner of war, Monsieur Potet, taught them French and drawing. 'Your little boys are very good', wrote Darwin, 'and I hope you will let them stay, till we write you word we are tired of them.'[15] Then the 'Etruscan' school moved back to the Potteries, taking Potet with them. Wedgwood worked out a detailed curriculum – English, French and drawing, a little Latin, accounts, riding and exercise, 'which will include gardening, Fossiling, experimenting etc. etc.'.[16] The girls had extra English lessons, Kitty 'as much French as she can bear', and instead of 'Fossiling' Sukey spent extra time on her music.

Radical as he was in some ways, in this respect Wedgwood's views on the needs of boys and girls were solidly conventional. A year earlier, when he was planning to commission Stubbs to paint two companion pieces, he put this in almost cartoon-like images. One picture would show his daughters, the other his sons:

> Sukey playing on her harpsichord, with Kitty singing to her which she often does, and Sally and Mary Ann upon the carpet in some employment suitable to their ages. This to be one picture. The pendant to be Jack standing at a table making fixable air with his glass apparatus etc.; and his two brothers accompanying him. Tom jumping up and clapping his hands in joy and surprise at seeing the stream of bubbles rise up just as Jack has put a little chalk to the acid. Joss with the chemical Dictionary before him in a thoughtful mood, which actions will be exactly descriptive of their respective characters.

(Sukey got her own back on her father, declaring airily a few years later that she could not write a good letter because she was 'surrounded by gauze, Ribbon &c: which naturally lead young ladies to think of balls, dancing &c – which are very flighty things & as unintelligible to you, as canals and improvements upon the River Trent are to me –'.)[17]

On the whole, women were not excluded from natural philosophy. Annie Watt knew enough chemistry to receive technical letters from her husband, especially on her family's trade of bleaching; Sally helped Wedgwood with his experiment books; Mary Priestley put up happily with keeping mice warm on the mantelpiece and providing laundry tubs for experiments; Edgeworth shared all his interests with Honora.[18] All the Lunar men encouraged their daughters to study botany and chemistry and mechanics. Edgeworth and Darwin both stressed this when they wrote on education, and years later Keir wrote a 'Dialogue on Chemistry between a father and daughter'. Mary Anne Galton, whose father Samuel joined the Lunar group in the 1780s, remembered that he 'wished me to be a philosopher; and used to instruct me in the rudiments of science':

> I had a pretty little monkey called jack, a dog, a cat, a rabbit, and other animals. It was my delight to hear my father explain the Linnean Orders; and to have him show me the teeth and claws of my various pets, classifying them, from the Primate Jack to the Brutum Sus.[19]

Her mother too would sit her on her knee and read Buffon's *Natural History*.

So the girls grew up in houses full of instruments and inventions and experiments, but the difference was that for women this could be only a hobby. While the Lunar men saw their sons as potential captains of industry or engineers or chemists or doctors, they saw their girls plainly as future wives and mothers – almost a more vital role. It was this attitude that lay behind Whitehurst's remark, when he was hunting on Boulton's behalf for a boarding school for Anne in London, that girls' education 'appears to me of more real importance than that of boys'.[20] They did not think girls had different innate capacities or that their subordinate position was fixed by God. But in subtle ways their scientific culture worked to reinforce custom. There was a drive, at the time, to identify a *natural* order underlying the social; comparative anatomical studies had suggested – surprise surprise – that women were 'formed' to bear & suckle children; new notions of 'instinct' talked up the maternal role. All this encouraged the ideas of complementarity that peaked in writers such as Hannah More: 'Women have generally quicker perception; men have juster sentiments . . . Women often speak to shine or to please; men, to convince or confute. Women admire what is brilliant, men what is solid.'[21] And so on.

The Lunar men were not immune to this, nor to set ideas about class. In the early 1790s, Darwin showed this even with regard to his own family, when he was planning a future for his daughters Mary and Susan Parker. In 1791 he wrote that with an annuity from him, and

> some employment as Lady's Maid or teacher of work they may be happier than my other girls, who will have not much more than double or treble that sum and brought up in genteel life – for I think that happiness consists much in being *well* in one's situation in life, and not in that situation being higher or lower.[22]

This was tough realism. But the Parker girls *were* happy: after they had taught in local boarding schools, Darwin set them up in a school of their own in a square brick house bordering on Brooke Boothby's estate in Ashbourne, Derbyshire. When it opened in June 1794, with their half-sisters Violetta and Emma among the first boarders, Darwin wrote them a prospectus, *A Plan for the Conduct of Female Education in Boarding Schools.* This incorporated all the latest thinking (including Whitehurst's views on ventilation), stressing the cultivation of gentleness, tenderness, sympathy and 'the charms which enchant all hearts!', while offering a broad curriculum, with a good dose of natural philosophy and botany.[23] The *Plan* placed Darwin firmly in the strata that Mary Wollstonecraft inhabited – at once progressive and fiercely supportive of women's right to education, yet socially conservative.[24]

The Lunar fathers saw education as linked to future roles more than to individual fulfilment. James Keir later put this strongly in his memoir of Day. To Rousseau, Keir noted, society was an 'unnatural state in which all the genuine worth of the human species is perverted' so he recommended that children be kept separate, unpolluted by its artifice. Keir thought this absurd as well as impractical: an education

> which has not society in view must be defective, not only in that instruction which ought to explain our duties and relations, but also in the acquisition of our most important habits, particularly that of controlling our selfish impulses for the sake of general order and happiness.[25]

While Wedgwood was rearranging desks and harpsichords in Etruria that erstwhile Rousseauian, Richard Edgeworth, was also grappling with a growing family. The two men were constantly in touch: Sukey stayed with the Edgeworths over Christmas in 1777, and had been very happy there:

> It is impossible to be otherwise in Mrs Edgeworth's company. I think Mr E. is very different to what I always thought him to be. I took him to be a very grave sedate man & now I think him to be just the contrary, he sings all the day from morning till night.[26]

Edgeworth was a natural educator, brimming with ideas about which he was 'neither bossy nor proprietorial', and a great reader, who could quote literature with enthusiasm from the Greeks to the present day.[27] He was also deeply intrigued by how children learned, starting with the acquisition of language, and was fascinated by the 'wild boy', Peter of Hanover. Peter had been discovered abandoned in a forest and lived for some time near the Edgeworths in Hertfordshire where 'we had many opportunities of trying experiments upon him'.[28] Although he could say a few phrases (including 'King George', accompanied by an imitation of the coronation bells), he never learned to talk properly; 'though his head, as Mr Wedgwood and many others remarked, resembled that of Socrates, he was an idiot'. Edgeworth visited him in 1779, just when Wedgwood and Darwin were setting up their school, and found his plight a sad argument against Rousseau's disdain of words and praise of the 'natural' life, especially when set against his tender observations of his own family.

By now the three oldest children of Edgeworth's first marriage, Richard, Maria and Emmeline, were all – more or less unhappily – at boarding school: this year Richard went to sea, a restless soul all his short life. At the family home at Northchurch Anna (the fourth and last surviving child of Anna Maria) was five, Honora was four and Lovell was three. Trying to teach the girls to read, Edgeworth looked through over forty books supposedly for children, but could find nothing until he came across *Lessons for Children from Two to Three Years Old*, by Anna Laetitia Barbauld, the daughter of Priestley's Warrington friend, Dr Aikin. (She too was a disciple of Rousseau, and had made a disastrous marriage to a Frenchman, which her brother blamed entirely on her reading of *La Nouvelle Heloïse*.) Responding to Barbauld's simple vocabulary and down-to-earth tales, Edgeworth's little girls learned to read in six weeks. He was thrilled, and by December 1779 Wedgwood too was a fan, telling Bentley, 'I should like to have Mrs Barbauld's books (all of them) for children. Our little lasses are quite out of *proper* books to say their lessons in.'[29]

He also reported the Edgeworths' latest plan, to write a succession of books for children themselves. Edgeworth was inspired both by Barbauld and by Priestley's abridgement of Hartley's *Observations on Man*.[30] This seemed to provide the theoretical justification for an entirely new approach to teaching small children: he and Honora, he decided, would work out a new system based on ordering and arranging the association of impressions, feelings and ideas from the earliest days. Edgeworth later claimed that Honora was the first to establish that 'the art of education should be considered as an experimental science'.[31] The first step was to ask some fundamental questions: What could children take in at different ages? What would they remember? How did learning take place in the mind? As in any new field – whether it be chemistry or volcanoes – the collecting of facts had to come before the decreeing of any system, and so Honora began a notebook charting the children's reactions, hoping to collect a body of evidence on the impact of different kinds of teaching.

The Edgeworths believed that children's books should be fun, to keep their attention, with crystal-clear and familiar subjects. They planned a programme of lessons for children from four to ten and, as a start, Honora wrote one story called 'Harry and Lucy'. Sending this to Priestley in early 1780, Edgeworth asked if she might dedicate it to him, as it was prompted by his writing and added that she hoped with the help of friends, to introduce into her work 'the first principles of many sciences or rather the facts upon which those principles are founded'.[32] This ambitious programme would finally be completed by Maria Edgeworth's *Early Lessons* in 1801, where 'science' is introduced from the start, as Harry and Lucy, watching their father shave, receive a lesson on heat and the use of the thermometer. In 1780, however, the Edgeworths' plan was abruptly halted by tragedy.

The illness that had brought them back to Honora's family in Lichfield in the spring of 1779 had now developed into consumption. She had shown symptoms of tuberculosis at fifteen, and her mother and three sisters had died of the disease. Edgeworth feared the worst. Through the summer and autumn he watched her weaken while Darwin tended her cautiously, but daily losing hope. From Day's house in Hertfordshire she went to consult the famous London specialist Dr Heberden, who gently broke the news that her case was hopeless. Back at Lichfield she seemed to revive, and Edgeworth took a small house across the border in Shropshire, but she was sinking fast. Edgeworth sat with her nightly. At the end of April he was called up to her room and he found her in violent convulsions. He felt her pulse and whispered, 'You are not dying,' but she begged him to sit with her and he took out his pencil, partly to note what she said but even more to have something to do, as a barrier to grief.[33] Next day she dozed and talked of the children and 'recommended it to me in the strongest manner to marry her sister Elizabeth'. All the next night, he sat up, waiting. On May Day, at six in the morning, her sister Charlotte called him. Propped on the pillows, her arm flung round Charlotte's shoulders, Honora turned to look at him with 'utmost tenderness'.

> At this moment I heard something fall on the floor. It was her wedding ring, which she had held on her wasted finger to the last instant – remembering, with fond superstition, the vow she had made, never again to lose that ring but with life. She never moved again, nor did she seem to suffer any struggle.[34]

Edgeworth's plain account, though shot through with emotion, is no Victorian deathbed scene. No tears or kneeling or prayers, no heavenly light, or angels, or afterlife, or duty. Instead Honora was practical, defiant of convention. But Edgeworth's lively spirits were doused; he could not speak about his feelings. He wrote to Wedgwood:

> Let me address this letter to the firm of W & B to ask whether I can have 12 profiles of my dear Mrs Edgeworth done in white on pale blue from a profile by Mrs Harrington & an excellent picture by Smart – I lost her Sunday – & you both know she is a real loss to
> yr. frd. R.L.E.[35]

A few days after the funeral, numbed and vacant, he went to Lichfield, and then took his two youngest daughters, Anna and Honora, to stay with Day in Essex. He listened and replied mechanically, he tried to read, he appreciated Day's kindness, but felt nothing except exhaustion.

Against his own inclinations he followed Honora's tough advice, returned to Lichfield and proposed to Elizabeth – Day's old love, who had sent him to France to learn the graces. She was 'distinctly surprised' to put it mildly, declaring he 'was the last man of her acquaintance, that *she* should have thought of for a husband'.[36] But that summer, worried that

Elizabeth was showing similar symptoms to Honora, Darwin suggested the family try sea bathing and Edgeworth went with the Sneyds to Scarborough. By the end of the stay his proposal had been accepted. The conservative forces of Lichfield Close reacted with horror. Major Sneyd and Canon Seward both objected; Edgeworth took on the Bishop of Lichfield publicly, with articles in the Birmingham papers, replies and rejoinders and much abuse – in which Edgeworth turned miserably to Boulton, who had also married his wife's sister in the teeth of opposition. At first the couple fled to Cheshire, but the Bishop forbade any clergyman in his diocese to marry them. Wedgwood heard that they had gone to Scotland but in fact Elizabeth retreated to Bath while Edgeworth took his little girls to London where he had the banns read in St Andrew's Church, Holborn. They were married there on Christmas Day 1780, with Thomas Day and Paul Elers (Edgeworth's first wife's brother) as witnesses.

This was a marriage of convenience, for the sake of the children, but although Edgeworth's affection never matched the intensity of his love for Honora, it was a successful match, emotionally and sexually: over the next fourteen years nine children were born, all but one of whom survived.

In 1780, the stir caused in Lichfield by this romance was matched by Anna Seward's sudden burst into fame with her *Elegy on Captain Cook*, killed in Hawaii the previous year. Most of Seward's poem, claimed Edgeworth, was actually written by Darwin – both he and Brooke Boothby swore they had heard him read it at the literary society, when Anna was present. But Darwin made no claims and worked hard to prevent a bad notice in the *Monthly Review*.

Darwin was undoubtedly flirting with Anna. In November they engaged in a whimsical feline correspondence, in which his cat, the Persian Snow Grimalkin, sent passionate addresses to Anna's Miss Po Felina, 'Dear Miss pussy', seen washing her 'beautiful round face' and 'whisking about with graceful sinuosity your meandering tail'.[37] Anna replied in kind – and at length. But this correspondence was, perhaps, a smoke-screen for his real love for Elizabeth Pole. On 27 November 1780, aged sixty-three, Colonel Pole died. What hope did Darwin have? Elizabeth was thirty-three, dashing, witty and rich: half the young men in the county were after her.

Darwin was forty-nine, stout, stuttering and lame, with two grown sons, and two illegitimate daughters. Early in her widowhood, said Anna, Elizabeth was teased about him in company, and asked what she would do with her 'captive philosopher':

'He is not very fond of churches, I believe, and if he would go there for my sake, I shall scarcely follow him. He is too old for me' – 'Nay madam, what are fifteen years on the right side?' She replied, with an arch smile, 'I have had so *much* of that right side.'[38]

Yet in the end Elizabeth Pole plumped for Darwin, laying down one hard condition – that he leave Lichfield and live at Radburn. After they were married at Radburn Church on 6 March 1781, Erasmus and his children left the house he had built so proudly and the town where he had been so prominent. Elizabeth whirled him out of his routine and within weeks the portly doctor was spending six weeks in London, his first visit for nearly thirty years. Erasmus junior moved to Derby to begin practising law, and the Radburn household included the three Pole children, as well as fifteen-year-old Robert Darwin and Susan and Mary Parker, now nine and seven. (Their mother Mary married a Birmingham merchant the following year but stayed close to Darwin and the girls.) In the spring of 1782 Darwin and Elizabeth had their first child, Edward: three more boys and three girls would follow in the next eight years.[39]

For Joseph Priestley, too, 1780 brought changes. All summer he was ill, probably with the first of several attacks of gallstones. This year, perhaps feeling that his religious work might benefit from greater independence, he left Shelburne's employment. Supporters rallied to fund his research, and his brother-in-law John Wilkinson found him a large house (belonging to the Quaker banker Sampson Lloyd) at Fair Hill, Birmingham. The family moved there in September 1780. It had everything Priestley needed except a laboratory, which he duly built.

At Fair Hill, Priestley was always keen to encourage the children around him. He let them play with the second-hand air-gun the lecturer Adam Walker had found for him, firing it occasionally and loading it for the boys, 'which he pleasantly did, though he seemed very much occupied in his laboratory, which was close at hand' (they particularly liked shooting at the wig block used to dress Priestley's wig, which they put on the coal-hole as a target).[40] Although he was furious when someone uncorked a bottle of special air, and devastated when his daughter Sally lovingly cleaned his lab and washed out *all* his bottles, his children were allowed to run in and out as he worked, an odd figure in his short laboratory apron. They vividly remembered the library with its screen for magic-lantern shows and shocks from the electrical machine.[41]

All the Lunar children stayed with each other, knew each other's parents and houses and interests. On the whole, the approach was informal: only in the Watts' house were lips tightened if no one wiped their feet. Each family tried out different types of education, but it was the childless Thomas Day who arguably had the greatest influence. Day too was now caught up in domesticity. Two years before, he had married Esther Milnes, possessor of a fortune inherited from her father, a Chesterfield merchant. In the 1770s she was living in Wakefield, where her cloth-making relations Robert and John were powerful merchant princes, known for their liberal, Dissenting politics.[42] Small had long ago suggested Esther as a possible wife, perhaps knowing her through connections between the Milnes family and Roebucks of Sheffield.[43] She was a perfect fit for Day. They moved first to Hampstead, where Edgeworth and Honora found the once frail Esther tramping the Heath in the snow with evident enjoyment. They were relieved to find her strength of opinion matched her husband's: rarely did a pair talk so much, or so vehemently. In early 1779 Day took a small estate and manor house at Stapleford-Abbot in Essex where he would 'have the pleasure of having my hands full of dirty works', as he was immediately going to enlarge it, he told young Erasmus Darwin.[44] He bought 'Ware's Architecture' at a bookstall, buried himself in it for three or four weeks, then tired of it, and was bothered and bewildered when masons and carpenters demanded door-sills, lintels and other trivia. Edgeworth was staying when the builder asked where the window of Esther's new dressing room should go. Deep in a treatise by a French agriculturalist, Day waved the question aside: why couldn't the wall be built first, then the window cut? Although the mason pointed out that the usual practice was the opposite, the room was built, and stayed windowless.

Despite their reclusiveness the Days always welcomed their friends' children. Esther's sister Elizabeth had died in 1769, leaving two sons, who often came to stay: the youngest, Thomas Lowndes, lived here almost permanently from 1783. As a bachelor a few years earlier, Day had looked after the eighteen-year-old Erasmus Darwin on a trip to London, prompting a sly verse comment to his younger brother Robert:

> Mr Day too was there, who was reckon'd you know
> A Man who had travel'd & rather a beau
> The very first moment we enter'd the Town
> Good Lord! I discovere'd he was but a Clown;
> Though he powders his Hair, & strives to look gay
> But I charge you don't tell him a word that I say.[45]

After the Days' marriage, Maria Edgeworth, at boarding school in London, stayed with them in the summer of 1781. She always remembered the impact of Day's 'romantic character', his metaphysics and 'the icy strength of his system', even when he forced her to drink Bishop Berkeley's tar water and write dialogues on happiness. Day was later appalled at the idea she should write professionally, yet she stayed fond of him.[46] She also remembered his odd sense of humour. Writing to her aunt years later about a hunting accident, she noted that there was 'floundering and dirt enough' to delight even Mr Day's fancy, 'whom nothing diverted so much as people's falling down in the dirt, especially if they were ladies & gentlemen & well-dressed'.[47] Sukey Wedgwood also stayed with the Days in later years, and all the children accepted Thomas's eccentricities. As Keir noted, although grave in company, with friends he had 'a singular gaiety of temper, which rendered him particularly agreeable to young people and children, whom he was always fond of pleasing and instructing'.[48]

There were no real children's books in the 1770s, although in this new age of sensibility a few were just beginning to be written. Around 1780 Day took up the challenge. To begin with he planned a book of extracts from famous lives, classics and history or novels, but to leaven the extracts he added a framing narrative, the story of young Harry Sandford and Tommy Merton, and their teacher, Mr Barlow. Day had not totally turned his back on Rousseau, but he had realized his ideas must be modified. *The History of Sandford and Merton* owes much to the 'natural education' of *Emile*, with the reinstatement of the book-learning that Rousseau threw out, and Day also borrowed without qualms much of the patterning of Brooke's *Fool of Quality*.[49] The first problem that faces his Mr Barlow (convinced, like Rousseau, that society is fundamentally self-interested) is one that also bothered Day: even if you educate children perfectly, how will they cope with the corrupt world they meet? And how can you reconcile Spartan 'hardiness' with the virtues of sensibility and sympathy?[50]

Unlike Rousseau's Emile, Day's boys live in the 'world' from the start. Six-year-old Tommy Merton, the spoiled son of a Jamaica sugar baron, meets his match in the person of the farmer's son Harry Sandford (kind to animals, especially toads and nasty insects, always in a good temper; never tells a lie). Harry becomes an ally and teaches him to read, but poor Tommy still seems to undergo a form of moral torture: not allowed to eat unless he tends the vegetables, forced to listen to improving stories. Yet for children of the 1780s the stories were fun: the boys build a house of branches and brushwood; Mr Barlow coolly subdues a bear who breaks loose from some travelling entertainers; Tommy becomes the hero of the village by saving a poor family from the bailiffs. The book also contains a mass of tales, from Androcles and the Lion to the conquest of Mexico and the *Arabian Nights*, but the moral is practical, in tune with the Lunar Society's views: a sensible man will behave well to everything around him:

> because it is his duty to do it, because every benevolent person feels the greatest pleasure in doing good, and even because it is in his own interest to make as many friends as possible. No one can tell, however secure his present position may appear, how soon it may alter, and he may have occasion for the compassion of those who are now infinitely below him.

When the first volume of *Sandford and Merton* appeared in 1783, Day found himself with a surprise bestseller. The second volume was published in 1786 and the third in 1789, on the eve of the French Revolution. Apt timing, since beneath the do-goody tone Day's democratic fires burned still: his clearest message of all is that the parasitic rich get fat on the work of the poor; it is they who provide the true wealth of the nation. Saccharine though it seems today, *Sandford and Merton* fed the imagination of the coming Romantic generation, and of the Victorians – running through a hundred and forty editions before 1870.

28: BRINGING ON THE ARTISTS

The end of 1780, which had seen the death of Honora Edgeworth, Priestley's move to Birmingham and Darwin's engagement to Elizabeth Pole, brought one final wrenching change: the death of Thomas Bentley. His health had been poor ever since the move to London: perhaps, it was said, because the Greek Street showroom had formerly been a dissecting room and was never properly fumigated – many who worked there fell ill. In the summer of 1780 he and Mary had gone to Margate and then to Etruria. The old correspondence with Wedgwood continued, with all Bentley's letters carefully bound in a great thick book, teasingly called 'Josiah's Bible'. On 12 November, Wedgwood wrote briskly, 'I have too many irons in the fire to write much now. I am contriving some vases for bodies, and bodies for vases, and I need not tell you that *new* things and *good* are very difficult to make.'[1]

Just over a fortnight later a rushed note arrived from Ralph Griffiths:

> Our poor friend yet breathes: but alas! it is such breathing as promises but a short continuance. Almost every hope seems to have forsaken us! I dread the thought of what will be the content of my next! Adieu! R.G.[2]

Wedgwood had no idea Bentley was so ill. He dashed south, to discover that his friend had died a day after Griffiths wrote, on 26 November. He stayed in London until Bentley's funeral and burial in the vault of Chiswick Church. He commissioned James 'Athenian' Stuart to erect a

John Flaxman, portrait medallion by himself

monument, and organized the management of Mary's future income from the firm. Tom Byerley took over as manager at Greek Street, and in an extravagant gesture of closure Wedgwood disposed of the whole of the Wedgwood and Bentley inventory: over two thousand lots were auctioned by Christie and Ansell in early December, in a great sale spread over eleven days. In the next few weeks, Wedgwood's friends thought he aged ten years.

Bentley had introduced him to high art and found so many of the artists for Etruria. Urbanely supervising the London artists and running the showroom, he embodied the balance of idealism and pragmatism that marked many of the Lunar projects. But for most of the group there was never a hard line between art and experiment; when Edgeworth remembered his Midlands friends he saw them all as equally 'men of literature and science'. For Wedgwood and Boulton, art was the stuff of their manufacture: the appeal of their goods lay in the combination of the artist's imagination, the craftsman's skill and the latest techniques of reproduction. But here – as in the coterie poetry of Lichfield, or the collective work on inventions or experiments – the issue of originality and 'ownership' of ideas often arose.

Wedgwood's relationship with artists involved a constant negotiation of this problem. He was proud of his own skill in shaping pots and in appreciating fine Sèvres porcelain or antique vases, but he never pretended to be an artist himself. He did, however, recognize quality. When he and Bentley first began hunting for modellers and painters, he scathingly dismissed the rococo models of the unfortunate Mr Tebo as 'like the head of a drowned puppy', and 'full as like Pigs as Hares'.[3] Conversely, he nurtured talent lovingly and several men and women stayed with him all their working days. One such was William Hackwood, hired as an 'ingenious boy' of twelve in 1769, and allowed a crack at a portrait medallion a couple of years later. In his teens Hackwood adapted busts and reliefs that Wedgwood found in London, and remodelled the borrowed gems and figures that were the base of the jasper cameo designs. He made a stream of portrait busts, from Augustus Caesar, Brutus, Cato, Homer and Horace to Andrea Palladio, Inigo Jones and Hermann Boerhaave; he modelled the first Wedgwood medallions, of George III and Queen Charlotte, and created fine reliefs such as the *Birth of Bacchus* in 1776, Wedgwood's largest jasper tablet so far. He soon became indispensable, perfecting the delicate undercutting required for bas-reliefs of fingers and faces, waving hair and near-translucent drapery. By the time the young man was nineteen, Wedgwood said he wished he had half a dozen more Hackwoods.

Briefly, tensions arose when Hackwood began to put a higher price on his work, which Wedgwood found exorbitant. But the real rub was that he saw himself as an 'artist' and wanted to claim his work: he began to sign his name beneath the shoulders on busts of Shakespeare and Garrick in 1777. Swiftly and firmly, this was stopped. Hackwood had to be satisfied, not with fame, but with high pay and high regard: Wedgwood entrusted him with the portrait medallions of the men he most admired: of his brother-in-law and old tutor, William Willett, who died in 1778; of Darwin (based on Wright's portrait); of Franklin and Priestley and Bentley. Hackwood stayed at Etruria through Wedgwood's lifetime and beyond, resisting Josiah Spode's attempt to lure him away in 1802 and finally retiring in 1832, after sixty-three years with the company.

Many small artistic treasures are found in Wedgwood's design history, such as the witty, minutely observed landscape vignettes by Thomas Bewick on some sets of Queensware. But Wedgwood had also cultivated amateur artists, particularly among well-bred women. He produced special 'paint chests' for them made from terracotta stoneware, black basalt and jasper, with a tray and twelve small pots for colours: Christie's catalogue of 1781 explains that these 'contain in a very small compass, a neat Apparatus for a Lady's painting, or for Use of young People learning to draw and colour Prints'.[4] By this date, such was the glamour of the pottery that some of these ladies were themselves approaching Wedgwood as would-be designers. Chief among them was Elizabeth, Lady Templetown, whose studies of women and children in the 1780s were perfect for the new 'feminine' market. She sent her designs in pencil, or 'beautiful cut Indian paper',[5] and from them Hackwood modelled graceful, informal reliefs – *Domestic Employment, Family School, Maternal Affection, Sportive Love*. In 1783, Wedgwood wrote coyly to express his delight that she liked his 'attempt to copy in Bas relief the charming groups of little figures', and requesting more, 'with the most perfect submission to Lady Templetown's pleasure'.[6] Following the shifting tide of taste, he reproduced her melancholy *Poor Maria* (based on Joseph Wright's painting of a scene from Sterne's

Sentimental Journey), followed only four years later by her 'Charlotte at the tomb of Werther', the first stirrings of Romanticism provoked by Goethe's epoch-making novel *The Sorrows of Young Werther*. Acknowledging the pulling power of her grand name (the opposite to Hackwood's), he even credited her with some of the designs in his 1779 catalogue.

Templetown's designs were followed in the later 1780s by those of Emma Crewe, from *Bacchanalian Children* to a whole series of *Domestic Employments*, sweetly adapted for medallions and vases, teapots and brooches. Here too, Wedgwood tapped a source of influence as well as profit: Emma's father was the Cheshire magnate John Crewe and her mother, one of Wedgwood's best customers, was a fashionable beauty and famous Whig hostess, the friend of Fox and Sheridan and Burke.

Another aristocratic designer, though from a far more raffish set, was Lady Diana Beauclerk, the eldest daughter of the second Duke of Marlborough.[7] Lady Diana's first marriage to the chronically unfaithful Lord Bolingbroke had exploded in a scandalous divorce in which he sued her lover and the father of her latest child, Samuel Johnson's rakish friend, Topham Beauclerk. Shunned by high society, she then shone as hostess to 'the Club', Johnson and Reynolds, Garrick and Boswell, and to make an income turned to her hobby of painting. Her success was boosted by a commission from Horace Walpole and a bestselling print of her cousin, Georgiana, Duchess of Devonshire, and after the death of her husband in 1780 she was free to work professionally. When Charles James Fox sent her drawings to Wedgwood in 1783, he jumped at her talent – and her contacts. Her lively designs still featured children, but more robust than sentimental – cheerily innocent Bacchanalian boys, tumbling amid flowing bowls, panther skins and swags of vines.

At the opposite end of the class spectrum from Diana Beauclerk, part of whose childhood was spent in the echoing halls of Blenheim, was Wedgwood's most famous jobbing artist, John Flaxman, who grew up among the models and casts in his father's Covent Garden shop. Visitors remembered a 'puny, ailing little boy', with a curvature of the spine, propped on a high chair with cushions behind his father's workbench. Moved by the sight of him trying to learn Latin from books found on a stall, the Reverend Anthony Matthew of Percy Chapel in Charlotte Street virtually adopted him, drawing him into the milieu, so familiar to Wedgwood and Bentley, of the cultured Radical Dissenters. The blue-stocking Harriet Matthews and her friend Anna Laetitia Barbauld read Homer and Virgil aloud while Flaxman drew scenes: at ten he was selling his commissioned drawings; at eleven and thirteen he won prizes for modelling from the Society of Arts; at fourteen, working alongside older artists such as John Bacon and James Tassie (also both associated with Wedgwood) he won a silver medal at the Royal Academy Schools.

Through Bentley, Wedgwood began ordering models from Flaxman in early 1775 for chimney-piece tablets, vases and bas-reliefs, satyrs and classical gods. Soon he commissioned portrait busts of Joseph Banks and Daniel Solander, and by the winter he was lamenting that Flaxman's Antony and Cleopatras were so fine that he could hardly bear to part with them. Flaxman was now receiving substantial commissions for individual sculptures and tombs, but Wedgwood provided a staple income and in return he produced many fine classical reliefs. One of his finest pieces was the *Apotheosis of Homer*, imaginatively adapted from a vase in Sir William Hamilton's collection in the British Museum, which prompted an ecstatic response when a tablet was sent to Hamilton in Naples.

In the early 1780s, newly married and determined to finance a trip to Rome, Flaxman wore himself out producing cameos, busts and portraits: of Rousseau and Sterne, Johnson and Sarah Siddons, Herschel and Captain Cook. In the end he sculpted at least twenty-two portraits of 'illustrious moderns', including two of himself and one of his wife. He often stayed at Etruria and he and his wife Anne became good friends of the family. In 1781 he was designing lavish mouldings and allegorical friezes for the drawing-room ceiling at Etruria Hall, and a year later, after teaching the Wedgwood boys drawing, he refused to set a price on his thirty-three lessons – a fair stint of work – 'because I fear in so short a time they could not profit much'.[8] When Wedgwood, touched, paid him generously, Flaxman responded by sending two small reliefs as a present, 'which you may copy in your excellent bisque for the manufactory if you think them worthy'.[9]

Flaxman completely understood the flexible adaptation of classical originals to current taste. As Wedgwood put it to Darwin in 1779, 'I only pretend to have attempted to copy the fine antique forms, but not with absolute servility. I have endeavoured to preserve the style and spirit, or if

The Dancing Hours, a reproduction of the jasper plaque, white on pale blue, designed by John Flaxman, *c.* 1775

you please, the elegant simplicity of the antique forms.'[10] Flaxman's *métier* was 'elegant simplicity'. He was also practical and careful, always giving precise details, and he understood the needs for adaptable models – his *Six Muses* of 1777 were designed so that they could appear as a frieze, or as individual medallions, or separated into matching bas-relief panels, or set around a vase.[11] He made perfect panels for chimney-pieces, such as his fluent *Dancing Hours* or playful *Cupids with a Goat.* He could imitate the pure antique taste that Hamilton so admired, but he could also dance easily to fashion – soon he too was producing groups of children, skipping, playing blind man's buff or marbles. And it was Flaxman who prompted Wedgwood to undertake his most inspired project, the copying of the Portland Vase. 'I wish you may soon come to town', he wrote in February 1784, 'to see Sir William Hamilton's Vase':

> It is the finest production of art that has been brought to England, and seems to be the very apex of perfection to which you are endeavouring to bring your Bisque and jasper. It is of the kind called 'Murrinan' by Pliny, made of dark blue glass with white enamel figures. The vase is about a foot high, and the figures between five and six inches, engraved in the same manner as a cameo, and of the grandest and most perfect Greek sculpture.[12]

At the same time, Flaxman pushed Wedgwood to try a very different style, the English Gothic, when he modelled a set of 'medieval' chessmen. Chess was increasingly popular, and silver and ivory sets were selling well, so Wedgwood decided to produce relatively cheap sets, at 5 guineas a time. He left the choice of style largely to Flaxman, with the result that politics

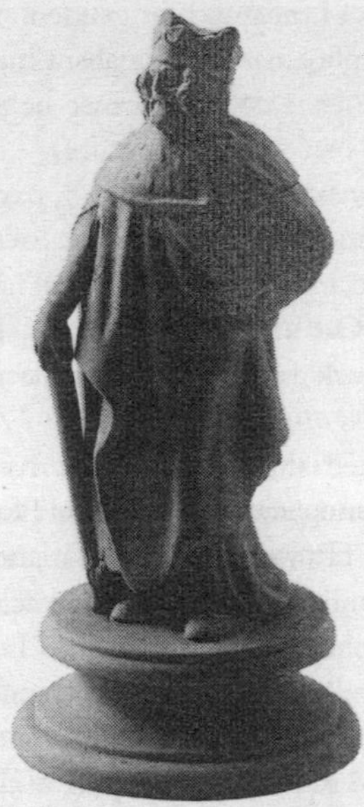

The King from Flaxman's chess set of 1783 said to show the actor John Kemble, in a Shakespearean role.

and taste and patriotism all overlapped in this relatively minor piece of work: a veneration for a neglected British history fused with radical fervour.

The mid-1770s saw a vogue for British antiquarianism. On his trip to the West Country Wedgwood himself had explored Stonehenge, noted the barrows on Salisbury Plain, and had exclaimed in awe at the grandeur of Maiden Castle and the strange standing stones of Cornwall. Enthusiasm flowered for a 'native' British past, for Celtic myths and legends, for the medieval ballads of Percy's *Reliques* and for the grand forgeries of 'Ossian' (James Macpherson) and 'Thomas Rowley' (Thomas Chatterton, whose suicide made him an icon in 1777). With his friends Thomas Stothard and William Blake, Flaxman was a great admirer of the 'Monumental effigies' of the Middle Ages, especially those in Westminster Abbey, and his lively chessmen closely echoed such sculpture (with a modern twist, since Kemble was said to have modelled for the kings, and Sarah Siddons, as Lady Macbeth, for the queens). The little figures carried a weight of meaning: in gentlemen's studies and libraries they brought popular antiquarian nationalism into the world of classical busts and dancing cherubs. Perhaps because he saw the clash, the chess pieces were Wedgwood's sole Gothic venture.

Flaxman eventually went to Rome in 1787, partly subsidized by Wedgwood. In Italy, where Canova became his close friend, he made a surprising impact with his sculpture and with engravings of his illustrations to Homer, Dante and Aeschylus. On his return in 1794, busy with monumental sculpture, he did little work for Etruria, yet remained a close family friend.

All these artists accepted that they were surrendering their work to Wedgwood's materials and production, just as Boulton's modellers did to his ormolu and silver. But in one intriguing instance Wedgwood worked to meet an artist's demands – those of George Stubbs. Stubbs was fifty in 1774, heavy, patient, obstinate in sticking to his own vision. His self-portrait shows him as bull-necked and double-chinned, scorning a wig; a wary, broadcloth man lacking all pretension. Like Hogarth, he preferred to work from nature and scorned the copying of the academy schools. (In complete contrast to Flaxman, when Stubbs went to Rome in 1754, he was back within a year, deciding he had nothing to learn.) The son of a Liverpool leather merchant, he was in every way a self-taught artist, apart from a few weeks as assistant to a minor portrait painter. He was already dissecting dogs and horses in his youth, and in his twenties, in York, where he painted portraits and studied anatomy at the hospital, he produced the drawings and engravings for the *Complete New System of Midwifery* by John Burton (the model for Mr Slop, Sterne's man-midwife in *Tristram Shandy*). In the late 1750s, with his redoubtable common-law wife Mary Spencer – an earlier Mrs Stubbs and two sons having vanished into the mist – he retreated to a Lincolnshire village. The remoteness was essential to avoid 'inconvenience' to neighbours while he worked on his astounding *Anatomy of the Horse*, eventually published in 1766. The air was thick with the smell of decomposing horseflesh. Each horse hung in tackle for six or seven weeks while he and Mary stripped the flesh off layer by layer so that he could draw the fat beneath the skin, the muscles and veins, tendons and ligaments, joints and bones and final drooping skeleton.

Wedgwood and Watt – two similar obsessives – would have understood this long, unnerving, fiercely penetrating gaze. And as with their own long labours and experiments, Stubbs's stinking, maggot-filled years bore fruit: in London in the 1760s he became known as a painter of horses and hunting scenes, alert to the inner beauty of form, the rippling muscles, the skull beneath the skin. By now he had two more compulsive themes – strings of mares and foals, as graceful as Flaxman's *Dancing Hours*, and the four-part series, endlessly reworked, of a lion attacking a horse, his own violent, disturbing adaptation of classical sculpture to the animal world.[13] His interests fitted this period where love of the sublime and the exotic jostled with scientific observation. With scrupulous attention to detail he painted animals such as the zebra brought from the Cape of Good Hope and presented to Queen Charlotte in 1762, and the cheetah given to George III in 1764, with its two Indian keepers. He painted the blackbuck antelope for William Hunter, and the Duke of Richmond's bull moose from Quebec. He drew Marmaduke Tunstall's mouse lemur, and for Joseph Banks he painted a kangaroo, and a dingo – animals newly discovered by Cook. All these works were used as accurate records by natural historians.

Stubbs, like Wedgwood, was a professional who knew his market and worked his contacts. He was a stalwart of the Society of Arts and was its President for a year in 1771. Also like Wedgwood, he was an inveterate technical experimenter – many of his canvases have flaked beyond repair due to his unique blend of 'pine-resin and beeswax and non-drying oils and fats'.[14] The experiment that brought the men together, however, was his attempt to work in enamels, which he thought might give even greater luminosity: after working on the chemistry of enamel pigments (a subject dear to Wedgwood's heart) he developed a palette of nineteen different tints, colours that would stay true after firing. But he also needed a ground that could withstand heat. For most enamel work, such as miniatures or snuff-box lids, the ground was usually metal and Stubbs began by painting large ovals on copper. But thick copper was heavy, while thin sheets would warp and bend in the heat. Reaching out for things to hand – in the familiar, impatient tradition of Watt and Priestley – in 1777 he cut out the bottom of a large Wedgwood dish and, to fit the oval, painted his wonderfully curled-up *Sleeping Leopard*.[15] After trying various potteries and the Coade stone-makers in vain, he approached Wedgwood and Bentley to make him oval plaques in biscuitware.

Wedgwood liked such challenges and he thought that if Stubbs succeeded, other painters might follow, so that this could become a profitable line. The manufacture was simple, but it was difficult to fire a larger size without its cracking. He tried out different clays, and resisted the painter's impatience: 'My compliments to Mr Stubbs,' he wrote firmly. 'He shall be gratified but large tablets are not the work of a day.'[16] The cost of the experiments worried him, and the alterations needed to the kilns, but after producing some successful small plaques, in May 1779 Wedgwood could report 'certainty & success' in firing larger ones, 'perhaps ultimately up to 36 inches by 24, but that is at present in the offing & I wo. not mention to Mr Stubbs beyond 30 at present.'[17]

This was why Stubbs was at Etruria in 1779, paying off his bills with the enamel ovals of Josiah and Sally and Sally's father, and with the conversation piece in oils of the family on the terrace. For Wedgwood this was a new experience, a partnership outside his work and – in the case of the paintings – a chance to play patron. He had liked Stubbs's light-hearted *Labourers (Lord Torrington's Bricklayers at Southill)*, and in 1781 ordered a copy on an enamel plaque, paying a hefty £340. But when Stubbs painted his family, he was baffled by the way the true likeness came and went at different stages; he followed poor Stubbs round like a curious child, reporting every step to Bentley: 'Our picture proceeds very slowly, but we have begun to make the horses sit this morning and I write by Mr. S in the new stable which is to be my study while he is painting here.'[18] The care required stunned him. 'Time and patience in large doses, are absolutely necessary in these cases', he wrote,

> & me thinks I would not be a portrait painter upon any conditions whatever. We are all heartily tired of the business & I think the painter has more reason than any of us to be so.[19]

In 1782 Wedgwood and Sally were also painted, very finely, by Joshua Reynolds in London, without such complaints. But for a while Stubbs was Wedgwood's 'own' painter, proudly introduced to local dignitaries: 'Nobody suspects Mr Stubs of painting anything but horses & lions, or dogs & tigers', he admitted, 'and I can scarcely make anybody believe that he ever attempted a human figure.'[20] In fact Stubbs's clear, dispassionate work had a profoundly democratic spirit – and if his aristocrats sometimes look wooden, his portraits of grooms and jockeys, African servants and

Indian animal-keepers are fresh, sympathetic and unsentimental.[21] Stubbs and Mary became part of the wide Etruria clan. Stubbs taught the boys perspective, just as Flaxman taught them drawing; Darwin came over to see about Mary's sore throat, and Stubbs painted his portrait in enamels in 1783. Stubbs did make some models for Etruria, but he stuck firmly to his own line – an inevitable *Lion and Horse*. 'He objected to every other subject so I gave it up', sighed Wedgwood, '& he is now laying in the horse whilst I write a few letters this good Sunday morning.'[22] This and a companion piece, *Phaeton and the Chariot of the Sun*, were both produced as plaques, while small cameos of individual horses seem to have been modelled from his designs, and some of his shooting scenes were made into transfers for creamware mugs and punch-pots. He used his large Wedgwood plaques from the late 1770s until 1795, but they proved an expensive undertaking: no other artists followed his lead and Stubbs was hurt by the Royal Academy's cool response to his bright colours, declared 'hurtful' to neighbouring paintings. Many plaques were unsold and were auctioned for a trifle at his death in 1806. To the end, he kept his anatomical curiosity – his last uncompleted project, begun in 1795 when he was seventy-one, was the almost surreal *Comparative Anatomical Exposition of the Structure of the Human Body with that of a Tiger and a Common Fowl*.

While Wedgwood chose Stubbs instead of Joseph Wright for the family portrait, he sensed that it was Wright who could best celebrate his craft. Wright clearly understood the power of science and industry, but he could also suggest the mysteries of the imagination and the force of inspiration from the past, as in paintings such as his *Virgil's Tomb* of 1779. When Wedgwood commissioned a painting to symbolize the history of his own 'mechanic art' he did not want realism, potters' wheels and sweat. Instead he wanted an allegory to please his high-born patrons, especially the women, so he set out to feminize and idealize his industry. In two companion pieces painted for Wedgwood, *The Corinthian Maid* and *Penelope Unravelling her Web by Lamplight* (spinning being the other great Midlands industry), 1783–85, Wright linked manufacturing with myth and domestic virtue.[23] Both are stories of fidelity showing the woman as artist and maker, but keeping the 'making' firmly out of sight.

Wright's confidant in these tricky commissions was the poet William Hayley, friend of Anna Seward, and supporter of Flaxman. Hayley reassured him that with 'historical or rather poetical subjects, you may take any liberties you please', but with regard to Penelope, Wright confessed he was worried: Ovid said she used to work at her loom, but a loom was so mundane; perhaps it could be hidden by drapery – which it was.[24] Even more bothering was the nudity. In painting the statue of Ulysses, Wright had tried to follow the ancients in making their statues naked, 'but being nearly seen in profile, the private parts become too conspicuous, for the chamber of the *chaste* Penelope'.[25] His solution of having Ulysses rest on his bow with his quiver over the offending genitals was hooted down by Hayley, who pointed out that this might well make 'some prophane Wag exclaim, "Happy is the man that hath his Quiver full."'[26]

The Corinthian Maid though, not Penelope, was the legend most apt to Wedgwood.[27] Hayley had told the story, derived from Pliny, in his poem *An Essay on Painting* in 1778, explaining how the maid, distraught that her lover was about to leave the country, had traced his profile on the wall, outlining his shadow as he slept in the lamplight. Thus the art of portraiture was born. More than that, her father was a potter, and he fixed the image for her permanently by working clay into the outline and firing it in his kiln: ceramic portraiture becomes the origin of likeness, both graphic and sculptural. This is a myth that justifies 'imitation'; and it shows an art born not of commercial greed but of a woman's love.

> The line she trac'd with fond precision true,
> And, drawing, doated on the form she drew . . .
> Nor, as she glow'd with no forbidden fire,
> Conceal'd the simple picture for her sire:
> His kindred fancy, still to nature just,
> Copied her line, and form'd the mimic bust.
> Thus from the pen, inspiring LOVE, we trace
> The modell'd image, and the pencill'd face.[28]

From Wright's first suggestion in 1778 the painting developed slowly, and from the start Wedgwood fretted – he wanted the potter's kiln in the background (represented by a fiery glow), but could not see logically how Wright's suggestion of a room filled with 'elegant Earthen Vessels' seen through an arch would work, since this was supposed to be the 'infancy of the Potter's Art'.[29] And at a late stage in April 1784, he took some women friends to see the picture and found it made them rather uncomfortable:

The objections were to the division of the posteriors appearing too plain through the drapery and its sticking so close, tho' truly Grecian, as you justly observe, gave that part a heavy hanging-like (if I may use a new term) appearance, as if it wanted a little shove up . . .[30]

Wright duly added more drapery, though the 'heavy hanging-like appearance' is still slightly comic, but as a whole this is a relatively austere neo-classical piece. Perhaps by design, perhaps by chance, the flowing duo of sleeping man and active woman, with their balanced limbs and sloping profiles, bears an uncanny resemblance to the figures on a Wedgwood bas-relief.

Yet this cool, gently lit composition actually denied the work itself – just as Wedgwood suppressed the names of his artists, and hid his technical secrets. The glowing surface is all: the product is admired, the bones of the enterprise ignored. A very different painting makes a similar point, Wright's evocative *Arkwright's Cotton Mills by Night*, in which the candle-lit mill is an industrial volcano, smouldering in Derbyshire dales, but the actual work is hidden, the busy night-shift workers with their whirring machines reduced to the glint of countless windows.[31]

The artists Wedgwood knew are like arrows, flying outward to the many networks that invigorated the Lunar world. The products of Etruria had an unnerving homogeneity, beautiful yet the same: we identify them all as 'Wedgwood', yet beneath this lies the individual skill of Hackwood and his fellow modellers; the high-born women sending designs; the idealism of Flaxman. Outside the factory, Wedgwood faced the gaze of Stubbs and then of Wright,[32] and while Stubbs anatomized and Wright mythologized, both pointed unerringly to the pride and possessiveness that drove this competitive era of experiment, luxury and trade.

29 : FROM THE NATION TO THE LAND

Day was never short of words. 'Surely,' he exclaimed, as the decade turned, 'there has never been a period in our history which more required the exertion of every talent and every virtue than the present, a period which stands unrivalled and alone for the excess of national calamity which has burst upon us, and threatens our destruction.'[1] He liked images of excess – tempestuous oceans, ancient enemies, internal conspiracies – but in 1780 Britain's outlook was indeed dark. In America the war seemed catastrophic; at sea the British Navy was battered by the French and Spanish fleets; in Ireland agitators were battling against British embargoes on colonial trade and even arguing for independence. The Irish Volunteer movement began and with the stirring example of America a rebellion seemed more than likely.[2]

At home the wartime slump, the pace of enclosure on the land and industrialization in the towns brought waves of unrest. Riots took place in Lancashire in the autumn of 1779: these were the ones that Wedgwood dodged when he was fetching Jack home from school in Bolton, with the mobs 'breaking the windows & destroying the machinery of the first mill they attacked, but the next, the machinery being taken away, they pull'd down the building & broke the mill wheel to pieces'.[3] To a manufacturer these were chilling scenes. In Lancashire the militia was called to disperse a huge crowd, over three thousand people, demonstrating against Arkwright's machines at Chorley.

Farm carts designed by Edgeworth

Amid this tide of troubles, bullied by George III, North hung on in power, but hardly in control. All the varied opposition forces were lining up against him; the Rockingham Whigs, with Fox and Burke as their eloquent spokesmen; the independent country MPs, both Whig and Tory; the Wilkesite radicals and republicans and the Rational Dissenters. In December 1779, a massive meeting organized by the Reverend Christopher Wyvill in York was followed by a nationwide petition against corruption and inefficiency. As with the Wilkesite furore of the late 1760s, this was not a mass movement but a protest of the gentry and bourgeoisie. Wyvill's first supporters were solid Yorkshire landowners. Esther Day's young relative James Milnes (currently building a huge new house, complete with Wedgwood panels for the fireplace, but later known locally as 'Jacky Milnes the Democrat') was on the committee of the Yorkshire Association.[4] A rash of provincial associations joined radical groups in London to demand 'economical reform'.

Day was jolted out of his armchair, away from his agricultural treatises, in the final weeks of 1779. But his politics, which look progressive and radical, were at base profoundly, poetically conservative. They harked back to Rousseau, who contrasted the modern material world with a purer past; to writers such as Adam Ferguson who compared the complex modern state to simpler organizations such as the Highland clan or Roman republic; to Saxon rights lost under the Norman yoke.[5] A disillusioned idealist, Day saw a Britain betrayed by its government, where 'all principles of natural equity and reason' had been subverted.[6] This winter he joined John Jebb's radical freeholders of Middlesex, exhorting men to join the cause:

> We are reduced to the very brink of despair, nor is there a gleam of hope, except from the rising spirit of the people, exasperated by an unexampled series of provocations and disgraces . . . Are we merely sheep to be fattened for the profit, and slaughtered at the pleasure of our masters?

This speech by Day, bound together with Fox's equally fiery appeal to the householders of Westminster, was published in February 1780. The following month Day was sharing a platform with Wilkes in Cambridge, attacking the iniquities of George III's Government – the Riot Act, the Game Laws, the 'public robbery' of the Excise and the standing army –

and demanding wider representation and annual parliaments. 'After a calm of the longest duration,' declared Day, 'the spirit of the people is at length excited, and I see a storm gathering which may be fatal to its enemies.'[7] The public must seize the gathering tide and hurl it 'full against the loftiest bulwarks of oppression; they will not resist its rage, they will be levelled with the ground, and leave you an easy victory'.

Day took part in the founding of the Society for Constitutional Information in April, organized by John Cartwright to bring together documents that would 'prove' the existence of a historic, enlightened Saxon democracy. Among those present were the campaigner John Jebb, Day's friend William Jones, the slavery campaigner Granville Sharp, the playwright Richard Brinsley Sheridan, and Thomas Bentley. Until Bentley's death, Wedgwood followed events closely through his friend's letters, arguing about 'representation', reading Day's speeches and backing him to the hilt. This month, the MP John Dunning launched a bravura attack in the Commons: in ringing tones, before an enthralled and silent House, he declaimed the famous resolution: 'The influence of the Crown has increased, is increasing and ought to be diminished.' Order papers were waved, MPs cheered, shouted, jeered. In the chaos the vote was narrowly in favour. North threatened to resign but the King made him stay – and after a week's fraught delay the Association's petition was rejected. The mood was changing. The Association's bold rhetoric had put off many moderate supporters. Burke's Economical Reform Bill was defeated, but Fox's cousin the Duke of Richmond introduced a Bill into the Lords for 'restoring' universal male suffrage and annual parliaments. And then, on 5 June, at the very moment that the Lords were debating Richmond's proposals, Lord George Gordon marched on Parliament at the head of a clamouring mob, carrying a petition from the Protestant Association against relief granted to Catholics.

As night fell, the demonstration turned to full-scale riot. Catholic chapels were burned; Catholic homes attacked. Drunken men, armed with clubs, roared through the streets. By next morning carriages and carts were jamming the roads out to the country. Day after day chaos reigned. Newgate was burst open and prisoners released; the King's Bench was burned; breweries and distilleries were plundered and the water supply of Holborn ran with spirits. The Bank of England was besieged, Burke's house surrounded, Saville's threatened, Lord Mansfield's looted and his famous library sacked. In these dark five days whole areas of London were ravaged, and though figures are unclear, nearly five hundred people were killed or badly wounded. Watt wrote to Black from London, 'We have had terrible doings here but I hope it is now over, I never saw or knew so many people overawed by a most despicable mob.'[8]

The all too evident power of the people helped to scuttle the reformist ship, already holed by uncertainty and divisions. Bizarrely, Wilkes now found himself on the side of authority, trying to calm the riots. Elizabeth Montagu, powerful coal-owner that she was, breathed a sigh of relief. Lord George Gordon, she wrote, had defused the discontent, 'and I hope in a great degree cured the epidemical democratick madness. The word petition now obtains nowhere, the word association cannot assemble a dozen people. We are coming to our right senses.'[9]

Day, with good cause, blamed the violence and deaths largely on the panic of the authorities, not the people. But from now on he stood back from politics, although Jebb tried to persuade him to stand as MP for Southwark, opposing the brewer Henry Thrale, Johnson's friend. Day declined. 'I need not tell my dear friend that the farce now acting upon all the stages in Great Britain is the *World in Uproar*,' Wedgwood wrote to Bentley in September, 'though in our borough you will suppose all is very quiet. We all sit still saying the will of the lord be done & there is an end of our farce.'[10] That was very much the general pattern. Although Fox was triumphantly elected for Westminster ('much to the honor of the electors', thought Wedgwood), North kept his majority. It seemed the radical energy of 1780 had dissipated like mist.

By the end of this year Britain's overstretched forces were fighting across the globe: against the French in India, the West Indies, North America and Africa; against the Spanish in Gibraltar, the Balearic Islands, the Caribbean and Florida; against the Dutch in Ceylon and the West Indies – and in America against the Colonists. When General Cornwallis surrendered at Yorktown in October 1781, North allegedly wept, crying out, 'Oh God! It is all over.'

Although Day's democratic sentiment seeped into the pages of *Sandford and Merton*, from now on he retreated increasingly from public life.

Romantically, he saw himself as a latter-day Cincinnatus, the upright man ready to step forward if his country needed him. He was not interested, he said, in fame, but only in the general good. Darwin, equally cynical about the political process, wrote teasingly to him in May 1781, 'Pray my good friend, why did you not contribute to the *benevolent* designs of Providence by *buying* a seat in Parliament?'

Mankind will not be served without first being *pleased* or tickled. They take the present pleasure of *getting drunk* with their candidate, as an earnest or proof, that he will contribute to their *future good*; as some men think the goodness of the Lord to us mortals in this world, his temporary goodness, is a proof of his future and eternal goodness to us.

Now you wrap your talent up in a napkin, and instead of speaking in the assembly of the nation, and pleading the cause of America and Africa, you are sowing turnips, in which every farmer can equal or excel you.

Briefly, the following year, Day was drawn back into the public arena. After the Commons finally voted to end the war in America on 27 February 1782, North resigned: in April a furious George III reluctantly accepted a Rockingham Whig ministry with Charles James Fox as Foreign Secretary and Shelburne as Home Secretary. Day's involvement in the peace process came through the American Henry Laurens, whose son John (killed in the last days of the war) had been his fellow student at the Inner Temple. Laurens senior had been captured on a Dutch ship in 1780 and clapped in the Tower until now, and after his release Day acted as his unofficial secretary until he joined Franklin and Adams among the American Peace Commissioners in Paris.

In May, Day wrote to Boulton, 'I sincerely hope and have some reason to believe, that we are now upon the eve of peace, which I hope will give new life and vigour to everything.'[11] But in July, Rockingham died. When Shelburne succeeded him as leader, Fox resigned in protest and the twenty-three-year-old William Pitt became Chancellor of the Exchequer. Bracing himself to meet the new regime, Day fired off a hundred-page tract, *Reflexions upon the Present State of England and the Independence of America*, and went on to write more diatribes, which later appeared as *Letters from Marius*.

All the Lunar stories overlap. But running beneath them all is the grand narrative of the nation. Everyone shared the feelings of humiliation during the final throes of the war; the jolting sense that Britain's dominions had shrunk, that its empire had turned on its axis, no longer facing west to the Americas but eastward to India and the new lands of Terra Australis revealed by Cook's voyages. The repercussions of war and its links to the drive for reform were felt very strongly in Ireland. When Edgeworth came back in May 1782, the Irish Volunteers – formed to 'protect' Ireland when English armies were in America – were at their height. Oppressive laws had been repealed (including the old penal laws against the Catholics), and the mood was buoyant, even aggressive. When the Act removing the control of the Privy Council and House of Lords was passed in May 1782, the Irish Parliament became technically 'independent'.

Edgeworth returned amid the national rejoicing in early June and immediately embraced the popular movement. He sat in the gallery of Parliament while Grattan was applauded and wrote an address to the Volunteer Corps of Longford, demanding complete reform of the Irish House of Commons, which was almost wholly under the thumb of ministers in Westminster who bought seats, offered places and handed out wholesale bribes. Edgeworth was one of those who first called for a national conference of Volunteers, and he attended it when it was held in Dublin in November 1783. But after the American peace was signed the Volunteer movement lost its leverage and many began to feel that only force would bring change. By late 1783, Edgeworth found himself among the moderates, working to damp down the plans of men like the militant Bishop of Derry.

Many of the Irish gentry, frightened by the surge of feeling, backed away from reform, whose banner was taken up by the lawyers and other professionals, men without land to protect. Edgeworth, however, continued to campaign. In 1784, he attended a Congress in Dublin with several future leaders of the Irish revolutionary movement. And he would always continue to plead his country's cause with anyone who seemed to threaten it, including his good Lunar friends. By now he was arguing fiercely with Wedgwood over Pitt's plans to end trade restrictions to Ireland but they were still entirely at one over parliamentary reform. In early 1784, Wedgwood reminded him of his own feelings:

respecting our House of Commons (that it was indeed a representation of certain lords and other Borough mongers, but it was a fallacy too gross for the shallowest of understanding to suppose it anything like a representation of the people, and I have not yet found any reason to alter this opinion.) I must therefore wish for a reform, and I would have it a thorough one. If you get the start of us, I hope we shall be wise enough to follow a good example.[12]

Edgeworth would continue his political involvement until the end of his life, but his reforming zeal, like Day's, also flowed in other channels. Their early devotion to the ideals of Rousseau had sparked their questioning of authority. But how could his teachings, which valued simplicity and attacked materialism, be reconciled with progressive science and industry?

Edgeworth managed this by sleight of hand, because his life was dominated by a parallel passion for mechanics and technical improvement. Day, by contrast, became increasingly reactionary and eventually a fierce opponent of the rise of industry. But there were two ways, perhaps, in which Rousseau could be reconciled with improvement: through education, and through farming – returning to the land. It is not surprising, therefore, that the two younger Lunar men were involved in both these fields. Teaching the young offered a blank slate on which to build a new consciousness, a new world. Both men had a romantic conviction that the labourers and peasants who worked on their land were somehow 'simpler', better material – they had less to 'un-learn'. Edgeworth would use a revealing image a few years hence, of paper-making, the creation of that desired blank sheet, and the printing of a new text. He was writing to Darwin about the fear of insurrection in Ireland,

And yet the people here are altogether better than in England. The higher classes are far worse; the middling classes far inferior to yours, very far indeed; but the peasants, though cruel, are generally docile, and of the strongest powers, both of body and mind.

A good government may make this a great country, because the raw material is good and simple. In England to make a carte-blanche fit to receive a proper impression, you must grind down all the old rags to purify them.[13]

Day and Edgeworth held typical 'enlightened' views on the importance of agriculture. For years now, more and more acres had been enclosed; farming practices had been improved with better breeding of livestock and rotation of crops; new ploughs and machines, such as Jethro Tull's seed-drill, were slowly introduced; new crops such as clover and turnips were grown; the Leicestershire stock-breeder Robert Bakewell bred sheep and cattle and pigs to produce more meat. Since the late 1760s, Arthur Young had travelled the country evangelizing the universal benefit of new techniques: his influential *Annals of Agriculture* began publication in 1784. From the late 1770s on it became fashionable for the landed gentry – and even for the King, 'farmer George' himself – to cultivate an interest in farming, to build model farms, to have their portraits painted not in their drawing rooms or landscape gardens but with their cattle and sheep in their farmyards and fields. There were imperatives other than fashion: the rising population meant an increase in demand (in the 1780s Britain ceased to be a corn-exporting country and began importing wheat). Farmers were urged by men such as William Marshall, whose *Minutes of Agriculture* and *Experiments and Observations* appeared in 1778 and 1779, to see themselves as mini-manufacturers, keeping their books neatly and managing their labour force efficiently. But even 'improvers' could still act like feudal lords: in 1786 the Earl of Dorchester would sweep away a ruined abbey, build a new mansion in its place and create a model village discreetly out of sight.

The 1780s and 1790s were critical years on the land, and agriculture inevitably entered the circle of Lunar interests. Priestley helped Young with his experiments on plants, but the appliance of science to the soil appealed especially to Darwin and Edgeworth, with their passion for mechanics and chemistry or botany. Darwin wrote expansively to Edgeworth about ploughs, manures and drainage and his own Derby garden, where the 82-foot-long hot-house produced 'abundance of Kidney-beans, cucumbers, Melons and Grapes'.[14] He was no farmer, but his last book, *Phytologia*, would deal with plant growth and cycles and methods of boosting production, increasing man's mastery of nature. He had old connections with the Edinburgh chemists and doctors, Cullen, Black and William Fordyce, for whom farming had long been a central issue: one of the books that made most impact in the 1770s was Lord Kames's *The Gentleman Farmer; being an Attempt to Improve Agriculture by Subjecting it to the Test of Rational Principles*.[15] He was also inspired by his friend James Hutton, who was not only 'the father of geology' but a practising farmer and agricultural theorist, writing a thousand-page manuscript, 'Principles of Agriculture'. Day and Edgeworth, however, were rather different from other improving farmers. Their reforming politics and old ideals made them focus not only on the land but the lives of the people who worked it. As Day found, this could be a thankless task. In a way, one feels his efforts were doomed from the start, informed by an unrealistic nostalgia for rural innocence. In 1781 he bought one of the most unpromising patches of land possible, two hundred acres of Surrey heath attached to Anningsley House at Ottershaw, near Woking. He settled there the following year. 'The soil I have taken in hand, I am convinced, is one of the most completely barren in England,' he wrote, with morose relish.[16] He set to work reclaiming marshland and planting woodland and employing the local labourers at humane, uneconomic rates, keeping them on in the slack winter days. He dosed them with medicine, and took religious services, while Esther ran a Sunday school and a knitting group to keep the poor in socks. He never made a profit – indeed he made a substantial loss, around £300 a year, he later told Edgeworth – partly, perhaps, because his altruism, which extended to all living creatures, rather hampered production. In *Sandford and Merton*, when Mr Barlow points out the flock of larks to Tommy, he says with pride:

> These little fellows are trespassing upon my turnips in such numbers, that in a short time they will destroy every bit of green about the field; yet I would not hurt them on any account . . . they find a scanty subsistence, and though they do me some mischief they are welcome to what they can find. In the spring they will enliven our walks by their agreeable songs.[17]

The local gentry raised their eyebrows at such eccentricities. Only Samuel Cobbett, farming (profitably) not far away in Cobham, openly admired Day's philanthropy and gave sympathetic advice. Day grew daily glummer, morosely accepting that recognition and gratitude were not on offer, and that virtue was its own cold reward.

> What then is the proper employment of benevolence below? In my opinion, to rectify as far as we are able (and that is very little indeed) the evils which proceed from the unequal distribution of property – by relieving those, first, who are in absolute want of the necessities of life, and particularly those who want them without their own fault.[18]

But there was no cause for optimism: 'for there will in every country be more than a sufficient crop of gentlemen and ladies, growing up like thistles among the corn.'

*

Edgeworthstown

Day grew more and more misanthropic; but in Ireland Edgeworth was still full of faith in human nature. Edgeworthstown was isolated, their nearest gentry neighbours being the Earl of Granard, eight miles away. The Pakenhams, friends of Edgeworth's childhood, were a bit further off, but 'there was a vast Serbonian bog between us; with a bad road, an awkward ferry, and a country so frightful, and so over run by yellow weeds, that it was aptly called by Mrs Greville, "the yellow dwarf's country"'. Much of their time was spent in the old Elizabethan house, with its book-lined rooms, clocks and maps, and 'innumerable ingenious mechanical devices'.[19] On one side of the hall was the library, where all the children worked, the older ones teaching the younger; on the other side was the workshop where everyone had a go at carpentry. Above was a maze of bedrooms and attics, below were low-roofed kitchens where cocks and hens pecked freely on the mud floors.

In this out-of-the-way spot, to the amazement of some old friends, Edgeworth became an innovative landlord, winning the lasting loyalty of his tenants. To begin with things were hard, as his daughter Maria acknowledged. She was thirteen when the family arrived in Edgworthstown and remembered the chaos that assailed them, as it did every Irish absentee landlord coming home: 'Wherever he turned his eyes, in or out of the house, damp, dilapidation, waste! appeared, Painting, glazing, roofing, fencing, finishing – all were wanting.' The yard was full of waiting petitioners, agents and tenants, and widows and orphans with heart-rending tales of distress. But as Edgeworth worked through the complainants, although he 'rated them roundly' at times, they 'saw into his character, almost as soon as he understood theirs. The first remark which I heard whispered aside among the people, with congratulatory looks at each other, was – "His Honor, any way, is *good pay.*"'[20]

But Edgeworth was no naïve outsider with his hand in his pocket. He had been elected to the Royal Society for his knowledge of mechanics in 1781, and now he brought the paternal organization and contractual firmness of Soho and Etruria, and the scientific farming of the Scottish school, to rural Ireland. He did away with the agent and ran the place himself, getting to know each individual tenant. To keep control of the land, and improve it as he wished, he issued only short leases and stopped subletting and family subdivisions. He abolished old feudal dues, like 'duty work' and 'duty fowl', and offered standard rates and cheap, unusually good, farm cottages – a great inducement. Far-sightedly, seeing the hardship that insecure land-holding brought (a bitter source of unrest in the century to come), he granted the right of tenure to holders who improved their land or worked hard. He would not accept defaulters, although he gave his tenants six months more grace than other landlords, and his schemes could seem downright foolish (Darwin ingeniously suggested he could use balloons to carry manure up difficult hillsides), but he was always an emotional soul, touched by distress, generous in need. The Edgeworths lent their tenants money, 'found jobs for their children, and kept in touch with them as far as America after they went away'.[21]

FOURTH QUARTER

For the FIRE hath this property that it reduces a thing till finally it is not.

Christopher Smart, *Rejoice in the Lamb*

30: FIRE

The issue of combustion smouldered at the heart of the 'chemical revolution' of the next two decades, swirling around the small, combative figure of Joseph Priestley in his dusty black coat. Priestley's arrival in Birmingham in 1780 gave all the Lunar men a new impetus. Wedgwood and Boulton discussed plans to raise extra funds for his research and Darwin agreed it would be a pity if 'Dr Priestley should have any cares or cramps to interrupt him in the fine vein of experiments he is in the midst of'.[1] Lunar meetings moved to Mondays, so as not to interfere with his church duties; Darwin invited him warmly to visit if passing through Derby; Wedgwood supplied him with retorts, tubes and crucibles. Soon Withering was reporting to the Liverpool merchant Rathbone on Priestley's work on nitrous air, telling him that 'He has lately been experimenting upon a kind of Air that suffers a Candle to burn freely, nay with an enlarged flame, & yet is instantly fatal to breathing Animals.'[2]

Priestley genuinely relished working with others. In detailed letters, he told Wedgwood about all his experiments. He took advice from Withering, and consulted with Keir. And if Watt was scathing at first about his serendipitous approach to experiment, he soon became a trusted confidant. When Watt was in Birmingham, Priestley's house-boy ran over almost daily to Harper's Hill with notes. One of these – about getting inflammable and fixed air from charcoal – contains a typical PS, inviting himself and his wife and daughter to 'take a dish of tea', clearly to thrash

Lavoisier's and Meusnier's equipment for obtaining hydrogen

out things further. Won over by his gentleness – and his astonishing openness with his results – Watt stopped grumbling at his slapdash ways and was soon sending Joseph Black neat lists of Priestley's 'beautiful' experiments.

After his arrival Boulton too leapt back into chemical research. Early in 1781 Logan Henderson sent him five boxes of ores and spars, or 'two horse-loads' of material from Cornwall. 'Chemistry has for some time been my hobby-horse,' he told Henderson, and he was attacking it with gusto:

I have made great progress since I saw you and am almost an adept in metallurgical moist chemistry . . . I have annihilated Mr. Murdock's bed-chamber, having taken away the floor, and made the chicken-kitchen into one high-room with shelves, and these I have filled with chemical apparatus. I have likewise set up a Priestleyan water-tub, and likewise a mercurial tub for experiments on gases, vapours etc, and next year I shall annex to these a laboratory with furnaces of all sorts, and all other utensils for dry chemistry.[3]

These were grand plans, very Boultonian, and probably unrealized but they were typical of the excitement in early 1781. Amid January frost and fog, Watt urged Darwin to come to their next meeting: 'I beg that you would impress on your memory the idea that you promised to dine with sundry men of learning at my house on Monday next,' he wrote. He was sure he could tempt him:

For your encouragement, there is a new book to be cut up; and it is determined whether or not heat is a compound of phlogiston and empyreal air, and whether a mirror can reflect the heat of a fire. I give you a friendly warning that you may be found wanting whichever opinion you adopt in the latter question; therefore be cautious. If you are meek and humble, perhaps you may be told what light is made of and also how to make it, and the theory proved by both synthesis and analysis.[4]

The touching try at matching Darwin's wit was scuppered by the latter's practice – although he phrased it in far from practical terms:

You know there is a perpetual war carried on between the Devil and all holy men. Sometimes one prevails in an odd skirmish or so, and sometimes the other. Now you must know this said devil has play'd me a slippery trick, and I fear prevented me from coming to join the holy men at your house, by sending the measles with peripneumony amongst nine beautiful children of Lord Paget's. For I must suppose it is a work of the Devil? Surely the Lord could never think of amusing himself by setting nine innocent little animals to cough their hearts up? Pray ask your learned society if this partial evil contributes to any public good? – if this pain is necessary to establish the subordination of different links in the chain of animation? If one was to be weaker and less perfect than another, must he therefore have pain as a part of his portion? Pray inquire of your Philosophs and rescue me from Manicheaeism.[5]

Here was a deliberate tease for Priestley, that believer in the happiness of the many and the ultimate providence of God. But it was a heartfelt question none the less: how could one justify the presence of misery in a world where all was for the best?[6]

Whatever they were working on, the Lunar men always consulted each other. They all had their laboratories, their libraries, their mahogany cabinets with wide shallow drawers of shells and fossils, insects and plants, pebbles and minerals. But with Darwin in Derby, Whitehurst in London and Day settled in Surrey, the dynamic of the group began to change. Soon Edgeworth too left, returning to Ireland in 1782. Letters now became almost as vital as meetings, and just as demanding. In a few years' time, Darwin would appeal to Edgeworth in comic despair:

You write short scrawling letters full of questions, which take up one line, and expect me to send you dissertations in return, on academies, hot-houses, philosophers, attorneys, etc, etc – and then . . . you tell me, after I have been laborious in enquiry, that you knew all this before.[7]

Long letters, often highly technical, flew between all the friends, working out problems, commenting and offering advice.

Darwin was happily building up his new practice and buying a house in Derby to save the journeys through Radburn lanes, but he missed the Lunar conversation. Feeling the financial pinch caused by his house-buying and the final closure of the Wychnor ironworks, and amiably pressing Boulton to repay a long-outstanding loan (he was still asking two years later), he launched into mock-baroque Lunar-speak:

Whether you are dead, and breathing inflammable air below, or dephlogisticated air above; or whether you continue to crawl upon this miry globe, measuring its surface with your legs instead of compasses, and boring long galleries, as you pass along, through its dense heterogeneous atmosphere; – as I am alive, now, I can not recollect how I ment to finish this long period, so here we'l leave it . . .[8]

He gave up, took a breath, and started again.

Darwin was joking, but not altogether. If Heaven was pure oxygen, humanity still lived in a 'heterogeneous atmosphere', a patchwork of good and bad air. Now that 'pure' air could be manufactured, could it not be piped into factories and ships, bottled to cure diseases? In 1782, roused by this possibility, Priestley fired off letters to a number of correspondents including Edmund Burke – who would later turn his idealism, and his aerial imagery, sharply against him.

In the coming decade the group widened, bringing in new younger members. The first was the young Samuel Galton, son of a leading Quaker gun-maker, who joined in 1781, strengthening Lunar ties with the web of Quaker merchants, bankers and manufacturers – Darbys, Gurneys, Lloyds, Barclays and Pembertons. Galton and his father both funded Priestley's work, and Samuel possessed a solid independent fortune (Watt eyed him keenly as a potential investor in steam-engines). A former Warrington student, now keenly building up his library and stock of instruments, telescopes and electrical machines, he became particularly close to Priestley, and their wives were the best of friends.

Galton was an amateur enthusiast in the old Darwin–Edgeworth mould, whose interests leapt from subject to subject like grasshoppers in summer. 'My father's insatiable thirst for knowledge', wrote Mary Ann Galton:

> made his books, his laboratory and other appliances for scientific research more attractive to him than general society. He had a large folio blank book in which he was wont to set down stray pieces of knowledge . . . Information on diet, on training, on pugilism, on horses, on building, the various resistances of timber &c, &c he notes in the book which was entitled the Book of Knowledge; it was alphabetically arranged, and formed many volumes . . .[9]

This multi-volume tome has been lost, and apart from a children's book on birds Galton's only recorded contribution, a fascinating one, was in the area of optics, light and colour. His paper on 'Experiments on the Prismatic Colours' showed that if the colours of Newton's prism were drawn on a wheel in the proportions he suggests, when the wheel was whirled round they would create white – an invention commonly dated a quarter of a century later.[10]

Two other younger men joined the group in this decade. One was the botanist Jonathan Stokes, who helped Withering from 1783, but was also remembered for his chronic absent-mindedness. At one meeting of the 'Lunatics', as the Galtons' butler (and they themselves) called them, 'we were astonished,' said Mary Anne,

> by hearing a sudden hissing noise, and seeing a large and beautiful yellow and black snake rushing about the room. My dear mother, who saw it was not venomous, said to me: 'Mary Anne, go and catch that snake'; which, after some trouble, and thinking all the while of little Harry Sandford and Tommy Merton I succeeded in accomplishing.[11]

Stokes explained that he had 'seen the poor animal frozen on a bank and put it in his pocket to dissect, but the snake had thawed and escaped'. He praised Mary Anne for her prowess, and gave her the snake as a present.

The other recruit, in 1787, was Robert Augustus Johnson, another useful connection for Boulton, and stalwart supporter of Priestley, despite his Tory loyalties. After a time in the Army, Johnson had married the strong-minded youngest sister of the Warwickshire landowner Lord Craven and had settled on her family estates at Coombe Abbey near Coventry. (Lady Craven described Johnson as 'a mild and good man, but entirely governed by his wife'.)[12] Visitors, too, often came to Lunar gatherings, but the result was not always comfortable. In the autumn of 1782, when the diplomatic Boulton was away, Smeaton came to a meeting at Watt's house at the same time as the blind lecturer Henry Moyes. In a tense discussion of rotative engines – clearly set up to pump Smeaton for ideas – the two guests clashed, which 'brought on a dispute which lost us the information we hoped for and took away all the pleasure from the meeting, which lasted two hours without coming half an inch nearer to the point'.[13]

Although outsiders such as Moyes and Smeaton were invited, it has to be said that the Lunar Society was rather a tight, exclusive circle. None of the talented Soho employees was brought in, although William Murdoch and John Southern (who became Watt's draughtsman in 1782) would have brought a vivid dash of technical imagination. And there were certainly no women. Following the Smeaton row, Moyes sent Withering some notes on Bergman's chemical essays (adding a PS from John Hunter who 'wou'd be exceedingly glad to have as much of the skeleton of the

Cat as you can conveniently send him'). Then he handed over to 'one who I find is determind to pour out her whole soul to you'. There followed a screed from the redoubtable Quaker Molly Knowles, a Staffordshire beauty who had married a top London doctor, and had strong views on women's right to learning. Molly had skirmished boldly with Samuel Johnson and was now set on demolishing the 'sly Philosopher' who had, she said, evidently made up his mind on one subject:

> *Women* to possess understandings of 'masculine strength', is an idea intolerable to most men bred up amongst each other in the proud confines of a College. There indeed they seem to monopolize *learning*, but happily *intellect* cannot be confined there; and as general education increases, Scholars will more & more discover to the confusion of their pride, that genius is shower'd down on heads, as seemeth to Heaven good, whether drest in caps of gauze or velvet – in large grey wiggs, or small silk bonnets . . .[14]

Molly was no fool. She had strayed from her point, she said, 'to express to dear Dr Withering some part of the pleasure I feel in the consciousness of being esteem'd by *him* – but to be *praised* by him, is *too* great a tryal for my humility'. Her sarcasm fell on deaf ears.

A few years later, Withering carried on an interesting correspondence with Lady Catherine Wright, a diplomat's wife from Plymouth who wanted to learn chemistry (ignoring Withering's advice that she choose botany, a more feminine subject). Catherine sharply declared that 'the Generalty of men have Agreed that Women ought to be kept in perpetual Ignorance & the most profound Darkness, respecting every part of Literature beyond a Book of Cookery'.[15] Although she acknowledged that 'a Married Woman cannot allways act just as she pleases', she bought a hydrostatic balance, made her own experiments and became absorbed by animal magnetism.[16]

Despite such contacts there would never be any silk bonnets at Lunar meetings. Indeed Faujas St Fond subtly defined the women's role in the Birmingham group, when he looked back on days spent here,

> in the midst of the arts and industries, and in the society of enlightened men and amiable women. Nothing can equal so peaceful a charm; the mind is fed and inspirited; the head is filled with facts, and the heart with gratitude. Such was our experience in the town which we could not leave without regret.[17]

The Lunar men always nurtured contacts with outsiders such as St Fond, welcoming visitors from abroad, and keeping up their contacts with societies in London and others that were springing up elsewhere. But one of the ways a group consolidates itself is in opposition. So far, they had fought many campaigns: for the canals, for their patents, for the rights of the provinces against the metropolis. Now, in the 1780s, they ranged themselves firmly – mistakenly and damagingly for their future reputation – behind Priestley in his fight against 'the new French chemistry'. The arguments circulated around language, measurement, and methods of experiment as well as underlying theory. The battle was profound and its significance was well recognized at the time: in 1796 Priestley himself acknowledged that 'there have been few, if any, revolutions in science so great, so sudden, and so general, as the prevalence of what is now usually termed the *new system of chemistry*, or that of the Antiphlogistians, over the doctrine of Stahl'.[18]

As contemporaries saw, Antoine Lavoisier was no sole genius, but was seizing on an approach already favoured in France, where a key component was the insistence on measuring change in weight, before and after experiments. Through this, Lavoisier and his colleagues gradually built up a radically new table of chemical compositions, 'simple substances' that could not, at the time, be broken down further. It was this startling re-mapping of the chemical boundaries that upset so many workers in the field.[19]

After Priestley had discussed 'dephlogisticated air' in Paris in 1774, Lavoisier had done a series of elegant experiments. Everything he tried suggested that in the calcining of metals and in the combustion of all materials one factor was the increase in weight: and he felt sure that this increase was due to the combination, the 'fixation' with air. In August 1778 he was ready to announce that it was the combination of Priestley's 'dephlogisticated air' with a metal that formed a calx during heating, or ash when wood or charcoal burned. Similarly, this was the air that humans and animals inhaled and used up, and 'fixed air' was exhaled: even human respiration could be subject to quantitative research. And the name of this air, he decided, should be 'oxygen' (*oxy* – acid, and *gen* – generator, because he wrongly thought all acids contained it). Combustion was not the release of phlogiston: it was 'oxidization'.

Priestley was rattled, returning fixedly to the subject, like a dog worrying a bone. His fierce insistence persuaded waverers such as Boulton, who confessed to Wedgwood that 'We have long talked about phlogiston without

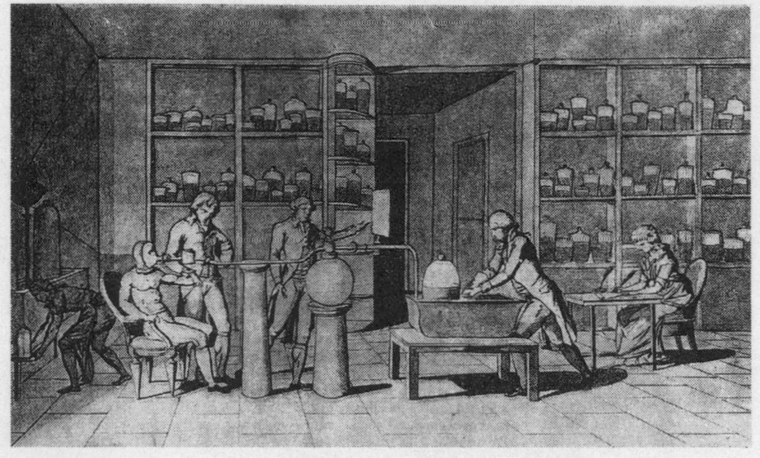

'Experiments on respiration', drawing by Mme Lavoisier, 1785.
Her self-portrait is on the right.

knowing what we talked about, but now Dr Priestly had brought the matter to *light*.'[20] The 'Goddess of levity', he joked, could now be weighed and measured. Wedgwood replied, just as light-heartedly, 'I am quite delighted with the resurrection of poor Phlogiston, as we had been old friends & I could not at my time of life supply his place with another.'[21] The group was suspicious of the terms the French chemists were employing: Lavoisier and his allies such as Guyton de Morveau wanted a new language, free of the old conceptual baggage and metaphoric weight, to express their new findings and ideas.[22] And the Lunar men were wary too of Lavoisier's use of extremely elaborate, expensive equipment for measuring, which required training to use and replaced the old technique of careful judging by the senses – another example, in Priestley's eyes, of taking power from the people and returning it to a new élite.[23]

The ironies are multiple – in the next decade Priestley would be firmly bracketed in the public mind with these same French experimenters. And in the early 1780s, in many ways the Lunar group *wanted* shared ground with fellow researchers in Europe. The more they compared their work to that of others, the clearer became the need for a common language and common standards of measurement. The morass of terminology was becoming impossible. Old words from alchemy, such as 'calx', clung on in use amid new terms coined by French chemists and quite different ones used by the English, or the Germans. Medicine and mining and other specialisms all had their own diction. Chemical symbols, the shorthand of experiment, were equally variable. Watt had been bothered by this, and had himself suggested a new system, in a paper praised by the French chemist Berthollet.[24] Keir had argued earlier that the speech of chemistry had become divorced from ordinary words, and the lack of common terms was holding science back. What he objected to now was that the new French terms carried an inbuilt theory – if you accepted the language you had to accept the implied ideas.[25]

Theoretical concerns apart, growing complexity demanded ever more precise language. (This was true of trade, as much as chemistry: ten years later Matthew Boulton would write to his London agent, 'As you and I shall often have occasion to speak of forms and proportions of buckles, it is necessary we should settle a distinct language that our definitions may be precise.'[26]) In science, standard units of measurement were needed, as well as language. Whitehurst and Watt were both now working on replacing local, traditional weights and measures by universal standards. And Watt was struggling (as if there was no war, no antipathy) to make French and British experiments 'speak the same language', trying to work out precise decimal units for 'the Philosophical pound'.[27]

At the same time, Wedgwood was working on measuring heat in his kilns. No mercury thermometer could withstand their ferocious temperatures and first he tried to judge these by the changes in colour that some clays underwent in firing, creating a graduated colour thermoscope: a mix of clay and metallic oxide, for example, would range from pink to dark brown or black. Then after Bentley's death he hired the experienced Alexander Chisholm as his scientific adviser and tutor to the boys, and with his support he developed a more reliable pyrometer based on the shrinkage of clay at certain temperatures.[28] Ingeniously, he tried to relate his scale to the conventional Fahrenheit scales, making tests at the London Mint to find the melting points of brass, silver, gold, copper, cast iron. His pyrometer was not perfect (and his extrapolation of the Fahrenheit scale was way off beam), but it was the best so far, used for a generation to come.[29] In March 1782 Boulton was talking of Wedgwood's 'curious and valuable thermometer' at a Lunar meeting.[30] In May, Wedgwood heard his paper read in London, and the following January he was elected to Fellowship of the Society – his powerful, mixed credentials suggested by the fact that it was James 'Athenian' Stuart who put his name forward.

Wedgwood could now feel that he was part of an international body of natural philosophers. The French chemists had also been working hard on the quantification of heat: the assumed element '*caloric*' was of immense importance to Lavoisier's theories – in many ways as fallacious a general principle as phlogiston. In comparing his pyrometer with other methods Wedgwood stumbled on a fatal variability in the 'calorimeter' developed by Lavoisier and the mathematician Laplace in 1781, which measured the heat generated in different reactions by running them in an ice calorimeter (a steel 'bomb' immersed in crushed ice and water), and checking how much ice melted. Furthermore, Wedgwood also noted the astonishing speed at which water tended to freeze again, almost at the same time as it was melting. Watt, Priestley and Joseph Black were all excited by this, and Darwin added some imaginative speculations, working out – for the first time – how air cools when it expands, and applying this to explain how clouds form when air rises and expands in the less pressurized higher reaches of the atmosphere, cooling and condensing in the process.[31]

Linked to the phlogiston row was the thorny question of who could claim the credit for discovering the composition of water – Priestley, Cavendish, Watt or Lavoisier. The story, it has been said, 'conforms perfectly to the well-established pattern of British discovery and French explanation'.[32] When Darwin wrote to Watt in January 1781, he tossed out some intriguing speculations. 'As to material philosophy, I can tell you some secrets in exchange for yours,' he promised, one of these being that water 'is composed of aqueous gas, which is displaced from its earth by oil of vitriol.'[33] So far, no one had made this connection between the constituents of air and water, which was still thought to be a single element.

Darwin's words may have nudged Watt into thought. But events were really set in train at a Lunar Society meeting soon afterwards, when Priestley carried out what he called 'a mere random experiment, made to entertain a few philosophical friends, who had formed themselves into a private society, of which they had done me the honour to make me a member'.[34] What had happened was that when he mixed inflammable air (hydrogen)

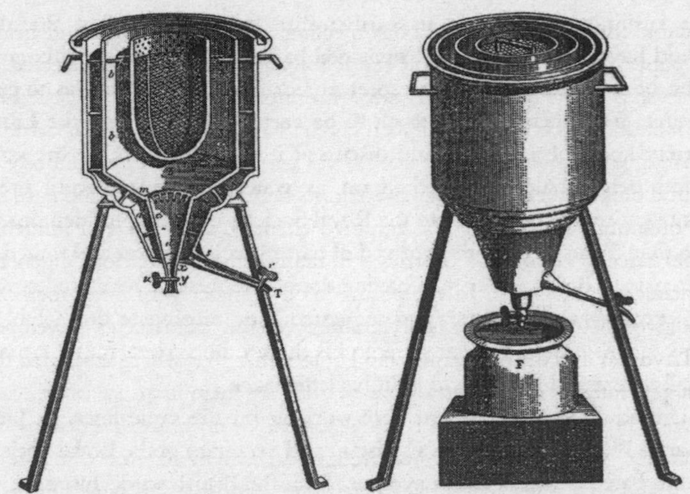

The ice calorimeter of Laplace and Lavoisier, sketched by Mme Lavoisier for the *Traité Elémentaire de la Chimique*, 1789

and 'dephlogisticated air' (oxygen), in a glass flask and sparked it with an electric charge, he found that the inside of the flask was covered with moisture. This March, Warltire was lecturing in Birmingham and came to some Lunar meetings, and, with Priestley and Withering, tried the sparking experiment with both copper and glass vessels. Again, the glass surface was dewy, but Warltire deduced that the moisture was already in the air.

Meanwhile Henry Cavendish saw the sparking experiment at one of Warltire's lectures.[35] He quickly adapted the Priestley–Warltire apparatus to produce more moisture and concluded that 'this dew is plain water, and consequently that almost all the inflammable air, and about one fifth of the common air, are turned into pure water' – in later terms, H_2O.[36] Although he did not publish yet, Cavendish told Priestley straight away. But, at the same time, Watt's curiosity had also been alerted by Priestley's findings, and he reached the same conclusions by a quite different set of experiments.

After talking to Watt, and hearing from Cavendish, in April 1783 Priestley worked on new experiments, in which it appeared that water heated fast in an earthenware vessel was turned not into steam, but directly into air. Many confusions arose. Wedgwood sensibly suggested that the air was rushing into the porous earthenware from outside, but Priestley would have none of this, and remained baffled: 'I have a tolerably good habit of circumspection with respect to *facts*,' he admitted, 'but as to conclusions from them, I am not apt to be very confident.'[37] All the Lunar Society knew of this work and discussed it in their letters. In the same month that Priestley was working on his 'water to air' ideas, April 1783, Watt sent him a draft letter to the Royal Society, including his demonstration that 'Water is pure air deprived of part of its latent heat and united to phlogiston.' But Priestley was having doubts because of his mistake over the earthenware container, and suggested they reconvene the 'Club' to think again. Although Watt stuck to his theory, he agreed, rather crossly. Fatally, he asked Priestley to hold his letter back.

Unknown to Watt, others were working on the same lines. In June, Charles Blagden, Cavendish's assistant and Secretary of the Royal Society, was in Paris, where he told Lavoisier about the British work. Instantly, on 28 June 1783, Lavoisier and Laplace improvised similar experiments, recording the creation of pure water in the French Academy's registers the next day. Watt was too cautious, too slow. By the time he sent a revised draft of his paper to Jean de Luc in early 1784, he had already been scooped. Lavoisier's paper, proving that water 'is not properly speaking an element, but can be decomposed and recombined', had been heard at the French Academy on 12 November 1783, and Cavendish's at the Royal Society on 15 January 1784. When Joseph Banks sent Lavoisier's paper to Priestley, he generously directed him to Watt, 'For the idea of water consisting of *pure air and phlogiston* was his, I believe, before I knew him.'[38] But Watt's earlier paper was not heard until May. Angry at being outmanœuvred he accused Cavendish of plagiarism and smarted under what he saw as class arrogance. Writing of Lavoisier and Cavendish to the Quaker Joseph Fry of Bristol – physician, type-founder and chocolate-maker – Watt grimaced, 'The one is a French financier; and the other a member of the illustrious house of Cavendish, worth above £100,000, and does not spend £1000 a year. Rich men may do mean actions.'[39]

The progress of this vital discovery is like the flow of water itself, the accumulated knowledge building up, finding an entry, running down different channels and finally bursting out. If Priestley did not actually 'discover' the constituents of water it was undoubtedly his innocent, eager gift for communication that brought it about – in England and, much to Lunar chagrin, also in France. Priestley and Watt, of course, couched their analysis of water in terms of phlogiston, but the language in which they wrote was already becoming obsolete. In 1787 Lavoisier and Laplace would publish their *Method of Chemical Nomenclature*, shortly followed by Lavoisier's *Elementary Treatise on Chemistry*, a textbook in which he defined the elements that formed the basis of his new terminology. Here logic and rationality were the key: using Greek roots, each compound would be named after the elements they were made up of, their names depicting the reactions that formed them: adding 'hydrochloric acid' to zinc would create 'zinc chloride', releasing the gas 'hydrogen' in the process.[40] The system was clear and appealing. For a time, the Lunar men stood out against it. Darwin was the first to be persuaded, and tried to win over Watt and Keir, who was now working on a new *Dictionary of Chemistry*. 'I am much obliged to you for your advice to me to be converted to the true faith in chemistry,' Keir wrote sardonically, joking that Darwin's main argument was 'not that it is true, but that it is becoming fashionable. This argument is of great consequence to an author, I grant, more than its truth.'[41] To Keir chemistry was an experimental and above all exploratory science: 'In fact, I neither believe in phlogiston nor in oxygene,' he wrote. Phlogiston was a mere 'mode of explanation', and his objections to the French were that they were dogmatic, pedantic and exclusive. If you adopted the terminology, you implicitly adopted all the unproved ideas that went with it – all the 'oxygene, hydrogene, calorique and carbone, all which are imaginary or at least hypothetical beings'. For the time being, he would stick to the old forms. Eventually, however, he stood down, abandoning his *Dictionary* after the end of the letter A.

31: BOULTON'S BLAZING SKIES

The one person who made no claim to any starring role in the 'water controversy', as Priestley called it, was Erasmus Darwin, even though he played a key role in starting it. But his joint work and friendships did not lapse. In the summer of 1782, for example, he had been fervently trying to interest Wedgwood in his 'great steam-wheel', having discussed his plans with John Michell, who had passed through the Midlands like 'a comet of the first magnitude'.[1] His steam-wheel was rather like Watt's attempt to make a turbine and busy as he was Darwin still thought wistfully of the old free-wheeling, fast-talking Lunar meetings. On Boxing Day, as the year neared its end, he lamented to Boulton that in Derby he was 'cut off from the milk of science, which flows in such redundant streams from your learned lunations; which, I can assure you, is a very great regret to me'.[2] It was sad to think of what he was missing, and he asked Boulton 'to make my devoirs to the learned insane of your Society'. Darwin and Elizabeth now had a daughter Violetta, born on 23 April. Just before this, Darwin had started 'an infant philosophical Society' in Derby. They would not presume, he told Boulton, 'to compare it to your well-grown gigantic philosophers at Birmingham. Perhaps, like the free mason societies, we may sometime make your society a visit.'[3] And, he added, 'I wish you would bring a party of your Society and hold one Moon at our house.'

Boulton, however, had other things on his mind. The adventures of individual members of the Lunar group ran between their joint work, like

Matthew Boulton, sketch for medallion by Rouw

solos or duets in a long choral work, and one long duet was the battle of Boulton and Watt to hold on to the lead in the field of steam. Watt's engine patent, which had driven invention forward so fast, was now becoming a brake to hold it back; other engineers were perpetually thinking of improvements and to beat them they too must find new designs. Up to now, all their engines had been used for pumping – for mines, canals, town water supplies – simply driving a pump rod up and down. But for years Boulton had seen that a far bigger fortune was in the offing if only they could make their engines drive wheels, so that they could power the mills and the machines for the new factories. With this in mind now he reminded Watt of his old work on rotary motion.

On Midsummer Day 1781 Boulton had sent a long, good-humoured letter to Cosgarne. Watt must work on rotation and not worry if some of the pumping engines were delayed. 'Don't vex about it, for what can't be cured must be endured':

> I think I understand all the elements of the horozontal, Eliptical, Ecliptical, Comical, Conical, Rotative, Mill; & a good Mill it is therefore I beg neither it nor any of the others which we have made (and nobody else) may be lost for want of Secureing, as the people in London, Manchester, & Birmingham are Steam Mill Mad, and therefore let us be wise and take the advantage. I do think in the course of a Month or two, we should determine to take out a patent for certain methods of producing Rotative Motions from the vibrating or reciprocating Motion of the Fire Engine, – remembering that we have 4 Months to discribe the particulars of the invention.[4]

This booted Watt into action. That summer he worked fast and hard, grumbling all the while. Rotative engines were a will-o'-the-wisp. Surely pumping would be enough? What about the Fens as a new market? Impatiently, Boulton repeated that there was no other Cornwall to be found. The profits would be tiny, muttered Watt, it would be dropping the substance to catch the shadow. Only Boulton could see that if the premiums were smaller, the market would be vast, and growing every minute: 'The Manchester folk will now erect Cotton Mills enough', he wrote, 'but want engines to work them.'[5] Moreover, if the mills wanted small engines, they could save time 'by making a pattern card of them . . . and confining ourselves to these sorts and sizes' – the beginning of standardization in the engine business.[6]

Finally Watt had tossed aside the simple crank method of getting rotary movement, holding that the uneven stroke of the huge engines would make this unusable (in fact the opposite was true since the crank actually controlled the lunging of the piston). But he was furious in 1780 when the Bristol engineer Matthew Wasborough and the button-manufacturer James Pickard took this up, and swore that their new gearing idea had been stolen from Soho, passed on casually over a pint in the Wagon and Horses by their employee Richard Cartwright. Watt patented five devices that would achieve rotary motion, substitutes for the disdained yet effective crank. The only one that really worked, however, was the 'Sun and Planet motion', and this was not his idea but that of William Murdoch.[7] A second umbrella patent the following spring included Watt's double-acting engine, where the steam pressed alternately on the opposite sides of the piston, making it push as well as pull.[8] This year, too, young John Southern – a fine engineer himself – was employed as Watt's draughtsman at Soho, taking much of the load of detailed work. But the complexities were still endless. Watt's engines were intricate and complex – far too much so, critics said, and every engine was custom built, with its own set of drawings. And now that his double-acting engine was working he had to tackle the most difficult problem of all: how to get tension in the connection to the beam on the piston's upward push – where the conventional chain would go slack – as well as on the pull. On 30 June 1784 he wrote with rare excitement, 'I have started a new hare! I have got a glimpse of causing a piston rod to move up and down perpendicularly by only fixing it to a piece of iron upon the beam . . .'[9] He had developed a clever arrangement of flexible bars and levers, and gradually he now worked out his ingenious, elegant 'parallel motion' – the invention of which he was most proud. The combination of sun-and-planet gear, double-acting engine and parallel motion made Boulton & Watt's rotary engines, like their pumping engines, the best in the world.

Yet, for a long time, steam brought no profits. In 1781 the engine company did not even have the cash 'to pay their Xmas balances nor their workmen's wages, but have had money from B&F on account for those purposes'.[10] A further bitter blow came when Boulton had to sack his partner Fothergill in November 1781 claiming that he was being cheated. Fothergill died bankrupt the following year leaving a wife and seven children.

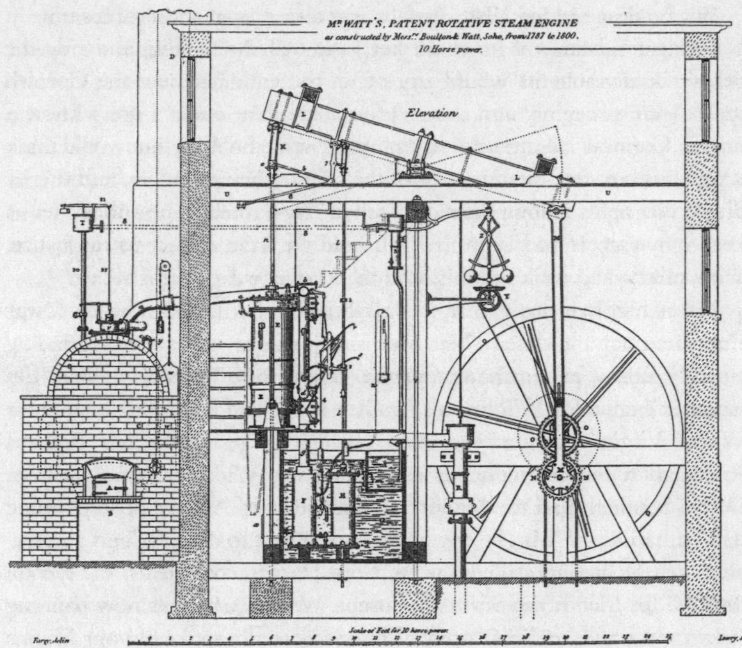

Watt's double-acting rotative engine, 1787–1800, engraved in 1827 for Farey's *Treatise on the Steam Engine.* The 'sun-and-planet' motion can be seen on the right, with the centrifugal governor above.

Without hesitating, Boulton stepped in to help, and also settled an annuity on Mrs Swellingrebel, who had lent Fothergill £6,000 for his initial investment in Soho so long ago, and had never been repaid.

'I feel I could call myself Mr. P, I pay all,' he joked. Meanwhile, thinking of their huge borrowings, Watt was burying his head in his hands, convinced he would die a pauper's death and that Annie and the children would starve. Finally, Boulton gritted his teeth and asked his agent and banker Matthews to get a loan, at any cost, to let his partner off the hook. For the first and only time he burst out at Watt's meanness.[11] Matthews, equally exasperated, cautioned Boulton tartly about ever talking about money to a man who would always moan, 'altho dragged from obscurity where he might have remain'd had you not stretch'd out an Arm to do it'.[12]

But Boulton needed Watt. And he was also a jump ahead of his time in believing firmly that if he could keep the cash-flow going, and ride the debts, the investments would pay off in the end. But now the Cornish mines were struggling and they had to abate their dues: 'I don't know a man in Cornwall amongst the adventurers,' wrote Boulton, 'but would think it patriotism to free the mines from the tribute they pay to us, and therby divide our rights among their dear selves.' Nevertheless, he added, 'let us keep our tempers and keep the firm hold we have got; let us do justice, show mercy, and walk humbly, and all, I hope, will be right at last.'[13]

It was hardly in his nature to walk humbly, but he knew when it was shrewd to bend a knee. Their first rotary engine was already grinding corn at Ketley in Northamptonshire. They were building a huge tilt-hammer engine for Wilkinson's Bradley Forge and a winding engine for Wheal Virgin. Darwin was even angling to get mill-owner Richard Arkwright to buy an engine for a small mill near Wirksworth, not far from Derby. Southern had tried before without success, Arkwright responding that he had heard 'Mr B's engine was so subject to disorder and so complex' that he had not called him in.[14] But Darwin could reach the tycoon through his friends the Strutts or Joseph Wright, who was now painting Arkwright's mills and his family. 'I have repeatedly spoke of your Engins to Arkwright's friends', wrote Darwin, 'and hope you may still be employ'd but I know very little at present of the matter.'[15]

Although nothing came of this, other big orders did arrive. By the spring of 1783 they could relax. Watt's letters were as concerned with heat and water as they were with engines. Boulton could look forward to holidays at home with his family. The previous year, John Wesley noted in his journal, he had 'walked through Mr Bolton's curious works', which employed five hundred people at Soho, and commented on how 'he has carried everything which he takes in hand to a high degree of perfection':

> His gardens, running along the side of a hill, are delightful indeed; having a large piece of water at the bottom, in which are two well-wooded islands. If faith and love dwell here, then there may be happiness too. Otherwise all these beautiful things are as unsatisfactory as straws and feathers.[16]

There was plenty of faith and love, but these were ominous words. In this summer of 1783, Boulton's world would change. The sky itself seemed full of portents. Towards the end of June came a great fall of honeydew and a creeping, hazy mist. Day after day the temperature rose, and a perpetual bluish fog clung low to the ground. No record so far showed a heat to equal this. Throughout July the air grew more and more sultry. 'The sun, at noon, looked as blank as a clouded moon', wrote Gilbert White, 'and shed a rust coloured light on the ground, and floors of rooms; but was particularly lurid and blood-coloured at rising and setting.'[17] Leaves fell early from the trees; fresh meat turned bad in a day; swarms of flies maddened the horses and cattle; the country people were full of fears and superstitions. The rusty light, noted the Rutland squire Thomas Barker, was 'very like Virgil's description of the summer after J. Caesar's death'.[18] The haze lasted for weeks, Barker remembered, 'and neither rain nor fair, wind nor calm, East or West winds took it away; and it was as extensive as it was common, for it was the same all over Europe, and even to the top of the Alps'. The same effects were noticed across Europe and the cause, it was later realized, was the devastating volcanic eruption in Iceland, which continued through June to July, sending lava flowing down the Skafta river and covering the countryside with toxic ash. The 'Skafta fires', and the earthquakes and famine that followed, killed livestock and caused the deaths of ten thousand people, a fifth of the island's population. For a time no one knew this, although the apocalyptic mood was heightened by news of a new volcano suddenly appearing off Norway and of earthquakes and eruptions in Calabria and Sicily.

July was the hottest month ever recorded. Whitehurst, staying with Darwin, could make nothing of the dry, misty weather. Darwin himself thought the dull glow was due to the red rays penetrating more powerfully than other colours. The whole of England was affected. Mist hung over the copper mines in Cornwall and over Keir's alkali factory in Tipton. It reached the Edgeworths in Ireland and curled round the trees on Day's estate at Anningsley. The bloody sun glinted off the windows of Withering's consulting room and off Priestley's powerful burning glass at Fair Hill. Many people fell ill.

On 11 July, in the midst of this terrible, sweltering, misty weather, this month of omens and fears, Anne Boulton stepped across the gravel on to the lawns of Soho House. Men were haymaking near by and the sky was overcast. Boulton had lavished care on his gardens, filling notebooks with details of trees and underplanting, grottoes and fountains. Following the style of Capability Brown, he built a cascade, tumbling down the hillside from a little pool in the woodland to the lake below, with its islands and boat house, wooded headlands and grassy banks. The lake still served the factory as its millpond, but the little pool above, with an ornamental shell feature added in 1778, was purely decorative, cool under the trees, a refuge from the dark sun. Above it, on the slope, stood a little temple of Flora.

A woman carrying beer to the haymakers saw Anne walking by the pool; ten minutes later she returned and found her 'upon her Face on the Water in a shallow part of it'. Desperate efforts were made to revive her, but it was clear that she was dead. The children, home from school for the holidays, were scurried away. Boulton was in Coventry, 'but my heart was at home'. He returned in the early evening 'and was met in the garden by my distressed Children with the fatal news. I cannot say more on that subject. The scene is not discribable by pen or tongue.'[19] Suicide was talked of, as bailiffs had been seen at the house two days before. But Anne had seemed happy, planning a trip with Boulton and the children to Buxton. Several letters mention her headaches and fainting fits and it seems that she may have suffered from periodic bouts of dizziness and had finally collapsed with a major stroke. She had been ill a little while before, but was thought to be recovering. As Garbett told a Lichfield family friend, Mrs Barker, she 'had often complained of a Giddiness in her head, was very liable to fall in Consequence of that & weak Ankles & high Heel shoes'.[20]

For all his flirtations Boulton had been quite sincere when he told Anne he loved her tenderly, as his wife, the mother of his children, and 'as my bosom friend, whom Providence hath decreed to go hand in hand with me through life'.[21] He had shared with her all the trivial, domestic pleasures and pains – worry over Matt's toothache and Nancy's writing master; having to eat breakfast by candlelight because of London fogs; his delight in the jars of tea sent all the way from Peking via Moscow; the problem of cleaning a new wig. It seemed only a short time before that he had ordered a 'genteel Cap &c. for a young lady about 40', and had brought her a silk dress from London, with a special note, 'May this easy cheerfull sprightly Dress be emblematical of the heart, health & spirits of the wearer. Mrs Boulton is requested to be that wearer for the sake & pleasure of him whose choice it is.'[22]

Anne was buried in the Robinson family vault at Whittington, near Lichfield, where Mary Boulton already lay. She disappears from the record. No tributes from friends, no consolations from Lunar colleagues. Perhaps Boulton burned them. 'Light sorrows speak, great griefs are dumb,' he would write later, of a friend in mourning for his father.[23]

Boulton wrote no poem to put in Anne's coffin as he had done for Mary; he designed no epitaph to place in the garden, as he had for Small. But after Anne's death his health collapsed, and all his schemes lay in abeyance. Darwin and Withering both advised him that the best thing he could do, once the children were back at school, would be to get out of Birmingham altogether.

In mid-August Boulton packed his bags and set off for Dublin and then Edinburgh, where Black and Robison introduced him to a young millwright called Rennie (a name he put down in his notebook for future reference). He discussed alkalis with Black; stayed a month with Roebuck at Carron, experimenting on iron ores, and in Perthshire he descended on Lord Dundonald, who had just taken out a patent for getting tar from coal (four years later he would illuminate Culross Abbey with coal-gas flares, the first gas lighting). As Boulton travelled, his saddle-bags grew heavier and heavier with the minerals he collected, and everywhere he went he gleaned information, or took orders. Watt begged him, if he came back via Manchester, 'please not to take any orders for cotton-mill engines': he had heard that they were springing up so fast on the powerful northern streams, that many would go to the wall, but Boulton thought differently: 'I think Fire Mills will in the end rival Water Mills,' he mused.[24]

After a whirlwind trip to London he was back in Cornwall. 'I think if we could but keep our spirits up and be active we might vanquish all the host,' he wrote. 'But I must own I have been low-spirited ever since I have been here – have been indolent and feel as if the springs of life were let down.'[25]

Towards the end of August, as Boulton headed north, the suffocating atmosphere was cleared by a series of violent thunderstorms. But then came an even greater mystery, a flaming meteor that shot across the sky on 18 August, roughly tracing the line of the east coast. At Radburn Darwin watched the fireball pass, showering sparks, and accurately worked out its height – fifty-eight miles up from the earth – from his own informal calculations, based on the height of his roof, and the angle of the house to the compass points. In the winter that followed snow fell on Christmas Day and lay until April, the deep drifts halting the stage-coaches; the next summer, too, was bizarre, bringing widespread floods and flailing hailstorms with stones as big as cockleshells, breaking windows and crushing crops.

In the years ahead Darwin would become increasingly absorbed by meteorology. From the late 1770s, all over Europe, ideas about the link between climate and health had encouraged organized observation. And Darwin's practice, which took him out in all weathers, inspired ingenious devices, such as a weather-forecasting device – a pointer on his study ceiling, connected to a weather-vane above (probably made by Whitehurst), which showed him the direction if not the force of the wind. He was becoming more and more interested in the winds themselves: in the sudden change from cold to warm as a south-west breeze replaced a northerly; in the formation of clouds and the patterns of weather. While his pointer showed the sudden changes, another instrument, an open tube attached to a chimney, with a windmill sail and a series of cogged wheels to show the revolutions, let him meter the north–south airflow. Soon he would develop his own theories of the global movement of the winds, making a series of significant discoveries, including the identification of warm and cold fronts, not rediscovered until the 1920s. His language and imagery remained that of his day, although his thoughts raced ahead. The air was 'rarified', he thought, on a line south of the British Isles, '& rises in a heap, which moulders down on the lower parts of the atmosphere, like the sand in an hour glass'. A current flowed from the cold Poles to the Equator, until the 'heap' at the line was heavier than the polar air, '& then the air oscillates, being very elastic & the upper part having no friction, like a pendulum'.[26] The movement from west to east, as opposed to north and south, he thought, was a function of the movement of the earth itself: eddies in the great air mass swirled and crashed, monsoons were born, trade winds arose, zephyrs played and thunder roared.

To the Lunar men the world was theirs to explore: its plants and minerals, its earths and airs, the fire beneath and the circling winds above. But Darwin's growing interest in the atmosphere was rivalled by a passion that now engulfed the whole group, and indeed most of the nation. In April, and again in June 1783, the Montgolfier brothers had sent the first successful hot-air balloons soaring into the air above France – and they later explicitly acknowledged the work of Black, Cavendish and Priestley as their inspiration. In August, J. A. C. Charles launched a hydrogen balloon, and a month later the Montgolfiers sent up their first passengers, a sheep, a cock and a duck. All came down safely. Franklin sent detailed reports from Paris to Joseph Banks, and to Priestley's friend the Reverend Richard Price.[27] 'The whole world is full of these flying balls at present,' wrote Watt, telling Black of the French Academy's plan to send up a huge balloon, 'and 4 or 5 Criminals tyed to it'.[28] The competition grew intense: in November De Rozier made the first manned ascent in a Montgolfier balloon, swiftly followed within a fortnight by Charles, piloting his own hydrogen balloon. In the same month the young Swiss inventor Ami Argand, a friend of the Montgolfiers, gave demonstrations at the English court.

The Lunar men were in the vanguard of all new developments, especially with hydrogen balloons: Darwin was the first Englishman to fly a large hydrogen balloon, from Derby, on 26 December 1783. In January 1784, despite the terrific falls of snow, Darwin also instigated one of the first attempts at 'balloon post', telling Boulton that his Derby Philosophical Society had tried to send a balloon to the Birmingham group which was calculated to fall in the garden at Soho, but 'the wicked wind' carried it to the garden of Sir Edward Littleton, MP for Stafford.[29] Boulton was already entranced, while Watt read the technical details of Charles's flight, published in France, and Boulton soon acquired the two-volume account of the Montgolfiers' experiences by Faujas de St Fond.[30] When St Fond visited Birmingham this summer, Priestley showed him a new apparatus for making an extremely light inflammable gas 'extracted from iron and water reduced to vapour', which he thought might be enlarged to fill 'aerostatic balloons' more cheaply, without the need for hydrogen.[31] Soon after this Watt sent Banks some early notes by John Southern, whose *Treatise upon Aerostatic Machines* would be one of the first and most practical books on the subject.

On 15 September, in front of a great crowd including the Prince of Wales, Vincenzo Lunardi set off from Chelsea with a cat, a dog and a pigeon, toppling to the ground at Ware in Hertfordshire. (The following summer, in front of a crowd of a hundred and fifty thousand, he arranged to take the famous beauty and actress Mrs Sage, and their friend George Biggins – amateur chemist and inventor of a coffee percolator – into the 'blue Paradisial skies', as shown on the cover of this book.[32]) In October 1784 Jean Blanchard took to the air, floating across southern England in fine autumn weather. In Selborne, Hampshire, Gilbert White roused the villagers to see his balloon pass, a speck dropping from the sky, floating above the church weathercock and the village maypole and on over his great walnut tree. In ten minutes it was lost to sight:

> The machine looked mostly of a dark blue colour; but some times reflected the rays of the sun, and appeared of a bright yellow. With a telescope I could discern the boat, and the ropes that supported it. To my eye this vast balloon appeared no bigger than a large tea-urn.[33]

Everyone now tried making balloons. White's young nephew made one from thin paper, heated 'by a cotton plug of wooll wetted with spirits of wine; & set on fire by a candle'. Fire-balloons became a craze for boys and a hazard for haystacks; Birmingham newspapers offered a guinea reward for news of offenders. In Ireland, Edgeworth's children also had a go. Writing a glowing letter of fatherly encouragement to Maria, he ended, '– I am interrupted – Enter a Baloon – Exit Daughter'.[34] Like kite-flying, there was something magical about the enterprise that brought old and young together. When Matthew and Anne came down to stay with Boulton in Cornwall, he put mining business aside to make a huge paper balloon and a vat of hydrogen gas: 'After great preparations, the balloon was made and filled, and sent up in the field behind the house, to the great delight of the makers, and all the villagers around Cosgarne.'[35]

Tremendously pleased with himself, Boulton decided to try this on a larger scale, as Watt later explained when he told 'the history of Mr Boulton's explosive balloon' to James Lind.[36] He made a balloon of thin paper, about five feet in diameter, coated with varnish and filled with a mixture of roughly one part common air and two parts hydrogen. In the neck of the balloon Boulton had tied a firework, 'a common squib, or serpent, to which was fastened a match of about two feet long . . . When the balloon was filled, the match was lighted and the balloon was launched.' The night was dark and calm, but the fuse was so long that it was about five minutes before it set the squib off, by which time the balloon was a couple of miles away. The huge mob that had gathered to see the experiment were sure the fuse had gone out so when the firework eventually caught they whooped with a 'large general shout'. The noise completely drowned that of the exploding balloon, apparently so loud that people near by 'took the balloon for a meteor, and the explosion for real thunder'. There had, allegedly, been a point to all this. Their intention, explained Watt, 'was to determine whether the growling of thunder is owing to echoes, or to successive explosions; but by means of that ill timed shout the question could not be solved'. He had been at home about three miles away and all he could see was 'that the explosion was very vivid and instantaneous; it seemed to last about one second, and the materials of the balloon taking fire, exhibited a fine fire-work for a few seconds more'.

No one was hurt by fiery falling debris, and if little was learned about thunder a good time was had by all. The fun continued when the first English balloonist, James Sadler, arrived in Birmingham and displayed his Grand Balloon in the New Theatre. An ascent was announced, the prospective aeronaut being named as Mr Harper. The balloon was moved to the Tennis Court in Coleshill Street, and a few days after Christmas crowds poured into town, 'in carriages, upon horses, and on foot', in 'incredible numbers'.[37] Something, however, went wrong. When the launch was postponed the disappointed mob scaled the scaffolding put up for spectators and finally managed to pull it down. In the scuffles the constables arrested four angry punters, one with a fractured skull.

Balloonists were setting records every day. Four days after Harper's bungled launch, Blanchard and the American John Jeffries crossed from Dover to Calais. Birmingham would not be cheated of its own triumphant flight. A second attempt was set for the 10th and this time the Birmingham philosophers were there to help. Withering, Southern and 'other scientific gentlemen' superintended the filling of the balloon. Despite fog and rain, a massive crowd turned out. Two ladies presented Harper with flags and then launched the balloon. 'A sublime and pleasing sight', it rose, hovered, dipped dangerously down towards the crowd. Then Harper suddenly shot upwards and northwards, piercing the cloud: 'and with a pure sun shining upon him, he passed through the purest ether, making such observations as his philosophical friends had suggested'.[38]

The barometer was low, the rain heavy, the air chill: in half an hour Harper rose 4,000 feet. Some time later, over Lord Gower's estate at Trentham, he swooped down to hail a labourer with a speaking trumpet and ask where he was. After almost an hour and a half in the air, he drifted down towards Newcastle-under-Lyme, fifty miles from Birmingham, hitting a tree and careering through bushes before the local blacksmith dragged the balloon's boat to a halt. During the trip Harper had collected bottles of air at different altitudes for Priestley but all were smashed. None the less, after a night at Lichfield he came back to Birmingham to a hero's welcome.

An enamel snuff-box was made commemorating the first Birmingham balloon flight. This was public science with a vengeance, the old combination of flamboyant demonstrator and admiring crowd. It was also, in a way, Lunar science. Ballooning could teach many things – about the power of gases and the resistance of materials, the strength of winds, the variations of the atmosphere. But above all it took men into another element, and promised a new future. Darwin's imagination was certainly stirred. He was already writing his *Loves of the Plants*, and found room to fit in a hymn to the balloonists in their 'silken castle', gliding on high, 'Bright as a meteor through the azure tides', propelling his readers towards gravity-defying inter-galactic flight itself:

Rise, great Montgolfier! urge thy venturous flight
High o'er the Moon's pale ice-reflected light;
High o'er the pearly Star, whose beamy horn
Hangs in the east, gay harbinger of morn . . .
Shun with strong oars the Sun's attractive throne,
The sparkling zodiac, and the milky zone;
Where headlong comets with increasing force
Through other systems bend their blazing course.[39]

*

Balloons were fun, but business called. The rotative engines were in demand, and with his usual scrupulousness Watt issued a detailed set of twenty-four directions for their maintenance. All parts must be oiled regularly, the piston packing changed weekly, the boiler cleaned monthly, and the water kept always at the same height, 'as carelessness in this point may cause the most sudden destruction of the boiler and consequent stopping of the works' (against which one waggish apprentice scribbled 'And hasty general holiday').[40] For the first time, the company of Boulton & Watt was in profit. Watt was credited (on paper at least) with over £4,000 up to the end of 1783. Immediately he transferred the whole sum to Scotland, and when Boulton appealed to him at a later moment of crisis, he said firmly that he had no funds to spare.

Restless, Boulton looked around for more new projects. One seemed a sure winner – the improved oil lamp invented by the Genevan Ami Argand, with a tubular wick and glass chimney which would cut out the smoke and smell.[41] When Argand came to London in 1783 to find skilled workers, he met de Luc, who introduced him to the court where he thrilled the royal family with demonstrations of balloons. Through Samuel Moore, the Secretary of the Society of Arts, he was put in touch with Boulton, who had been interested in lamps ever since the designs of Franklin and William Small twenty years before. In May 1784, a patent was obtained and it was stated that Soho had the exclusive licence. It was not to be. Wondrous samples were made, including an Adam-designed lamp presented to Louis XVI, but the workshops were swamped by Argand's constant improvements, and the market by imitations. Argand's letters became ever more desperate, and Boulton's more evasive. A couple of years later the patent was declared invalid on the ground that the design had already been made known: 'Poor Argand is oppressed with grief,' Boulton wrote ruefully.[42] He did make Argand lamps – silver-plated mantel lamps, brass wall-hanging lamps, Sheffield-plated double burners, sometimes with Wedgwood jasperware bases. The whole group was involved: pages of Darwin's commonplace book and reams of letters to Wedgwood are filled with lamp designs, while Watt also designed carriage lamps. In the end, in 1787, Boulton also began making a new 'hydrostatic lamp' to another design – Argand set up his own lamp factory in Geneva, but died in 1803 in poverty and distress.

Some Boulton ventures succeeded, such as Watt's copying machine. Others failed or faded, like the Argand lamp and Keir's copper alloy. But into all of them he poured his whole heart and effort. And even while the Argand lamp briefly promised to revitalize Soho, he still needed to prove he could truly sell power to the world. Where was power most needed? Every day, for all the people? Surely in providing the staples of life – beer and bread. And where could his grand engines be shown off best? Not in Cornwall, land of sea and fogs, but in London, the great Metropolis. The answer was to make engines for London breweries and to build a London corn mill, a publicity stunt *par excellence*. By late 1783 Boulton &Watt were making engines for two brewers, Goodwyn and Whitbread. Here the engines replaced horses, so there was no saving of coal, and Watt had to work out a new way of calculating royalties. Eventually he estimated that a horse could raise 33,000 lbs of water one foot per minute. 'Horsepower' has been the unit of power measurement for engines ever since.

The brewers were delighted: they could grind without accident or interruption.[43] George III visited Whitbread's and was properly impressed. But the real showpiece was the famous Albion corn mill on the Surrey bank near Blackfriars Bridge, the largest steam mill in the world so far. Boulton and Watt were connected with it from the start, in 1782–83. Watt designed two of his biggest double-acting rotative engines, each powering ten sets of millstones, which ground 150 bushels of wheat an hour and ran hoists for loading and unloading. Samuel Wyatt designed the building and young John Rennie was hired to oversee the mill work. (Watt, perhaps smarting at the outspoken Murdoch, wrote to Robison for a reference, one of his questions being 'whether he is free from that kind of self conceit that convinces a man that his own opinion is better than his masters'.)[44] Rennie's mill work made his name, using iron shafts and gears instead of wood for the first time. And the great London mill brought one final new invention. On a visit there in 1788 Boulton spotted a new device for regulating the distance between the millstones and therefore the speed of grinding, produced by the centrifugal force of two lead weights which rose to horizontal when in motion and fell down when the speed slowed. He often made suggestions, many of them good, and Watt usually dismissed them or claimed he had invented them already. This was different. With Southern's help Watt created the centrifugal governor, two balls flying out as the engine speed increased, working a lever that opened and shut the throttle valve, adjusting the engine's performance. It was his last significant invention.

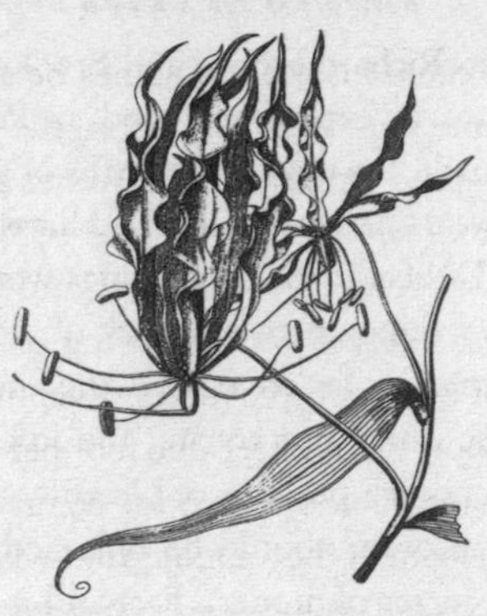

32: THE LINNAEAN ROW

Darwin was shouldering his way fast into Derby life. He became closer to old acquaintances such as Joseph Wright, saw Whitehurst when he was back from London, and made new friends, like the doctor John Beridge who introduced him to the poet William Hayley. A few miles north, at Cromford, stood Arkwright's two mighty cotton mills: he would build a third, even larger, at the entrance to Matlock Dale in 1783. Downstream was the hosiery factory of Arkwright's former partner, Jedediah Strutt.[1]

Strutt's son William – a lifelong friend of Darwin, and through him of Richard and Maria Edgeworth – became a stalwart member of the Derby Philosophical Society.[2] To begin with this was small, meeting in the members' homes with the women dancing attendance, sometimes reluctantly. The Miss Strutts, Sukey Wedgwood reported, had to serve tea 'and did not like this at all', but Darwin, 'with his usual politeness made it very agreeable to them by shewing several entertaining experiments adapted to the capacities of young women; one was roasting a tube, which turned round itself'.[3] Soon this informal group was put on a firmer footing. On 7 August 1784, they adopted their own rules, arranging to meet at the great old coaching inn, the King's Head, and to buy books, with a proper catalogue and strict fines for late return of books and missed meetings.[4] During the 1780s this was an important meeting place for local scientists, doctors, landowners and entrepreneurs, its members including the Strutts, the porcelain manufacturer William Duesbury, the paper-maker Robert

'Gloriosa superba', from *The Botanic Garden*, 1791

Bage, and gentry such as Richard French and Brooke Boothby. At the opening meeting Darwin was in expansive mood as President, preaching the power of science to strangle 'the monstrous births of superstitious ignorance'.

Within a couple of years he had established himself as a force in Derby, just as he had been in Lichfield. His own doings were various and curious: at one moment he was writing on an outbreak of virulent cattle disease for the Derby *Weekly Entertainer*: he advocated the then radical measure of killing and burying the cattle and tracing the infection to its source, as opposed to the contemporary practice of bleeding and purging, and later suggested a government order should be obtained to kill cattle within a five-mile radius – a precurser of modern foot-and-mouth policy.[5] A month or so later he was advising the Duke of Devonshire (nephew of his school-friend George Cavendish, and husband of the gorgeous Georgiana) on diet, drinking and gout. With regard to his new house in Full Street, right in the centre of the town, he was conducting a furious row with a neighbour about a new window looking into his yard. And to improve the water supply before his family moved here in early 1784, he drilled an artesian well (a name not coined for another fifty years), and wrote a pioneering paper on the engineering and the underlying geological principles.[6]

Yet there was something rather strained about Darwin's ceaseless activity. Zestful as he was, a telling note sounds in his definition of one of the diseases later included in *Zoonomia*:

> *Taedium Vitae.* Ennui. Irksomeness of life. The inanity of sublunary things has afforded a theme to philosophers, moralists and divines, from the earliest records of antiquity: 'vanity of vanities', says the preacher, 'all is vanity!' Nor is there anyone, I suppose, who has passed the meridian of life, who has not at some moments felt the nihility of all things.[7]

His recommended cures included 'the agreeable cares of a matrimonial life' and the cultivation of science, 'as of chemistry, natural philosophy, natural history, which supplies an inexhaustible source of pleasurable novelty, and relieves ennui by the exertions it occasions'.

He took his own medicine. Almost as soon as he was married, in quiet hours at Radburn, with its spreading lawns and trees, its hot-houses and old walled gardens, he had begun writing furiously on several subjects simultaneously. Two of these harked back to the creation of his botanic garden in Lichfield when he had mentioned to Anna Seward the possibility of turning Linnaeus into verse, and had also jotted down a note about a possible translation. Before he left Lichfield he had founded a small botanical society – very small, in that it consisted only of himself, Brooke Boothby and a cathedral proctor, William Jackson. Seward was scathing, regarding it as another of Darwin's ego-building enterprises: William Jackson was a man 'sprung from the lowest possible origin', she said, who fawned on Boothby and 'worshipped and *aped* Dr Darwin' – nothing but 'a would-be philosopher, a turgid and solemn Coxcomb, whose morals were not the best, and who was vain of lancing his pointless sneers at Revealed Religion'.

However sycophantic, Jackson was a useful aide. After Darwin left, he cared for the Lichfield botanic garden, and was of help when Darwin finally decided to tackle the translation of Linnaeus's *Systema Vegetabilium*. This idea was perhaps fed by his lingering rivalry with Withering, and prompted by his brother Robert's *Principia Botanica*, an introduction to Linnaeus, already available in manuscript and eventually published in 1787. For Darwin's translation to be successful he knew he would have to co-opt these networks. Letters and more letters flew from Radburn. He sent sample pages to forty or more leading botanists, including Carl Linnaeus the younger (who had succeeded his father as Professor of Botany in Uppsala in 1778) whom Wedgwood had told him was in London.[8]

Darwin was proceeding according to the practice of the time. Many well-known books, such as Thomas Pennant's antiquarian surveys of England or Gilbert White's *Natural History of Selborne* were almost collaborative works, built out of letters and enquiries over several years. Darwin's correspondents included the rural clergyman John Lightfoot, a Fellow of the Royal Society, whose *Flora Scotia* had appeared two years before; the Norfolk surgeon James Crowe, an expert on mosses, fungi and the willows that hung over the broad East Anglian streams; the Dorset doctor Richard Pulteney, another FRS, who wrote on botany for the *Gentleman's Magazine* and had just published a popularizing *General View of the Writings of Linnaeus*.[9]

Thirty-three correspondents wrote back with suggestions, sent samples of rare plants from their neighbourhoods, and argued whether the botanical terms should have Latin or English suffixes. Should 'Petalum', for example, be written as 'Petal'? Darwin deluged Banks with letters and enquiries about books. The linguistic problem was the most immediate, and in some ways the most fascinating. Some decisions were easy: he stuck firmly to Linnaeus's 'stamina' (which he was the first to call 'stamens') and 'pistils' in preference to Withering's prissy 'chives' and 'pointals'. But Darwin was setting out to build a new botanic language, creating vernacular compounds in English as Linnaeus had done in Latin. English, he felt, might be even more suited, its compounds more vivid and precise:

> so the words awl-pointed, for acuminatum; and bristle-pointed for cuspidatum are more expressive than the latin-words. And so end-hollowed for retusum; end-notch'd for emarginatum, edge-hollow'd for sinuatum; scollop'd for repandum; wire-creeping for sarmentosus . . .[10]

The nuances of word creation and transcription could be baffling in their complexity:

> . . . Some of our friends think the word eggshap*e* should be written eggshap*ed* – we thought sword-form and eggshape and halbert-shape etc might be used adjectively better – as a sisser-form*ed* leafe, would be a leaf cut out by sissers, whereas a sisser-form leaf would mean one in the form of sissers. So thread-form*ed* styles, would be styles formed of thread; and thread-form style, one of the form of a thread.

He would, he said (no doubt swallowing some pride), try to get the opinion of Dr Johnson on this, and write it 'eggshap*ed*' if the Great Cham thought it better. And, he added ruefully in his next letter, he was trying hard to avoid any terms that were ridiculous or obscene ('grinning' seemed to be causing him problems).

Joseph Banks, President of the Royal Society for the past three years, was flattered by Darwin's dedication of the work to himself and more than ready to help. He checked Darwin's translation of terms and lent precious books from his own library, such as Murray's edition of the *Systema* and Carl Linnaeus the younger's newly published *Supplementum Plantarum*. By now Brooke Boothby had been to see Banks personally and the work was driving ahead. As soon as the first part appeared in 1783 Darwin was thinking ahead. Perhaps they should embark on a second translation, this time of Linnaeus's *Genera et Species Plantarum*. But was the field clear? According to Withering, a member (unnamed) of the Lichfield Botanical Society approached him in the spring of 1783 to see if he was thinking of undertaking this. Flattered, Withering said that he was too busy with the next edition of his *Botanical Arrangement* to undertake it himself, although he would be happy to help. The next year, he was brought some proof sheets, and as they talked this over Withering brought out some pages of his new edition, illustrating his new method of accents to help pronunciation. He also mentioned that he had employed a friend to translate the *Systema Vegetabilium.*

Throughout the strange summer of 1783 and the balloon mania of that winter the *System of Vegetables* was being published in four parts, re-issued in two volumes, amounting to almost a thousand pages, describing 1,444 plants, in twenty-four classes. A notably helpful feature was a system of pronunciation, using accents. In the Preface, Dr Johnson and all the correspondents were thanked: but one botanist, whose help was not acknowledged, received nothing but scorn:

> Dr Withering has given a Flora Anglica under the title of Botanical Arrangements . . . but has intirely omitted the sexual distinctions, which are essential to the philosophy of the system; and has . . . rendered many parts of his work unintelligible to the latin Botanist.[11]

When the *System of Vegetables* received a glowing notice in Griffiths's *Monthly Review* in 1785, Withering could hold back no longer.[12] He was stung when the reviewer described his own book as a 'translation of Hudson' and sneered at his 'over-strained notions of delicacy'. He was already incensed by Darwin's use of the foxglove cases in Charles's thesis, a crime now compounded by Darwin's paper on foxglove and dropsies (carefully timed to be read at the Royal Society of Physicians in January 1785 just before Withering's own *Account of the Foxglove* appeared). Withering wrote angrily to Ralph Griffiths. He described the visits of the Lichfield Botanical Society and accused them of deliberately sabotaging his own planned translation and of stealing his pronunciation scheme: he would have given permission if asked, but now it would look as though *he* were the copier:

> After such a breach of confidence, I did not wonder that Dr Darwin, in the preface to the work, which I know was written by him, should leave me out of the honourable list of persons they consulted, or that he should in that preface evidently step out of his way to attack the Botan: Arrangment; for I have often remarked that

the man who has injured me finds it more difficult to justify himself, than I should do to forgive him.[13]

The correspondence raged throughout the summer, with Griffiths firmly against publishing his letter, since every aggrieved author would then demand space, and Withering bitterly holding forth upon the correct role of a critic.

Darwin simply shrugged off the dispute as if it was not worth bothering with. His next translation was ready for the press so soon that he had to thrust late material from Banks's librarian Jonas Dryander into an appendix. It was published in two hefty volumes in 1787 as *The Families of Plants, with their natural characters* . . . translated from the last edition of the *Genera Plantarum.* Once again, Darwin used the Preface to goad Withering, as a 'pseudo-botanist' given to inventing 'uncouth' English botanical names.

Botany had become the arena for Lunar competition rather than co-operation. Darwin knew perfectly well that Withering was working painstakingly on his second edition of the *Arrangement* and had even enrolled the help of their new Lunar member, Jonathan Stokes. Withering had known him since childhood and in 1776 Stokes, as a medical student, had reported on the foxglove cases to the Medical Society of Edinburgh. He was bright and well connected: in 1781 he went with John Hunter and others to Royal Society meetings in London where he met both Banks and Carl Linnaeus.[14] Several Lunar men were at the same meetings, Priestley and Whitehurst as Fellows, Boulton and Wedgwood as guests. They spotted him as a rising star: his Edinburgh dissertation (dedicated to Withering) was on dephlogisticated air, quoting Priestley, Keir and Darwin; he had catalogued the plants of the botanist John Fothergill and had attempted a new classification of minerals and earths. Furthermore, he had toured Europe, meeting famous botanists, and was already building up his own plant collection. In May 1783 he was going to the Royal Society as Priestley's guest, and by midsummer he had a job as a physician at Stourbridge:

I am within an hour & ½'s ride of the philosophical circle consistg. of Drs Priestley and Withering & Messres Boulton, Watt, Keir & Galton who meet once a month to converse on Philosophical subjects. They have for title the Lunar Society & I have been regularly invited to their meetings.[15]

He worked hard on the new *Botanical Arrangement*. When its two volumes appeared in 1787 (the same year as the Lichfield *Families*) it included many new plants and descriptions of habitats, plus a long bibliography by Stokes with pithy annotations. The final volume, devoted to *Cryptogamia* (fungi, mosses, etc.), came out five years later. By that time, Withering had fallen out with his helper and Stokes moved away, allegedly taking all Withering's precious books. When he eventually returned them many pages had been cut out as 'illustrations', to Withering's horror. Stokes – who went on to do interesting botanical work of his own – was cold-shouldered from now on.[16]

Withering's long-drawn-out work involved all the Lunar members to some degree, even the children. Mary Anne Galton remembered that during walks with her father, they 'were constantly occupied in looking for new plants, which my French governess would afterwards draw under his direction'. They were encouraged to make natural-history collections and supplied with books and microscopes but were also forced, as she saw it, to help Withering collect fungi, to find new species. To children, this was staggeringly boring so they took nature into their own hands and spent hours 'painting over the fungi in sundry methods' with various chemicals. She was convinced that Withering was fooled.[17]

The two doctors and botanists were sharply different. Darwin was bold and imaginative, Withering pernickety and precise – his books included instructions for readers on how to build their plant-collecting boxes to within a tenth of an inch, and provided little spaces to fill in meticulously when a specimen was spotted. He seemed more interested in bagging species than in the plants themselves. Darwin, by contrast, took a personal delight in his plants. In one later notebook of scuffed brown leather, obviously shoved into a jacket pocket in all weathers, he listed all the hardy plants in his garden, each individual clump he passed on his daily walk down to the Derwent and back: delphiniums and phlox, saxifrage and antirrhinums, peonies and cistus.[18] The notebook's sections are not botanical but domestic: 'Beyond the shed', 'Beyond the fish-pond', 'Turn up towards the summer house'. Pencilled corrections mark sadly, 'gone' (or the even more poignant 'gone?'). His fruit-trees have little asides, such as 'Early nonsuch – a fine apple indeed'.

Darwin could not limit his interest to the translations alone, which were dry, slogging work once the language issues had been solved. He turned more and more to the plants themselves. Gardening forced its way into his inventions; he designed a 'melon-ometer or a brazen gardener', a balanced metal tube with two copper globes, one filled with inflammable air (hydrogen) and the other with mercury, which would swing up and down according to the temperature, and could be put at the back of a hot-bed, or a greenhouse, to hold it open 'as long as the sun-shines'.[19] In his notebook his language sometimes becomes a seamless mix of precise 'scientific' observation and poetic apprehension which looks forward to Coleridge, or to Dorothy Wordsworth. One January morning, he noted:

as I was riding out the trees were exceedingly beautiful, incrusted with so much rhime – about nine in the morning, about 1/2 or 3/4 of an hour before the setting of the moon a warm wind arose & in about 3 or 4 minutes all the rhime drop'd, & the whole face of nature was changed.[20]

Darwin never managed to keep the subjects he was working on apart; his concern was for 'the whole face of nature'. With insouciant confidence he saw his Linnaean translation as a springboard for wider explorations: when he asked Banks for the loan of two books by French and Swiss naturalists in 1783, he explained that he wished to see 'if the physiology of plants can be brought to anything like science – or be compared with that of animals'.[21] He was already experimenting, using coloured dyes to trace the veins in leaves to see if they 'breathed' through tiny pores:

As those insects which have many spiracula, or breathing apertures, as wasps and flies, are immediately suffocated by pouring oil over them, in the year 1783 I carefully covered with oil the several leaves of phlomis, of Portugal laurel, and balsams; and though it would not regularly adhere, I found them all die in a day or two, which shows another similitude between the lungs of animals and the leaves of vegetables.[22]

Many of these investigations eventually found their way into *Phytologia*, his vast study of vegetable life. Still more observations are buried in the notes to *The Loves of the Plants*, of which a first draft was probably ready by 1784. Nothing expresses more clearly Darwin's powerful blend of enthusiasm, wayward imagination and scholarship. It is as if the copious word-making of the translation (leading to the coinage of many now

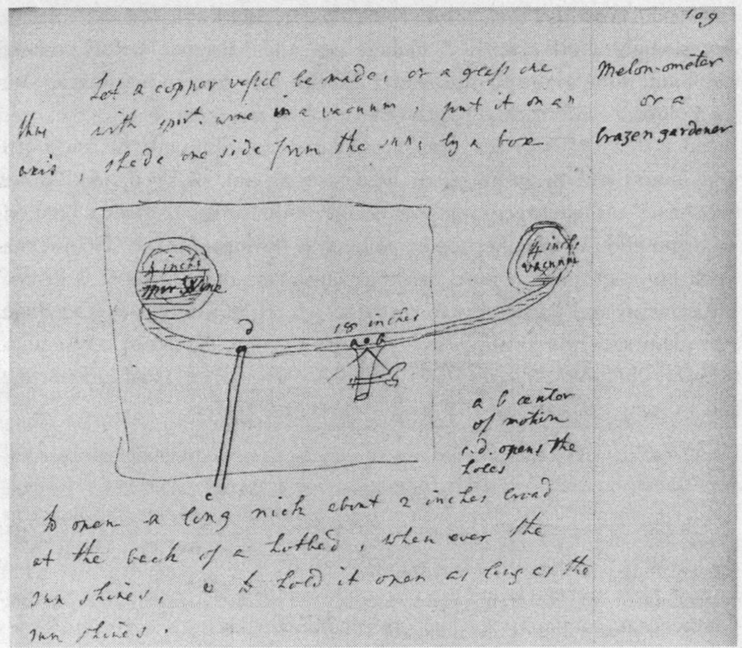

Darwin's melonometer, from his commonplace book for 1781

common words such as 'bracts', 'anthers', 'florets') had triggered some creative impulse. The poem's origins lay in Lichfield, when he had idly suggested to Anna Seward that the Linnaean system was 'a happy subject for the muse'. At the same time he had also dashed down a note in his commonplace book: 'Linnaeus might certainly be translated into english without losing his sexual terms, or other metaphors, & yet avoiding any indecent idea.' In the margin he wrote 'Classes':

> 1 one male (beau). 2 two males &c. 13 many males. 14 two masters (lords). 15 four masters. 16 one brotherhood. 17 two brotherhoods. 18 many brotherhoods. 19 contemporaries. 20 male coquetts. 21 one house (separate beds). 22 two houses (separate houses). 23 polygamies. 24 clandestine marriages.[23]

And after the heading 'Orders', he began with 'one lady (wife belle), two ladies &c (wives)'. The words in brackets in this quote were afterthoughts scribbled above the first word. Even so, he could still see that certain things might shock. He tried 'male ladies' and 'viragoes' before settling for 'male coquetts', and brusquely crossed out 'many marriages' and 'cuckoldoms', before choosing the more sober 'polygamies'.

Since then, perhaps spurred by his desire for Elizabeth, he had been working on and off on his poem. The opening note of *The Botanic Garden* harks back to Linnaeus to write a kind of Ovidian *Metamorphosis* in reverse. To begin with, at least, the undertaking was far from serious. He offers a 'CAMERA OBSCURA' where shadows dance on a white canvas, a 'trivial amusement', an 'INCHANTED GARDEN'. Here is Darwin as showman again, adopting the jovially ironic delivery of some master of ceremonies at family charades. Whereas Ovid, 'a great necromancer', had turned men and women, gods and goddesses into trees and flowers,

> I have undertaken by similar art to restore some of them to their original animality, after having remained prisoners so long in their respective vegetable mansions; and have here exhibited them before thee. Which thou may'st contemplate as divers little pictures suspended over the chimney of a Lady's dressing room, connected only by a slight festoon of ribbons. And which, though thou may'st not be acquainted with the originals, may amuse thee by the beauty of their persons, their graceful attitudes, or the brilliancy of their dress.[24]

The setting is equally playful, a whimsically sexual fairy-tale scene, a *Midsummer Night's Dream* land of ancient oaks, love-sick violets, virgin lilies and jealous cowslips, of glow-worms and spiders and horned snails. But then each flower is introduced, with the action of their stamens and pistils in the 'nuptial bed' of the calyx, the fantasy translated back into botanic terms in the notes that cluster below. Although the flowery entanglements involve every possible combination of males and females, this is a gentle sexual world, free from violence or assault, presided over by a female deity.

In an Interlude the author is challenged by a bookseller. 'Your verses, Mr Botanist,' he charges him, 'consist of *pure description*; I hope there is *sense* in the notes.' The poet claims that his poetry is visual, meant to amuse, while science demands prose, 'as its mode of reasoning is from stricter analogies than metaphors or similes'. Surrendering to poetry or painting is like unreflective dreaming, 'during which time, however short, the objects themselves appear to exist before us'. There is something in *The Loves of the Plants* of a flip-book of images, or a modern animated cartoon. And while the verse is full of fanciful apparatus, it is in the prose, where Darwin speaks in his own voice, that the 'poetic' suggestiveness of his botanic vision emerges. His notes mix Linnaean definitions with his own observations, and with those of authorities ranging from Ray and Derham and Hans Sloane to travel books and histories and the letters of friends. The contrast is obvious even in the shortest stanza, as in this on *Collinsonia.*

> Two brother swains, of Collin's gentle name,
> The same their features, and their forms the same,
> With rival love for fair Collinia sigh,
> Knit the dark brow, and roll the unsteady eye.
> With sweet concern the pitying beauty mourns,
> And soothes with smiles the jealous pair by turns.[25]

The note runs as follows:

> Collinsonia. Two males one female. I have lately observed a very singular circumstance in this flower; the two males stand widely diverging from each other, and the female bends herself into contact first with one of them, and after some time leaves this and applies herself to the other. It is probable one of the anthers may be mature before the other . . . The female of the rose bay willow herb bends down amongst the males for several days, and becomes upright again when impregnated.

Here the comment is matter of fact and to the point, though still heavily anthropomorphic. But often the notes meander to other subjects: to insccts or animals, the use of cypress in myth and history, the insulating function of 'downy clothing' on some vegetables, and even the Osticaks of Russia who manage to mix their mushrooms to produce 'intoxication for 12 or 16 hours'.

Darwin himself is very much present, the weighty doctor in his stout boots bending with rapt attention over a tiny plant or peering down his microscope at stamens and pistils and petal formation. He tells his readers of the holly trees in Needwood forest, prickly at their base but smooth-leaved on the higher branches above the reach of animals. He stops on his rounds on an autumn morning to look at *Tremella*, the fungus found particularly in nearby Dovedale. The verse swoops to a lurid botanical sublime, as the lonely *Tremella* faints amid dark Derbyshire crags, lamenting her secret love:

Through her numb'd limbs the chill sensations dart,
And the keen ice-bolt trembles at her heart.
'I sink, I fall! oh, help, me help!' she cries,
Her stiffening tongue the unfinish'd sound denies.[26]

Once again, Darwin's prosaic note has far more genuine, 'natural' pathos: 'I have frequently observed fungusses of this genus on old rails and on the ground to become a transparent jelly, after they had been frozen in autumnal mornings.'

The notes often have a tone of absorbed wonder, marvelling at the curious variations in plant life and at the clever adaptation of animals, birds and fishes, from the swallow to the whale. And sometimes Darwin's prose, even when its subject is very plain, attains an unselfconscious biblical rhythm:

Caterpillars which feed on leaves are generally green, and earth-worms the colour of the earth which they inhabit; butterflies which frequent flowers are coloured like them; small birds which frequent hedges have greenish backs like the leaves, and light coloured bellies like the sky, and are hence less visible to the hawk, who passes under them or over them . . . The lark, partridge, hare, are the colour of dry vegetables or earth, on which they rest. And frogs vary their colour with the mud of the streams which they frequent; and those which live on trees are green.[27]

Inevitably, as in his commonplace book, his notes are packed with queries. Writing on mimosa, he wonders if sensitive plants fall asleep, like animals. Noting how lichen covers the rocks with tapestry, dying off to create a loam which then supports other plants, he wonders if this was how the whole earth became covered by vegetation 'after it was raised out of the primeval ocean by subterraneous fires'. The sterile stamens of *Curcuma* (turmeric) prompt thoughts of the rudimentary wings on insects or the teats on male animals. Several animals

have marks of having in a long process of time undergone changes in some parts of their bodies, which may have been effected to accommodate them to new ways of procuring their food . . . Perhaps all the productions of nature are in their progress to greater perfection – an idea countenanced by the modern discoveries and deductions concerning the progressive formation of the terraquous globe, and consonant to the dignity of the Creator of all things.[28]

As contemporary knowledge and deductions are brought together here, they provoke a kind of speculation familiar to Lunar meetings. In his eventual Advertisement for *The Botanic Garden*, Darwin would state his aim as 'to inlist Imagination under the banner of Science', leading his readers from sensual experience and poetic pleasure to Linnaean theory and 'the ratiocination of philosophy'. But in 1781 he was more modest, keeping his authorship quiet – understandably, as both poetry and botany were very much women's territory. That October Darwin thanked Banks for commenting on a poem, claiming only to have 'written several of the notes, and corrected some of the verse, a part of which was written by Miss Seward, and a part by a Mr Sayle'. The poem's design, he said, 'was to induce ladies and other unemploy'd scholars to study Botany, by putting many of the agreeable facts into the notes'.[29]

Over the next two years the poem grew. Then in May 1784, on the recommendation of Henry Fuseli, who had offered to illustrate it, Darwin approached the publisher Joseph Johnson. He explained that the draft he was sending would be the second part, and that he was still working on the first, *The Economy of Vegetation*, and made two provisos: that he keep the copyright so that he could amend the poem later, and that it appear anonymously. 'I would not have my name affix'd to this work on any account', he wrote, 'as I think it would be injurious to me in my medical practise, as it has been to all other physicians who have published poetry.'[30] If it could be printed by the Lichfield printer John Jackson (who had just published Anna Seward's verse novel, *Louisa*) he could revise it in the press.

Johnson was keen but still Darwin hesitated. The Lichfield Society translations and Withering's new *Botanical Arrangement* would be well thumbed and read long before *The Loves of the Plants* ever disturbed the dust on the booksellers' shelves. By that time, this poem of the sex life of plants would be miraculously expanded to fire off in a myriad directions, hymning all the achievements of the day, from canals to button-making.

33: PROTECTING OUR INTERESTS

Piracy, politics, patents, spies. The empires of the new industrialists seemed assailed on all sides. Wedgwood's chief concern, as always, was Etruria. The high cost of wheat during the American War and the lack of paid work caused frequent food riots and in March 1783 a crowd mobbed a barge laden with corn on the canal outside Etruria, believing it was being taken to Manchester to be sold at inflated prices. The militia was called and the mob was told that if they did not disperse within an hour, the order would be given to fire. An hour passed. 'Dr Falkener had got the word "fire" in his mouth,' twelve-year-old Tom Wedgwood told his father, 'when two men dropt down by accident, which stopt him, and he considered about it more. The women were much worse than the men.'[1]

Wedgwood was shaken by the violence on his doorstep and swiftly published a pamphlet, *An Address to the Young Inhabitants of the Pottery*, urging them not to follow the incitements of their elders. Expressing his belief (as genuine and heartfelt as it was blinkered) that industry did not exploit the workers, but brought benefits all round, he looked back to the time when local houses were 'miserable huts', the lands poorly cultivated, the roads impassable, whereas the district now showed the effects 'of the most pleasing and rapid improvement' – high wages, good houses, roads and canals, and a name known in countries that had never heard of its existence before.[2]

At the time the barge was captured Wedgwood was in London, following

Sir Richard Arkwright, after the painting by Joseph Wright the Commons debates on the treaties with America and France. In April 1783, he wrote to Charles James Fox and the Treasury Commissioner Sir Grey Cooper on behalf of the Staffordshire potters, protesting that their exports were being damaged by restrictions abroad, while low British duties let imports flood in. The lure of newly re-opened borders also made him anxious, prompting a second pamphlet, *An Address to the Workmen in the Pottery on the Subject of Entering into the Service of Foreign Manufacturors.* 'Liberty' had its limits: this was a scaremongering tract, full of horror stories of emigration, of epidemics, bankruptcies and disasters in America, and of the perils of living in France, the lack of a poor law, the difficulty of ever getting home again. To cap it, he offered fifty guineas for information about people trying to suborn local workmen – and in a passage in the draft, later crossed out, suggested setting up a national intelligence network.[3] He also suggested to Watt that manufacturers might combine to form a society to prevent emigration and pressure the Government to represent the interests of their newly powerful class.

One difficulty was that manufacturers had so little voice in Parliament. Up to now many industrialists had steered clear of politics, finding it suited them better to be free from party ties. With odd exceptions, they were curiously indifferent about reform, but inevitably they began to be pulled into the public sphere. Boulton and Samuel Garbett, for example, were both leaders in the formation of a General Commercial Committee in Birmingham in September 1783, created to combat plans to repeal laws against exporting brass, but also holding a broader brief to protect local interests. Another problem, to begin with, was that Government itself was in turmoil. Soon after the end of the American War, almost in recognition of the re-orientation of empire from west to east, Fox brought in a much needed Bill to reshape the East India Company, bedevilled by stories of corruption and violence. To opponents, Fox's proposed 'Commission' looked like jobs for the boys, a quick way to feather Whig nests. Once the Bill passed the Commons, Pitt and the King (who loathed Fox) got together and when it reached the Lords an open letter from the monarch was circulated denouncing the Bill's supporters as his foes. Fox was trumped, the coalition was defeated and next morning William Pitt was Prime Minister.

In the General Election of April 1784 Pitt confirmed his victory.

Boulton composed a jubilant loyal address to King George: like most manufacturers, he was banking on the twenty-four-year-old Pitt. But Pitt faced a fearsome national debt, which he hoped to ease by increasing taxes, cutting spending and rationalizing trade restrictions. To the alarm of the manufacturers he seemed to have no clue about their interests: one of his first proposals was to tax raw materials, in particular dyed cotton and linen. New licences had to be taken out, excisemen could enter a factory day or night, detailed information was demanded and machinery could be confiscated if arrears went unpaid. The cotton interests were appalled by this 'fustian tax' and called for support from other industries. Boulton burst out angrily to the Chacewater mine-owner Thomas Wilson, 'Let taxes be laid on luxuries, upon vices, and if you like upon property; tax riches when got and the expenditure of them, but not the means of getting them: of all things don't cut open the hen that lays the golden eggs.'[4]

Pitt met the most vocal manufacturers personally, including Boulton, who quickly sensed that 'our young Minister' was not listening and would soon pledge himself to carry even more obnoxious measures.[5] He was right. Blind to the forces ranged against him, Pitt stirred more opposition by his proposed commercial agreement with the new Dublin Parliament. His 'Irish Resolutions' were a genuinely well-intentioned measure, meant to heal wounds, ease tensions and generate loyalty to the Crown by removing old protective regulations. But to English manufacturers (cleverly manipulated by opponents of Pitt, including Fox, Lord Sheffield and William Eden, former Whig Commissioner of Trade) this looked like a system whereby English imports into Ireland had to face a tariff, but Irish goods came in free. The cotton-spinners, the potters and the iron-masters combined to fight the treaty, roping in other trades to back them.

Wedgwood stepped in to organize the campaign. What they needed, he believed, was a national organization (an interestingly different form of association from the earlier radical activities of Day). In the icy February of 1785 he got in touch with Boulton and travelled to Birmingham to meet Garbett and others: 'I mean to recommend them the measure of a Committee of Delegates from all the main factories and places in England and Scotland, to meet and sit in London all the time the Irish commercial affairs are pending.'[6] Slowly a General Chamber of Commerce of Manufacturers of Great Britain was formed. Even Watt grudgingly became involved, writing a paper on the iron trade.

Wedgwood spent months in London this year. The old god of rationality was invoked in the cause of British commerce: 'You well know', he wrote impatiently to Boulton on May Day, 'we have to deal with *unreasonable houses*, with *unreasonable people*, with *unreasonable propositions*.' Part of his frustration was with the timidity of the manufacturers themselves, in fear of government and in thrall to their patrons. The iron-masters, convinced they would be ruined, had promised to come to a meeting but not one of them had made it:

> The principal glover in this town has a contract under government, so he cannot come. The button maker makes buttons for his majesty, and so he is tied fast to his Majestie's minister's button hole. In short the Minister has found so many buttons and loop holes to fasten them to himself, that few of the principal manufacturers are left at liberty to serve their country . . .

Another source of fury was the pettiness. Committees, then as now, got mired in detail. The debates in the Commons seemed endless, Wedgwood told Priestley. Yet the ferocity of the movement had startled the Government: sixty petitions swamped Westminster and so many amendments were tabled that although the mangled Bill passed both houses in London, in Dublin it won such a small majority that it had to be effectively abandoned. Exhausted, but feeling he had won, Wedgwood took a rare holiday, spending August on a tour of Derbyshire.

Wedgwood's position had been curious, quite at odds with his declared belief in free trade. He had believed the alarmist rumours of the opposition and thrown his considerable presence behind an unworthy and illiberal measure. The Irish – including Edgeworth – felt they desperately needed to nurture their growing industries and prevent distress. A blockade began and feelings ran high: in 1784 a Dublin merchant was tarred and feathered for breaking the embargo on importing English goods. Trying to explain to Edgeworth his conviction that complete union with England would be a far better option than protective tariffs, Wedgwood admitted they were not of one mind. But in politics, he said, as in religion, hardly any two people who thought at all, thought exactly alike on everything. The main thing was '*to agree to differ*, to agree in impartial investigation and candid argument'.[7] There had been nothing impartial about Wedgwood's battle, but he wrote long and passionately to support his point.

Politics was one problem, patents another, particularly for Boulton and Watt, who had already faced more than one challenge. In 1781 Jonathan Hornblower junior, backed by a Bristol capitalist, had taken out a patent for a two-cylinder 'compound' engine working with expansive steam.[8] Boulton was chipper and unworried by any Hornblowers:

> If we do suffer our selves to be pissed upon by that family we deserve to be sh–t upon. For Gods sake dont loose a moments peace upon that head, but proceed, get the Engines erected & Money in our pockets.[9]

The Hornblowers went ahead, setting up their first compound engine at Radstock Colliery near Bath. 'I am rouzed', Watt roared uncharacteristically, 'and shall prepare myself to meet the worst and not lie down to have my throat cut – I beg you would summon up your resolutions & not lose the battle before you fight it.'[10] Their engine erector James Law tried to spy on the Radstock engine-house but could only squeeze a glimpse through a window; Watt then set out himself to Bristol where he learned that the chief partner, Major Tucker, was in Bath. Having followed him there and heard that he was out hunting he finally cornered him in the village pub, 'a potato-faced, chuckle-headed fellow with a scar on the pupil of one eye. In short I did not like his physiog.'[11] The major suavely promised to bring the matter up with his partners and Watt had to be content with putting a notice in the Bristol paper about infringement of his patent.

The panic had its farcical elements, yet Watt was right to be nervous. Not long before, Arkwright's patent for a carding machine had been challenged on the grounds that it was void, because obscure and incomplete. 'I don't like the precedent of setting aside patents through default of specification,' Watt wrote anxiously. 'I fear for our own . . . all the bells in Cornwall would be rung at our overthrow.'[12] The divisions of interests became clear in 1783 in yet more Arkwright battles.

If Boulton was king of the West Midlands, Arkwright was emperor in the East. Born in Preston in 1732, he had been a barber, a peruke-maker and innkeeper before experimenting with spinning machines. After establishing his patent in 1769, he set up in business, partly financed by Jedediah Strutt and his partner, and soon grew enormously rich, building his own mills and licensing his power-driven water-frame and carding machine to others. We can still see him vividly in Joseph Wright's brilliant portrait and in the description Thomas Carlyle drew from this: 'a plain, almost gross, bag-cheeked, pot-bellied man, with an air of painful reflection, yet almost copious free digestion'.[13] It was widely claimed that the key elements of the water-frame, the source of his fortune, had been based on the designs of others, including Lewis Paul and John Wyatt, whom Boulton had pulled out of poverty twenty years earlier. Perhaps because of this, as well as Arkwright's bold-faced assumption that he knew more about engines than they did, Boulton and Watt found him hard to take. When he lost his patent in 1781, Boulton reported that everyone who knew him agreed he was 'a Tyrant & more absolute than a Bashaw & tis thought his disappointment will kill him. If he had been a man of sense & reason he wd. not have lost his patent.'[14]

Darwin, however, living just downstream from Cromford had become entranced by the mill's machinery, cramming his commonplace book with designs for spinning machines and stocking-frames. He added a hymn to Arkwright's spinning process in *The Loves of the Plants*, under *Gossypia*, the cotton plant, accompanied by detailed notes:

> Next moves the *iron hand* with fingers fine,
> Combs the wide card, and forms the eternal line;
> Slow, with soft lips, the *whirling Can* acquires
> The tender skeins, and wraps in rising spires;
> With quicken'd pace *successive rollers* move
> And these retain, and those extend the *rove*;
> Then fly the spooles, the rapid axles glow; –
> And slowly circumvolves the labouring wheel below.[15]

His determination to turn everything into images was too much for anyone with a sense of the ridiculous and later parodists would gleefully turn carding to cooking:

> The spiral *grooves* in smooth meanders flow
> Drags the long *chain*, the polish'd axles glow,
> While slowly circumvolves the piece of Beef below.[16]

Darwin's enthusiasm was genuine, however, and in January 1785 he appealed to Boulton to help Arkwright get back the carding patent. He had been asked to look over the 1781 case and thought the decision unjust. Arkwright had made many improvements to the machine since, 'all of which I am master of, and could make more improvements myself'.[17] And if this patent had been overthrown so easily, was Boulton's own secure? This was a common cause: it would be in his interest to give evidence at Arkwright's trial, and might also lead to profit –

> and you make fire-engines for cotton-works in future; which I understand from spinning wool also, will increase tenfold . . .
>
> Tho' your inventions, by draining Cornwall, have supply'd work to ten thousands – Mr Arkwright has employ'd his thousands – I think you should defend each other from the ingratitude of mankind. His case comes on the 14th of February, [so] that the time is urgent.

The arguments bore weight. Watt agreed to go with Darwin to London, braving the freezing weather. But when he did get in the witness box his evidence was so guarded as to be virtually useless, while Darwin's stutter gave the court great trouble. None the less, after the long day in court, the verdict went to Arkwright.

His triumph did not last. Since 1781 Lancashire cotton-spinners had spent a fortune on buildings and machines, employing around thirty thousand people – men, women and children. They could not afford to become his licensees at prohibitive rates and in April they applied to have the decision annulled. The momentous trial of *Rex* v. Arkwright took place in June, and once again Darwin and Watt jolted to London. The argument lasted till one in the morning, when the jury brought in a verdict for the Crown. A jubilant Manchester broadsheet crowed that 'the old Fox is at last caught by his over-grown beard in his own trap'.[18]

In September Wedgwood suggested that Watt should meet Arkwright to discuss the whole issue of patents. They did meet, and drew up the 'Heads of a Bill', but there matters stayed.[19] In the event, Arkwright's demands for a new trial were refused, and the King's Bench ruled that the letters patent be cancelled.

It was often argued, and rightly, that patents held back other inventors. This was true of no one so much as Watt, who was ruthless in this respect.

23. Blue John mounted ewers, Soho, c. 1772
24. Matthew Boulton's Sidereal clock, Soho, 1771
25. Jasper 'Triton' candlestick by John Flaxman, c. 1775
26. Jasper medallions mounted in cut steel by Boulton

27. *The Wedgwood Family*, by George Stubbs, 1780.
On horseback: Thomas, Susannah, Josiah II, John; in foreground: Mary Ann and Sarah, Catherine, Sally and Josiah. There is a distant view of Burslem to the right.
28. *The Corinthian Maid*, by Joseph Wright, c. 1782–5

29. *George Stubbs*, by Ozias Humphrey, c. 1775
30. *The Fall of Phaeton*, jasper plaque designed by George Stubbs, c. 1785
31. The Portland Vase, blue-black jasper, 1790
32. *The Apotheosis of Homer*, jasper plaque designed by John Flaxman, 1778
33. Jasper tea-set 'Domestic Employment' designed by Lady Templetown, modelled by William Hackwood, c. 1783

34. *The Edgeworth Family*, by Adam Buck, 1787
From left to right: Maria, Emmeline, Henry, Charlotte, Sneyd, Lovell, Richard, Anna, Bessy, Elizabeth holding William, Honora.
35. *Richard Lovell Edgeworth*, by Hugh Douglas Hamilton
36. Wedgwood creamware teapot for the Volunteer Societies of Ireland, c. 1775

37. *Matthew Boulton*, by C. F. von Breda, 1792 38. *James Watt*, by C. F. von Breda, 1792
39. *William Withering*, by C. F. von Breda, 1792 40. *Samuel Galton*, by Longastre

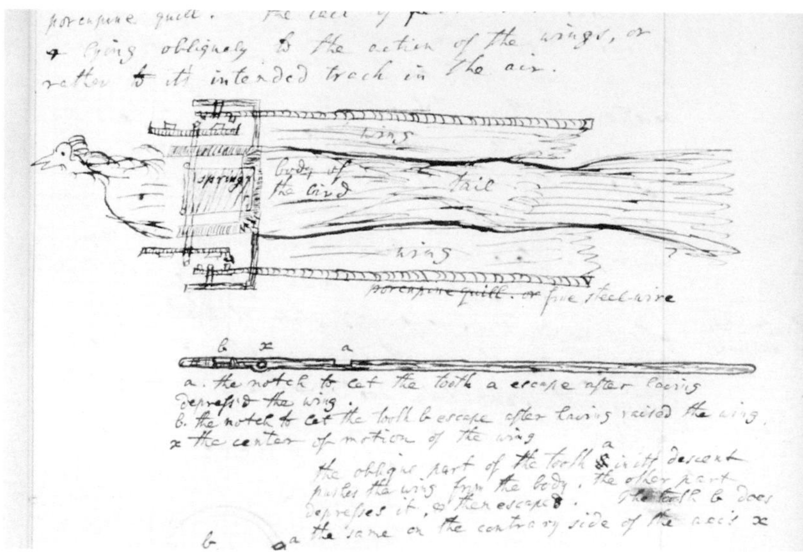

41. Darwin's Mechanical Bird, from his Commonplace Book
42. *Erasmus Darwin*, by Joseph Wright, 1792
43. *Cupid Inspiring the Plants with Love*, by Philip Reinagle, from R. J. Thornton, *The Temple of Flora*, 1805

4. *The Hopes of the Party prior to July 14th*, by James Gillray, 1791. George III being beheaded: Horne Tooke on left and Priestley in black on right of executioner

45. Priestley's house, from *Views of the Houses destroyed during the Riots at Birmingham*, 1791

46. *Joseph Priestley*, by Ellen Sharples, 1797

47. *The Face of the Moon*, pastel by John Russell, 1795
now in Soho House

The umbrella patent of 1784, for example, contained a special clause laying claim to any use of steam on a wheeled carriage, something William Murdoch had been working on. He was already running a little three-wheeled engine round his living room in Redruth; in 1776 he made another model 'Travil a Mile or two in a Circle' in the Assembly Rooms at the King's Head in Truro, carrying a fire shovel, poker and tongs; he even ran the small fiery monster down a lane at night, terrifying a local vicar who thought it was the Devil (although a partisan Trevithick swore it 'wobbled like a lame duck').[20] But Watt's patent blocked Murdoch developing this any further – and once again, it was Trevithick, in the 1790s, who successfully took up the idea.

Manufacturers also had to deal with other threats, from industrial spies. Watt had been wary ever since 1779, when he showed two Prussian visitors round Soho and took them home to dinner, only to learn that they had built engines of their own on his designs as soon as they got home. Wedgwood's fear of espionage had also grown over the years and at a Lichfield meeting in 1785 he reported that 'three sets of spies' were now known to be operating in the Midlands.[21] And Boulton had made things worse by introducing to Wedgwood the Danish businessman Ljungberg, who settled in Birmingham in 1784, even though it was an open secret that he had been sent by the Danish Government to gather information. When Ljungberg left the country five years later, his bags were found to be crammed with drawings of tools and machines and models and even samples of clay, won by bribing Wedgwood's workmen. The Danish Embassy paid £300 to get him out of gaol, and as soon as the cash was handed over he fled home, his secrets well worth the bail money.

Boulton often juggled the virtues of secrets against the value of contacts and in this case the bet paid off: a few years later, when Ljungberg was a high Danish official, Soho received the largest engine order from abroad that ever came its way. He made the same kind of hopeful calculation with Baron Stein, a Prussian mining official who inveigled his way into seeing the new Watt engine at a London brewery in 1787, and was found, Boulton reported to Joseph Banks, 'with his 2 assistants making their attacks with Bribery & Blacklead pencils'.[22] As Boulton wrote ruefully, 'If I have anything to reproach myself of in my conduct towards strangers it is

'The Brewhouse', from Croker's *Complete Dictionary of the Arts and Sciences*, 1764–66

in haveing been too liberal & too unguarded in showing those things which my duty to my Country & to my self required more reservation in.'

It was true, too, that Boulton liked to display his power. He liked to rule. And at Soho, at least to outsiders, he looked like an emperor. The visitors' book included counts from Gottenburg, *Académiciens* from Paris, merchants from Amsterdam. The Baron de Montesquieu made the tour and so did the Viscount de St Chaman; 'Mr Selva, a Venetian architect' rolled up in his carriage, and 'His excellency Allen Fitzherbert, Minister to St Petersburg'.[23] And he still had to pander to demanding patrons such as Elizabeth Montagu, now bombarding him with letters about her new house in Portman Square, plate-glass windows and all. 'A Philosopher wd laugh at my reckoning a house an important Object,' she wrote skittishly, but, 'I am not a Philosopher, & whoever is not, is apt to consider things according to their bulk, & I am sure in that view my House is not a bagatelle.'[24] Boulton had not lost his power to flirt, nor to pull the strings of patronage. When his friend, the needlework artist Mary Linwood, was planning an embroidery exhibition in London, he crooned, 'If my dear Miss Linwood would but confine me in her silken fetters at Soho . . .' and managed to get her an introduction to show her work to the Queen.[25]

In 1786 Boulton was busy showing off the Albion Mill to his grand clientele. Watt was aghast at the rumour that the grand opening in March 1786 was to be celebrated with a masquerade: 'What have Dukes, Lords and ladies to do with masquerading in a flour-mill?' he asked with horror. Everything that made them look conspicuous should be avoided: 'Let us content ourselves with *doing*.'[26] Boulton retorted that he had shown only three people round himself, but when important men such as Eden, Lord Lansdowne and Lord Penrhyn ('who wants an engine'), the Indian Directors ('who will be our customers'), and the President of the Royal Society asked to go round, he was not going to refuse them. But Watt had a point. When Boulton was escorting one group, he suddenly realized that a distinguished *Académicien*, the Marquis de Coulomb, had disappeared: he was discovered in the dark hole where the rotary wheels were, hastily covering a sketch with his handkerchief.

Throughout the 1780s, the cat-and-mouse game continued. If manufacturers could get hold of foreign knowledge or patterns, that was fair game, 'patriotic even'. If men from other countries tried to get theirs, these were 'knavish foreign tricks' that must be stopped. Sometimes the profit motive was refreshingly open: John Wilkinson's brother William had been running a huge foundry in France since the 1770s, and continued to do so even when the countries were at war: the Wilkinsons would be ruled by no one.

The big men could do their own power-broking. In 1785, anxious about the falling price of copper ore due to over-production and to undercutting by the booming Anglesey mines, Boulton helped to set up the Cornish Metal Company, pledged to buy a certain tonnage annually from the Cornish mines (in effect, producing a copper mountain). Boulton had no time for tedious negotiations with any 'set of large wig'd committee gentlemen in solemn dulness'.[27] Instead he called on the combined forces of the fearsome John Wilkinson and the Anglesey tycoon Thomas Williams, who confronted the adventurers at a huge public meeting: 'The Cornishmen would not have submitted to have been kicked and piss'd upon by me as they have been by them . . . Mr Wilkinson and Mr Williams have rose very high in my estimation as men of great abilities.'[28] More mildly, he told his daughter that he had no doubt 'but the revolution we have made will be the Salvation of Cornwall from ruin'.[29]

In 1785 all the Lunar businessmen had shown their strength in public in different ways. They could also be pleased with their rising status in the scientific world: Boulton, Watt and Withering, and then Keir and Galton, were all elected Fellows of the Royal Society this year. Now Wedgwood joyfully accepted Edgeworth's suggestion to lay 'an *embargo* upon politics, wishing it however not to be *temporary*, but perpetual', and turned with relief to other subjects: thermometers and canal barges and the merits of his new air-closet, whose air-flow ingeniously burned crap to ash and answered 'each intention perfectly, the least offensive smell not having been perceived, that I know of, by the nicest nose'.[30]

Yet by February 1786, while late snowdrifts lay heavy on the hedges, Wedgwood was already back in London, working with the General Chamber on the new commercial treaty with France. After the Irish skirmishes, Pitt's arrogance had turned to a wary respect and he tactfully gave the job of negotiating this treaty to William Eden – who had now changed sides and taken a government post. Wedgwood was delighted with the eventual French treaty, signed on 26 September. It was a triumph for British interests, phrased in terms that made it an early landmark of *laissez-faire* economics – even a glimmering anticipation of some form of common European market – providing for equal rights of travel, residence, purchase of goods and religious toleration, 'freely and securely, without licence or passport, general or special, by land or sea'.[31] In Birmingham it was hoped, somewhat optimistically, that 'our buttons will barter for pipes of Champaigne'.[32]

In November Boulton and Watt sailed for France, taking Matt with them. John Wedgwood and Tom Byerley were also on board (John sending back graphic accounts of Boulton's seasickness). The Birmingham men had been invited to inspect the great mill at Marly which supplied drinking water to Versailles from the Seine. They also hoped that the Foreign Minister Vergennes and the Controller General Calonne might both now look kindly on granting them a special licence to build their engines. To smooth the process, they had agreed to advise Calonne's brother on buying the ironworks at La Charité, set up long ago by the Birmingham renegade Michael Alcock.

In Paris, they were greeted by ministers, scientists and bankers including the influential Delesserts family. 'We shall be extreamly busy every day next week in delivering our Letters & makeing visits of Ceremony,' Boulton told his daughter Anne, adding a PS, 'I actually did not know Matt this morng. when I first met him after his hair was dressed a la francois – youl be astonished at the alteration – we dined today at Mr De Lasserts who live magnificently in a palace . . .'[33] He himself dressed from top to toe *à la mode*, and carried a sword. (But what did Watt wear?) The visit was packed. And although they won no definite order and soon Calonne himself was out of a job and their hope of a 'privelege' was lost, yet their trip did bring gains: Watt learned from Berthollet the new technique of chlorine-bleaching, which he passed on to his linen-maker father-in-law in Glasgow.[34] And Boulton – who had enjoyed himself hugely – came back inspired by dreams of another venture, the minting of coinage at Soho.

Others were less pleased by the French treaty. While the thrusting northern mills and factories wanted free trade, the older industries such as the clothiers wanted protection. Wedgwood was pilloried by many for changing tack: arguing for restrictions against Ireland and then for free trade with France. Murmurs of corruption were heard, and accusations of arrogance were rife. The latter were true. It was ludicrous, Wedgwood felt, that 'a man who should get a delegation from the tooth-brush makers of London would have a vote equal with a delegate sent from Birmingham or Manchester'.[35] The Chamber soon split apart.

The alliance of politicians and manufacturers was also uneasy: they came together only to further their own goals. When Sheffield taunted Eden for his sudden friendship with men such as Wedgwood, Eden replied that he knew manufacturers were 'wavering and fickle'; that those who gain 'are shy and sly, and snug and silent, that those who do not gain are disingenuous, and sullen and suspicious; and those who either lose or think they may lose, are confoundedly noisy, and absurd and mischievous'.[36] There was a long way to go to make any common cause. Watt was probably right when he told a French correspondent in 1787 that 'our landed gentlemen' reckon 'us poor mechanics no better than the slaves who cultivate their vineyards'.[37] But in its short life the General Chamber had shown the power of industrial muscle as never before, and had developed methods of organizing and lobbying that would set the pattern for the century to come.

This display of strength ruffled opinion, especially among groups with the older entrenched interests. The farmers were furious when the clothiers' desire to protect their failing trade led to them introducing a Bill to prevent the export of wool. This, as Arthur Young and others saw it, set manufacturing directly against agriculture. At this point another Lunar voice was heard, that of Thomas Day. In his *Letter to Arthur Young*, he was full of praise for the improvements that had followed industry. 'What morasses have been drained,' he cried, 'what arid deserts cultivated, what an increase of animals, of happiness and population!'[38] But he now felt that commerce had become swollen-headed, boorish, unstoppable: 'In its last stage, I fear, it is apt to become an impervious torrent, that threatens destruction in its course, and bears away liberty, public spirit, and every manly virtue.'[39] Hyperbole ruled: the manufacturers were reducing the farmers to slavery, with no more feeling than Liverpool traders. The battle-lines of the future were being drawn up. 'The manufacturers', he wrote, with more passion than accuracy:

> compose an orderly and collected body, closely united under the common standard of interest, ready to profit by the negligence of their enemies, to improve every advantage, and though often repulsed, returning ever to the charge. Their adversaries, on the contrary, though formidable by their numbers, weight and influence, seem to be incapable of union or organization.[40]

Day's apocalyptic outburst was lurid. But, as so often, he was a fore-runner. Many people in the years to come would share his fear and revulsion at the mighty phalanx of industrial power, a new and growing force, marching across this green and pleasant land.

34: FAMILY & FEELING

The various Lunar friendships had now stretched over many years. Writing to Boulton, Darwin justly referred to 'the true spirit of our long and ancient friendship; which I dare say will not cease on either side, till the earthy tenement of our minds becomes decomposed'.[1] Distance made little difference to their affection, which was compounded of working partnerships, shared interests and rivalries. Their children brought them still closer. Matt Boulton, James and Gregory Watt and the Wedgwood and Priestley boys came to know each other well. The Edgeworth girls stayed with the Days and the Wedgwoods and the Keirs. The Wedgwoods and Darwins became fonder than ever. 'Mrs Darwin says she hears your whole family are going to town in a body, like a caravan going to Mecca,' the doctor wrote, '& we therefore hope you will make Derby a resting place, & recruit yourselves & your Camels for a few days, after having travell'd over the burning sands of Cheadle and Uttoxeter.'[2]

At twenty-two, Sukey Wedgwood (now called Susannah or Susan, appropriate to her years) was urging Wedgwood to share the coach to London with Darwin: 'Do consider how happy you will be with the Doctor all the way!! Such a deal of talk.'[3] She often stayed at Derby, while the Pole girls went with her family to London. She was a good source of gossip, reporting home how tired Elizabeth Darwin was, and how she declared she would have no more children after this one. Susan even tried to teach Darwin music. He loved it – and her. 'Your medallion I have not

Susannah (Sukey) Wedgwood

Drawing by Francis Darwin's daughter, showing the Darwin family panicking at the thought of rabies when Francis, who had been bitten by a dog, pretended to bark. On the left of the stairs, from the bottom upwards are Elizabeth Darwin and Susan and Mary Parker, while on the right are the surgeon and Violetta and Emma: Darwin himself comes out of his study to see what the fuss is all about.

yet seen,' he confessed to Wedgwood. 'It is covered over with so many strata of caps & ruffles & Miss Wedgwood is whirled off to a card-party – seen and vanished like a shooting star.'[4] Watching Robert Darwin and Susan together, the fathers put their heads together and made plans.

In 1786 Darwin and Elizabeth had another son, Francis, a lively mischievous boy who would be the centre of many family adventures in years to come. In Ireland the Edgeworths also had a new baby, named Charles. But the older children, now nearly grown, were also on their parents' minds – it seemed a long time since the days of the 'Etruscan school' and some of the Lunar men were now pushing their sons, perhaps too hard, to follow in their footsteps. Of the tall Darwin boys (both well over six foot), Erasmus had settled into a lawyer's life in Derby and Robert, who had spent the past three years at Edinburgh and Leyden, was now setting up as a doctor in Shrewsbury. He was immediately successful, acquiring patients, money and self-importance. To cap this, Darwin made sure that Robert's thesis 'On the Ocular Spectra of Light' (bearing a suspicious resemblance to Darwin's own work, and to Galton's) was published in the *Philosophical Transactions*, and managed to manœuvre his election to the Royal Society – three of the signatures supporting Robert in November 1787 were those of Josiah Wedgwood, Matthew Boulton and James Watt.

These three were also plotting for their children. John Wedgwood was now twenty. He had spent a year in Edinburgh – where Jos and Tom followed him in 1787. But none of the boys liked the city, complaining of the dirt, the moors, the wild wind, the smoky chimneys, the mediocre plays and assemblies.[5] In 1787 John was sent to France, and then to Switzerland and Italy to glean ideas for Etruria. He made the most of his father's contacts and was courteously welcomed in Naples by Sir William Hamilton – and it was the social life, not the commercial opportunities, that he really liked.

Boulton and Watt also wanted their sons to become thoroughly European in outlook, and proficient in French and German. James Watt was sent abroad at fifteen, partly as a drastic solution to family tension: ever since the birth of his children with Annie, Watt's son and daughter by his first marriage had been cold-shouldered. Watt refused to go to Margaret's wedding in Scotland, and regarded James's visiting her as sentimental.[6] James himself was a constant target of Annie's criticism, and of Watt's Calvinistic stress on work and the woes of the world. After he finished two years at Wilkinson's foundry, learning carpentry and machine-drawing, Watt seems to have given up hope of him becoming a good engineer. In late 1784 he went to Geneva (where young Joseph Priestley was already), to study geometry, algebra and drawing under the supervision of de Luc. Even here, Watt rebuked him for finding the lectures of the great Saussure 'pompous', for reading too many novels and going to the theatre, and advised him to 'take care that you do not acquire a taste for nice eating', while Annie taunted him with comparisons. Joseph Priestley had managed to come back without turning into a coxcomb, 'No long bills to barbers and hairdressers for powder and perfumes was in his charges,' she sniffed.[7] James soon moved on; first to Stadtfelt in North Germany to learn German and mathematics, then to Freiberg for mineralogy before returning in 1787 to a chill reception. Annie made it clear that she did not want him in the house. Could he be found a job in the Albion Mill in London? Instead he was apprenticed to a firm of Manchester calico-makers and printers. A short while later, when he got into debt, he appealed in heartfelt terms to Boulton rather than to his father: 'Never having been a young man himself he is unacquainted with the inevitable expences which attend my time of life, when one is obliged to keep good company and does not wish to act totally differently from other young men.'[8] Boulton slipped him £50 straight away, and a brief avuncular lecture.

At the end of the decade, Matt Boulton would follow James to Germany. But in 1787 the seventeen-year-old Matt was in France. He had stayed on after his father's Paris visit, first with a tutor at Versailles and then, to his delight, in the capital itself, studying French science, reading Voltaire and Fanny Burney – and sending home reports on French button-makers. Boulton was more indulgent than Watt, happy for Matt to read novels and curl his hair, writing him letters that mix encouragement and tenderness with great expectations. Nevertheless, even amid the distractions of Paris, Boulton hoped that he was getting his hands dirty. 'A man will never make a good Chymist,' he wrote, 'unless he acquires a dexterity, & neatness in making experiments, even down to pulverising in a Morter, or blowing the Bellows.' This taught him the essential virtues: 'Distinctness, order, regularity, neatness, exactness & Cleanliness are necessary in the Laboratory, in the manufactory, & in the Counting house.'[9]

For a while Darwin, Boulton, Watt and Wedgwood could concentrate on their private lives. But as they did so, another member of the Lunar group, Joseph Priestley, was being pushed into the limelight, and this time the issues were not science or trade, but religion and politics and Priestley's whole vision of the world. Since 1780 he had won a name as a preacher in Birmingham and his congregation had swelled yearly. Catherine Hutton, daughter of the bookseller William Hutton, was among those who joined and became devoted to him. 'I look upon his character as a preacher to be amiable,' she told a friend, 'as his character as a philosopher is great.'

In the pulpit he is mild, persuasive and unaffected, as his sermons are full of good reasoning and sound sense. He is not what is called an orator; he uses no action, no declamation; but his voice and manner are those of one friend speaking to another.[10]

Not everyone in Birmingham found his manner reasonable. The Anglicans were growing increasingly nervous of the strength of Dissent in the town – of Unitarians, Quakers, Methodists and Baptists. In hindsight, this was a time for tact, a quality unknown to Priestley. Within a year of his coming he had swung into action. He set up three kinds of Sunday schools, as he had done at Leeds: one for children under fourteen, another for older children and a third for young men. He taught the principles of Dissent, but also literacy, history, chemistry, and briefly, from 1784, the Sunday schools grew into a local interdenominational movement. If these schools reached out to the working population, his next move – to reorganize the library – was a bourgeois venture. It was founded in 1779, and when Priestley arrived it was just a small collection of books in a cupboard, opened for an hour once a day for a small circle wealthy enough to pay the guinea membership fee and annual subscription of six shillings. Priestley drew up rules (again based on his experience in Leeds, where he had founded a public library), and advertised its existence in *Aris's Birmingham Gazette*. A proper library room was found, opening hours extended, a librarian hired: next came a proposal from 'some of the members to form themselves into a separate society, for the Purchase of Books of science, and especially Foreign Publications of that Class'.[11]

Priestley never voted for the purchase of his own books and always for those of his opponents, but he simply could not avoid the inevitable rows. By now he was also a national figure, both as a scientist and controversial theologian. The rumblings had begun with his *History of the Corruptions of Christianity* in 1782, where he had deliberately courted opposition with Archdeacon Horsley, the editor of Newton's works and eventually Bishop of St Davids. (It became a running joke that any ambitious cleric simply had to take on Priestley, and a mitre would be theirs.) Part of the problem was Priestley's highly individual, millenarian view that progress had to be aided by 'particular providence', the possibility of direct Divine intervention. And such a dramatic change in the world's fortunes might well be accompanied by apparently random suffering, dramatic incidents, calamities.

It was in this mood that on 5 November 1785, the anniversary of the Gunpowder Plot, Priestley preached a sermon, published as a pamphlet, *The Importance and Extent of Free Enquiry*. His subject was actually the slow spread of Unitarian churches and he was arguing that the sect's supporters should not be discouraged:

> We are, as it were, laying gunpowder, grain by grain, under the old building of error and superstition, which a single spark may hereafter inflame, so as to produce an instantaneous explosion; in consequence of which that edifice, the erection of which has been the work of ages, may be overturned in a moment, and so effectually as that the same foundation can never be built upon again . . .

He was careful to stress that his was a gradualist approach: 'Till things are properly ripe for such a revolution, it would be absurd to expect it, and in vain to attempt it.' But the damage was done: the words 'instantaneous explosion' and 'revolution' appearing a few lines apart were quite enough to brand him an insurgent. And he had chosen his words wilfully, ignoring the advice of friends like Wedgwood, who had suggested he drop the 'gunpowder'. From now on, he was nicknamed 'Gunpowder Joe'. (By a curious twist, the nickname should really have gone to 'Gunpowder Antoine' Lavoisier, who had been Director of the Gunpowder Administration for years, and whose improvements put French munitions well ahead of Britain's in the American war.)

After Priestley's incendiary sermon his opponents were quick to move. He had been using such rhetoric for years – one has only to look back to his 'air pump' simile of twenty years before – and it was easy to take extracts from his writings to demonstrate his subversion. Threats were heard, implying that since his meeting house and that of Theophilus Lindsey were not properly licensed they could be closed down. In the meantime, wrote Archdeacon Horsley ominously, one could trust to 'the trade of the good town of Birmingham and the wise connivance of the magistrate', who had so far turned a blind eye, 'to nip Dr Priestley's goodly projects in the bud' before they ripened to dangerous effect.[12]

Then came a comparative lull. During 1786, none of the group exactly relaxed but they felt they could get back to something like normality. Despite the rows boiling around him, Priestley published the sixth and last volume of his *Experiments and Observations*. Between the trips to London, Wedgwood turned back to art, enthralled by the Portland Vase. The spinning machines in Darwin's notebook were succeeded by a flurry of lamp designs. He was polishing off *The Families of Plants*, and tinkering with *The Loves of the Plants*, Arkwright's '*iron hand*' and all. And this year, although Boulton was caught up with the Albion Mill, his diary was sprinkled with Lunar meetings: Monday 15 March at Watt's house (marked with a little crescent moon); 15 May at Galton's; 12 June at Dr Stokes's.[13] Writing to Edgeworth on Christmas Eve 1786, Wedgwood returned to amicable discussions of cobalt, and of china pie dishes (an Edgeworth suggestion that proved a lasting hit, the shortage of flour for pastry increasing the appeal of crusty-looking china lids). The only damage lingering on from his campaign over the Irish trade treaty, he reported, was that his Dublin agent refused to buy any more of his 'double damned ware', after one Irish lord swore that if any was landed in the country he would personally smash it up.[14]

Beneath all their activity the ground was shaky. The fuse lit by Priestley in Birmingham burned slow, but it burned steady. Every aspect of knowledge had begun to take on a political tinge, even language itself: Joseph Spence advocated reforming the alphabet, to make spelling easier for the 'laborious part of the people'; Horne Tooke discussed the evolution of language, defying divisions between the 'vulgar' and élite.[15] This cause too was taken up by the Lunar men, especially Priestley and Darwin.

It was in the realm of language, books and education that Priestley met his first challenge. In Birmingham in 1787, a storm blew up over books of theological controversy in the library, and another over the running of the Sunday schools. The timing of the arguments was not random. This year saw the first motion in the House of Commons to repeal the Corporation and Test Acts which debarred Dissenters from civil office and the universities. Priestley went to the Commons to hear the debates and corresponded directly with Pitt. The motion was defeated, but it could not stop the drive for reform. And as every month passed it also became clear that unrest in Ireland simmered in shadows and corners and that the troubles in France, beneath the glamour that so seduced Boulton, ran far deeper than treaties or trade.

*

The increasingly passionate tones of the reformers reflected the way that the politics of feeling now pervaded all aspects of British life. It could be seen in Day's benevolence at Anningsley and in *Sandford and Merton*, and in numerous drives to help orphans, 'fallen women', prisoners or innocent natives of newly discovered and colonized lands.[16] The influence of Rousseau and other European writers was coloured by Eastern philosophies brought back from India and beyond: animals had souls and even atoms had sensations. From mid-century on, animals became objects of sympathy: Mrs Barbauld had long ago lamented the loneliness of her friend Priestley's mice, confined for his experiments on air, and in her poetry even a worm called for tenderness:

> Beware, lest in the worm you crush,
> A brother soul you find;
> And tremble lest thy luckless hand
> Dislodge a kindred mind.[17]

Such sentiments found plenty of satirists, but few doubted that sensibility was the mark of the civilized. Torrents of feeling poured through Gothic novels with frail heroines facing hopeless odds, and flowed afresh in old ballads rescued from the past. New writers sprang from the provinces, often radical in tone, from the poetry of Burns to the novels of Darwin's friend Robert Bage, which inevitably have a dig at aristocratic pride.[18]

Sometimes sensibility came into conflict with industry. Anna Seward's poem 'Colebrook Dale' of 1785 praises the flourishing expansion of 'polite' Birmingham but expresses horror at the fate of 'violated Colebrook', its glens and woods draped in black palls of sulphurous smoke.[19] But far more important than the rape of rural innocence was the exploitation of human beings. And the politics of feeling clashed most sharply with the politics of trade on one particular, newly ignited issue, that of slavery. Since Day's *Dying Negro* of 1773 the movement had been pushed forward by Wesley's heartfelt *Thoughts on Slavery*; by Adam Smith's curt argument that free labour came cheaper in the end; by the horrifying scandal of the slave ship *Zong* in 1781, whose owner threw 132 sick slaves overboard and claimed the insurance, arguing that his act was within the remit 'to jettison some cargo to save the rest'. In court, Lord Mansfield agreed, comparing the fate of the slaves to the disposal of a cargo of horses.

The cult of the noble savage had idealized the image of the slave, torn from the earth of Africa, and the flood of sympathy chimed with disapproval of the denial of natural rights. Slavery reduced people to brutes, in Priestley's words, 'so that they are deprived of every advantage of their rational nature'.[20] During and after the American war, the movement grew. In 1783 the Quakers presented an Abolitionist petition to Parliament. Four years later the Society for the Suppression of the Slave Trade was founded, joining the London Quaker committee with long-time Abolitionists such as Granville Sharp and powerful Anglican members, Thomas Clarkson and William Wilberforce, MP for the West Riding and close associate of Pitt.

Wedgwood became a committee member, helping to draw up a petition in Stafford. He wrote to Watt, 'I take it for granted that you and I are on the same side of the question respecting the slave trade. I have joined my brethren here in a petition from the pottery for abolition of it, as I do not like a half-measure in this black business.'[21] One of the people he appealed to was Anna Seward, so appalled at the plight of nature yet full of reservations about the fate of slaves. One of her arguments was that abolition would damage West Indian commerce, while another was that slaves would only change their masters, giving their free labour to foreign rivals. There were splits in the movement already, Wedgwood admitted, between those who wanted to abolish and those who merely wanted to regulate the traffic. But even a change of master 'would be a blessing of no small magnitude'. The Spanish system allowed slaves to work towards freedom and at least this gave some hope. 'Contrast this chearing state but for a moment', he pleaded, 'with the absolute despair of a West Indian slave, wearing out by immoderate and incessant labor, with *known* and *calculated* certainty, in the course of a few years.'[22] He could not tell her 'an hundredth part of what has come to my knowledge of the accumulated distress brought upon millions of our fellow creatures by this inhuman traffic'. Anna was won over.

Wedgwood promoted the cause tirelessly, using all his lobbying skills, his mass of contacts, his brilliant grasp of advertising and publicity. He asked Hackwood to model a seal for the Society, showing a kneeling black slave raising his hands in supplication, inscribed with the plea, 'Am I not a man and a brother?' Wedgwood sent a packet to Franklin in Philadelphia, pleased to be fighting a common cause with the old man once again, and linking the individual cause to the wider struggle:

> This will be an epoch before unknown to the World, and while relief is given to millions of our fellow Creatures immediately the object of it, the subject of freedom will be more canvassed and better understood in the enlightened nations.[23]

Hundreds of cameos were produced, becoming popular symbols, like the ribbons worn for causes today – men displayed them as shirt pins and coat buttons or carried them set in snuff-box lids, women as bracelets, brooches, ornamental hairpins. Soon, remembered Clarkson, they were everywhere; for once fashion was promoting justice and humanity. Women transferred the emblem into needlework which they sold to raise funds, and later designed their own medal, 'Am I Not a Slave And A Sister?'[24]

Wedgwood threw himself fully into this new cause, and by the summer of 1788 he was tired. He had a throbbing, ghostly pain from his missing leg, and his eyes hurt – altogether he felt a lot less virile, and when Darwin suggested valerian for his eyes he jokingly approved the old herb remedy, alluding to the story of 'its tendence to make the animal fibre rigid, the very essence you know, of old age'.[25] Darwin was Wedgwood's ally in the anti-slavery campaign, just as he had been twenty years ago in the fight for the canals. Then the thrust had been material, a matter of prosperity of trade and the regions; now it was emotional and humanitarian, but the techniques of persuasion were much the same – marshalling public opinion, attacking Westminster, changing hearts and minds. Darwin recommended reprinting Defoe's *Life of Colonel Jacque*, which had condemned the trade as early as 1722, and put Wedgwood's medallion into his *Botanic Garden*:

> Hear, oh BRITANNIA! potent Queen of isles,
> On whom fair Art, and meek Religion smiles,
> How AFRIC'S coasts thy craftier sons invade
> With murder, rapine, theft, – and call it Trade!
> The SLAVE, in chains, on supplicating knee,
> Spreads his wide arms, and lifts his eyes to Thee;
> With hunger pale, with wounds and toil oppress'd,
> 'ARE WE NOT BRETHREN?' sorrow choaks the rest; –
> AIR! bear to heaven upon thy azure flood
> Their innocent cries! – EARTH! cover not their blood!'[26]

The Wedgwood Slave Emancipation Medallion, black on yellow jasper, modelled by William Hackwood in 1787

Slavery found its way, too, into his depiction of *Cassia*, the American cashewnut whose seeds were swept east by the Gulf Stream across the Atlantic, just as the slaves were carried on the opposite course. And though his appeal was to the legislators, who have 'power to save', it was also aimed at all those who engaged in or even countenanced the trade: 'He, who allows oppression, shares the crime.'[27] Darwin was appalled by what he learned. 'I have just heard', he wrote to Wedgwood, 'that there are muzzles or gags made at Birmingham for the slaves in our islands. If this be true, and such an instrument could be exhibited by a speaker in the house of commons, it might have great effect.'[28]

In Birmingham, as his note suggests, feelings were more equivocal. Following the lead of Manchester, the town soon had its own Anti-Slavery Committee, with Priestley a leading member, which declared that although the local people supported the commercial interests, they would oppose 'any commerce which always originates in violence and often terminates in cruelty'.[29] But resistance was strong. Unlike the great ports of Bristol or Liverpool, Birmingham was not directly involved in the trade but it still depended heavily on it. Leading manufacturers even petitioned Parliament against abolition, claiming that a large part of their wares were 'adapted to and disposed of for the African Trade, and are not saleable at any other Market; and that several Thousands are employed in, and maintained by, such'.[30] Even the Galtons sold their guns to Africa.

Galton, Boulton and Priestley all joined the deputation to welcome Olaudah Equiano, whose autobiography was such a focus of the anti-slavery campaign, when he came to speak in Birmingham in 1789. But Boulton was not above slippery thinking. A few years earlier he had been happy to dine with three plantation-owners 'who wish to see steam answer in lieu of horses', including one 'very amiable man with 10 or 12 Thousand £ a year' – the notorious Pennant, owner of the largest plantation in Jamaica, who went on to invest his sugar fortune in the slate quarries of North Wales.[31] Equiano's visit was hardly a Damascus road conversion: a year later Boulton and Watt were corresponding without a qualm with the slave-trader John Dawson of Liverpool about supplying engines for Trinidad.[32] Powerful interests in the House of Commons, backed by the voice of the King himself, blocked the Abolitionists' way. Their first Bill was roundly defeated and, although Wilberforce and his allies fought on year after year, it would be almost twenty years before the British slave trade was finally banned.

Despite all the setbacks, the language of rights and liberty was on everyone's lips. The Lunar men, optimistic about their children and full of plans for their own work, were in confident mood. Briefly, it seemed, the established order was beginning to bend and bow before the tide of reforming sentiment. But not always. Writing from London in April 1787 Sukey Wedgwood reported with glee that their acquaintance Lady Clive was not only upset at Wedgwood's politics of trade and reform, but:

> she has another charge against you – your religious principles cannot be right, or you would not be acquainted with such a man as –'– as – – Dr Darwin!! as for her ladyship she would rather die than have his advice if there was not another physician in the world.[33]

35: GRAND PROJECTS

The Lunar men talked, corresponded and collaborated constantly. But they were never a single unit, rolled into a ball surging forward towards a single goal. Each took their own separate paths. In their mid-fifties Darwin, Boulton and Wedgwood each had one venture dear to their hearts: for Wedgwood, it was the Portland Vase; for Darwin, *The Botanic Garden*; for Boulton, the new Soho Mint. Each was a visionary project. Each displayed the nature of the man. Each fused ideas of technology, technique and materials with different ideas of history and progress and with use of the poet's language or the artist's skill.

At first the idea of the Soho Mint sounds like self-parody – having got so far, where could the great, perpetually debt-bound entrepreneur go now? The answer, laughable in its bold simplicity, was to make money by making money.[1] Once again Boulton spotted a gap in the market and another golden chance for his new steam power. Making coins would combine the steam expertise of Boulton & Watt with the old metal crafts of the Soho workshops *and* use some of the dread Cornish copper mountain. (Mines were again threatened with closure because of over-production and in the inevitable riots Boulton had found himself the scapegoat.) Designing coins would call on the art of the sculptor and engraver; pouring them out by the ton would demonstrate the virtues of machine power. Even better, in Boulton's eyes, the venture was 'patriotic' and even philanthropic, preventing crime and saving potential felons.

Admission ticket to view Wedgwood's copy of the Portland Vase, 1790

It was true that British coinage was in chaos. As industry expanded and wages rose, small change was desperately needed. The Royal Mint was still using heavy screw-presses which needed two men to pull the heavy balls on the end of the levers, and hammer down a single coin at each jerk of the press. Slow and creaking, the Mint stamped its gold and silver pieces, one by one. The only issue of copper in the past thirty years had been a meagre batch of halfpence in 1770, and even silver was scarce. In places where weekly wages came in shillings, employers sometimes had to hand out one pound note between four or five men, who went together to the shop, or pub, to spend it.

Counterfeiting was inevitable, especially in Birmingham, with its skilled metal workers. If a machine could make an impressed button, it could clearly make a coin. By the end of the 1780s, as he jingled the change he got at tollgates, Boulton reckoned up to two-thirds of the copper coins were fakes.[2] As a trader with a big wage bill, he had been concerned with the lack of small coins for years, suggesting that Parliament should look urgently at the currency and talking of applying steam to minting when Watt first came to Soho. Twenty years later, he took up the idea again. He tinkered with new ideas – striking a coin in a collar to fix the diameter, standardizing the thickness, making raised, bevelled edges to stop them wearing down – but even though he minted his own guinea to prove Soho's capability, no one at Westminster would bite.

Instead he set up his own presses and in 1786 he struck 100 tons of copper coins for the East India Company. When he and Watt were in Paris they visited the mint and noticed some advanced ideas developed by the medallist Jean Pierre Droz, but not picked up by the French Government. These included a jointed collar, which could hold the coin at the moment of striking and allow an inscription to be struck on the edge. Back home, Boulton approached the Government again and organized petitions. All that he achieved was the setting up of a cumbersome Privy Council Committee, which pondered the issue for three years and finally made recommendations that were never carried through. But Boulton impressed them: after he gave evidence in January 1788 he was asked to submit patterns for a new halfpenny.[3] At first he had the pattern made on Droz's press in Paris, and then brought over Droz himself (exactly the kind of thing that British manufacturers got so hot about when the cross-Channel traffic went the other way). The Soho samples were ready by May, complete with their inscribed rim, declaring, 'Render unto Caesar the things that are Caesar's.'

Having invested capital he could hardly spare, Boulton now offered to undercut the Mint by half in the cost of production. The patterns were approved and it seemed the deal was on. But official wheels ground slow, especially given the vested interests in London. It would be ten years before a government order came. In the first flush of excitement, however, Boulton was sure he would make halfpence rain from heaven. In 1788 he built the first steam-powered mint in the world, tucked away near his manufactory behind the curving extension called the 'Fairy Farm' which held his Tea Room for visitors, his Fossil Room and Laboratory.[4] The steam-engines drove a mechanical contrivance that gave a twist to each of the coining-press screws with the help of a vacuum cylinder connected to each press. In the rolling-mill back at the manufactory he put in new rolls to turn the copper ingots into sheets and strips 'thinner than copper had ever been made before for coins', and in the lapping-mill next door, powered by the 'Lap engine', he put new machines for cutting out blanks. In the mint, he set up a plant for scaling and annealing the metal.

Soon he had six presses thumping and big orders on the stocks: copper coins for the American colonies and silver for the Sierra Leone Company. Over the years, more foreign business came his way – pennies for Bermuda, faluces for Madras. Another line was in local coinage and trade tokens, which now began to be issued by manufacturers (as they had been in a previous coin famine a century before). In 1787 Thomas Williams's Parys mine in Anglesey had issued penny pieces made of their own metal, struck in their own mint. John Wilkinson swiftly ordered a fine coin with his bust surrounded by the proud name 'JOHN WILKINSON IRON MASTER'; on the reverse was a smithy, with two large tilt-hammers. This was the first coin to bear a head other than the monarch's and, not surprisingly, it caused a considerable stir. The *London Magazine* carried a sharp 'Epigram on Mr Wilkinson's Copper Money', alluding to the classical heroes on antique medallions:

> So Wilkinson, from this example,
> Gives of himself a matchless sample!

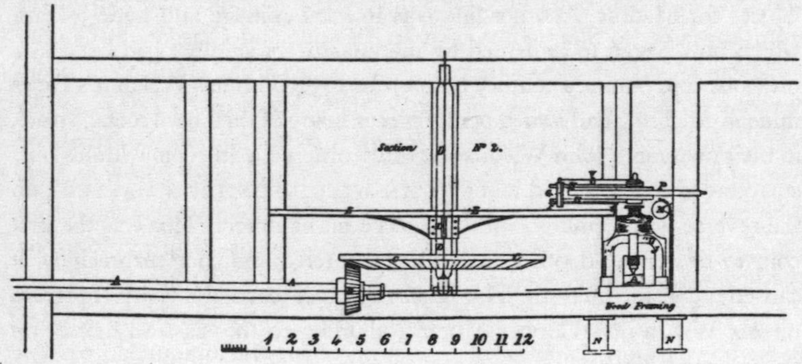

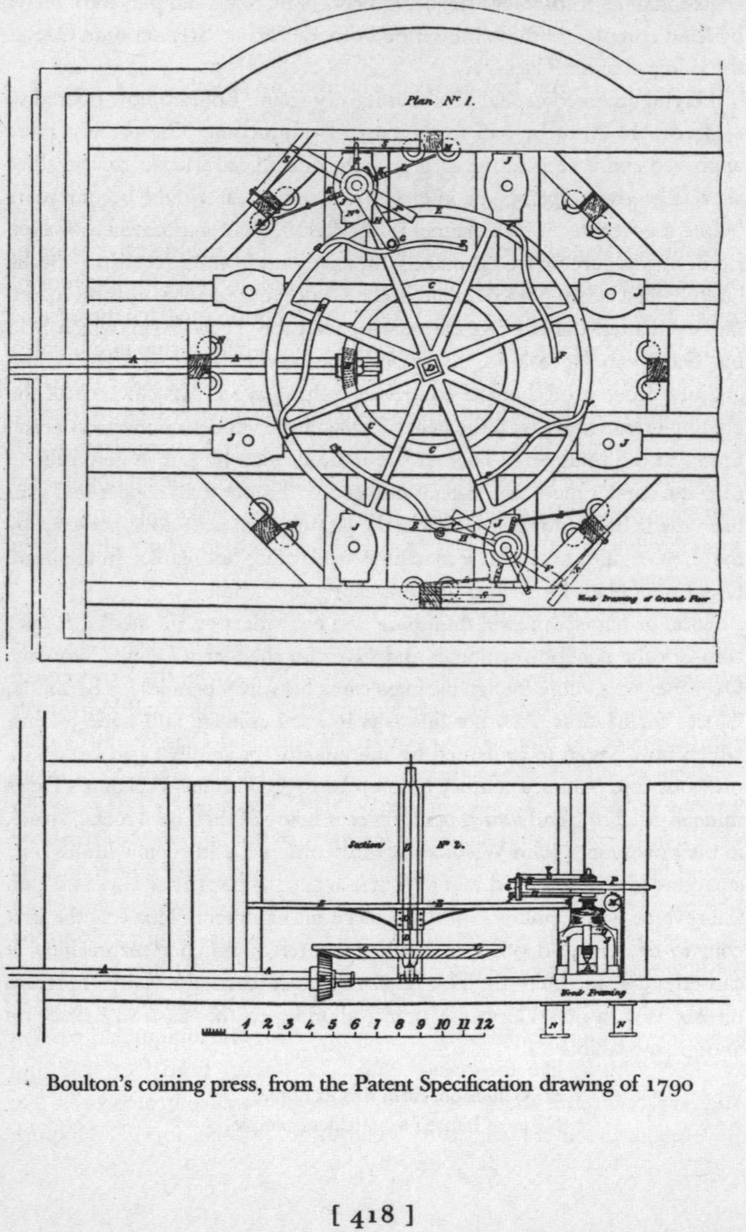

Boulton's coining press, from the Patent Specification drawing of 1790

And bids the *Iron* monarch pass
Like his own metal wrapt in *Brass*!
Which shows his modesty and sense,
And how, and where he made his pence!
As iron when 'tis brought in taction,
Collects the copper by attraction.
So, thus, in him, twas very proper
To stamp his brazen face on *Copper*.[5]

Boulton stamped several Wilkinson tokens over the next five years in vast amounts (one order alone, demanded 'speedily', was for five tons of coins). Every issue had his head on the front, but on the back the designs might be a woman in classical dress with a cog wheel, a ship in full sail, a nude figure of Vulcan at his anvil or an emblem of Justice. One run commemorated the launching of the *Trial* in 1787. Angered by the barge-builder's slowness in providing new boats to carry a vital contract of heavy guns for the Government, Wilkinson challenged the monopoly and built his own iron boat: hundreds crowded on the banks expecting to see 'John's Folly' sink, then halloed as she floated high, to a fiery salute from the foundry's own guns.

These coins, just as much as the success of Wedgwood's Chamber of Commerce, were a show of manufacturing power and independence. The iron-master's stern profile tossed defiance to the Crown, the Government and the old order. Boulton made tokens for companies and individuals from Lancaster to Southampton, from Dundee and Glasgow to County Offaly in Ireland. Holding them in your hand is like shuffling an iconography of the day. So many aspects of the age and of the Lunar project come together here: a profile of a thrusting merchant on one side might be backed by a rural scene, or an evocation of ancient ritual. The dignity of the provincial towns with their old coats of arms could be countered by a sharp selling slogan on the reverse. On the 'Leeds halfpenny', Bishop Blaize, patron of woolcombers, is offset by a scene of the new Mixed-Cloth Hall, and the slogan 'Success to Yorkshire Woollen Manufactures'. Some tokens show a polite neo-classicism; others the antiquarian passion for 'Old Albion', like the druids that adorn coins from Cornwall and Anglesey. Still more shout the virtue of the new transport networks, like the elegant token for the Sturt Navigation or that for the Holborn innkeeper Christopher Ibbertson, where St George and Dragon flips over to declare 'Mail & Post Coaches to all Parts of England'. Sometimes, too, the inscription hints at commercial values subverting old ways. The beautiful Glasgow token with its arms and its reclining river god is emblazoned 'LET GLASGOW FLOURISH' – a telling truncation of the full city motto, 'Let Glasgow flourish . . . through the preaching of Thy Word.'

Mundane as they were, the tokens were a decorative art, and Boulton used fine engravers such as the temperamental Droz, and later Dumarest and Conrad Heinrich Küchler. And with each order, his technical experience grew. In July 1790, he took out a patent for applying steam power to coining 'in place of men's labour'.[6] As he explained two years later, his mint had eight large-screw coining-presses, all linked together, but each with its own clutch and gears, so that eight different sizes of money could be made at the same time, from tiny French sous to English crowns. The presses had to be arranged in a circle and the whole arrangement was like a heavy children's roundabout, whirling incessantly, and pouring out a torrent of copper, from fifty to a hundred and twenty coins a minute. This was the apotheosis of mechanical 'copying', always a theme of Boulton's work. And these presses themselves were so smart that the operation was literally child's play: 'Each machine requires the attendance of one boy of only twelve years of age, and he has no labour to perform. He can stop his press one instant, and set it going again in the next.'

He limited his boys to ten hours, much less than many factories, and dressed them proudly in white jacket and trousers, washed once a week. And his contemporaries could even praise him as a philanthropist, a true man of sensibility. In his notes to *The Botanic Garden*, Darwin suggested Boulton 'should be covered with garlands of oak' for saving potential forgers from the gallows. He also gave the impression that the machine almost worked on its own, for the glory of Boulton – and his country:

> Descending screws with ponderous fly-wheels wound
> The tawny plates, the new medallions round;
> Hard dyes of steel the cupreous circles cramp,
> And with quick fall his massy hammers stamp.
> The Harp, the Lily and the Lion join,
> And GEORGE and BRITAIN guard the sterling coin.[7]

From coining Boulton moved to medals. One of his earliest was to commemorate the recovery of George III in 1789, after his year-long bout of madness: the first cast was presented to the Queen by her Reader, and Boulton's friend, André de Luc.

Wedgwood too produced a 'restoration medallion'. He was always quick to cash in on topical events, making a medal in 1787, for example, to mark the commercial treaty with France, and another this year to celebrate the first Australian colony, at Sydney Cove, using clay that Banks had sent him from this distant bay. But Wedgwood still burned to prove that the new pottery to which he had given his life could match the treasures of the past – and one of the greatest of these to reach Britain was the Barberini Vase. This was a marvel of dark-blue glass with white cameo figures, probably made in Alexandria in the time of Augustus, at the start of the millennium.[8] In the early seventeenth century it came into the possession of Cardinal Francesco Barberini and stayed in the family's palace in Rome, reproduced in countless works of engravings, including Montfaucon's *L'Antiquitée Expliquée*, one of Wedgwood's chief sources. In 1780, the last of the Barberini line, ruined by card debts, sold it to the Scottish antiquary and dealer James Byres, who had plaster-of-Paris replicas made by the gem-engraver James Tassie before selling it on to William Hamilton for £1,000. In 1784 Hamilton too sold it, with elaborate secrecy, to the Duchess of Portland ('a simple woman', Horace Walpole called her, 'but perfectly sober and intoxicated only by *empty* vases'),[9] and after her death a year later it was put up for auction with the rest of her vast collection, known saucily as 'the Portland Museum'. The winning bid, on 7 June 1786, was made on behalf of her own son, the third Duke.

All sorts of rumours surrounded the vase: it was thought to be made of agate, not glass; it was said to be a funerary urn that had once held the ashes of the Emperor Alexander Serverus and his mother. The motif itself still remains mysterious, perhaps connected with the story of Peleus and Thetis. For years Wedgwood had been trying to copy it in jasper from prints and immediately after the sale he wrote to the Duke of Portland to ask if he could borrow the original: on 10 June 1786 it arrived at Etruria. Wedgwood wrote at once to Hamilton, caught between elation and despair. Working from Montfaucon's prints, he had felt sure he could equal the original, 'or excell if permitted', but now that he had seen it, he said, 'my crest is fallen, and I should scarcely muster sufficient resolution to proceed if I had not, too precipitately perhaps, pledged myself to many of my friends to attempt it in the best manner I am able'.[10] He would use the best modellers and would make it in black jasper, which was harder than glass and could be cut like agate and polished on a lathe. The most difficult challenge would be to imitate the fine undercutting, so lacy and subtle that it was more like painting than sculpture.

He was so nervous that he insisted he would take subscriptions only on the understanding that the purchasers need not take the copies if they did not succeed. The challenge called into play a lifetime's experience, all his technique, his gathering of a perfect team, his experiments with ceramic bodies and with firing. He was helped by Henry Webber, and by his son Jos, as well as Hackwood and William Wood, but perfecting his copy – first in black and later in deep blue – took the next four years.

Wedgwood was disturbed by the shape of the vase itself, which was much fatter and squatter than his idea of the beautiful, and indeed at times he thought of transferring the figures to a different, finer shape altogether – and of using them on other works, and in other forms. All the time, he kept in touch with Hamilton, writing often and at length for his advice. He thanked him for copies of antique busts and told him of his chimney-pieces and vases and figures: 'But my great work is the Portland Vase.'[11] By midsummer 1787, he had finished a third version of the figures and was worrying about how he could ever come near to the delicacy of the original, where the artist had been able to pare down the white cameo glass to near transparency so that the dark-blue ground gave the effect of shading. This he got round by applying shading. He was also anxious about problems in firing: a little more or less heat, a little more or less time in the kiln, and the effect could be entirely different. His final, and most difficult, problem was finding a jasper dip that would match the intense dark blue of the vase, without breaking down in the kiln.

By the following spring he had produced two or three copies of the entire vase but none that satisfied him fully. The first, almost perfect, copy was sent to Darwin for inspection in October 1789 – perhaps the highest mark of Wedgwood's trust. He made him swear to keep it private, but Darwin inevitably went his own way: 'I have disobeyed you,' he wrote, 'and shewn your Vase to two or three, but they were philosophers, not cogniscenti. How can I possess a jewel, and not communicate the pleasure to a few Derby Philosophes?'[12]

Wedgwood was also collecting all the commentaries so far written on the vase, so that he could print a selection for his purchasers.[13] One interpretation came from Darwin, and was later reprinted in the notes to *The Economy of Vegetation.* Darwin was convinced that the vase depicted the initiation into the Greek Eleusinian mysteries.[14] He asserted, as the Masons did, that the true wisdom of the ancient world had belonged to the Egyptian magi, perhaps received by them from India, and that this had been written in their hieroglyphic code, later interpreted as myth. In the Eleusinian mysteries the nub of the great wisdom remained: the belief that in the universal cycle of nature organic matter could not be destroyed, but only transformed – a belief reflected in Darwin's own evolutionary theories.

In the early spring of 1790 Wedgwood was ready. (Fifty copies were allegedly made, although only sixteen or so survive.) The Queen had a private view on May Day, before a private showing at Joseph Banks's house. By happy chance, the Duke of Portland owned the land on which Wedgwood's London showroom stood, so the Portland Vase was appropriately exhibited at Portland House in Greek Street. Sir Joshua Reynolds bestowed a formal certificate declaring it to be 'a correct and faithful imitation both in the general effect and the most minute details of the part'.[15]

In 1789, when Wedgwood asked Darwin's opinion on his first precious copy, he was returning a compliment. Earlier in the year Darwin had sent him *The Loves of the Plants*, a month before it was published. This poem took even longer to see daylight than Wedgwood's vase: in 1787 Darwin had finally accepted Joseph Johnson's offer but even so he would not put his name to the poem lest it looked as though he thought it important.

In his late fifties, like Boulton and Wedgwood, he was beginning to feel the effects of middle age, taking calomel for gout and suffering the odd sleepless night. In 1788 he was reading the proof pages of Keir's new *Dictionary of Chemistry*, having the pages sent to him four at a time, so that he could read them like a newspaper. He was still learning, determined 'to improve myself by all methods'. In some quarters, though, he found it hard to be open-minded. Another row with Withering erupted, this time waged by Robert Darwin from Shrewsbury, after Withering intervened to change the treatment of one of his patients. Furious, near slanderous exchanges followed, in which Robert (egged on by Darwin) accused Withering of greed and incompetence.[16]

In the gaps of time, morning and night, Darwin took up his pen. He wrote indoors, balancing a cushion on the arm of his chair and resting his writing board on it; in the heat of summer he wrote in his summer-house, its windows opening over the river; on the way to see patients, he wrote in his carriage, the front of which 'was occupied by a receptacle for writing paper and pencils, likewise for a knife, fork and spoon; on one side was a pile of books reaching from the floor to nearly the front window . . . on the other, a hamper containing fruit and sweetmeats, cream and sugar'.[17] As well as countless letters, Darwin was writing on clouds and spa waters and piling up material for *Zoonomia*, his ever expanding work on diseases. And all the time his poem lay on his desk, acquiring new touches and more notes. Finally, in April 1789, *The Loves of the Plants* was published, almost overshadowed by the birth of the Darwins' sixth baby, Henry. Immediately, Darwin began work on a second edition, and then on a third (including new verses on *Cannabis*, noting with awe that a plant grown in England shot from seed to fourteen feet in five months). And in the autumn of 1789 he canvassed details of more inventions, quizzing Boulton about his engines and coining-machines. Spotting good publicity, Boulton cheerfully sent a modest 'Catalogue of Facts':

> Convenience and properties of B&W improved Engines.
> a. it is the most *powerful* Machine in the World
> b. it is the most *tractable*
> c. it is the most *regular* and hath the property in itself of conforming to the least work.[18]

Watt too sent a long note, with a friendly wry grumble: 'I know not how steam-engines come among the plants; I cannot find them in the Systema Naturae, by which I should conclude they are neither plants, animals nor fossils, otherwise they would not have escaped the notice of Linnaeus. However if they belong to *your* system, no matter about the Swede.'[19]

All the information was fed into Darwin's poem by the end of 1790. The only hold-up in the publication of *The Economy of Vegetation* came from

The frontispiece to *The Loves of the Plants*, 1791
designed by Emma Crewe

his determination to include an illustration of the Portland Vase – eventually engraved by William Blake. (He did not know his name, but 'Johnson sais he is capable of doing any thing well,' wrote Darwin hopefully.[20]) Meanwhile, the critical response to *The Loves of the Plants* was rapturous. Darwin's rhyming couplets may have been old-fashioned but his subject matter was entirely novel: his notes were like an encyclopaedia of the new technology and advances in science. The words 'pleasure' and 'instruction' were mixed in equal amounts in the reviews, while the ageing Horace Walpole was weak with enthusiasm for 'the most delicious poem on earth'.[21] Darwin crowed with pride, telling Keir and Edgeworth and other friends exactly how much he was earning. When the Poet Laureate Thomas Warton died in May 1790, he even hinted to a friend who had the ear of William Pitt that perhaps he might be put forward for the laureateship, to help his 'large and increasing family'.

In odd ways, Darwin did speak for the nation. One can understand how young poets such as Wordsworth and Coleridge and chemists such as Humphry Davy were equally bowled over. In its final version *The Botanic Garden* explodes into life with a vision of the creation of the universe, presaging the 'Big Bang' theory. Milton meets Herschel, as Chaos thrills to the command, 'Let there be Light!'

> Through all his realms the kindling Ether runs,
> And the mass starts into a million suns;
> Earths round each sun with quick explosions burst,
> And second planets issue from the first;
> Bend as they journey with projectile force,
> In bright ellipses their reluctant course;
> Orbs wheel in orbs, round centres centres roll,
> And form, self-balanced, one revolving Whole.

It was astoundingly up to date, the first literary vehicle for the new language of French science. Its opening notes leap from the 'Rosicrucian machinery' drawn from Pope's *Rape of the Lock*, of Gnomes, Sylphs and Nymphs, with their sense of hidden powers and alchemical mysteries, to the newest ideas on heat and gases, even employing the term 'calorique'. 'You are such an infidel in religion that you cannot believe in transubstantiation,' wrote Keir, 'yet you can believe that apples and pears, hay and oats, bread and wine, sugar, oil, and vinegar are nothing but water and charcoal, and that it is a great improvement in language to call all these things by one word, oxyde hydro-carbonneux'.[22]

Brand new words, such as 'oxygen' and 'hydrogen', and Darwin's own lasting coinages such as 'iridescent' share the page with the Cyclops who dwell in 'Etna's rocky womb'. In the crammed notes, Darwin's own observations on phosphorescent evening light and warm springs jostle against Cook's voyages, Virgil's *Georgics* and Priestley's discovery of 'pure air'. (Darwin even imagines that this might be applied to the 'diving bell', and looks forward to a time when adventurers will 'journey beneath the ocean in large inverted ships or diving balloons'.[23]) Contemporary readers enjoyed the vision of a sexualized plant world ('sexual reproduction is the *chef d'œuvre*, the master-piece of nature,' Darwin wrote[24]), but they were even more entranced by the heroic, industrial poetry. Sometimes *The Botanic Garden* itself seems like a great Lunar meeting, a playful, disputatious conversation with forebears and friends: the poem is a prism, catching the reflected light of all their concerns.[25]

As its title suggests, *The Economy of Vegetation*, echoing Buffon, emphasized the self-regulating economy of the natural world. It was also, to many people, very shocking, replacing Genesis with a pagan version of the Creation, imbued with sex:

> When Love Divine, with brooding wings unfurl'd
> Call'd from the rude abyss the living world . . .[26]

It smacked worryingly of the atheistical ideas of those French *philosophes* who presented the universe as entirely material, an active entity displaying life in ever-changing combinations.[27] Even more significantly, Darwin described the great variety of devices gradually evolved for survival and managed to suggest that the arts, invention and industry were somehow the latest stage of this 'natural' historical progress.

He took these ideas much further in his next book, *Zoonomia*. Keir had wisely suggested that he divide this into two volumes and the first appeared in 1794, the second two years later. And just as *The Botanic Garden* made him, briefly, England's leading poet, so this made him Britain's leading medical writer. Darwin's medicine embraced the whole person, physical, social and psychological, and the afflictions he listed included not only influenza and colic, cancer and toothache but depression, anger, religious mania, delusion. Living beings were a bundle of fibres, nerve endings that could contract in response to stimulation, linked in consciousness in the 'sensorium' – the organ both of thinking and feeling. This would respond to four main causes: 'irritation', the response to external stimuli; 'sensation', excited by pleasure or pain; 'volition', the ability to act in reaction to desire or aversion, and 'association'.[28] Volition and association led to the growth of non-selfish emotions, the ability to sympathize with others, to appreciate beauty or to spy out injustice. And through reproduction, Darwin controversially suggested, the imprinted patterns of experience were passed on to each new generation, which progressed beyond, in its turn.

There was much here that echoed the great authorities of Darwin's youth, and of his later years.[29] But no one had formulated a theory of the development of life, free from the guiding hand of the Creator, as Darwin did in his long chapter 'On Generation'. 'Too old and hardened to fear a little abuse', he was no longer cowed by the wrath of Canon Seward and his ilk.[30] 'Would it be too bold to imagine', he asked,

> that in the great length of time since the earth began to exist, perhaps millions of ages before the commencement of the history of mankind, would it be too bold to imagine that all warm-blooded animals have arisen from one living filament, which THE GREAT FIRST CAUSE endowed with animality, with the power of acquiring new parts, attended with new propensities, directed by irritations, sensations, volitions and associations; and thus possessing the faculty of continuing to improve by its own inherent activity, and of delivering down those improvements by generation to its posterity, world without end![31]

The Temple of Nature, published after his death, broadcast these bold ideas even more clearly. The universe was formed by 'chemic dissolution' producing heat, then by the forces of 'repulsion' (the exploding mass), 'attraction' (the cohering force of gravity) and contraction. As the earth's surface changed, so organic life developed beneath the sea: 'everything from shells'. When he was a child, Darwin's imagination had been fired by electricity and the Leyden jar. Now Galvani had discovered that a muscle in a dissected frog contracted when a spark was fired from an electrical machine.[32] Was the whole body an electrical circuit? And even apparently inanimate matter had the potential for life. In his first note to *The Temple of Nature*, Darwin described how he had made a paste of flour and water and how the tiny animalcules or 'eels' had grown in a sealed glass jar.[33]

Years later, in the cold and rainy August of 1816, Shelley and Byron discussed galvanism and Darwin's experiments, inspiring the listening Mary Shelley to the idea that the scientist might not only observe life, but also create it: 'Perhaps the component parts of a creature might be manufactured, brought together, and endued with vital warmth.'[34] Darwin himself had no such hubris. He was merely, he declared, recording the real. Metamorphosis was the stuff of life as well as story: the butterfly from the caterpillar, the frog from the tadpole. Everywhere around them his readers could see how animals adapted to their surroundings; the fossil record showed how many species had become extinct; animal breeding showed how characteristics could be moulded through generations. Lust, hunger and the need for safety ensured that the strongest animals won the fight to propagate the species, developing features that let them fight for their quarry, escape predators, or find food in a crowded, competitive environment:

> Some birds have acquired harder beaks to crack nuts, as the parrot. Others have acquired beaks adapted to break the harder seeds, as sparrows. Others for the softer seeds of flowers, or the buds of trees, as the finches.[35]

But if Nature was red in tooth and claw – or as Darwin put it in *The Temple of Nature*, 'One great Slaughter-House the warring World' – this was the inevitable cost of development. And this was equally true of the competitive world of trade, or of politics. At times Darwin's vision of competition came close to that of Malthus's *Essay on Population* (1798). But while Malthus induced only gloom, Darwin continued to project the optimism that had inspired the whole Lunar project, borne out by the struggle of life on the planet itself:

> Shout round the globe, how reproduction strives
> With vanquished Death – and Happiness survives;
> How Life increasing peoples every clime,
> And young renascent Nature conquers Time.[36]

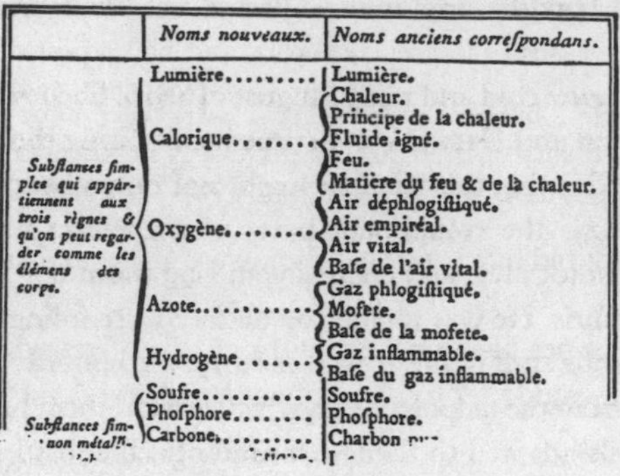

	Noms nouveaux.	*Noms anciens correſpondans.*
Subſtances ſimples qui appártiennent aux trois règnes & qu'on peut regarder comme les élémens des corps.	Lumière	Lumière.
	Calorique	Chaleur. Principe de la chaleur. Fluide igné. Feu. Matière du feu & de la chaleur.
	Oxygène	Air déphlogiſtiqué. Air empiréal. Air vital. Baſe de l'air vital.
	Azote	Gaz phlogiſtiqué. Moſete. Baſe de la moſete.
	Hydrogène	Gaz inflammable. Baſe du gaz inflammable.
Subſtances ſim- ... non métalli- ...	Soufre	Soufre.
	Phoſphore	Phoſphore.
	Carbone	Charbon pu- ...

Lavoisier's 'Table of Simple Substances', 1789

36: TURNING

In February 1788, aged seventy-four, John Whitehurst dies. The friends are beginning to face the onslaught of time.

Mary Anne Galton is ten, a devout, solemn little girl. With astonishment she watches the famous Dr Darwin squeezing his massive bulk out of his carriage when he comes to treat her mother at Great Barr. His head is almost buried in his shoulders, his old scratch wig tied in a little bobtail behind. He stammers, his eye is keen, he hates 'fermented liquors' and believes in eating largely, especially sweet things. The Galtons set out hot-house fruit and West India sweetmeats, clotted cream, Stilton cheese. Darwin eats it all. He tells stories of a duchess poisoned with lead face-paint and he quotes unintelligible Greek. He is a monster. And he sneers at sacred Scriptures.

Dr Darwin often used to say, 'Man is an eating animal, a drinking animal and a sleeping animal, and one placed in a material world, which alone furnishes all the human animal can desire. He is gifted besides with knowing faculties, practically to explore and to apply the resources of this world to his use. These are realities. All else is nothing; conscience and sentiment are mere figures of the imagination.'

Among his men friends, some have noted, he may not have said 'sleeping'.[1]

In May Edgeworth commiserates with Wedgwood, who has been ill. Edgeworth confesses that he himself has been 'reduced to a dreadful state by overwork & anxiety' but has restored his health by relaxing, and by taking '3 or 4 glasses of Claret, not Port, and half a pint of Porter'.

I am really sober . . . but Dr Darwin thinks I am inclined to be a toper – I therefore consider his theory upon that subject as very doubtful and trust rather to the large Experiment of thousands, who are in excellent health around me.[2]

But Darwin had told him roundly, 'I drink water *only* and am always well.'[3]

Matt Boulton comes back on holiday from Paris, dressed in high French fashion. 'It was wonderful to me', remembers Mary Anne:

to see Dr Priestley, Dr Withering, Mr Watt, Mr Boulton himself, and Mr Keir, manifest the most intense interest, each according to his prevailing characteristics, as they almost hung upon his words; and it was impossible to mistake the indications of deep anxiety, hope, fear, curiosity, ardent zeal, or thoughtful gravity . . . My ears caught the words, 'Marie Antoinette', 'The Cardinal de Rohan', 'diamond necklace', 'famine', 'discontent among the people', 'sullen silence instead of shouts of "Vive le Roi"'.[4]

In Paris, David paints Lavoisier and his wife. Impeccable in curled wig and white cravat, black knee breeches and buckled shoes, the scientist sits at a table draped in red velvet, on which stand shining glass tubes and apparatus. Quill in hand, he looks up at Marie-Anne, whose hand rests gently on his shoulder like that of a Muse. She wears silver-white taffeta tied with blue ribbon. This is a still, safe world of science, reason and affection which no rough wind can tumble.

The phlogiston debate rumbles on, with Darwin the first to defect. In November 1788 he challenges Watt to read the Introduction to Fourcroy's *Elements*, 'and tell me if the facts are in general true – if they be, the theory holds them nicely together':

I shall wait with Patience to see this great dispute decided, which involves so great a part of the theory of chemistry – and thank the Lord, that chemical Faith is not propagated by fire and sword, – at present I am inclined to the heterodox side of the question.[5]

Watt succumbs. Keir hesitates. Yet he is troubled by the French chemists: he has rheumatism, his own *Dictionary of Chemistry* is going slowly. In

Antoine Laurent de Lavoisier and his Wife, by Jacques-Louis David, 1788

January 1789, Berthollet tells him that the Paris philosophers are swamped with new terms: 'Nous sommes ici inondés de *carbonate* du *sulfate*, de *nitrate* &c.' Keir writes to Priestley,

I wish Mr Berthollet and his associates would relate their facts in plain prose, that all men might understand them, and reserve their poetry of the new nomenclature for their theoretical commentaries on the facts . . .[6]

Priestley replies, calling on his time yet again:

I wish very much to see you, haveing got a quantity of *green liquor*, by air from the *lead ore*, which I wish you to examine with me. I have also other things to *show*, and to *tell* you, especially what I think a *coup de grace* to the new doctrines.[7]

He protests too much.

In the new year, Darwin writes to Boulton, anxious to smooth over any tensions from Robert's row with Withering:

I often lament being at so great a distance from you – your fire-engine for coining money gave me great pleasure for a week after I saw it – and the centrifugal way of regulating the fire. If you ever come near Derby pray give me a day of your company . . .

Adieu, from, dear Boulton,
Your old affectionate friend
E. Darwin.[8]

In February 1789 the country cheers George III's recovery from madness. Wedgwood writes, teasingly, to Darwin: 'I hope you are all very loyal at Derby and drink heartily for his majesty's sake as we do here and shall continue to do for some time yet.'[9] In Birmingham (where the town elders have already expressed their loyalty to the Regent elect, just in case), there are fireworks and by the Navigation Office a massive bonfire burns three tons of coal. In winds and freezing rain, crowds fill the streets. All the public buildings bear brilliant transparencies and emblems and at Soho, the manufactory, house and grounds are 'completely and grandly illuminated with many thousand lamps of various colours, most judiciously displayed'. Later a huge Montgolfier balloon glides over the town 'in a very majestic and pleasing manner'.[10]

In July, as the evening shadows lengthen after a glorious day, Mary Anne sees a carriage thundering towards the Galtons' house:

After some minutes the door of the drawing room opened and in burst Harry Priestley, a youth of sixteen or seventeen, waving his hat, and crying out, 'Hurrah! Liberty, Reason, Brotherly Love for ever! Down with kingcraft and priestcraft. The majesty of the people for ever! France is free, the Bastille is taken: William was there, and helping. I have just got a letter from him. He has put up the picture of the Bastille, and two stones from its ruins, for you,' (addressing himself to me,) 'which you will soon receive; but come, you must hear his letter.' We all stood thunderstruck.[11]

Two months later, Keir writes to Day, congratulating him on the coming final volume of *Sandford and Merton*. But Day never reads this kindly letter or sees his published book. At Anningsley he has a colt which he refuses to break in by force, trusting to the sympathy between them. On 28 September 1789 he sets out to meet Esther but his colt takes fright at a winnowing-vane by the edge of the road and plunges and bolts, throwing him head down on the stones. Carried to a nearby house, he dies before the surgeon can reach him, martyr to his own philosophy of kindness.

By the start of the twentieth century even his grave cannot be found. One rumour about him lingers on – that he had buried £25,000 somewhere about his house in Anningsley. Labourers in the 1920s are still arguing as to which of the great oaks it lies beneath.[12]

'Sir', writes Edgeworth to Day's nephew,

I have an excellent picture of Mr Day by Wright; if Mrs Day should wish for it, it is at her service: she may be sure no other person shall have it, but one of my own children . . . I have called my new born son Thomas Day, and I hope his name may excite him to imitate the virtues of my excellent friend.[13]

The baby lives only three days. Esther turns her face to the wall. Two years later she too is dead.

The winter brings iron frosts, drifts of snow and rimy mists. Ah, Darwin sighs: 'The loss of ones friends is one great evil of growing old.'[14]

37: RIOTS

The Fall of the Bastille in July 1789 thrilled British Radicals. Dandies flaunted the *bonnets rouges* of the revolutionaries, odes filled the press, men and women danced in the streets. Among the younger generation, the sixteen-year-old Coleridge imagined the spread of intellectual liberty: 'No fetter vile the mind shall know,/And eloquence shall fearless grow'.[1] Wordsworth rejoiced, and was in Paris the following year to toast the anniversary. In a benighted London, Blake cried to his fellow citizens:

The fire, the fire, is falling! Look up! look up! O citizen of London, enlarge thy countenance . . .

Spurning the clouds written with curses, stamps the stony law to dust, loosing the eternal horses from the dens of night, crying Empire is no more! and now the lion and the wolf shall cease.[2]

Among the Lunar friends, Priestley, of course, was wholeheartedly enthusiastic. So was Wedgwood. 'I know you will rejoice with me', he wrote to Darwin,

in the glorious revolution which has taken place in France. The politicians tell me that as a manufacturer I shall be ruined if France has her liberty, but I am willing to take my chance in that respect, nor do I yet see that the happiness of one nation includes in it the misery of its next neighbour.[3]

Darwin agreed. In *The Economy of Vegetation* he inserted a long section, imagining the French people bound like Gulliver in Lilliput by the myriad

Detail from James Gillray, *Repeal of the Test Act*, February 1790 strings 'of Confessors and Kings', bursting their bonds when touched by the 'patriot flame'.[4] At the start of 1790 he asked Watt, 'Do you not congratulate your grand-children on the dawn of universal liberty? I feel myself becoming all french both in chemistry and politics.'[5]

As far as Watt was concerned, the political excitement was nothing but a nuisance, causing philosophical correspondence from France to shrink to a trickle, 'so that I have lately heard nothing interesting in chemistry'.[6] Among the other Birmingham men, Boulton, as usual, remained cannily detached, supporting Priestley and his Dissenting friends, while making the right pro-Government noises and keeping a keen eye for a potential new market for his coinage. Withering, however, was open in his hope for the new regime, while Keir saw the Revolution as a straightforward victory against despotism, following the model of the Glorious Revolution. It was the 'sole *triumph of Reason*':

> having been the effect of the gradual illumination of the human mind over a whole nation, by *Philosophy* shewing that the true end of Government is the happiness of the Many and dispelling those baneful *prejudices* which established the tyranny of the *Few*, and which were the relics of the ignorance of barbarous ages.[7]

Over the next two years opinion in Britain became more sharply divided. From 1789 onwards campaigners made much of the contrast between the religious freedoms claimed in France and the position of Dissenters at home: their second attempt to repeal the Test Acts had been defeated this May, by a mere twenty votes.

The Dissenters' campaign was close to the hearts of the Lunar Society, and in Birmingham, where so many prominent businessmen were Unitarians, Quakers or Baptists, they followed the progress of the fight with eagerness and concern. In October 1789 Priestley's friend William Russell persuaded the different Birmingham sects to set up a joint committee, with the aim of creating a national organization. The following month the Dissenting minister Richard Price defied old age and illness to preach a ringing sermon, 'On the Love of Our Country', to the London Revolution Society, a club of liberal Whigs named in honour of the Glorious Revolution of 1688. Dinners had been held across the nation the year before to celebrate the centenary – Boulton himself had hosted the Birmingham feast – but Price's burden was that the legacy of 1688 was far more radical than people supposed. It was a charter endowing them with the right to choose their governors and cashier them if they did wrong. He had lived, he said, to see knowledge undermine superstition, the rights of men made clear. Calling for instant repeal of the Test Acts and for equal representation, he hailed the new regime in America and the Revolution in France:

> You cannot now hold the world in darkness. Struggle no longer against increasing light and liberality. Restore to mankind their rights; and consent to the correction of abuses, before they and you are destroyed together.[8]

Price's sermon provoked a powerful opponent, Edmund Burke, former supporter of the American rebellion and of many liberal causes. In *Reflections on the Revolution in France, and on the Proceedings in certain Societies in London relative to that Event*, published the following year, Burke presented the Revolution not as a beginning but an end, the wrecking of all institutions, the vandalism of culture, the replacement of humanity by the rule of 'calculators and economists'. But in late 1789, Burke was in the minority – many people shared Price's optimism and enthusiasm for reform grew month by month. Unfortunately, as it did so, the opposition to Dissenters hardened. In Birmingham in early 1790 Priestley found himself under increasing pressure. All the old arguments were rehashed over the library and the Sunday schools. He lost his voice and his stammer became worse. His house was attacked, and his maid fired on when she opened the window to see who was breaking in. Sermons were preached against him in St Philip's, one a blatant personal attack by the rector, Spencer Madan. At a county meeting in Warwick, chaired by the Earl of Aylesford (former pupil of Priestley's enemy Bishop Horsley), the Church party ominously considered the 'proper' action to defend the constitution. Try as he would to defend himself in 'Familiar Letters to the Townsfolk', the slogan 'Damn Priestley' was 'chalked on every blank wall in Birmingham'.[9] In late February, a week before the third debate on the Test Acts, Russell told him he had heard that if the Bill was defeated the Church leaders had ordered the bells to be rung, and it was feared the mob might be deliberately provoked against the Dissenters. They were thinking of appealing to Garbett, Boulton and others to prevent this.

By the time the Bill was read, the rift in the Commons between Burke on one side and Fox and Sheridan on the other was absolute and irretrievable. Defending the *status quo*, Burke fiercely attacked the Bill, citing 'subversive' extracts from Priestley's work and Price's revolutionary sermon as evidence of their intention to destroy the Established Church. Although Fox spoke until he was hoarse and the debate wore on until three in the morning, defeat was inevitable.

Priestley was in London that spring, often consulting with Fox. By now he had been dubbed the 'arch-priest of Pandaemonium liberty' in the *Gentleman's Magazine* and was a familiar figure in political cartoons, like 'Dr Phlogiston: The Priestley politician or the Political priest', with his gunpowder tract in one pocket and 'Revolution toasts' in the other. (His family thought these hilarious, and his daughter Sally made a special collection of them.) Back in Birmingham, papers and letters poured out against him, some in the spoof form of *Very Familiar Letters* from 'John Nott, Button Burnisher', much relished locally. His own supporters, under the names 'Abel Sharp, Spur maker' and 'Alexander Armstrong', sprang to his defence, the latter contrasting the complacent contempt of the Anglican clergy with Priestley's assertion of the working man's worth and right to independent views: 'As the saying is, every man for himself and God for us all. We may ask questions and give answers of *our own making*.'[10]

When Burke's *Reflections* finally appeared in November this year it provoked thirty-eight replies, including Mary Wollstonecraft's *Vindication of the Rights of Man*, and Tom Paine's *Rights of Man*, published in two parts in 1791 and 1792. 'Have you seen Mr Paine's answer to Mr Burke?' Priestley asked Wedgwood. 'It is most excellent, and the boldest publication that I have ever seen.'[11] One of the first responses had come from Priestley himself. He admitted that revolution might be 'calamitous to many, perhaps to many innocent persons', but this would be justified by the end, 'eventually most glorious and happy'.[12] He sketched a millenarian blue print for revolution and government for the 'greater good' – a state that would not interfere with religion; where men would enjoy their 'natural rights' and there would be no Established Church, no state rituals, no standing armies. Expenditure on defence and armies could be devoted to public works. Universal goodwill would prevail, and prejudice fade away. Colonies would vanish, 'no part of America, Africa or Asia, will be

Dr Phlogiston, cartoon of 1 July 1791

held in subjection to any part of Europe'. In fact the ideal state promised by Revelation and the Scriptures would be indistinguishable from that 'which good sense, and the prevailing spirit of commerce, aided by Christianity, and true philosophy, cannot fail to effect in time'.

In April 1791 Richard Price died, and on May Day Priestley preached his funeral address at Hackney. The visionary tone remained. A new light was dawning, a new wind blowing, which was showing humanity the true nature of rights, and of government: 'While so favourable a wind is abroad, let every young mind expand itself, catch the rising gale, and partake of the glorious enthusiasm.'[13] Reformers, Priestley said, must prepare for persecution, and remember that 'the world may bear down particular men, but they cannot bear down a good cause'.

Early in May he returned home. A whole brew of trouble was stirring, showing how hard it is ever to assign a single cause to events. Birmingham trade was suffering as the new fashions dropped buckles in favour of laced shoes and anger turned on the big manufacturers, many of them Dissenters. Quite separately, there were rows over the local Court of Requests, which dealt with small debts; over the rising parish rates; the rating of small properties, and the accusation that the citizens were being forced to pay for a new Police Act.[14] All these quarrels became jumbled together as a 'presbyterian conspiracy' to wrest control of the town.

Priestley and his friends were being undermined on all sides. And as midsummer came events in France roused still more conservative panic: in June Louis XVI and Marie Antoinette escaped from the Tuileries, only to be recaptured at Varennes. At the end of that tense month, Priestley told his London friends that Birmingham was to form its own Constitutional Society. Its first act would be to hold a public dinner for 'Any Friend to Freedom', on 14 July, the anniversary of the storming of the Bastille. Keir was invited to be Chairman at the dinner and accepted without hesitation, 'never conceiving', he wrote later,

> that a peaceable meeting for the purpose of rejoicing that twenty-six millions of our fellow creatures were rescued from despotism, and made as free and happy as we Britons are, could be misinterpreted as being offensive to a government, whose greatest boast is liberty, or to any who profess the Christian religion, which orders us to love our neighbours as ourselves.[15]

The dinner was announced in the *Gazette*, quickly followed by a brief announcement that the names of all who attended would be printed in a halfpenny pamphlet: the threat was clear. Next, copies of a handbill were found, declaring the dinner was to celebrate the 'second year of Gallic Liberty', and denouncing Parliament as venal, Pitt as hypocritical, the clergy as 'legal oppressors', the Royal Family as extravagant, the Crown 'too weighty for the head that bears it', and ending by calling on sympathizers to keep the peace but only 'till the Majority shall exclaim, *The Peace of Slavery is worse than the War of Freedom.* Of that moment let Tyrants Beware.'[16] The broadsheet was hot stuff, and could equally well have been written by fierce Radicals or planted by the Church and King party to discredit the Birmingham Dissenters. The latter group immediately disowned it, and offered a hundred guineas for the name of its author – but then so did the magistrates, and so did the Government. As the day of the dinner drew near, Priestley was wisely persuaded not to go.

On the day the mood was nervous. At one point the dinner was postponed, but the landlord persuaded them to carry on. At three o'clock, eighty or so men arrived at the Royal Hotel, to be met by a small group of hecklers (some of them blithely shouting 'No Popery', perhaps in memory of riots long past, much to the consternation of Presbyterian diners). The dinner itself was quiet though hardly sober, with the men raising their glasses eighteen times, first to 'The King and Constitution', then to 'The National Assembly of the Patriots of France', 'The Rights of Man' and 'The United States of America', and going on to embrace the revolution in Poland, and Hampden and Sydney and the Whig martyrs.[17]

Down the road, the Anti-Jacobins were having their own meeting, breaking up in time to see the dinner guests leave, around five o'clock, and to pelt them desultorily with mud and stones. Around three hours later a larger, more determined mob arrived. Furious that the dinner was long over they broke all the hotel windows and then turned, as if directed, straight to Priestley's New Meeting House. The gate and doors were breached, pulpit and pews torn out, cushions hurled out and burned, the Meeting House library ripped up and the building itself set alight. Another detachment reached the Old Meeting House, burning the furnishings in the burial ground before they torched the building. The fire-engines were slow (the Vicar of St Martin's had apparently ordered the key to their shed to be removed), and when they did come the crowd let them turn their hoses on neighbouring buildings but not on the Meeting House itself.

The next target was Priestley's house, a mile away at Fair Hill. Priestley and Mary were playing backgammon when Samuel Ryland rushed in and persuaded them to take refuge at the Russells' near by. William Priestley and some of Priestley's students stayed on, determined to defend the house, when news came that the Russells' house would be next. William Russell, Justice of the Peace, merchant and Unitarian leader – deaf as a post but utterly fearless – took his pistols and rode back to Fair Hill to warn William to move manuscripts and valuables.

When the first wave of attackers arrived at Fair Hill, drink and sharp words persuaded them to leave. Russell and William held off a second mob until the stones flew, when they fled. The crowd moved in, breaking down the doors and flinging furniture from the windows, tearing and burning all the books and manuscripts in the library. Priestley saw it all:

> It being remarkably calm; and clear moon-light, we could see to a considerable distance, and being upon rising ground we distinctly heard all that passed at the house, every shout of the mob, and almost every stroke of the instruments they had provided for breaking the doors and furniture. For they could not get any fire, though one of them was heard to offer two guineas for a lighted candle . . . I afterwards heard that much pains were taken, but without effect, to get fire from my large electrical machine, which stood in the library.[18]

At last the house began to blaze. As the Russells, Priestleys and friends walked the moonlit roads to yet another refuge, the night was bright with flames and loud with noise. To Catherine Hutton, Priestley was a martyr, and his steadfastness and calm had an otherworldly quality: 'No human being could, in my opinion, appear in any trial more like divine, or show a nearer resemblance to our Saviour, than he did then.'[19]

Behind them, the garden of Fair Hill was vandalized, its trees and shrubs uprooted. Some men had found the cellar, breaking the necks off bottles for lack of corkscrews, standing ankle deep in wine. Others discovered the laboratory, behind the house. All Priestley's instruments were destroyed – his mortars made by Wedgwood, his retorts, his flasks, his machines. There were yells and cheers. The walls were razed, the flames roared. One man was killed by a falling cornice. Dr Spencer, a Justice of the Peace, was said to be present, warning rioters not to hurt each other as the house and laboratory burned. By four o'clock in the morning, the fields near by were strewn with exhausted rioters, still armed with bludgeons.

As dawn broke, the refugees returned to the Russells' house. Here they tried to sleep until rumours arrived that the mob was regrouping, shouting for Priestley's life. Hurriedly Mary and Priestley dressed and drove to their daughter and that afternoon Priestley left town, losing his way as darkness fell, trekking forlornly and circuitously to Worcester before catching the night mail to London. Here he took sure refuge, at last, with Theophilus Lindsey.

The Birmingham authorities at first seemed unconcerned: most of the gentry had spent the day at a Staffordshire archery contest. On Saturday morning, however, Lord Aylesford arrived at the smouldering ruins of Fair Hill and persuaded the mob – who hailed him as a good Church and King man – to return to the city. This was an error (or perhaps not), since the houses of several Dissenters lay on their route. All the doors and window-shutters in Birmingham seemed to have 'Church and King for ever' scrawled upon them. Every shop was shut and many people had fled.[20] At several points the mob was fed information clearly designed to incite it: rumours of treasonous toasts at the banquet, of subversive leaflets at Fair Hill, of letters implying that in August the Dissenters planned to blow up churches and assassinate the King.

The propaganda had its effect. At two in the afternoon of this second day the crowd stormed Baskerville's old house, now the home of John Ryland, a merchant and banker. This too was burned: seven men died in the cellar, drinking while the blazing roof collapsed in on them. Even so there was no official action against the rioters until the Birmingham prisons were opened and their occupants freed; then a hundred or more special constables were sworn in, but this hastily assembled force was quickly routed in a pitched battle, with one constable later dying of his injuries.

The mob next turned on other targets, including William Hutton's house and bookshop. As morning broke, they marched out to the country, where seven more Dissenters' houses and farms and a small meeting house were ransacked. Each was carefully chosen. All through these days, people remarked not on the madness but the clean efficiency of the operation. As *The Times* put it, the demolition workers 'laboured in as cool and orderly a manner as if they had been employed by the owner at so much per day'.[21] Withering reported that 'the mob had lists of the devoted that were to be attacked' and added that these were 'occasionally altered and constantly added to'.[22]

Withering was an Anglican – 'who perhaps never heard a Presbyterian sermon' – and he had not been to the dinner, although he had walked cautiously through the crowd on his way home on the 14th. He was, however, a known friend of Priestley, and of Keir and the dangerous 'philosophers' and wealthy manufacturers of the Lunar Society. In the days that followed his servants at Edgbaston Hall kept a nightly watch. Boulton and Watt, equally anxious, armed their Soho workmen and mounted a permanent guard. On Friday night, Withering's gardener reported that his house was to be attacked, but he did not know when. On Saturday morning a message came that it was to be burned the next day, and this gave Withering time to remove 'three carts and two wagon loads' of his most valuable furniture, books and specimens. Driving these away, they met a crowd of about three hundred men bent on plunder – such gangs often came ahead of the incendiaries, knowing the family would be on the run with their possessions – one wagon had to be hidden in a haystack and covered with hay while Withering turned off the crowd with money and drink. A second, 'more determined set of scoundrels' was driven out with more bribes and liquor and by evening, most of Withering's goods were stored in the nearby church. At eight the incendiaries arrived, about three hundred strong. Withering was in Birmingham, awaiting the arrival of the military and at Edgbaston again his stout friends – a baker, a clerk, a coachmaker – turned off the mob, reinforcing their defence party with 'some of the grinders and some famous fighters from Birmingham hired on purpose'. In the fight that followed, at least one attacker was killed.

As the rioters regrouped at Edgbaston Hall about ten o'clock that night, a troop of dragoons finally arrived from Nottingham. The streets were illuminated and the people cheered: the danger seemed past. A large gang of colliers poured into town the next day, still hoping for a fight, and bringing new fears – 'It is apprehended the whole town wil be laid in ashes' – but by that time the dragoons had been reinforced by cavalry from Lichfield and soon order was more or less restored.[23] Odd skirmishes and 'depredations' continued in the countryside but by 20 July, a week after the banquet, 'not one rioter was met with, and all the manufactories are at work, as if no interruption had taken place'.[24]

The riots were over. Four meeting houses and twenty-seven houses had been attacked; at least eight rioters and one special constable had died, and many more had been injured. In the aftermath, the horrified Lunar men counted their losses, while Boulton gave vent to some hearty self-interest. Soho at least had been unharmed. Writing to a French friend, he pretended to utter detachment – to be above all the fuss:

> The Town of Birmgm's quite unharmonied party spirit and Rancour tears all good neighbourhood to pieces but I am happy in living alone in the Country & am almost silent upon this disonant Subject. By minding my own business I live peacably & securely amidst the Flames, Rapin, plunder, anarchy & confusion of these Unitarians, Trinitarians, predestinarians and tarians of all sorts.[25]

Breathing a sigh of relief now that the fuss was over, he acknowledged that the thing that mattered most to him was his factory. He had appealed to his workmen not to join 'the Court of King Mob', but had set cannon in place, just in case. 'I should have suffered them to have burnt & destroyed my house rather than fired on them,' he confessed, 'but I was solemnly determined to have poured destruction on their heads the moment they had attempted my manufacture, the destruction of which would have put 1,000 persons out of employ.'[26]

The official response was ambivalent. During the riots, *The Times* was plied with false information (later publicly retracted), reporting, for example, that treasonous toasts had been made. James Gillray published *A Birmingham toast, as given on the 14th of July, by the Revolution Society*. None of those he drew – including Sheridan, Fox, Horne Tooke or Priestley (holding a Communion dish and calling out 'The [King's] Head, here!') – had been there, but Gillray ineradicably linked the diners to the levelling 'king-choppers', and to secret feasts celebrating the death of Charles I. A second print, *The Hopes of the Party*, showed the same group outside the Crown and Anchor in the Strand, preparing to cut off George III's head, with Priestley mouthing consolation to the bewildered fat king: 'A man ought to be glad of the opportunity of dying if by that means he can serve his country in bringing about a Glorious Revolution.'[27]

A Birmingham Toast, by James Gillray, 14 July 1791. Priestley holds up the platter, crying, 'The – Head here.'

Not surprisingly, George III declared he could not but feel pleased that Priestley had suffered 'for the doctrines he and his party have instilled, and that the people see them in their true light'.[28] Pitt stayed silent. Burke was openly delighted. On the other hand Dundas, the Home Secretary, percipiently saw the riot itself, not the dinner, as showing the dangerous 'levelling principle' at work: the mob's cause could swing easily from pro- to anti-Government. (Two years later the same Birmingham men were indeed crying 'Tom Paine for ever'.) Dundas's urging that the local magistrates search out the organizers was quietly ignored – hardly surprisingly since the justices themselves, like the Anglican clergy and local landlords, had undoubtedly fostered the outbreak. When the cases finally came before the Warwickshire Assizes the following April, seventeen people out of the hundreds involved were prosecuted.[29] With a jury perhaps anxious to show support for Church and King, only four were convicted, of whom two were hanged and two pardoned.

Priestley's Lunar friends rallied round. Galton took in Mary Priestley immediately after the riots, courting danger himself. Wedgwood wrote quickly, 'Can I be of any use or service to you upon the present occasion? Assure yourself, my good friend, that I most earnestly wish it. Believe this of me – act accordingly, instruct me in the means of doing it & I shall esteem it as one of the strongest instances of your friendship.'[30] In September, Galton pressed Priestley to say when he might come back to the town, and promised he would meet him, happy to show his public attachment to a person whose friendship did him great honour.[31] Furthermore, the next Lunar meeting would be the following Monday at Barr – might Priestley come?

Darwin wrote to Wedgwood, 'The Birmingham riots are a disgrace to mankind.'[32] The Philosophical Society of Derby sent an 'Address to Dr Priestley' after their meeting on 3 September:

> Almost all great minds in all ages of the world, who have endeavoured to benefit mankind, have been persecuted by them; Galileo for his philosophical discoveries was imprisoned by the inquisition; and Socrates found a cup of hemlock his reward for teaching 'there is one God'. Your enemies, unable to conquer your arguments by reason, have had recourse to violence; they have halloo'd upon you the dogs of unfeeling ignorance, and of frantic fanaticism; they have kindled fires, like those of the inquisition, not to illuminate the truth, but, like the dark lantern of the assassin, to light the murderer to his prey.[33]

But there was a sting in the tail of this letter which went on to express the hope that Priestley would not risk his life again among these people, and would abandon polemic and cultivate science. Then his already widespread fame would 'rise like a Phenix from the flames of your elaboratory with renovated vigour'. And not all the Derby Society agreed: the Reverend Charles Hope complained to the local press that not enough members were present: Darwin promptly had him thrown out.

That message of sympathy but 'stay away' was even clearer in Watt's response. Galton's (vain) hope that Priestley might come to the September Lunar meeting filled him with alarm and he set off to Barr with pistols in his pockets. Watt had not written to Priestley at all until he heard he was thinking of returning to Birmingham to preach at the temporary 'Union Meeting House', to be opened in November. The news made Watt snatch up his pen to remind the doctor that he had duties 'to your family, to your friends, and to humanity in general, that should direct you not to risk a life so valuable to them all'.[34]

Priestley's supporters, in London and in Europe, were vociferous in their support: Lansdowne, Stanhope, Sheridan and Fox stood by him; Dissenting congregations and constitutional societies across the land took his side; French patriots offered him a house in Paris; his brother-in-law John Wilkinson arranged for funds to be placed in France and America in case he had to emigrate. Meanwhile, Wilkinson also took a house for Priestley in Clapton, and he took over as Price's successor as minister in nearby Hackney. He settled in that September and old Lunar friends immediately helped him to set up a new laboratory: Wedgwood and his son Tom sent money, mortars and apparatus; Galton and Withering sent money; Watt sent a copying-press. All advised him to tone down his *Appeal to the Public on the Subject of the Riots in Birmingham*, but all volunteered to act as witnesses for his claim for damages.[35] He missed his friends sadly, telling Watt that he never failed to think of the Lunar Society, 'at the usual times of your meeting'.[36]

In the following year, Keir took up his pen, writing his own pamphlet in Priestley's defence as 'Timothy Sobersides, Extinguisher-Maker'.[37] The Lunar Society had been fatally wounded by these riots – less by the damage to property, or even by the departure of Priestley, but by the slur attached to all the things they believed in: 'reason', 'science', 'freedom', 'experiment'. This was a riot against intellectualism, and its abiding image is of book-burning. One witness remembered that 'the highroads for full half a mile of the house were strewed with books, and that on entering the library there was not a dozen volumes on the shelves, while the floor was covered several inches deep with the torn manuscripts'.[38]

One slogan chalked on Birmingham walls was 'No philosophers – Church and King For Ever.' The reference was to the French *philosophes*, held to blame for igniting the fuse to revolution. But Birmingham had its own 'philosophers' near at hand, the Lunar Society. The vernacular pamphleteering had pointed a finger at the high-thinking, impractical men of learning (men with 'a fine deal of know'), and members of the Royal Society ('the King's Club').[39] Priestley's intellectualism itself became a focus: he was out of touch with the working man. Burke's chief *bête noire* was theory divorced from practicality. He declared the French uprising to be 'a revolution of doctrine and theoretic dogma'.[40] Elsewhere he defined this 'theoretic' mentality as embodied in the *philosophes* who were not just indifferent, but downright dangerous: 'These philosophers consider man in their experiments no more than they do mice in an air-pump, or in a recipient of mephitic gas.'[41] Only two years before, Keir had written proudly,

'The diffusion of a general knowledge, and of a taste for science, over all classes of men, in every nation of Europe, seems to be a characteristic feature of the present age.'[42] But now in Burke's words science was smeared with blood, its practitioners arrogant and uncaring. In the *Reflections*, attacking Richard Price's sermon, Burke described the old minister as linked with 'literary caballers, intriguing philosophers, political theologians and theological politicians', all of them wholly 'unacquainted with the world in which they are so fond of meddling, and inexperienced in all its affairs, on which they pronounce with so much confidence'.[43] Using terms that harked back to Priestley's soda-water, French chemistry and Helmont's 'gas' that burst out of its container, he yoked science to subversion:

When I see the spirit of liberty in action, I see a strong principle at work, and this, for a while, is all I can possibly know of it. The wild *gas*, the fixed air, is plainly broke loose: but we ought to suspend our judgment until the first effervescence is a little subsided and the liquor is cleared, and until we see something deeper than the agitation of a troubled and frothy surface.[44]

This was clever but also acute. Priestley had little sense of political action or administrative practice – God would decree the means of change and the methods. He genuinely believed that revolution could usher in the millennium, telling his old friend Thomas Belsham at the end of his life, 'You may probably live to see it; I shall not. It cannot, I think, be more than twenty years,'[45] and his mingling of reason and prophecy always seemed to carry a veiled threat of violence.

The linked attack on science and revolution was seen in the cartoons. And when Priestley's successor, the Reverend John Edwards, arrived at the New Meeting House in 1792, 'John Nott Button Burnisher' jumped on the opportunity to damn his predecessor's pretensions:

But Meister Edwards, shoodna you ha sumthin arter your nam? Caus plane J. Edwards loks as if yow want a mon of larnin' and ability as you say. Now, I was thinkin, spoze you was to put T.G., A.B., J.L., B.M, it whood stond for Tailors Gose, Are Balon, Jock o'Lonton and Bag a Muneshine. How much honsomer this whood look shoodna it than Meister Priestley's DD, F.R.S.?[46]

The Birmingham riots were a symbolic moment: the start of that long distrust of the British masses for the intellectual. For the first time, it became dangerous to have initials after your name.

38: HANDING ON

'The Hellish miscreants who committed so many outrages here, by banishing Dr. Priestley have almost broke up our Lunar Society,' groaned Watt to Joseph Black. 'At least when we meet we have more politics than Philosophy.'[1] And it was true that although meetings took place sporadically for the next few years, after Priestley left the light slowly faded from Lunar life. The French Revolution interposed itself, like the turning earth casting its shadow on the moon, covering the circle of talk and experiment until all that remained was a shadow, glowing red with reflected light.

This was not to say that individual achievements were at an end. In many ways the members of the group were at the height of their achievements. In 1792 Sir Joshua Reynolds directed the Swedish portrait painter Carl von Breda to Birmingham, where he painted Boulton with his minerals and his manufactory, Watt with his engine plans and Withering with his foxglove: the trio were exhibited at the next Royal Academy show. This year, too, Joseph Wright painted his last portrait of Erasmus Darwin, fatter than ever, his button straining to hold his fine broadcloth coat, a humorous, imposing figure with quill in hand.[2] Darwin's weighty prose writings, *Zoonomia* and *Phytologia*, his study of horticulture, were still to come; Boulton and Watt would win their patent battles; the Soho Mint would flourish; the Portland Vase be hailed as a national treasure. But there was a sense of change – and often of danger – a feeling of handing

Darwin playing chess with Erasmus, silhouette, *c.* 1790

over to a new generation who had different ways, different views, and who had 'grown out of' the Lunar movement, in both senses. The next few years would witness a fascinating kaleidoscope of struggle and tension between the older Lunatics and their sons, at once admiring and in revolt against their dominating fathers.

Already, in 1790, Wedgwood had taken his three sons, and his nephew and long-standing manager Tom Byerley, into the partnership. But the only son who was really committed to the firm was the second, Josiah, and even he felt it was below him to wait on customers in the London showrooms. He had been brought up for too long, he said, 'to feel equal to everybody, to bear the haughty manner of those who come into a shop'.[3] In 1792 he married Elizabeth Allen, daughter of a West Country gentry family. In the same year, his older brother John left the firm to become first a banker and then a country gentleman, when he married Elizabeth Allen's younger sister, Louisa.

At twenty, Tom Wedgwood – tall and elegant, brilliant, charming, hypochondriac – also went his own way. He still worked for the firm, designing patterns, experimenting, and corresponding with Priestley on his papers on light and heat.[4] A genuinely innovative chemist, Tom has a claim to be recognized as the real father of photography: taking up Scheele's experiments on the photosensitivity of silver salts – which turned black in the light – ten years later he managed to copy silhouettes and drawings on to glass, by putting them in bright light on to pale leather or white paper moistened with silver nitrate.[5] But he found no way to fix them – the salts kept on darkening in the light, and his creations were doomed for ever to the dark room. Tom was a proto-Romantic, a restless soul. In 1792 his Etruria years ended: that summer he set off for Paris to join James Watt junior. Darwin treated him for years for his many ailments, prescribing him in 1794: 'about 5 grains of rhubarb and 3/4 of a grain, or a grain, of opium, taken every night for many months, perhaps during the whole winter'.[6] This was the first push down a fatal slope, perhaps not only for Tom: at the time he was staying with his brother Josiah near Ottery St Mary, the home of the Coleridge family.

Handing over was not so easy. And if Tom Wedgwood moved into Radical–Romantic circles, befriending Wordsworth in Paris and Coleridge in Somerset, Jamie Watt's path, at least to begin with, was even more alarming to his father. In Manchester James had lived and breathed politics. His close friends were two leading Radicals, the barrister Thomas Cooper, and Thomas Walker, head of a big textile merchants: through them he joined the Manchester Lit. and Phil., and the Jacobin-friendly Constitutional Society, founded in 1790. With Cooper, Walker and others he argued hotly that the Lit. and Phil. should publicly support Priestley after the riots. When the committee refused, all the proposers walked out. James had almost finished his apprenticeship. The rift with his father and stepmother ruled out working at Soho, and they blocked his idea of setting up in business with young Joseph Priestley, who was equally at a loss. Finally, James became a European salesman for the Walker brothers, his political allies. In March 1792, allegedly to sell textiles, he sailed for Calais with two leisured 'friends of liberty', Cooper and John Tuffen.[7]

A fever was abroad in Britain. The second part of Thomas Paine's *Rights of Man* had just been published; Jacobin sympathizers cut their hair short, and wore it without powder; new clubs were born, among them the London Corresponding Society. As if in sympathy with their rising passions, the spring was unnaturally hot and thermometers soared to 82 °F in March, the month that James Watt sailed. When he reached Paris, he met Lavoisier and other chemists, many of them his father's correspondents. They were now all talking politics instead of philosophy. In April, James presented an address from the Manchester Constitutional Society to the Paris Jacobin Club, and he and Cooper marched, waving banners, in an international rally. Both were personally denounced by Burke in the House of Commons. (A newspaper report nearly caused Ann Watt to die of fright. She told her husband, 'Had I not known you were in London I certainly would have thought it was you.'[8])

Birmingham was being targeted as a nest of insurrection: at one point Burke flung a dagger on the floor of the House, claiming it was part of a Birmingham arms shipment for the French Government. A despairing Watt senior warned his son, 'The eyes of our Ministry are upon you.'[9] He was furious at being tarred with a Jacobin taint just as his patent cases were reaching a crisis. Yet as his letters to Annie showed, he had his own resentments. He was being treated in the House of Commons as an 'extortioner', claiming the inventions of others: 'It may so be deemed but if it is,

I hope to live to see the end of a corrupt aristocracy that has not the gratitude to protect its supporters, nor the sense to uphold its own decrees.'[10]

Other Lunar sons were in Paris this year. In June, William Priestley applied for naturalization there, and was introduced to the National Assembly. (*The Times*, reporting the Assembly's welcome of this 'Birmingham hero', joked that Priestley was about to follow him and had offered his own and Tom Paine's 'combustible' works as ammunition, a good idea as it might prove 'a prodigious saving in the article of gunpowder'.[11]) Tom Wedgwood arrived next, reporting in July that James was 'a furious democrat . . . Watt says that a new revolution must inevitably take place, and that it will in all probability be fatal to the King, Fayette, and some hundred others.'[12] By then the Government was already bracing itself for a Prussian invasion and in early September, when that invasion came, counter-revolutionaries were massacred by the thousand in the gaols of Paris and the provinces. English onlookers were horrified. Watt wrote to his wife that James had sent some account 'of the late atrocious murders of which he expresses his abhorrence, but remains of the same principles as before . . . I told him that no duty could urge him to espouse the cause of the French that his duty required him to follow his business and that he had no right to cover me with infamy.'[13] Watt had told James, he said, to give up all contact with his associates in Paris: 'If he does not he must do so with me.' By the time Watt's letter reached Paris, the leaders of the Commune, whipped up by Marat's rhetoric, had called for a new constitution. The French monarchy was formally abolished on 21 September, and next day France was declared a republic.

Shortly afterwards the new regime made several foreigners honorary French citizens, including Priestley, Paine, Jeremy Bentham, William Wilberforce and George Washington, and a month later Priestley was elected a member of the Convention. He declined the dangerous honour. He was the constant subject of attack, pilloried in pamphlets, speeches and cartoons and an accolade from France would not help him. He already found life in London difficult, 'most of the members of the Royal Society shunning me on account of my religious and political opinions'.[14] Unable to work, he looked back nostalgically to the days at Fair Hill, writing often about his sense of loss and promising to send papers to Lunar meetings. 'One of the things that I regret most on being expelled from Birmingham, is the loss of your company, and that of the rest of the lunar society,' he told Withering. 'My philosophical friends here are cold and distant.'[15]

At that time, Priestley was thinking of following his son to France, 'tho unwillingly'. One of the most touching moments in his voluminous correspondence comes in June 1792, when he writes to his arch rival, Lavoisier, that in case of more riots,

> I shall be glad to take refuge in your country, the liberties of which I hope will be established notwithstanding the present combination against you. I also hope the issue will be as favourable to science as to liberty . . .[16]

But in Paris itself, many British sympathizers were becoming troubled by the pace of events. James Watt, whose loyalties among the many factions, like those of Tom Paine, lay with the more moderate Rolandists, now found the lead being taken by Danton and Desmoulins, whose demagoguery he loathed. Stories later clustered about Watt's stay: it was said, without evidence, that he had intervened in a duel between Danton and Robespierre; had been denounced by Robespierre as an agent of Pitt, and that this was why he fled. The truth was less romantic. In October he left Paris on business, resuming his travelling salesman role, journeying south to Marseilles, Livorno and Naples. But even as he travelled his thoughts remained in Paris: he could even see the possible assassination of the King as no crime, 'when it may save the lives of millions of my fellow creatures'.[17] Over the coming year, his correspondence with his enraged father reached its nadir. 'I despair,' wrote Watt. 'You have imbibed notions of liberty that in my opinion are utterly incompatible with the happiness of mankind.'[18]

In these days, 'Everything rung and was connected with the Revolution in France', as Lord Cockburn remembered. 'Everything, not this thing or that thing, but literally everything, was soaked in this one event.'[19] Boulton, for example, had spotted a good business opportunity. In January 1791, hearing that the National Assembly planned to overhaul the French monetary system, he offered to strike an entire new French coinage for them. His contact was with the Monneron brothers and although the Assembly rejected his plan, the brothers eagerly launched a token coinage of their own. Boulton struck five-*sol* pieces, based on a medal by Augustin Dupré, showing the *fédères* taking their oath before France, who holds up the tablets of the Rights of Man. Around the design ran the motto 'Vivre Libres ou Mourir'. In money-starved Paris, the tokens were so popular that the crowd demanding them had to be dispersed by the cavalry. Three patriotic medals were also struck at Soho, showing the 'Serment du roi', 'Rousseau' and 'Lafayette', and although the Monnerons went bankrupt in March 1792, Augustin Monneron continued to trade, shipping over forty tons of tokens from Birmingham, before private coinage was banned in September 1793. Ever sensitive to the public mood (and capable of playing to two markets at once), for the past year Boulton had also been acknowledging the reaction in Britain, issuing poignant medallions of 'The Last Interview', showing Louis XVI torn from his family before his execution, and of Marie Antoinette on her way to the guillotine.

All the Lunar men felt the impact of the Revolution in different ways. In *The Economy of Vegetation* Darwin had delighted openly in the freeing of the 'Giant-form of Liberty in France', and in December 1791, with Joseph Strutt and Samuel Fox, he founded the Derby Society for Political Information, one of the earliest of the Radical societies formed in the wake of *The Rights of Man*. In 1792, Darwin was still optimistic. In the middle of a packed letter to his old schoolfriend Dixon (addressed with revolutionary humour as 'Richard Dixon Citizen'), he wrote, 'The success of the French against a confederacy of Kings gives me great pleasure, and I hope they will preserve their liberty and spread the holy flame of freedom over Europe.'[20] The manifesto of the Derby Society (one of its demands being full adult male suffrage) was published in July, presented to the National Assembly in November, and printed in the *Morning Chronicle* the following month.

The timing could not have been worse, and soon the *Chronicle*'s editors were arraigned for seditious libel. Even before the September massacres, waves of panic hit England, fostered by Pitt's Government: a widely disseminated propaganda print, *The Contrast*, engraved by Thomas Rowlandson, showed the dire consequences of levelling French tendencies. 'Church and King' mobs demonstrated against known Radicals, including Thomas Walker, whose Manchester warehouse was attacked. A proclamation against seditious writings was issued and Tom Paine was burned in effigy in many towns. Thomas Reeves, a Government placeman, founded the Association for Preserving Liberty and Property against Republicans and

The Contrast, designed by Lord George Murray and engraved by Thomas Rowlandson, one of the most popular prints of late 1792

Levellers, whose membership was soon double that of the Radical societies: 'Captain Commandant of the Spy-gang', Coleridge called him.

In January 1793 came the news of the execution of Louis XVI, and in February, as the revolutionary armies crossed the Dutch border, Britain and Holland were pulled into open war, joining the Austro-Prussian coalition. In London, as Robert Augustus Johnson told Withering (who was wintering in Lisbon for his health and had begged for political news), troops were recalled, rewards were offered for information on publishers of seditious papers and servants were bribed to report conversations: 'Reform and Revolution were termed synonymous'; all those who did not agree with the Government were labelled republicans '& treated with the most scurrilous abuse, & the grossest calumny'.[21] In short, 'A gloomy terror sat on almost every countenance.' William Priestley left France for America, and many British Jacobins emigrated, among them Watt's friend Cooper, and Joseph and Harry Priestley. 'Such is the spirit of bigotry,' Priestley told Withering, 'that great numbers are going to America, and among others all my sons, and my intention is that when they are settled, to follow them, and end my days there.'[22] He signed off miserably, 'Wishing tho hardly hoping for better times'.

It was a hot summer. The nation sweated and trembled. Boulton kept quiet, while his barrels of tokens were shipped across to Paris. France was now the enemy, embroiled in bloodshed: in September the Committee of Public Safety, under Robespierre, took control of the Government and declared terror to be the prime political weapon. There was fear of invasion and James Keir, while never losing his zeal for reform, was typical of many ex-army men whose patriotism now focused on military targets, publishing *An Essay on the Martial Character of Nations*, and a *Dictionary in the Art of War Ancient and Modern* (perhaps picking up the notes for the book which his publisher had thrown in the fire over thirty years before).

Throughout the year, as James Watt junior travelled slowly north through Switzerland and Germany, he was in a torment of indecision, thinking seriously about joining the revolutionary armies, although all his friends argued passionately against it. The following spring James slipped quietly back from Amsterdam, knocking on the door of Soho House early in March 1794. 'Bad as this country may be, it is the best I know,' he told a Dutch friend in April.[23] From now on, both he and Matthew Boulton junior shielded themselves from trouble in the cloak of industry: that year, both became partners in their fathers' firm. James never married. He lived alone with his Sicilian manservant, his cat and his dog, and immersed himself in work: the former republican became a doughty capitalist.

Over the past year, the atmosphere in Britain had become increasingly tense. In August 1793, in Edinburgh, the lawyer Thomas Muir had been sentenced to fourteen years' transportation, merely for advocating parliamentary reform, and the Unitarian clergyman Thomas Palmer for seven. Innumerable sedition cases were dragged before the courts, and finally, in December, the editors of the *Morning Chronicle* were tried for publishing the Derby Address, defended by the great Thomas Erskine, who had fought the case against Paine the year before. Despite the summing up by the judge, Lord Chief Justice Lord Kenyon, who attacked the 'wicked' intent of the Address, and declared it was clearly 'a gross and seditious libel', the jury seemed reluctant to convict. After five hours, they came back to say that though 'guilty of publishing', there was no malicious intent – a nonsense verdict. Sent back to the jury room they spent the night arguing, and at five in the morning recorded 'Not Guilty'.

Darwin and his fellow members of the Derby Society had a narrow escape: if the verdict had been otherwise they too might have been tried. It did not help Darwin that, in passing, Erskine had said that the Address was rumoured to come from the pen of an admired writer, 'the first poet of the age; – who has enlarged the circle of the pleasures of taste, and embellished with new flowers the regions of fancy'. Nor that during this year his stirring hymn to the Revolution, safely embedded in *The Economy of Vegetation*, was extracted and reprinted in the most outspoken and influential Radical weekly, *Politics for the People*, published by the tireless Radical bookseller Daniel Eaton (tried himself this year for distributing *The Rights of Man*).

Although Darwin's views were radical he was not a revolutionary. Family, property, law and order; the 'improvement' science could bring; the idea that nature itself offered a deep-laid pattern of order and progress – these were the staples of his belief. His local arena was still the Derby Philosophical Society, where one guest in 1792 remembered his 'wonderful sallies of imagination and wit, which kept us in perpetual laughter and astonishment',[24] but his scientific work was now recognized more widely, bringing a fellowship of the new Linnean Society in 1792, and election to the American Philosophical Society the following year. And his practice continued to flourish, despite the taint of sedition. Yet family affairs came first. One-year-old Harry had died in 1790 – bled as ruthlessly by his father as any adult patient. Soon a daughter was born, his namesake, Harriot. 'The worst thing I find now', he confided to Dixon, 'is this d–n'd old age, which creeps slily upon one, like moss upon a tree, and wrinkles one all over like a baked pear.'[25] At sixty, Darwin had six children under nine. But the house slowly began to empty: his three Pole stepchildren married and he would soon set up Mary and Susan Parker in their school at Ashbourne, where Violetta and Emma, aged eleven and ten, were their first boarders.

Even in the atmosphere of growing repression there were moments of light-heartedness. That summer Darwin wrote happily of swimming in the river with Elizabeth and the children. Patty Fothergill, daughter of Boulton's old partner, kept an exuberant diary this summer, with not a

Darwin's wire-drawn ferry across the Derwent, 1789, drawn from memory by Francis Darwin

syllable about war or revolution: with the Boultons she stayed at Derby, where Darwin 'took us through his Gardin (and across a river in a boat that goes by itself) to another Garden full of all the common Weeds which he collects and is very fond of'.[26] As Boulton's party went on to Derbyshire, meeting the Edgeworths in Matlock, this could have been any other summer – with visits to the Castleton caves and card parties and balls: 'After supper we danced as usual and as it was the last night we kept it up a good while and danced Over the Hills and Far Away which is one of the most romping dances I ever saw . . .'[27]

When Darwin found himself identified with insurrection he withdrew into science and, in his writing, he linked the hope of change with a biological vision of evolution, rather than with current politics. In 1791 Priestley had declined Darwin's advice. 'Excuse me', he responded courteously but firmly, 'if I still join theological to philosophical studies, and if I consider the former as greatly superior in importance to mankind to the latter.'[28] But by now the political climate had grown too hot, even for him. His sons had already emigrated, and in February 1794, with all his friends pressing him to flee to America, Joseph Priestley booked his passage. Carefully, he packed his apparatus and his books. On 30 March he preached his last sermon at the Gravel Pit in Hackney. Ten days later the boat sailed.

Before he left, Priestley learned that Lavoisier had been arrested for his work as a 'tax farmer' under the old regime. The Academy of Sciences, Lavoisier's last refuge, had been abolished in 1793. After months in prison, Lavoisier was tried on 5 May 1794. He went to the guillotine the same day, and his body was thrown into an unmarked grave. During his trial, one of the charges against him was his correspondence with Englishmen.

Priestley dedicated his last work to be published in England, *Experiments on the Generation of Air from Water*, to the Lunar Society, declaring that it had 'both encouraged and enlightened me; so that what I did there of a philosophical kind ought in justice to be attributed almost as much to you as to myself.'[29] ('Almost' is nice.) In his memoirs he noted, 'I consider my settlement at Birmingham as the happiest event of my life.'[30] But although they continued to correspond, to fund his research and supply equipment, to the Lunar friends Priestley left behind it was as if an era had ended. And they had their own cares. Watt's savagery in his dealings with James had been intensified by the pain of watching his daughter Jessie dying of consumption before his eyes. Darwin shared Watt's anguish, treating Jessie with opium and foxglove, and consoling him tenderly at her death in June 1794. As a doctor and a father, he had shared so much of this parental misery with his friends, a sense of helplessness that intensified their closeness. In his writings, however, he could fit these private sufferings into a larger pattern, and this was part of the burden of *Zoonomia*.

Even in the climate of reaction the wave of visionary radicalism swept on. Darwin was much admired by younger writers such as William Godwin, whose 1793 *Enquiry concerning the Principles of Political Justice* took empiricism, 'associationism' and social evolution to a powerfully anarchistic, if logical, conclusion. If individuals were educated to understand their social duties, freed from the tyrannical interference of government and law, society would run itself: mind would control matter, and even disease and death would fade away.

Another admirer who felt he could vanquish disease was the doctor Thomas Beddoes, whom Darwin first met in 1787.[31] Edinburgh-trained, a keen chemist and geologist, Beddoes was an eager convert to French chemistry and an equally fervent supporter of the Revolution: in 1793 he left Oxford, resigning from his lectureship in chemistry before he was pushed. Beddoes was a natural heir of the Lunar Society – almost a caricature – his current project was to see if inhaling oxygen, hydrogen and other gases could help chest diseases such as tuberculosis. Darwin advised him to move to Bristol, with its hot-wells and wealthy patients, and as he was a great admirer of Thomas Day, Keir introduced him to Edgeworth who was there with his family seeking a cure for his consumptive son Lovell.[32] The Edgeworth children loved his experiments, and he compiled the imaginative chapter on 'Toys' for *Practical Education.* Soon he proposed to the twenty-year-old Anna and to everyone's surprise – since he was thirty-three, short and fat and famously gawkish – she accepted.

Edgeworth had his doubts about his 'little fat Democrat' son-in-law, recognizing that 'if he will put off his political projects till he has accomplish'd his medical establishment he will succeed and make a fortune – But if he bloweth the trumpet of Sedition the aristocracy will rather go to hell with Satan than with any democratic Devil'.[33] But Beddoes faced more problems than reactionary duchesses. One was simply how to get the pure gases he needed and the most effective apparatus. Here the Lunar men helped again. After Jessie's death, Watt began designing new apparatus for making unpolluted gases in bulk. The subject filled his letters to Darwin: 'I shall wait for Dr Beddoes' next book, I think, before I carbonate or hydrogenate, or azotate, oxygenate anyone,' wrote Darwin.[34] He did not have to wait long: the first volume of Beddoes's *Medicinal Uses of Factitious Airs*, with Watt's contributions, appeared at the end of the year.[35] Soon Darwin was trying oxygen on his patients, including his poor daughter Emma, who was also swallowing 'bone-ashes, soda phosphoreta and powder of bark . . . so that the oxygene gas is but a part of the process'.[36]

When the Pneumatic Institution was finally founded in 1798, with almost all the Lunar men as subscribers, Tom Wedgwood and Gregory Watt recommended a genius as Beddoes's assistant – the teenage Humphry Davy, son of their landlady in Penzance, where they had both been staying for their health. Over the next few years, while Beddoes poured out books, didactic novels and pamphlets, the vivacious Anna held open house to an extraordinary circle of campaigners and idealists, including Davies Giddy (who later changed his name to Gilbert and became President of the Royal Society) as well as Robert Southey and Samuel Taylor Coleridge. In 1794 Southey and the nineteen-year-old Coleridge were hatching ecstatic plans for the 'Pantisocracy', a Utopian community of twelve couples on the banks of the Susquehanna. The scheme inevitably collapsed but for the next two years the spirits of Locke and Hartley, Priestley, Darwin and Godwin imbued Coleridge's writing in his *Religious Musings*, his lectures and the *Watchman*. Knowledge would bring progress; man and society were perfectible; Eden would be reclaimed.

Darwin's support of Beddoes and work on *Zoonomia* let him envisage the improvement of society through medicine, rather than politics. Most of his work was serious; some of it was fun, like his correspondence with Tom Wedgwood on making an air-bed. ('He thinks feathers always stink', Darwin told Watt with amusement, 'and wishes to rest on clouds, like the Gods and Goddesses, which you see sprawling on ceilings.'[37]) But to the powers that be, he still seemed mildly threatening. Darwin's neighbour, Mr Upton, with whom he had had a petty dispute about windows a decade earlier, turned out to be a Government spy. And later this year his work and ideas were parodied in *The Golden Age*, an alleged poetical epistle from Darwin to Beddoes, where wild fantasies of plants adapted to grow meat were linked to 'the mean muck of low Equality'.

The summer of 1794 was a low point in Lunar, and national, life. In May the Habeas Corpus was suspended and twelve reformers were imprisoned, to be tried for high treason in October. The first case brought was against the shoemaker Thomas Hardy, Secretary of the London Corresponding Society; the second against the veteran radical Horne Tooke. One by one, thanks largely to Thomas Erskine's hard-hitting defence, they were acquitted. The other cases were dismissed. But despite the victory the Radical wave had crested and broken: too many men had been hounded, imprisoned, transported or driven abroad.

The acquittals in the treason trials were cheered among the Lunar men: Wedgwood himself had been a marked man; his name, he had been told, 'stood high in the list' of those who would be arrested if the French invasion came.[38] But at Etruria the legal triumph meant little. In late November, Wedgwood fell ill. To help his palpitating heart, Darwin prescribed him alum and nutmeg, mixed with quinine and powdered rhubarb, but he seemed so much better after a trip to Buxton that in early December the doctor exuberantly advocated him 'to leave off the bark, and take no medecine at present'.[39] But then an infection had set in on the right side of his jaw, and as the pain increased he fell into a fever. Darwin had little or no hope from the beginning, but 'was with him as often, and as long, as possible'.[40] The three weeks' illness allowed Sally and the family to prepare for the worst, said his son Josiah, and they would never forget their obligation to Darwin: 'We are all persuaded that he kept him alive many days by his friendly and uncommon attention.'[41] For the last two days, according to his nephew Tom Byerley, Wedgwood was unconscious; but another story, circulated in the family for a hundred years, told things differently.[42] Darwin had kept his friend well enough to say goodbye to his family, but he had also given him a private supply of laudanum. On 2 January 1795 Wedgwood told Sally and Susannah not to come in as he was sure he would sleep soundly all night. Next morning – Sukey's thirtieth birthday – they found his room locked from the inside. The carpenter, Greaves, came running with a ladder and climbed in through the window to find him lying dead.

Wedgwood, who had started with £10 forty years before, left a fortune in his will. The following spring, as the two fathers had long wished and planned, Robert Darwin married Susannah Wedgwood, heiress to £25,000 – over a million in modern terms.

39: 'FIRE ENGINES & SUNDRY WORKS'

The winter of 1794–95 was the worst for almost fifty years, so cold that the milk froze in the milkmaids' pails. January was the coldest month ever recorded, with a −3.1 °C average in central England. In March there were food riots in the Midlands and in Cornwall. As the price of wheat soared the riots spread; in Birmingham and in the cotton towns, mills and warehouses were attacked. A terrible harvest added to the tension. At the end of the year, after a stone from a hissing mob hit the King's coach on the way to the opening of Parliament, Pitt passed his notorious Two Acts against 'Seditious Meetings' and 'Treasonable Practices': the former hit particularly at the intellectual societies, requiring them to be licensed and proscribing discussion of religion or politics. The repression deepened: reformers were attacked, juries were packed, working men were picked off by press-gangs to fill the Navy's ships. In 1797 came the mutinies at Spithead and the Nore. Many were imprisoned without trial. The year 1798 saw Nelson's victory at the battle of the Nile but the triumph made no difference to Government nerves: this year Darwin's publisher, Joseph Johnson, was gaoled for selling seditious works.

When Darwin, full of grief, wrote to Edgeworth in early 1795 he lamented the state of the world as much as Wedgwood's passing:

He is a public as well as private loss – we all grow old, but you! – when I think of dying it is always without pain or fear – this world was made for the demon you speak of, who seems daily to gain ground upon *the other gentleman*, by the assistance

Soho Foundry, by John Phillp

of Mr Pitt and our gracious – I dare not mention his name for fear that high treason may be in the sound . . .

America is the only place of safety – and what does a man past 50 (I don't mean you) want? potatoes and milk – nothing else. These may be had in America, untax'd by Kings and Priests.

Sometimes the harsh times and the personal tragedies made life near unbearable. Watt's older, married daughter Margaret died in Glasgow in 1796, and Darwin wondered, he said, what there was to make a man want to keep on living, especially when public affairs were so dismal. 'Activity of mind', he thought, was the only thing that could keep one from brooding over 'disagreeable events, which already exist, or are likely soon to exist, in England as well as in the other countries devoted to this bloody war!'[1]

Yet, as the war with France dragged on, the Lunar men still looked forward. And while the poor and the landed classes suffered, the industrialists did well out of the war, the Lunar men among them. Armies had to be fed, clothed and armed. Copper output rose, especially for the sheathing of ships. Coins were needed. Henry Cort's method of puddling and rolling iron, which Boulton had supported ten years before, at last began to be widely used. Furnaces and forges poured out steel and iron, and coal production soared. Even the cotton factories experienced a boom. The 1790s saw a spate of enclosure acts and a new canal mania, with speculators making vast fortunes. Just over the horizon lay the era of smoking factory chimneys and sweated labour – the age of the machine.

From the beginning, one response to the French Revolution was that it could not happen in England: the constitution was too sound, the rights of Englishmen secure; the poor infinitely better off than gruel-eating French peasants. This was a keynote of older Whig thinking, heard in Keir's modified cheer at the Fall of the Bastille: 'Happily the same necessity does not exist in this country.' It was also a theme of the Evangelicals, ringing through the speeches of William Wilberforce and the tracts of Hannah More, determined to open the eyes of the working people not to their needs but to their blessings.

This fervent advocacy of quietude was balm to manufacturers. And in 1796, in the middle of war, Boulton struck the same note:

As the Smith cannot do without his Striker so neither can the Master do without his Workmen. Let each perform his part well and do their Duty in that state to which it has pleased God to call them & this they will find to be the true ground of Equality.[2]

His notion of equality is actually a straightforward acceptance of hierarchy, rephrased in industrial rather than feudal terms. This was an unblushing depiction of men, women and children as the manufacturer's tools. It was only sensible to use them well. For all their talk of liberty Wedgwood and Boulton were leaders in the drive to subdue and control, epitomized a few years later by David Dale's and Robert Owen's model factory and village at New Lanark.

Boulton had always been genuinely concerned about working conditions, insisting that the walls of the Albion Mill be whitewashed, for example, to counteract the darkness and dirt, giving Christmas presents to all his workers, raising their wages each year, sorting out problems with their landlords. But in times of unrest such as these, the manufacturers had even more need of a loyal, quiet workforce to keep trade going. In 1791 there was a strike among engine-fitters at Soho: Boulton settled it quickly and there were no further problems for several years.[3] And in 1792, the year of *The Rights of Man*, the rules of the 'Insurance Society belonging to Soho Manufactory' were printed and circulated, with an elegant engraved plate, including emblems of Stability (the cube), Fidelity (the dog) as well as Art, Prudence, Industry and Plenty. 'The Flowers that are strewed over the Bee-hive', runs the explanation, 'represent the Sweets that Industry is ever crowned with.'[4] In the same year, as Treasurer of the new Birmingham Dispensary, Boulton announced grandly, 'If the funds of the institution are not sufficient for its support, I will make up the deficiency.'

There was plenty of other work around Birmingham, yet the workforce of Soho did stay loyal to Boulton through all his crises, and through all these insurrectionary years, as much because of his exuberant personality as his rules. His opportunism and constant flow of ideas sent the workshops into frenzy and made the men swear, but filled their pockets none the less. He behaved like the rumbustious ruler of a small empire: the adjective 'princely' was often used to describe him. Commenting on his coining efforts, Watt paid a double-edged tribute: 'He has conducted the whole more like a sovereign than a private man-

The plate decorating the Rules of Soho Insurance Society

ufacturer, and the love of fame has always been to him a greater stimulus than the love of gain.'[5]

Love of gain was still important though. In new conditions, Boulton looked to new markets: to France for his coinage, for example. But he also realized, as he neared retirement, that it was vital to cash in on their greatest commercial achievement, the steam-engine. Although orders rose in the three years from 1790, as more workers were taken on wages rose too, and in fact each year showed a greater and greater manufacturing loss. Two things had to be done: rationalize the production and gather in all the royalties they were owed. These tasks would fall largely to the next generation: in October 1794 their sons were formally admitted to the partnership, and the firm's name changed to Boulton, Watt & Sons, including Watt's brilliant younger son Gregory, now nineteen. To add a dose of experience, William Murdoch came back from Cornwall to help.

Ever since he first urged Watt to bend his thoughts to the rotative engine, Boulton had realized that it would mean new production methods. With the old reciprocating engines used in mines or on canals, there had always been engineers on the spot who were used to the old Newcomen engines and could erect the Soho machines from their many parts. This was not the case with mills and factories which had so far depended on water or horse-pulled wheels. Now the engines often had to be made completely at Soho and tested there before being re-erected at the site: they were smaller, supported on their own frame rather than being built into a huge engine-house, and amounted to a complete, movable machine. Looking at his falling profits and realizing that the costs to subcontractors such as Wilkinson were biting hard, Boulton decided to build a complete engine works at Soho, something he had dreamed of for thirty years, since his letter to Watt of 1769 when he said it would be well worth his while to make engines 'for all the world'. The catalyst was a final row between the fiery-tempered Wilkinson brothers, John and William, in 1795, which led to the closure of the Bersham ironworks. The Coalbrookdale Company could supply Soho with cylinders but not in enough quantity, and those made elsewhere proved hopeless.

A mile from his manufactory, Boulton began building the new Soho Foundry, on the bank of the Birmingham Canal. Complete with smithy, forge, boring mill, foundry and turning shops it was ready in three months. Several of Wilkinson's experienced workforce were taken on, and on 30 January 1796 it opened with a huge celebratory lunch for all the workmen and builders, and with a bullishly fantastical speech from Boulton, self-declared 'Father of Soho'.[6] *Aris's Birmingham Gazette* caught the tone, so much so that it reads like a press release from Boulton himself:

> Two fat sheep (the first fruits of the newly cultivated land of Soho) were sacrificed at the Altar of Vulcan and eaten by the Cyclops in the Great Hall of the Temple which is 46 feet wide and 100 feet long. These two great dishes were garnished with rumps and rounds of beef, legs of veal, and gammons of bacon, with innumerable meat pieces and plumb puddings, accompanied with a good band of Marian Music. When dinner was over, the Founder of Soho entered, and consecrated this new branch of it by sprinkling the walls with wine, and then in the name of *Vulcan*, and all the Gods and Goddesses of *Fire* and *Water*, pronounced the name of it SOHO FOUNDRY, and all the people cried amen . . . These ceremonies being ended, six cannon were discharged, and the band of Music struck up *God save the King*, which was sung in full chorus by two hundred loyal subjects.[7]

John Southern wrote tongue in cheek to James Watt junior, who missed the great event, that the whole thing went off without any mishap except 'a little or rather not a little thieving . . . Knives, Forks, Pepper boxes, mustard Pot, drinking cups – Jugs – Tablecloths, Knifecloths, Spoons and C.'[10]

Under the new partnership, especially with the keen involvement of James Watt junior and the eagle-eyed supervision of William Murdoch, the foundry gradually overcame its early difficulties with quality. The two old men, Boulton and Watt, had tackled one problem. But the foundry cost over £20,000, an investment not recouped for fifteen years, so it was even more urgent to pull in all available cash. When it came to raking in their royalties, they looked not only to the old engines, but also to all the pirate copies that infringed Watt's patent. A great many such engines had been built. The success of the rotative engine had been almost too great for Boulton and Watt to meet the demand, and many mill-owners and manufacturers naturally turned elsewhere, hoping to get other engineers to build them on the sly. For a long time Watt was reluctant to go to war because he feared that the terms of the patent were so wide that no court would uphold it. But now this seemed vital for practical reasons – less to protect the patent, which had only a few more years to run, than to collect unpaid royalties and ensure future payments. From 1794, the younger partners began hunting down infringements: the threat of a lawsuit was often enough to bring the royalties in.

But if the mill-owners grudgingly settled, the situation was different in Cornwall. This was where most of their engines were, where most money came from – and where their monopoly was most resented. At last Watt brought an action against Jonathan Hornblower, who had been erecting his double-action engines all over the duchy. The owners sourly paid up rather than go to court, and when Hornblower tried to get his own rival patent extended, Boulton & Watt opposed it energetically and finally won the day.

One rival was down and out, but others sprang up in his place. Edward Bull, their former engine erector, built 'inverted' engines for the western mine captains, with the help of the talented twenty-one-year-old Richard Trevithick. They won injunctions to stop Bull and Trevithick – the latter evading all attempts to catch him and serve it, until suddenly, with baffling panache, he turned up with Bull on the doorstep of Soho itself, and was cornered in the tap room of the nearest pub. The court never doubted that

Bull had infringed the patent. The problem – as Watt had feared – was the validity of the patent itself. For years this issue dragged on, being taken before the Appeal Court in 1795, again with no clear decision. Meanwhile the partners could continue to fight individual cases. One, in 1796, was against Jabez Hornblower and his partner Maberley, who had invented a 'pendulum steam engine', using Watt's separate condenser. When the jury decided for Watt there were great celebrations and firing of cannon at Soho, but Hornblower and Maberley made a desperate last stand, challenging Watt's patent as a whole. In January 1799 the Court of King's Bench unanimously decided for Watt: 'We have *won the cause* hollow,' he told Boulton.[9] James Watt junior commanded Soho to send forth its trumpeters: 'Tell it in Gath, and speak it in the streets of Ascalon. Maberley and all his hosts are put to flight!'[10]

In winning these cases Watt's reputation as a 'Philosopher' was of undoubted importance, and so were his carefully nurtured connections: half of the witnesses who testified for them were Fellows of the Royal Society, and the rest had at some stage been Soho employees.[11] But in many ways this was a symbolic victory, and an expensive one. The court costs were huge and the patent would run out in a year: Watt's engine was ageing, as he was. In 1799, their Cornish agent Thomas Wilson estimated, with not wholly believable precision, that royalties owing amounted to £162,052 3s 7d – roughly £9 million in today's terms. But although the furious Cornish mine captains were forced to pay up some of this, they certainly never paid all, and their resentment meant the end of the firm's business there. Even before the patent lapsed, Trevithick's powerful, high-pressure engines were steaming and sweating away. Despite a terrible explosion at Greenwich (exploited to the hilt by Soho propaganda) Trevithick's monsters were soon at work from Cornwall to Coalbrookdale.

When the sons entered the business in 1794, Boulton was sixty-six and Watt only fifty-eight. But while Boulton carried on in style, Watt concentrated on his experiments and his work with Beddoes. The engine royalties had freed him from his panic over money. In 1790 he and Ann had moved from the old Regent's Place on Harper's Hill, now submerged by Birmingham brick, to a new house designed for him by Samuel Wyatt, called Heathfield, on Handsworth Heath, twenty minutes' walk from Soho. When the heath was enclosed the next year, he bought forty acres, planting trees, building hot-houses and a walled kitchen garden.[12]

For a time, all looked well. The talented Gregory was sent to the Radical, progressive University of Glasgow, and his professors there were warmly invited to the Watts' home. But sadness found its way into Heathfield with Jessie's death in 1794, followed by a kind of estrangement from Gregory. A year later Annie was writing heart-rending letters, chiding him for his discontent and for not wanting to please his parents who 'would always be ready to fulfil his every wish', if only he would 'open your heart to them with freedom' and not greet them with such cold 'discreet reserve'.[13] The gap between the generations was wide. Over the next few years, Gregory and his friend William Creighton were zestfully exchanging letters in which serious discussions of minerals were interspersed with scatological verses and graphic drawings of Mam Tor and old caves like the Devil's Arse, as well as vivid cartoons of geologists in the field, with bags of rocks – and bottles of gin – a very different tone to the old Lunar explorations of Derbyshire. They recorded ludicrous royal patents, not for steam-engines, but for schoolboy-crude devices such as the 'Centrifugal Purge', and indulging in a happy parody of 'satanic' camaraderie, addressing each other as 'Most Enraged Saint', 'Beneficent Demon', 'Beautiful blooming Chermbocolos'.[14] At exactly the same time, Watt senior was sending Gregory affectionate, detailed accounts of the Cairngorms and the beauty of this 'most romantic valley'.[15] By contrast to his son, he seemed old, innocent and puritanical. But despite Gregory's brilliance and overflowing spirits, by now there were signs that he too was sickening.

In these years Watt often withdrew into the solace of work. In a long attic above the kitchen wing at Heathfield he created a workshop, placing his lathe and workbench by the window, building shelves for his fossils and other specimens, and arranging all the old tools he had brought with him years ago from Scotland. This untidy, dusty, chemical-smelling haven was allegedly a retreat from Ann Watt's obsessive tidiness, so much so that he carried up a frying pan and dutch oven to make his meals. Boulton, too, had been nest-making. In 1787 he had commissioned Samuel Wyatt to remodel Soho House. It was to be elegant, with a long pillared portico,

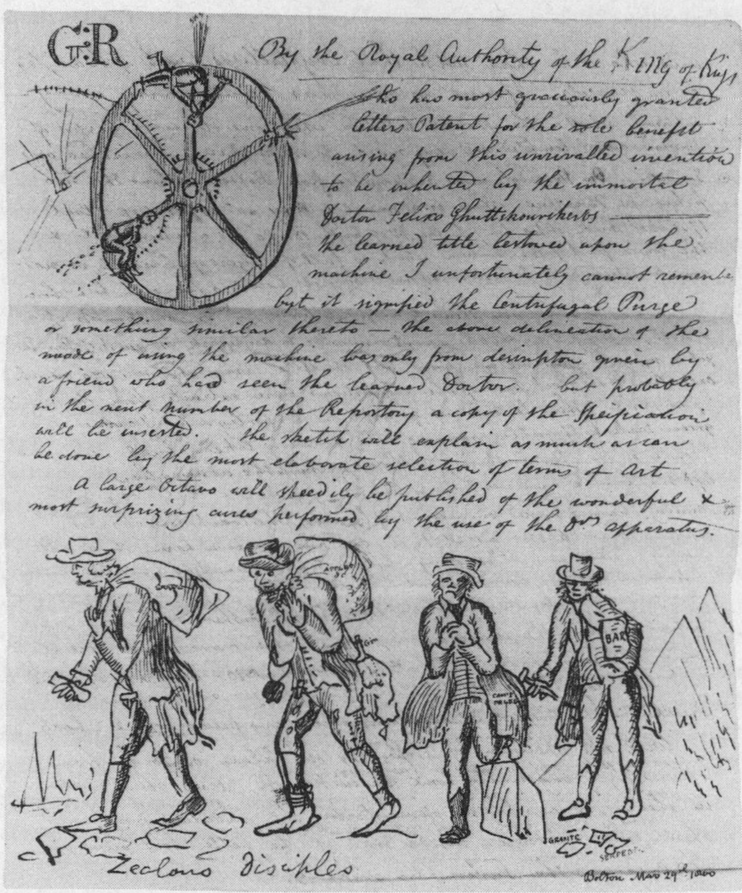

Drawings by Gregory Watt, in his letters to William Creighton, April 1800

marble pillars in the hall, marble fireplaces and mahogany doors. But it was also to be the house of a scientist: a year after the Wyatt plans an alternative set, never used, were drawn by John Rawsthorne, with rooms in the wings marked for 'Wet Chymistry', 'Dry Chymistry', 'Natural history', 'Botany – Green House', 'Astronomy'.[16] Boulton finally bought the freehold of his land in 1794, after making a careful listing of pros and cons, and immediately had new plans drawn up for aviaries and orangeries and tea-rooms. A final design by James Wyatt in 1796 added a classical architrave and round, Adam-style porch. In the end he quarrelled with James Wyatt and Samuel came back to finish it off.

No expense was spared, inside or out. Accounts for furnishings include festoons for windows, japanned chairs and 'french Hair cushions', white lace and sprigged calico, borders with roses and poppies, green fringed silk and large feather beds, as well as an Adam-style floor-cloth for the hall.[17] War, distress, riots and repression never touched this pursuit of domestic luxury. The garden, too, was overhauled, Boulton filling his notebooks with ideas and sketches, and discussing the use of steam to power cascades with the landscape designer Humphry Repton, who graciously accepted a copying machine as a present. (Repton was taken less by the Soho garden than by Boulton's buckles or 'latchets', asking if he could give him a pair, since bending to lace his boots was becoming very difficult to 'a man of my rotundity'.[18]) John Phillp's elegant drawings show the Soho garden at this date, and by the century's end Boulton's fine house was finished. It was technologically up to the minute, including piped hot water for the bath (the first ever heated by steam) and a system that let heated air flow through the house, filtered out through holes in the stairs. Grates and steam-heating had long been on the minds of both Watt and Boulton, and Franklin and Whitehurst before them. Boulton had corresponded with Lord Lansdowne about warming up his library, and had fitted heating in Edgbaston Hall for William Withering, who hoped it would help his bad chest, writing in November 1796 that the sharpness of the air made him 'think of your steam-warmed Atmosphere in which I promise myself much comfort during the Winter Months'.[19] In the end, though, the Witherings could not stand the smell of the soldered copper pipes and the whole system came back to Soho.

Although Watt and Boulton both now had time to think of creature comforts, they had always been opposites, and while Watt retreated, Boulton used his relative freedom from the engine business to concentrate on other things. He still led campaigns to control local standards, for example the gilding of buttons, in the Button Act of 1796 and took up new ideas, including a patent for hydraulic rams, worked on earlier by Whitehurst and Montgolfier. But his real concern was with the mint. To begin with this did not have an altogether easy ride. As soon as it looked successful the copper producers put up their prices; this affected other local manufacturers, including the button-makers, who promptly lowered their wages – once again, Boulton was blamed. In 1792, on the ready for riots, he wrote angrily to Thomas Wilson that he had done more for Birmingham than any other individual; he had nothing to do with white metal buttons, the trade most affected: 'I mix with no clubs, attend no public meetings am of no party, nor a zealot in religion.'[20] It made no difference. A week or so later he was reporting that 'workmen are parading the streets with cockades in their hats and assembled by beat of drum and headed by Ignorance and Envy with their eyes turned towards Soho'.[21]

Boulton armed his own men with clubs, but the uproar passed and the mint hammered on. When he refurnished Soho in grand style in the 1790s he even paid his supplier James Newton with casks of coins. Coinage was ever more in demand. The costly war brought inflation, and in 1797 the Bank of England suspended its obligation to convert paper money to gold, when required. This year Boulton finally won the longed for Government contract, for 480 tons of penny coins and 500 tons of twopenny pieces – hefty coins known as 'cartwheels', of which no more have been made since. 'Papa dined with Mr Boulton not many days ago,' wrote young Sarah Withering to her brother William, 'and brought home a precious collection of specimens for future coinage – the twopences are in circulation, sad clumsy things.'[22] This was only the beginning: from 1797 to 1799, Boulton recorded that the 'total quantity of peny pieces coined' weighed 1,266 tons, 'equal to 45,407.440 of pieces'.[23]

Full of excitement, he made improvements to the mint itself. In 1798–99 he chucked out the presses he had installed ten years before, damning them as violent and noisy. Instead, he built a new extension and set up new machines based on very different principles. From now on the steam-engine in Boulton mints would not drive wheels and shafts but pump a vacuum into a long cylinder, called 'the spirit pipe', which gave the main motive force to the coining presses, a brilliant invention of John Southern (son-in-law of that old inventive genius John Wyatt).

The new presses were a triumph. More and larger orders followed well into the new century, for pennies, halfpence and farthings, and even for a huge hoard of silver five-shilling pieces, recast from Spanish dollars from

Mexico, captured by the Navy. Month after month, Boulton's coins weighed down the barges heading south to the capital. And in 1805, when a new Royal Mint was finally planned on Tower Hill, Boulton was contracted to supply the machines and the steam-engine and plan the buildings. In his lifetime, machines went from Soho to supply the Royal Mints of Spain, Russia and Denmark.

The Soho mint offered adventure as well as profit. In 1799 Boulton took his own men on a raid against three local counterfeiters, engaging in a pitched battle, with people breaking down secret doors, jumping through trap-doors, and scurrying down ladders. It was as much public display as law-enforcing, and no one was prosecuted, but it made him feel young again, able to enjoy a fight. The following Christmas he found himself in another, when he heard rumours of a plot to rob the manufactory. Two nights running, he and young James Watt and a bunch of employees hid under the stairs armed with blunderbusses and sticks. The first night the burglars came but could not open the door: the second night they forced it, seized their plunder and were caught in a fierce scuffle. Staggering into bed in the early hours and hearing yet more noises, Boulton fired his gun into the darkness. 'In the morning,' he told Anne, 'I found our gray pony had been graseing under the windows & had been more frightened than I was. We have examined but find she is not hurt.'[24]

Boulton never grew out of showing off. ('I like Boulton, he is a brave man,' Walter Scott is supposed to have said when he heard of the robbery.) Even in his seventies he loved the chance to impress visitors. In 1794, Charlotte Matthews, who had been his banker and undertaken commissions for him in London since her husband died, admonished him for not being more sober. She noticed, she said, that for his new waistcoat he had chosen 'the silk pattern with the pink in it, now as I wear only black I cannot permit you to wear pink . . . at your age a similarity rather than a diversity of colours should be preferred'.[25] He rarely listened; this was the era of extravagant male fashion – high collars, stocks, tight-waisted jackets and tight breeches for Bond Street swells. A pink waistcoat, Boulton might well have thought, was a mere bagatelle.

In August 1799 he sent Charlotte Matthews a jubilant account of a visit to Soho by the Russian Ambassador during negotiations over the mint for St Petersburg. This was corporate hospitality on a lavish scale. With Anne and her friend Miss Mynd they saw Kemble playing Hamlet, and Macready's farce of *Bluebeard*; they had elegant dinners ('as I was well provided with dishes & a Man Cook for ye whole week'), musical evenings and daytime tours round the Birmingham workshops. The climax was a 'Secret expedition' on the canal in a specially upholstered barge. Another barge went ahead, crammed with musicians playing trumpets, French horns, clarinets, flutes, bassoons and a large double drum. After sailing into the foundry for a brief tour, they went on to another site where they saw 'one of our Engines Forging Steel, a Tilting Forge, grinding plates of Steel & Sad Irons, grinding Gun Barrells, Boreing them & also grinding Spectacle Glasses'. The *pièce de résistance*, which must have stunned the poor Ambassador absolutely, was the voyage home through Dudley Tunnel and into the huge limestone caverns under Dudley Castle Hill.

> We again entered our Boats & by the assistance of the Musick we increased the apparent population of the County immensely. After stoping several times to see Coal Mines, Iron Furnaces, Fire Engines & sundry works we at length enterd the regions of darkness wch I dispelled by 100 Torches wch I had provided. New & immense Caverns opend as we proceeded, through wch we often espied through openings the Inhabitants of ye Earth looking down upon us. The Band of Musick playing all the time & beautyfully echoed from the Salons of Erebus.[26]

This trip along the murky canal from Smethwick to Dudley must be one of the oddest of all eighteenth-century tourist trips, taking the industrial – and subterranean – sublime to a joyously ludicrous peak. It comes perilously close, too, to a Boultonic egotistical sublime, as if he had not only arranged the trip and the hundred torches and the music, but had also conjured up the factories and fire-engines and furnaces with some personal demonic force.

Despite Boulton's bursts of enjoyment and bravado, for much of the time from the mid-1790s he was far from well, stricken by kidney complaints, while Watt, the lifelong hypochondriac, now flourished in health. Keir was also often confined to home with rheumatism, but in his sixties he was still managing his factory, watching his daughter Amelia grow up in their big house at Hill Top. The real invalid was William Withering, not yet fifty, who was fighting persistent illness (allowing Boulton the woeful pun that 'the Flower of Physick was in such a Withering way as to render a transportation to Lisbon necessary –'[27]). He resigned from the General Hospital in 1792, and after his second winter abroad he gave up his medical practice – although even in Portugal he analysed hot springs and was made a member of the Portuguese Academy of Sciences.[28]

'I find foreign travel very inconvenient to a Physician as causing the spending and precluding the getting of Money,' he wrote drily.[29] But in fact Withering had no money worries, having bought a large slice of land for a song in the early 1780s, which was now right in the centre of expanding Birmingham. The rents, leases and sale of plots (including a large strip for the new Union Street), plus investments in Cornwall and canal companies, kept his family in style. And he kept busy, writing a short paper on the preservation of fungi,[30] and spending six hours a day preparing the third edition of his *Botany*. 'Shall I risk 2000 copies?' he asked his bookseller hopefully.[31] His advisers included all the experts of his day and this was the best of his books, published in four volumes in 1796 as *An Arrangement of British Plants*: four more editions were edited by his son and the last, eleventh edition, appeared in 1877.

Despite their common support of Beddoes, the one person Withering did not write to was Darwin. It seems sad that these two Lunar men should remain foes to the last: both, for example, were currently writing on hybrid plants and Spallanzani's experiments in producing 'mules'. Withering could be prickly, pompous and intransigent but those who knew him well were very fond of him and he tried hard, if hopelessly, to keep some of the Lunar spirit going. French chemistry, French wars and the British reaction to them had together blighted the joint Lunar work. 'Philosophical news I have none of any consequence,' wrote Watt to William Roebuck in 1795. 'These cursed French have murdered Philosophy & continue to torment all of Europe.'[32] Their links with international scientific circles were not only weakened in practical terms, but had begun to look politically dubious as well. And all those with industrial interests were working harder than ever, leaving little room for science. But these still often overlapped, Keir's work stopped him going to a Lunar meeting to meet Kirwan, when Tipton was nearly undermined by colliers driving seams from nearby pits. But when Keir and his partner Alexander Blair opened their own colliery at Tividale in 1794 they found that the coal was surrounded by basalt – in some cases lying beneath it – and this led directly to Keir's paper on local mineralogy for Stebbing Shaw's study of Staffordshire.[33]

Old controversies rumbled on, and new positions were taken. Darwin was busy in Derby and Edgeworth in Ireland, telling Keir and Darwin about his new inventions. Letters went back and forth – Darwin asking Boulton about Schweppe's soda-water, Boulton asking him good-humouredly for advice on his 'gravelly complaint'. He was too busy to get old: 'I am kept up like a top, by constant whipping.'[34] But although they exchanged affectionate letters the remaining Lunar men were each now working individually: the give and take and fast exchange had gone. In 1799, Withering lamented that it was some time since he had seen Robert Augustus Johnson, who had left Kenilworth to take up farming:

> We have a loss of him at our Lunar meeting, which has never flourished since the departure of Priestley. The Members are to a Man either too busy, too idle, or too much indisposed to do anything; and the interest wch everyone feels in the state of public affairs draws the conversation out of its proper course.[35]

40: 'TIME IS! TIME WAS! . . .'

In Ireland, public affairs threatened to overwhelm Edgeworth altogether. In 1794 he was fifty and still as passionate as ever, as Maria remembered:

> In the education of the heart, his warmth of approbation, or his strength of indignation, had powerful influence. The scorn in his countenance when he heard of any base conduct the pleasure that lighted up his eyes, when he heard of any generous action; the eloquence of his language, and vehemence of his emphasis, commanded the sympathy of all who could see, hear, feel, or understand.

Not all. The red-faced Irishman who laughed uproariously, and still liked to dance round a table, seemed ridiculous to Coleridge and Anna Edgeworth's sharp young set in Clifton. Although the Edgeworths' *Practical Education* created widespread interest in 1798, as the century turned it was Maria who was the star, not Richard: in 1800 she published her first full-length novel, *Ormond*. The London set found her father 'a boisterous bore'. That was Byron's phrase, although he later recanted, remembering him as 'a fine old fellow, of a clarety, elderly red complexion, but active, brisk and endless.' 'Edgeworth bounced about, and talked loud and long,' Byron wrote in his journal, 'but he seemed neither weakly nor decrepit, and hardly old.'[1]

By the time of Maria's success the Edgeworth family had lived through another revolution. Since the first rumours of a French invasion in 1794, Edgeworth had been trying in vain to get the Irish Government, and then the Admiralty, to take up his telegraph invention. In France, Claude

Edward, Emma and Violetta Darwin, c.1787

Chappe had already beaten him to the starting post, erecting sixteen stations between Paris and Lille in 1791, stone towers with arms pivoting off a pole: his simple code messages could be transmitted from Lille to the capital in two minutes. Edgeworth grew increasingly frustrated and through all the ups and downs, negotiations and displays and trials, he wrote thumping, excitable letters to Darwin. He wanted to build a line of twenty-foot towers, with huge triangular pointers turning in a circle, sending messages by different combinations of their positions. His system was numerical, creating a code of 7,000 possible different combinations, he explained, 'which refer to a dictionary of words'. In addition, it could be adapted to seven different vocabularies, 'lists of the Navy, Army, Militia, Lords, Commons, geographical and technical terms, &c.'[2]

Darwin duly responded to this vision. Edgeworth's telegraph, he thought,

> would be like a Giant wielding his long arms, and talking with his fingers, and those long arms might be covered with lamps in the night. You should place four or six such gigantic figures in a line, so that they should spell a whole word at once, and other such gigantic figures within sight of each other, round the coast of Ireland; and thus fortify yourselves, instead of Friar Bacon's wall of brass round England – with the brazen head, which spoke 'Time is! Time was! Time is past!'[3]

Edgeworth kept trying to get this adopted until 1804. But if he could not succeed as telegraph engineer, he could still express his patriotism in politics. In 1796 he became an Irish MP by buying a pocket borough, St John's Town, County Longford. In Parliament he campaigned strongly for better education for the poor and an end to local magistrates' prejudice against Catholics. 'The lowest order of the people has been long oppressed,' he told Darwin, 'and horrid calamities may ensue from their ignorance.'[4] His most important work in years to come would be for the Board of Enquiry into Education in Ireland.

In 1797, however, Edgeworth's telegraph experiments and his political life were both halted when Elizabeth died, of consumption like her sister, after seventeen years of marriage, aged forty-four. Within a year he had married a fourth wife, Frances Beaufort, a year younger than Maria. The marriage also a brought a new close friend in his wife's brother Francis Beaufort (later naval hydrographer, creator of the Beaufort wind scale, and instigator of the voyage of the *Beagle*, on which Darwin's grandson

Charles would sail). Frances was entranced by the household at Edgeworthstown, all 'chymists and mechanics, & lovers of literature & a more happy more accommodating more affectionate family never yet came under my observation'. The marriage was a great success, and over the years six more babies were added to the Edgeworth tribe. But in 1798 Ireland was on the verge of rebellion. On the way home from their wedding in May the couple saw a hanged man dangling from a cart, a cruel warning to others.

For some time now, the gentry had been recruiting troops of yeomanry to defend their homes. One reason for Edgeworth's reputation as a subversive was that while nearly all these troops were Protestant, the Edgeworthstown yeomanry – recruited as a reluctant last resort – mixed men of both creeds. When the French landed in County Mayo in August, his troop was ready but still had no arms, and he and his family were forced to flee to the Protestant-held town of Longford, an uncomfortable refuge: in the middle of the hysterical rejoicing at the news of the rebels' defeat an excited Protestant mob tried to lynch him as a French spy. Next morning the family returned home. It was an eerie moment. The village had been ravaged as the rebels moved through, doors broken, windows shattered. But within the house, wrote Maria,

> everything was as we had left it – a map that we had been consulting was still open on the library table, with pencils and slips of paper containing the first lessons in arithmetic . . . a pansy, in a glass of water, which one of the children had been copying, was still on the chimney piece.[5]

Like an island in the eye of the storm, the hurricane of rebellion had swirled around them and left their world intact.

The uprising was harshly suppressed, and in the aftermath there were new moves for an Act of Union, an idea that Edgeworth first supported, and then, at the last minute, voted against, feeling it was wrong 'to force this measure down the throats of the Irish, though five sixths of the nation are against it'.[6]

Meanwhile, with Frances and Maria, he had visited England, calling on Robert Darwin in Shrewsbury, staying with the Keirs and meeting all the Birmingham men. 'Watts and Bolton', wrote Frances enthusiastically, 'are as well worth seeing as the fire Engine that goes by their name.'[7]

Edgeworth and Maria also made a separate trip to see Darwin in Derby. It was the last time the two old friends would meet.

Darwin was still suffering the rollercoaster of fame and notoriety after the publication of *Zoonomia*. In this book, he had perhaps written his own apologia, a credo for Lunar science, in the entry on *Credulity*, one of his Diseases of Volition.

> *Credulitas.* Credulity. Life is short, opportunities of knowing rare; our senses are fallacious, our reasonings uncertain; man therefore struggles with perpetual error from the cradle to the coffin. He is necessitated to correct experiment by analogy, and analogy by experiment . . .[8]

Even so, no one should rest satisfied in the belief of facts until experiments could be repeated or confirmed by others. Ignorance and credulity had always marched together, he wrote, and had 'misled and enslaved mankind'. Philosophy, or experimental science,

> has in all ages endeavoured to oppose their progress, and to loosen the shackles they had imposed; philosophers have on this account been called unbelievers: unbelievers of what? of the fictions of fancy, of witchcraft, hobgoblins, apparitions, vampires, fairies; of the influence of the stars on human actions, miracles wrought by the bones of saints, the flight of ominous birds, the predictions from the bones of dying animals, expounders of dreams, fortune-tellers, conjurors, modern prophets, necromancy, cheiromancy, animal magnetism, metallic tractors, with endless variety of folly? These they have disbelieved and despised, but have ever bowed their hoary heads to Truth and Nature . . .[9]

Stirring stuff. Such words were a trumpet call to young idealists such as William Godwin, who passed through Derby in 1797, hoping to see Darwin. The doctor was away in Shrewsbury, so Godwin moved on. But he later confessed he was mortified to have missed him: 'I believe we were wrong,' he told Mary Wollstonecraft. 'So extraordinary a man, so truly a phenomenon as we should probably have found him, I think we should not have scrupled to sacrifice 36 hours.'[10] The year before, Coleridge had also been in the town, writing excitedly,

> Derby is full of curiosities, the cotton, the silk mills, Wright, the painter, and Dr Darwin, the everything, except the Christian! Dr Darwin possesses, perhaps, a greater range of knowledge than any other man in Europe, and is the most

inventive of philosophical men. He thinks in a *new* train on all subjects except religion. He bantered me on the subject of religion . . .[11]

Although Darwin teased him about his Unitarianism, to Coleridge he still seemed a 'wonderfully instructive and entertaining old man'. (But Coleridge did add a final note: 'I absolutely nauseate Darwin's poem.')

Coleridge's curiosity about Darwin was partly fostered by Tom Wedgwood. Tom was not only a fine chemist, but also a theorist, currently hypothesizing about the possibility of analysing sensations, both physical and moral, in a way clearly influenced by *Zoonomia*, tracing them from the associations linked to pain or pleasure. Almost like Godwin, he thought such knowledge would pave the way to moulding behaviour and developing a future generation imbued with goodwill. Sensation and goodwill of a different kind were also a by-product of Davy's experiments in Bristol. He was working on the effect of different proportions of oxygen and nitrogen, and in July 1799 Robert Southey wrote excitedly to Tom,

Davy has invented a new pleasure, for which language has no name . . . it makes one so strong and so happy! and without any after disability, but instead of it, increased strength of mind and body. O excellent air-bag! Tom I am sure the air in heaven must be this wonder-working gas of delight.[12]

The laughing-gas, nitrous oxide, could do no harm it seemed. But Darwin's prescribing could: over the next few years opium loomed large in the lives of Tom Wedgwood and Coleridge.

In a way, the laughing-gas itself caused lasting damage to Beddoes, and indirectly to the Lunar men. Humphry Davy often visited Birmingham, and earlier this year had spent ten enjoyable days arguing with Keir and Watt over phlogiston and calorique. 'I had a great deal of chemical conversation with them,' he wrote. 'Mr Keir is one of the best informed men I ever met with, and extremely agreeable.'[13] Excited by his new discovery of nitrous oxide, and following the democratic precept, Davy worked on apparatus, including an air-tight breathing chamber that would allow everyone to experience the euphoric new gas. He and his friends had all tried it out – even Edgeworth 'capered about the room without having power to restrain himself'.[14] But there were problems. The apparatus sometimes did not work and the gas was hard to prepare. (Matt Boulton and Gregory Watt were very disappointed when they tried to follow

Davy's instructions.) More significantly, the experiments – and Davy's shocking suggestion that emotions such as happiness and benevolence were created by body chemistry and not by God – were immediately pounced on by opponents. In cartoons and in the press Beddoes, Davy and Priestley were aligned with revolutionary frenzy and mad religious enthusiasm, and with other 'French' fads of the day, such as animal magnetism and Mesmerism.[15] Davy, a shrewd careerist, jumped ship as fast as he could. He did not abandon his old friendships, but he dumped their old politics and philosophy: in 1801 he moved to the newly founded, ultra-respectable Royal Institution in London, condemning the French Revolution, endorsing the social hierarchy, stressing the practical application of chemistry and electricity, and re-welding natural philosophy to a divinely ordered universe.

Among the friends he left behind, Lunar science was also being subverted. And some of the explorations of 'sensation' that Tom Wedgwood shared with Coleridge employed vast quantities of chemical and organic substances. Wordsworth's sailor brother John became a new source of supply, and the distinguished Sir Joseph Banks sent down a four-ounce bag of hemp, or *bhang*, for 'experiment'. 'We will have a fair trial of Bang,' Coleridge wrote gleefully to Tom. 'Do bring down some of the Hyoscamine pills, and I will give a fair trial of Opium, Henbane, and Nepenthe. By the by, I always considered Homer's account of Nepenthe as a banging lie.'[16] Losing weight drastically, Tom alternately fought and fed his addiction: 'The dullness of my life is absolutely unsupportable without it,' he told his brother.[17] He died in July 1805, aged thirty-four.

Gregory Watt had died the year before, in October 1804, of consumption. He too had shown great promise, and his first geological paper had been read at the Royal Society that April.[18] 'He was a noble fellow and would have been a great man,' Davy lamented. 'Oh! there was no reason for his dying – he ought not to have died.'[19] Watt collected his son's writings and drawings, from childhood on, and kept them in a trunk beside his attic workshop bench all his remaining years.

Darwin could not prevent these deaths, and in Tom's case he may even have contributed to it. He was not always a good doctor. But it is from the late 1790s that some of the lasting legends about him come, recorded by his grandson Charles. George III sent for him to be his personal physician, it was said, but he refused. A mysterious gentleman arrived from London, to consult him as 'the greatest physician in the world, to hear from you if there is any hope in my case'. Darwin examined him and declared the issue hopeless, then asked him why, if he came from London, had he not seen the famous Richard Warren, the senior royal physician? 'Alas! doctor,' came the reply, 'I am Dr Warren.'[20]

Darwin had his moments of glory and happiness: his first grandchild, Marianne, the daughter of Robert and Susannah, was born in April 1798. But he felt increasingly like a survivor of a thinning crew. Deaths had tolled through the decade: Smeaton in 1792, Arkwright in 1793, Wedgwood in 1795, Joseph Wright and James Hutton in 1797. In 1799 Joseph Black died. So did his old opponent Withering, at the age of fifty-seven, only a few days after he moved into a new house almost next door to Priestley's old home at Fair Hill.

Darwin also felt that opinion was slowly turning against him. He was bothered by a vast manuscript from a young student, Thomas Brown, with penetrating queries on *Zoonomia*, such as the relation of sensation and ideas – good questions and tough to answer.[21] Next he found himself bracketed with William Godwin and Richard Payne Knight – dilettante historian of priapic cults and author of *The Progress of Civil Society* – in a pro-Pitt satirical magazine run by the young George Canning, Under Secretary for Foreign Affairs. The *Anti-Jacobin*'s 'Loves of the Triangles' borrowed Darwin's phrasing but disdained botany for a more serious science, 'to enlist the IMAGINATION under the banner of GEOMETRY'. It was a wicked parody of *The Botanic Garden*, notes and all, complete with an introductory note, taking as its text 'the eternal and absolute PERFECTIBILITY OF MAN', and claiming that it was demonstrable that humanity had risen entirely by its own energies from the '*Cabbages of the field* to our present comparatively intelligent and dignified state of existence'.[22] It linked Darwin ineradicably with the French Revolution, with irreligious views on evolution, and with cranky ideas on electricity as a force for good. It was also, unfortunately, very funny.

Two months later the *Anti-Jacobin* carried Gillray's cartoon, *The New Morality*, showing Darwin staggering under a basket marked 'Zoonomia, or Jacobin Plants'. Gillray showed Darwin in the company of Priestley,

The New Morality, by James Gillray, in the *Anti-Jacobin*, August 1798, showing Darwin as an ape carrying a basket on his head labelled 'Zoonomia, or Jacobin Plants'

Gilbert Wakefield, Southey and Coleridge. But however much Coleridge and Wordsworth were influenced by his imagery and ideas, they too now turned their backs on him. The Preface to their *Lyrical Ballads* in 1798 firmly placed incidents from common life above myth and fancy and 'gaudy' language. From now on, thanks to the parody, Darwin's verse looked silly; thanks to the *Lyrical Ballads*, he also looked old-fashioned. This was true of his Radicalism as well as his style: 'I have snapped my squeaking baby-trumpet of Sedition,' wrote Coleridge this year, '& the fragments lie scattered in the lumber-room of Penitence.'[23]

Darwin shrugged off the attacks, taking refuge in his work on *The Temple of Nature*, in which scientists and industrialists were even more clearly presented as deities of the modern age. Another pantheon of agricultural innovators starred in the hefty *Phytologia; or the Philosophy of Agriculture and Gardening*. This book was firmly embedded in the ethos of scientific improvement, particularly the Scottish tradition. Darwin dedicated it to Sir John Sinclair, first President of the Board of Agriculture, who asked him to write it, and it was one of a flurry of such books, including Lord Dundonald's treatise on agriculture and chemistry and Richard Kirwan's study of manures.[24] It was also a response to the famines and bread shortages of the 1790s, their urgency heard in Darwin's call to use land for wheat not cattle, and grain for bread rather than brewing (drunkenness, he noted, also being a major factor in riots). But although the book was practical, Darwin's real aim was to systematize agriculture and gardening into a Science. And overlying this was an even grander theory, that plants, insects, animals, mankind and human knowledge itself were all interlinked, part of the rolling, progressing sphere of nature.

Because he was concerned with productivity, Darwin had to deal with the vulnerability of nature as well as its strengths, with pests and predators and declining species. Once again he fell back on the belief that nature was internally organized for survival – the pain of the individual was less important than the 'greatest happiness of the greatest number' (the Utilitarian formula recently expressed in social terms by Jeremy Bentham).[25] But while he saw that developments such as enclosure did increase production, he also felt the human cost and proposed giving enclosed land to more people, in a system of allotments that would feed the poor, not drive them to the cities for work.

Phytologia was imaginative as well as practical, particularly in the way it pushed the chemistry of plant life to new boundaries: in the analysis of photosynthesis, the detection of the key roles of nitrogen, carbon and phosphorus in plant nutrition. A section on plant physiology and structure was followed by one on the economy of vegetation – on seeds, photosynthesis, nutrition, manures and drainage – and finally a forward-looking section on increased productivity of flowers and crops, including those of the future such as sugarbeet. To Darwin the flow of the sap and the breathing of the leaves displayed a life as vivid as the flow of blood, the power of the lungs, the beating heart; plants too had muscles, nerves and 'brain'. As contemporary critics noted, his analysis was weakened by his habit of drawing analogies between plants and the body. Yet the sense of shared vitality also makes *Phytologia* wonderfully readable. Its bounding rhythms summon up a flow of discoveries and ideas, and its intense attention to small details makes every bud, every seed, every leaf intensely individual, astounding in their variety and ingenuity.

Darwin enjoyed writing *Phytologia*; he agreed with Edgeworth and Day, 'in thinking the cultivation of the earth the most engaging & most lasting & certainly most innocent amusement a man can follow.'[26] But even here he could not banish politics. Proposing the forestation of Britain's mountainous regions – another of the book's prophetic strands – and talking of cultivating timber, he turned aside to lash out at the present 'insane state of human society', where war dominated all, 'and mankind destroy or enslave each other with as little mercy as they destroy and enslave the bestial world'.[27]

It is rare to find Darwin so bitter. And personal grief, as well as political despair dogged him this winter. Although by and large his family was flourishing, from Robert in Shrewsbury to little Harriot at home, he had grown increasingly worried about Erasmus, now nearly forty. Erasmus was a good local lawyer, much relied on by the community and by his friends, including the older Lunar men, but he seems to have felt that his father's second family had shoved the first aside, complaining, for example, about the meanness of the memorial put up to his aunt Susannah, who had cared for them for so long.[28] He grew more and more withdrawn, with spells of the paralysed inactivity that marks depression, interspersed by high spirits and startling acts: suddenly, in the summer of 1799, he bought Breadsall Priory, five miles out of Derby, a 'large forlorn old Pile' complete with fishponds and cottages. Darwin was jovially dismissive, telling Robert of the house 'where Erasmus intends to live . . . and sleep his life away!'

The high was followed by a low, culminating in Erasmus's seemingly trivial but panic-stricken inability to finish his year-end accounts. On 29 December, after spending two days wrestling with his books in his old house in Derby, his clerk told him to go to bed. 'I cannot,' he answered, with his head in his hands, 'for I promised if I'm alive that the accounts should be sent in tomorrow.'[29] What happened next is unclear. Family stories, backed by the second-hand account of Anna Seward, describe Erasmus rushing from the house toward the icy, swollen river Derwent, leaving his hat and neckcloth on the bank and flinging himself into the water. Others, including Darwin, held that he slipped. As soon as his absence was noticed and his neckcloth found in the garden, rescue boats set out and Darwin was called. He paced the bank in the night and returned home in despair. When news came that Erasmus's body had been found, he staggered so much that his daughters made him sit, 'which he did, and leaned his head on his hand', staying silent for many minutes.[30]

He never quite recovered. He blamed himself, as others did, for not noticing Erasmus's distress. A family friend, Mrs Clive, wrote soon afterwards that she had loved him most dearly, and had 'often acted the part of a mother to him'; she felt really at fault herself, 'knowing how much he suffered *at times* but I supposed his father must know'.[31] Over the coming year Darwin wrestled with Erasmus's fateful accounts. Soon he and Elizabeth decided to move into the Priory themselves, setting to work on much needed alterations. He tinkered with his inventions and flickered into life with various patients, including Georgiana, Duchess of Devonshire, whom he instructed on how to build the new 'Galvanic pile' (in fact Volta's discovery), a battery of dissimilar metals, immersed in water.[32]

The Watts came to visit, and his books came out, first *Phytologia*, then a new edition of *Zoonomia*. But Darwin was ill and tired, struggling through at least one bout of pneumonia by dosing himself with the controversial digitalis. Early in 1802 he and Elizabeth finally moved into the Priory. After only two weeks, on 10 April, he fell ill with shivers and fevers. A few days later he was up, walking in the garden with Elizabeth's foster-mother Susan Mainwaring, joking about the building work he swore he would never see finished. On 17 April he began a letter to Edgeworth. 'I am glad to find, that you still amuse yourself with mechanism, in spite of the troubles of Ireland,' he began, before telling him of the new house with its talkative stream and its shady valley, like Petrarch's *Val Chiusa*.[33] 'I hope you like the description,' he wrote, 'and hope farther that yourself and any part of your family will sometime do us the pleasure of a visit.'

Halfway through writing his letter he slept. Next morning he woke shivering and went to the kitchen to warm himself and asked for a bowl of buttermilk to be brought to him in his study. There, among his books, he lay on the sofa and asked to be covered with his greatcoat.[34] He was fainting and sick and fading fast. The unfinished letter with its warm invitation arrived at Edgeworthstown with a note attached from Mrs

Mainwaring: 'Sir, this family is in the greatest affliction. I am truly grieved to inform you of the death of the invaluable Dr Darwin.'[35]

Darwin was buried in Breadsall church, by the side of Erasmus, as he had wished. Elizabeth stayed on at the Priory until her death in 1832, her greatest joy being in filling the house with her children, grandchildren and great-grandchildren. Darwin's spirit lived on here, typified by a bizarre, colourful coda: in 1877, when alterations to the church exposed the family graves, his granddaughter Elizabeth Wheler recorded, 'My grandfather's coffin had burst open and his remains were visible and in perfect preservation. He was dressed in a purple velvet dressing-gown and his features unchanged.'[36]

WANING

For lo! The New-moon winter-bright!
And overspread with phantom-light,
(With swimming phantom-light o'erspread
But rimm'd and circled by a silver thread)
I see the old Moon in her lap, foretelling
The coming on of rain and squally blast . . .

Coleridge, *Dejection: An Ode*

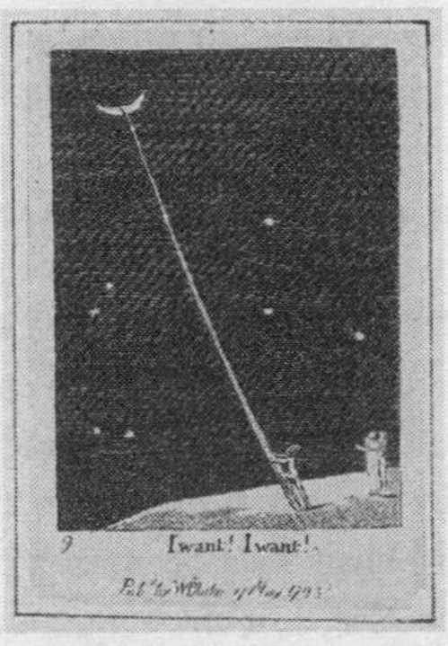

EPILOGUE: 'FIDDLEDUM, DIDDLEDUM'

In the summer of 1819 Maria Edgeworth was in the Midlands, carrying with her the memoirs of her father, who had died two years before. It was an awkward document, the first half full of Edgeworth's vivid, comic, egotistical early memories, the second half an account of the later years in Ireland, written from the notes he left her when he died. She had crossed the Irish Sea the year before to see the people it might embarrass – including the Sneyd family and Sabrina Bicknell, the subject of Day's long-ago experiment. But this year and the next, before leaving the manuscript with the publishers in London, she called on the families of her father's friends. She was especially close to Amelia Keir, who ten years earlier had married James Moilliet, a Swiss banker who had settled in Birmingham in 1789 with a recommendation from Boulton. When the *Memoirs* were published the following year Maria fled from the reviews to stay with the Moilliets in Switzerland. By the end of that year the Lunar giants were all gone, a chapter of deaths: Priestley in 1804, Boulton 1809, Edgeworth 1817, Watt 1819, Keir in 1820. Only Galton lived on to see the Great Reform Act of 1832.

The bonds the Lunar men had built stayed strong all their lives. In his years in America, defending his phlogiston theory until the end, Priestley constantly referred to the support of the Birmingham group. Despite the patronage of Jefferson, William Small's old pupil, who became President

William Blake, 'I want! I want', from *The Gates of Paradise*, 1793 in 1801, even in the New World Priestley was constantly attacked as 'subversive' and he faced disappointment and sadness, including the death of Mary and of his sixteen-year-old son Harry. But he kept up his lectures, his writing and experiments, addressing his Lunar friends in his mind, quarrelling, for example with Darwin's views on evolution in *The Temple of Nature*, which he thought 'atheistic'. His fluent pen never slowed – he was still correcting proofs in the week he died. And even at the last he remained optimistic, telling his grandchildren, 'I am going to sleep as well as you, for death is only a good long sleep in the grave, and we shall meet again.'[1]

In Birmingham the emphasis was on money rather than faith: on the future of the industrial empire, not the kingdom of salvation. Cash was still a priority. In 1801, after the redoubtable Charlotte Matthews died, the Soho companies set up their own London agency or bank, and much of the steam-engine business came through this.[2] Untroubled by the general industrial unrest over the Combination Acts of 1799 and 1800 (which ushered in twenty years of persecution of the young trades unions), Boulton triumphed with his mint and watched Soho flourish under his and Watt's sons and William Murdoch. For years, Murdoch had been working on the problem of gas lighting – discouraged, as usual, by Boulton and Watt who dissuaded him from taking out a patent. But when Gregory Watt saw a demonstration in Paris he pushed them into encouraging Murdoch's work. At the Peace of Amiens, which heralded the short-lived gap in the French wars from 1802 to 1803, the whole of Soho blazed with gaslight in a 'grand illumination'. Boulton rose from his sick-bed to see it. Within years, the streets of London and other cities, and many mills and factories across the land, were lit by gas.

'Soho Illuminated for the Peace of Amiens', in *Aris's Birmingham Gazette*, 1802

The Soho sons, however, were of the new generation where industrial strength was a possession to hang on to, not to display and share. Boulton's old habit of showing visitors round his great manufactory was banished. In 1802, his son and the Watt brothers placed this notice in local and London papers:

Soho Manufactory: The Public are requested to observe that this manufactory cannot be shewn in consequence of any Application or recommendation whatever. Motives, both of a public and private Nature have induced the Proprietors to adopt this Measure and they hope their Friends will spare them the painful task of a Refusal.[3]

Boulton never quite lost his energy, nor his love of show and grand gesture. In 1805 he had a special Nelson medallion struck, to be given to every man who fought at Trafalgar (in copper for officers, pewter for the rest). Two years later, nearly eighty, he led and won the fight to build a theatre in Birmingham, in the teeth of opposition from Nonconformists who damned it as immoral. (Despite bad rheumatism, Keir came over to Soho House to join 'Mrs Siddons and a haunch of venison', to celebrate her appearance.[4]) But by the spring of 1809 Boulton was writing to his daughter Anne with uncharacteristic self-pity and a shaking hand, 'I shall not be hapy until I clasp you in my arms for I am now very very mis[er]able . . . pray come soon for I am very ill.'[5] He died on 17 August. Old Soho hands wept at his funeral, and Watt and Keir wrote memoirs, Keir's amusement and Watt's irritation at Boulton's excesses and risk-taking filtering unstoppably through their solemn phrases.

Keir too lived on into his eighties, surviving the death of the once so beautiful Susanna in 1802; he uttered one cry of grief, his daughter said, despite his firmly stoical philosophy, and wore Susanna's ring round his neck on a black ribbon for many years. Five years later his stoicism was called on again when he had to rebuild his house after it was burned to the ground – to save himself the trip to Tipton he had installed a counting house at home, and his clerks managed to set fire to a beam in the cellar by raking the ashes of their fire on to the floor. Shrewd, laconic and urbane, Keir adored his six Moilliet grandchildren, for whom he wrote a tenderly parodic version of Thomas Gray's 'On a favourite cat', when his own cat Selima drowned in the pond:

Susanna weeps, James drops his stick and drum,
Amelia wails and merry John is dumb.
E'en ancient Gra, whose breast had erst been steel'd
With sights of misery in the martial field . . .[6]

He wrote more seriously, passing on to them some old Lunar knowledge in the splendidly clear 'Dialogues on Chemistry between a Father and his Daughter'. But he took to the grave his prime formula, the secret of the great alkali process that had made him rich.

Seven years earlier, in 1813, the old men had realized that the Lunar Society, at least as a semi-formal body, was now a hollow shell. It was time to call it a day. A lottery was held to dispose of their books and Samuel Galton walked away with the hoard. No one complained, although all of them would have liked the prize since they had no thoughts of giving up experimenting, especially Watt. As his old friends dropped off one by one, Watt seemed to cheer up distinctly. He had no money worries now. His health was better than it had been all his life, though he did get a little confused, dropping asleep while talking or wandering into unconnected anecdotes. His old thrift and anxiety vanished. He set his heart on a country estate, exploring Devon and Dorset and Monmouth before investing in a clutch of nearly thirty farms in mid-Wales, and making his chief base in the beautiful valley of the upper Wye, at Doldowlod, near Rhayader, in an old farmhouse with spreading acres. This was not his only summer retreat. The one-time workaholic now took holidays to spa towns and to Scotland. The old Puritan who had growled at his son James if he read novels, and refused to employ an assistant who might like music, now cried over Mrs Opie's books and went to the theatre to see *The Honeymoon* to cheer himself up. He and Ann even went to France, as the Edgeworths did too, during the brief peace of 1802–3.

Tall and thin, with a mass of white hair, in his old age Watt was fêted and honoured, revered by scientists such as Humphry Davy and John Dalton, and young engineers like John Rennie and Thomas Telford. Glasgow University gave him an honorary degree; Lord Liverpool offered him a baronetcy (he declined); the French Academy made him an associate. To Walter Scott, watching him shine among the literati of Edinburgh was like seeing a sorcerer from another age: 'This potent commander of the elements – this abridger of time and space – this magician, whose cloudy machinery has produced a change in the world the effects of which, extraordinary as they are, are perhaps only now beginning to be felt.' Remembering Watt arguing with a philologist, he noted that 'he talked with him on the origin of the alphabet as if he had been coeval with Cadmus'. (But he added, equally tellingly, that Watt turned out to be 'as shameless and obstinate a peruser of novels, as if he had been a very milliner's apprentice of eighteen . . .')

Watt died in 1819. His image was carefully promoted by his family and friends and he was the first of the group to be mythologized, the model of the heroic engineers of the Victorian age who would stride through the books of Samuel Smiles, rising by their genius and determination from humble beginnings. His memorial in Westminster Abbey, written by a fellow Scot, Henry Brougham, hailed him as 'among the most illustrious followers of science and the real benefactors of the world'.[7]

Lunar links continued. There were signs that James Watt junior wanted to marry Anne Boulton, but this came to nothing, and despite other suitors Anne stayed on at Soho House to look after her brother, moving out when he married Mary Ann, daughter of the iron-master William Wilkinson. Galton's son, Samuel Tertius, married Violetta Darwin: their brilliant son Francis founded the dubious statistical science of eugenics. In later generations more weddings linked Galtons, Edgeworths, Keirs, while Charles Darwin, son of Sukey and Robert, also married a Wedgwood, his cousin Emma. His own theories of evolution would reach the world in *On the Origin of Species* in 1859.

These children, however, inhabited a different world. Maria Edgeworth was right to be nervous about reviews. Her father had continued to work for education, particularly in Ireland, and had never ceased to suggest improvements to carriages, roads, agricultural practices, telegraph schemes, but by 1820 Edgeworth's mechanical – and matrimonial – adventures were something to be ridiculed, not admired.[8] And by and large in the first decades of the new century opinion turned against the whole mix of enlightenment, reason and democratic reform that the Lunar group had come to represent. Even before the century turned, the *Anti-Jacobin* had written this programme off as infantile gibberish:

Reason, Philosophy, 'fiddledum, diddledum'
Peace and Fraternity, higgledy, piggledy
Higgledy, piggledy 'fiddledum, diddledum'[9]

In the next generation, after twenty years of virtually continuous war from 1793 to 1815, reform and innovation were looked upon with even more suspicion, and satirists such as Rowlandson continued to mock the pretensions of science, even in Davy's respectable and fashionable lectures.

Detail from *Chemical Lectures* by Thomas Rowlandson

By then the informal networks that supported so much of the achievement of the Lunar men had vanished. Natural philosophy was becoming more specialized, retreating into the institutions. In chemistry Lavoisier's system was now the norm and ideas of 'phlogiston' seemed as remote as alchemy, while the climate of evangelicalism labelled Darwin's evolutionary ideas as the God-defying fantasies of a crank. From a different perspective, Romantic writers spurned the arid insights of reason and the denial of innate instincts, preferring an infant 'trailing clouds of glory' to the blank slate on which the model citizen could be imprinted. To them, the appeal of experiment lay in its revelation of hidden powers in nature, akin to the inspiration of 'genius'. When Shelley was a boy, he pursued his sister with much the same Leyden jar experiment that Erasmus Darwin had practised to entertain his siblings so many years before: Shelley's sister Hellen described his looming up with 'folded brown packing paper under his arm and a bit of wire and a bottle'. Her heart, she wrote, 'would sink with fear at his approach; but shame kept me silent, and with as many others as we could collect, we were placed hand-in-hand around the nursery table to be electrified'.[10] As a student in 1810 Shelley's rooms were cluttered with crucibles and phials and microscopes, electrical machine, air pump and galvanic trough: but this electrified, atheistical student bore less resemblance to the Lunar rationalists than to the Coleridgean visionary of 'Khubla Khan', of whom 'All should cry, "Beware! Beware!/His flashing eyes, his floating hair!"'

To its opponents – and some of its Romantic devotees – science was thus dangerous, demonic, revolutionary. For three decades or more, the springs of that inventive Lunar energy ran underground. The invention and industrial expansion, however, tore on apace. By 1800 Boulton & Watt had already sold nearly five hundred steam-engines, although this was still only a small part of the power used across the country.[11] Soon Trevithick tried out the first steam railway locomotive engine, and from the 1820s the railways would begin first to creep and then to charge over the land. The economic geography of Britain changed, with the population rising in the coalfields and industrial areas of the Midlands and the North, and moving from country to city: in 1700 one person in six lived in a town, by 1800 it was one in three.

The factories and the mills with their steam-engines boomed on, admired for the money they made. But they were increasingly regarded with distaste as well as awe, as they darkened the skies with smoke and devoured human labour. In the Midlands and the North the Luddite riots of 1812 brought home the distress of the hand-workers forced out of their livelihood. And some years later, Ebenezer Elliot laid the blame firmly on 'Watt! and his million-feeding engineery!', when he imagined guiding a blind man around the industrial scene:

> . . . thou can'st hear the unwearied crash and roar,
> Of iron powers, that urged by restless fire,
> Toil ceaseless, day and night, yet never tire
> Or say to greedy men, 'thou dost amiss'.[12]

Once the libertarian mood was lost, the 'rational' exploration of society hardened into the codes of the Utilitarians, Gradgrinds of the future. The old 'progressive' views were co-opted by a joyless crew. The circle turned in ways the Lunar men could not foresee in their hopeful prime of dreams, of energy, of passionate invention.

The group itself was more than an assortment of single beings. It had an evolving life of its own, becoming almost a mirror, or a microcosm, of the way that the different currents flowed together through the second half of the eighteenth century, conflicting and colliding to create a new society. And these appear differently in different lights. The Lunar group were bourgeois capitalists who constantly downplayed the role of labour and overstated the importance of leaders, thinkers, inventors; but they were also radicals, educators and firm believers in the democracy of knowledge. Buoyant, sparkling, self-made men, they used the old networks of patronage and class, but they also defied them, shifting the axis of power from metropolis to province, from the money men to industry, from Parliament to the people.

Hard-headed though they were and often driven by self-interest, they were also, in their way, romantics. In 1818, the year Maria Edgeworth completed her father's *Memoirs*, Mary Shelley published *Frankenstein, or the Modern Prometheus*. 'The ancient teachers of this science', her narrator declared, had 'promised impossibilities and performed nothing'. By contrast:

> The modern masters promise very little; they know that metals cannot be transmuted and that the elixir of life is a chimera. But these philosophers . . . have indeed performed miracles. They penetrate into the recesses of nature and show how she works in her hiding-places. They ascend into the heavens; they have discovered how the blood circulates, and the nature of the air we breathe. They have acquired new and almost unlimited powers; they can command the thunders of heaven, mimic the earthquake, and even mock the invisible world with its own shadows.

The Lunar men were no Frankensteins, but they did feel, confidently, that they could explore every element, and that science could eventually give men almost 'unlimited' powers.

It has been said that the Lunar Society kick-started the industrial revolution. No individual or group can be said to change a society in such a way, and time and again one can see that if they hadn't invented or discovered something, someone else would have done it. Yet this small group of friends really was at the leading edge of almost every movement of its time in science, in industry and in the arts, even in agriculture. They were pioneers of the turnpikes and canals and of the new factory system. They were the group who brought efficient steam power to the nation. They were in the white heat of the drive to catalogue and name plants, to study minerals, to detect and work out the history of the formation of the earth. The philosophers among them were keenly concerned with the nature of human knowledge itself, with the process of learning, and beyond this with enquiring into the origin and evolution of all organic life. And all of them, though not always with great success, applied their belief in experiment and their optimism about progress to personal life, and to the national life of politics and reform.

They formed a constellation of extraordinary individuals, a tangle of friendships and dependencies, arguments and loyalties. They were colourful, strong, idiosyncratic: Darwin squeezing into his carriage, Boulton showing off his machines, Watt despairingly holding his head, William Small soothing and planning, Whitehurst scouring the Derbyshire moors, Keir carefully watching how crystals form, Edgeworth inventing his wooden horse and weeping for Honora, Day trying to bring benevolence to ungrateful Surrey, Wedgwood fingering the rough, blue fineness of his jasper, Priestley picking up his glass of mint and musing on the magic greenness of its leaves. They felt the greatness of the cosmos and its limitless possibilities, the beauty of the infinitely small – the bud, the grain of quartz, the microscopic animalculae – and the grandeur of the vast, the thundering force of steam, the rolling clouds, the relentless flow of lava over aeons. They knew that knowledge was provisional, but they also understood that it brought power, and believed that this power should belong to us all. The legacy of the Lunar men is with us still, in the making of the modern world, and in the inspiring confidence with which all these friends, in their different ways, reached so eagerly for the moon.

CHRONOLOGY

1704	Newton's *Opticks*	
1706		John Baskerville born
1707	Act of Union with Scotland	
1709	Abraham Darby uses coke for smelting	
1713		John Whitehurst born
1714	Accession of George I	
1715	Jacobite uprising defeated	
1727	Accession of George II	
1728	Pope, *The Dunciad*	Matthew Boulton born
1730		Josiah Wedgwood born
1731		Erasmus Darwin born
		Thomas Bentley born
1732	Stephen Gray's experiments on electricity and combustion	
1733	John Kay invents the flying shuttle	Joseph Priestley born
	Pope, *Essay on Man*	
	Voltaire, *Lettres philosophiques*	
1734		William Small born
		Joseph Wright born
1735	John Harrison perfects first chronometer	James Keir born
1736		James Watt born
1739	David Hume, *A Treatise of Human Nature*	
1740–48	War of the Austrian Succession	
1741		William Withering born
1744	War with France	Richard Lovell Edgeworth born
	First documented cricket match Kent v All England	
1745–46	Second Jacobite uprising	Boulton enters father's business
1746	Discovery of the Leyden jar	

CHRONOLOGY

1748	Peace of Aix la Chapelle Hume, *Enquiry concerning Human Understanding* Excavations begin at Pompeii	Thomas Day born
1749	First volumes of Buffon, *Histoire naturelle*	Wedgwood ends apprenticeship Boulton marries Mary Robinson
1751		Priestley moves to Daventry
1753		Samuel Galton born
1754	Society of Arts founded	Watt goes to Glasgow Darwin studies in Edinburgh Wedgwood enters Whieldon's
1755	Johnson, *Dictionary of the English Language* Lisbon earthquake	Watt apprenticed in London
1756	Start of Seven Years War	Darwin moves to Lichfield
1758	John Dollond invents achromatic telescope	Franklin visits Midlands
1759	French defeated at Minden British capture Guadeloupe, Quebec Laurence Sterne, *Tristram Shandy*	Wedgwood opens Ivy House Works; develops creamware Mary Boulton dies
1760	Accession of George III	Boulton marries Anne Robinson
1761	Bute Secretary of State Duke of Bridgewater's canal opens	Wedgwood rents Brick House Works Priestley moves to Warrington Academy Darwin elected FRS
1761–65		Boulton builds Soho Manufactory
1762	Rousseau, *Nouvelle Héloïse*, *Emile* Catherine the Great becomes Empress of Russia	Boulton & Fothergill partnership Wedgwood meets Bentley, becomes 'Queen's Potter'
1763	Treaty of Paris ends Seven Years War Wilkes tried for seditious libel	
1764	Adam's designs for Kenwood James Hargreaves invents spinning-jenny	Watt repairs model of Newcomen engine Wedgwood marries Sally
1765	Stamp Act imposes tax in American colonies	Watt devises separate condenser Promotion of Grand Trunk Canal
1766	Pitt (Earl of Chatham) forms Government Hamilton's *Antiquités* Rousseau visits England	Priestley elected FRS Keir moves to Midlands Wright paints *The Orrery* Edgeworth comes to Lichfield
1767	Ferguson, *History of Civil Society*	Priestley moves to Leeds, *History of Electricity*
1768	'Wilkes and Liberty' riots Royal Academy founded Cook's first voyage	Watt visits Soho Priestley, *Essay on Government* Wright, *The Air-Pump*

1769		Watt's first patent
		Opening of Etruria
		Day adopts foundlings
1770	Lord North forms Government	Day moves to Lichfield
	Goldsmith, *The Deserted Village*	
	Cook lands at Botany Bay	
1771	Mackenzie, *The Man of Feeling*	Edgeworth visits Rousseau
	Smollett, *Humphry Clinker*	Keir *Dictionary of Chemistry*
1772	Failure of Scottish banker Fordyce	Roebuck ruined
	Somersset case; slaves free on reaching England	Priestley becomes librarian to Lord Shelburne
1773	North regulates East India Co.; Warren Hastings first Governor-General	Boulton acquires rights in steam-engine
		Opening of Assay Office in Birmingham
	Boston Tea Party protest	Day and Bicknell, *The Dying Negro*
1774	Continental Congress at Philadelphia	Watt moves to Birmingham
	Accession of Louis XVI	Wedgwood makes 'Frog Service'
		Stubbs works at Etruria
		Priestley isolates oxygen, *Experiments on Air*
1775	War of American Independence	Boulton & Watt form partnership
		Watt obtains 25-year extension of patent
		Small dies; Baskerville dies
1776	14 July American Declaration of Independence	Boulton & Watt build first engine; begin work in Cornwall
	Tom Paine, *Common Sense*	Withering, *A Botanical Arrangement*
	Adam Smith, *The Wealth of Nations*	
	Gibbon, *Decline and Fall of the Roman Empire*	
1777	Burgoyne defeated at Saratoga	Keir, *Treatise on Gases*
1778	Pitt the Elder (Chatham) dies	Day marries Esther Milnes
	War with France and Spain	Whitehurst, *Inquiry into Formation of the Earth*
	Catholic Relief Act in Ireland	
	Lavoisier's new theory of combustion	Wedgwood perfects 'jasper'
1779	Cook killed in Hawaii	Iron bridge at Coalbrookdale
	Crompton's spinning mule	Priestley, *Experiments and Observations* (3 vols, 1779–86)
	Samuel Johnson, *Lives of the Poets*	
1780	Yorkshire Petition for reform	Priestley moves to Birmingham
	Gordon riots in London	Watt patents copying-press
	North grants Irish free trade	Honora Edgeworth dies
		Bentley dies
1781	October, Cornwallis surrenders	Arkwright patent overturned
		Watt patents rotary motion
		Galton joins Lunar Society
1782	Rockingham ministry with Charles James Fox and Shelburne	Priestley, *History of the Corruptions of Christianity*
	Ireland granted Dublin Parliament	

CHRONOLOGY

1783	Fox–North coalition	Anne Boulton dies
	Treaty of Versailles	Rivalry over composition of water
	Pitt becomes PM (until 1801)	Lunar work on balloons
	Cort's 'puddling' process for iron	Day, *Sandford and Merton* I
	Montgolfier brothers launch balloon	Darwin, *A System of Vegetables*
1784		Darwin starts Derby Philosophical Society
1785	Edmund Cartwright invents power loom	Wedgwood organizer of General Chamber of Commerce
		Priestley's 'Gunpowder' sermon
		Withering, *An Account of the Foxglove*
1786	Anglo-Irish Trade Bill abandoned	Opening of Albion Mill
	Commercial treaty with France	Wedgwood begins copying the Portland Vase
	Mozart, *The Marriage of Figaro*	
1787	Impeachment of Warren Hastings	Wedgwood makes anti-slavery medallion
	Anglo-French commercial treaty	Darwin, translation of Linnaeus *The Families of Plants*
	Society for Suppression of Slave Trade founded	Lavoisier and Laplace, *Chemical Nomenclature*
		Lavoisier, *Elementary Treatise on Chemistry*
1788	Proposal to abolish slave trade defeated	Whitehurst dies
	George III's madness: Regency crisis	Boulton opens Soho Mint
	Gilbert White, *Natural History of Selborne*	
1789	Storming of the Bastille	Day dies
	Washington elected President United States	Darwin, *The Loves of the Plants*
	Blake, *Songs of Innocence*	
1790	Motion to repeal Test Acts defeated	
	Burke's *Reflections on the French Revolution*	Portland Vase displayed
1791	March, Thomas Paine, *Rights of Man*	14 July, Church and King riots
		Priestley leaves Birmingham
1792	London Corresponding Society founded	Darwin, *The Economy of Vegetation*
	Thomas Paine, *Rights of Man*	James Watt junior goes to Paris
	Mary Wollstonecraft, *Vindication of the Rights of Women*	
	September Massacres in Paris;	
1793	Execution of Louis XVI	Thomas Beddoes practises pneumatic medicine in Bristol
	France declares war	
	Scottish Treason trials	
	Godwin, *Political Justice*	
1794	Hardy and others acquitted of treason	Priestley emigrates to America
	Lavoisier guillotined in Paris	Darwin, *Zoonomia*
1795	Treasonable Practices Acts	Wedgwood dies
1796	Spain declares war	Withering, *An Arrangement of British Plants*
	Jenner discovers smallpox vaccine	

1797	*Anti-Jacobin* founded	Boulton makes coins for Royal Mint
		Darwin's *Female Education*
1798	Publication of *Lyrical Ballads*	Edgeworth elected MP in Irish Parliament
	Malthus, *Essay on the Principle of Population*	Edgeworth, *Practical Education*
	Uprising in Ireland crushed	Beddoes Pneumatic Institution
1799	Radical groups suppressed	Withering dies
1800–01	Act of Union with Ireland	Darwin, *Phytologia*
		Maria Edgeworth, *Castle Rackrent*
1801	Pitt resigns	
	'Dalton's law' on partial pressure	
	Trevithick's passenger steam-carriage	
1802	Treaty of Amiens	Darwin dies
1803	War resumes with France	Darwin, *The Temple of Nature*
1804	Napoleon proclaimed Emperor	Priestley dies
1805	Battle of Trafalgar, death of Nelson	
1807	Abolition of Slave Trade	
1808	Peninsular War begins	
1809		Boulton dies
1812	Napoleon invades Russia	
1813	Anglo–American war (1813–15)	Edgeworth, *Construction of Roads and Carriages*
	Jane Austen, *Pride and Prejudice*	
1814	Napoleon abdicates and is exiled to Elba	Steam-press used to print *The Times*
1815	Napoleon returns; Battle of Waterloo	
1817		Edgeworth dies
1818	Mary Shelley, *Frankenstein*	
1819	Peterloo Massacre in Manchester	Watt dies
1820	Shelley, *Prometheus Unbound*	Keir dies
		Edgeworth, *Memoirs*
1829	Stephenson's 'Rocket'	
1832	Great Reform Act	Samuel Galton dies
1859	Charles Darwin, *On the Origin of Species*	

ACKNOWLEDGEMENTS

Anyone working on the Lunar Society must be grateful to the pioneering scholars of the late twentieth century whose meticulous research laid the groundwork for all of us who have followed: I am thinking particularly of Robert Schofield, Eric Robinson, A. E. Musson, D. McKie, Jennifer Tann and John Money: my debts to all these, and many others, are reflected in the notes to this book. I would, however, particularly like to thank Desmond King-Hele, biographer of Erasmus Darwin and editor of his letters, who has encouraged me from the start, has commented on the draft and has been unfailingly generous with his time.

Because the group comprises so many people and diverse interests, my thanks are multiple. Among historians of science I have benefited from the comments of Patricia Fara, Jan Golinski and Simon Schaffer, and especially from the enthusiastic advice of the late and much missed Roy Porter. Maxine Berg has been a tremendous help on the economic background to the luxury trades (as have other members of the Luxury Project at the University of Warwick). Val Loggie, Curator of Soho House, Birmingham, has also been a constant source of encouragement and George Demidowicz has given expert advice on the Soho Manufactory, Foundry and Mint. Sir Nicholas Goodison kindly shared his great knowledge of Boulton's ormolu, and corrected my mistakes; Marilyn Butler has enriched my understanding of Richard Lovell Edgeworth, while Judith Egerton and Stephen Daniels have enlightened me on the work of George Stubbs and Joseph Wright. Stella Tillyard found the lovely cover picture, and has been fun to talk to throughout my work, and Alison Samuel has given me invaluable last minute comments and boosted my confidence. For their help at different points, I would also like to thank Madeleine Budgen, Norma Clarke, Maurice Crosland, Pauline Diamond, Brian Dolan, Peter Jones, Adam Hart-Davis, Janet Hunter

(Wakefield Library), David Kynaston, Martin Claggett, Rita McClean, Susan Marling, Marion Roberts, Anne Secord, Barbara M. D. Smith and Francis Spufford.

I have been fortunate in having access to many different archives, but I would especially like to thank Sian Roberts and the staff of Birmingham City Archives, notably Adam Green, Fiona Tait and Tim Procter of the Soho Project Team; and also Gaye Blake-Davis, Director of the Wedgwood Museum, Barlaston, and Helen Burton at the Wedgwood Archive, University of Keele. For the use of copyright material I am grateful to the following individuals and institutions: The Trustees of Birmingham Assay Office Charitable Trust; Birmingham City Archives; Birmingham University Library; The Syndics of Cambridge University Library; Cambridge University Press; The Darwin Collection at Down House (English Heritage), Downe, Kent; The Wedgwood Trust, Barlaston; The National Library of Ireland; The Royal Society of Medicine; Dr Johnson's Birthplace Trust, Lichfield; Williamsburg University Library, Virginia, and The Wellcome Library, London.

I am also grateful to the institutions and individuals who have given permission to reproduce visual material, as named in the List of Illustrations. It has been a particular delight to establish contact with David Craig, descendant of William Small's brother Robert, who kindly sent me a hitherto unseen portrait and with Michael Butler who provided the Adam Buck sketch of the Edgeworth family.

On the publishing side, I have been marvellously supported by Deborah Rogers and by Jonathan Galassi of Farrar, Straus and Giroux, and by Stephen Page and Joanna Mackle at Faber. I owe a special debt to my shrewd, humorous and encouraging editor, Julian Loose, to the designer Ron Costley and copy-editor and typesetter Jill Burrows for their careful work, and to Kate Ward, Angus Cargill, Anna Pallai and the whole Faber team. I would also like to thank Reginald Pigott for his map.

This is a book about friendship, and one of the great pleasures in writing it has been working with Shena Mason, historian of the Birmingham Jewellery Quarter. From the beginning, out of sheer generosity and interest she has found things, sent things and checked things – and her vivid letters have always made me laugh and spurred me on.

Two people, above all, deserve my gratitude. I could not have written this, or any other book, without the constant interest and warm support of my friend Hermione Lee. And my final thanks go to Steve Uglow, whose own research uncovered several surprising details and in whose company, with immense enjoyment, I have visited so many of the places described.

ABBREVIATIONS, SOURCES AND NOTES

Some archives and sources are abbreviated throughout, as given below. Otherwise a full reference is given the first time a work or source is listed in each chapter; thereafter it is in a short form.

AB	Anne Boulton (née Robinson; second wife of Matthew Boulton)
AnneB	Anne Boulton (daughter of Matthew Boulton)
AR	Anne Robinson (later Boulton; *see* AB)
AS	Anna Seward
AW	Ann (Annie) Watt (née Macgregor; second wife of James Watt)
BF	Benjamin Franklin
CD	Charles Darwin (grandson of Erasmus Darwin)
ChasD	Charles Darwin (son of Erasmus Darwin)
ED	Erasmus Darwin
EDjnr	Erasmus Darwin (son of Erasmus Darwin)
EM	Elizabeth Montagu
JB	Joseph Black
JF	John Fothergill
JK	James Keir
JnWe	John Wedgwood (brother to Josiah)
JP	Joseph Priestley
JR	John Roebuck
JS	Jonathan Stokes
JW	James Watt
JWjnr	James Watt junior (son of James Watt)
JWe	Josiah Wedgwood
JWejnr	Josiah Wedgwood II (son of Josiah Wedgwood)
JWh	John Whitehurst
JWr	Joseph Wright
LH	Logan Henderson
MB	Matthew Boulton
MD	Mary (Polly) Darwin (née Howard; wife of Erasmus Darwin)

ME	Maria Edgeworth (daughter of Richard Lovell Edgeworth)
MH	Mary Howard (later Darwin; *see* MD)
MRB	Matthew (Matt) Robinson Boulton (son of Matthew Boulton)
MW	Margaret (Peggy) Watt (first wife of James Watt)
RAJ	Robert Augustus Johnson
RLE	Richard Lovell Edgeworth
RWD	Robert Waring Darwin (son of Erasmus Darwin)
SG	Samuel Galton
SWe	Sarah (Sally) Wedgwood (wife of Josiah Wedgwood)
SuWe	Susannah (Sukey) Wedgwood (daughter of Josiah Wedgwood)
TB	Thomas Bentley
TD	Thomas Day
TWe	Tom Wedgwood (son of Josiah Wedgwood)
WM	William Matthews
WS	William Small
WW	William Withering

ARCHIVES AND UNPUBLISHED SOURCES

BCA	Birmingham City Archives
B&F	Boulton & Fothergill Papers BCA
B&W	Boulton & Watt Papers BCA
BL	British Library, London
DAR	Darwin Papers, University Library, Cambridge
Down	Darwin Archive, Down House, Downe, Kent
ED Commonplace Book	Erasmus Darwin, Commonplace Book (ms), on loan to the Erasmus Darwin Centre, Lichfield
JBM	Johnson Birthplace Museum, Lichfield
JWP	James Watt Papers, Birmingham City Archives
Keir Memorandum (1802)	James Keir, Memorandum on Erasmus Darwin, sent to RWD 12 May 1802, DAR 227.6:76
Keir Memorandum (1809)	James Keir, Memorandum on Matthew Boulton, sent to MRB, 3 December 1809, MBP 290/112.
MBP	Matthew Boulton Papers, Birmingham City Archives
NLI	National Library of Ireland, Dublin
RS	The Royal Society, London
RSA	The Royal Society of Arts, London
RSM	The Royal Society of Medicine, London
UCL	Galton–Pearson Papers, University College, London
W.	The Wedgwood Archives, Special Collections, University of Keele
Watt Memorandum (1809)	James Watt, ms memoir of Matthew Boulton, Glasgow, 17 September 1809; BCA
Wellcome	The Wellcome Library, London
WWP	William Withering Papers, Special Collections, Birmingham University Library

ABBREVIATIONS, SOURCES AND NOTES

PUBLISHED SOURCES

Unless otherwise specified, place of publication is London.

Anderson and Lawrence   R. G. W. Anderson and Christopher Lawrence, *Science Medicine and Dissent: Joseph Priestley (1733–1804)*, Wellcome Foundation catalogue, 1987

Annals   *Annals of Science*

Arts and Sciences United   *Josiah Wedgwood: 'The Arts and Sciences United'*, Exhibition Catalogue, Science Museum, London, 1978

BG Ec. Veg.   Erasmus Darwin, *The Botanic Garden* part I *The Economy of Vegetation* (1791)

BG Plants   Erasmus Darwin, *The Botanic Garden* part II *The Loves of the Plants* (Lichfield, 1789)

Barker-Benfield   G. J. Barker-Benfield, *The Culture of Sensibility: Sex and Society in Eighteenth-century England* (Chicago, 1992)

Berg *Manufactures*   Maxine Berg, *The Age of Manufactures: Industry, innovation and work in Britain, 1700–1820* (1985)

Birmingham Life   J. A. Langford, *A Century of Birmingham Life: A Chronicle of Local Events, 1741–1841*, 2 vols (1841, rev. 1870)

BJECS   *British Journal of Eighteenth-Century Studies*

BJHS   *British Journal for the History of Science*

Boswell *Johnson*   *Boswell's Life of Johnson*, edited by G. B. Hill, revised and enlarged by L. F. Powell, 6 vols (Oxford, 1934–50)

Botanical Arrangement   William Withering, *A Botanical Arrangement of all the Vegetables Naturally Growing in Great Britain* (Birmingham, 1776)

Brewer   John Brewer, *The Pleasures of the Imagination: English Culture in the Eighteenth Century* (1997)

Butler   Marilyn Butler, *Maria Edgeworth: A Literary Biography* (Oxford, 1972)

Butler *Romantics*   Marilyn Butler, *Romantics, Rebels and Revolutionaries: English Literature and its Background, 1760–1830* (Oxford, 1981)

Clow and Clow   A. and N. Clow, *The Chemical Revolution* (1952)

Constantine   David Constantine, *Fields of Fire: A Life of Sir William Hamilton* (2001)

Consumer Society   N. McKendrick, J. Brewer and J. H. Plumb, *The Birth of a Consumer Society: The Commercialization of Eighteenth-Century England* (1982)

Craven   Maxwell Craven, *John Whitehurst of Derby, Clockmaker & Scientist, 1713–88* (Ashbourne, 1996)

Cultures of Natural History   N. Jardine, J. A. Secord and E. C. Spay (eds), *Cultures of Natural History* (Cambridge, 1996)

Daniels   Stephen Daniels, *Joseph Wright* (1998)

Darwin   Charles Darwin, *The Life of Erasmus Darwin, being an introduction to an Essay on his Scientific Works, by Ernest Krause*, 2nd edn (1887)

Delieb   Eric Delieb, *The Great Silver Manufactory: Matthew Boulton and the Birmingham Silversmiths, 1760–1790* (1971)

Dent   Robert K. Dent, *Old and New Birmingham, A History of the Town and Its People* (Birmingham, 1880)

Dickinson *Boulton*   H. W. Dickinson, *Matthew Boulton* (Cambridge 1936) ['Watt Memorandum (1809)' included as Appendix I]

Dickinson *Watt*   H. W. Dickinson, *James Watt, Craftsman and Engineer* (1935)

Dickinson and Jenkins H. W. Dickinson and Rhys Jenkins, *James Watt and the Steam Engine* (Oxford, 1927)

Edgeworth *Memoirs* *Memoirs of Richard Lovell Edgeworth Esq.* (1820); text used is 3rd edn (1844)

EDL Desmond King-Hele (ed.), *The Letters of Erasmus Darwin* (Cambridge, 1981)

EHR *English Historical Review*

Electricity Joseph Priestley, *The History and Present State of Electricity* (1767)

Enlightenment Roy Porter, *Enlightenment: Britain and the Creation of the Modern World* (2000)

Essential Writings Desmond King-Hele (ed.), *Essential Writings of Erasmus Darwin* (1968)

Experiments on Air Joseph Priestley, *Experiments and Observations on Different Kinds of Air*, 3 vols (1774, 1775, 1777)

Experiments and Observations Joseph Priestley, *Experiments and Observations relating to various Branches of Natural Philosophy*, 2 vols (1779; Birmingham, 1781)

Fara Patricia Fara, *Sympathetic Attractions: Magnetic Practices, Beliefs, and Symbolism in Eighteenth-Century England* (Princeton, 1996)

Farrer K. E. Farrer (ed.), *Letters of Josiah Wedgwood*, 3 vols (Manchester, 1903–6)

Fitton R. S. Fitton, *The Arkwrights, Spinners of Fortune* (Manchester, 1989)

Foxglove William Withering, *An Account of the Foxglove* (Birmingham, 1785)

Franklin Papers Leonard W. Labaree (ed.), *Papers of Benjamin Franklin* (Newhaven, 1965)

Galton Karl Pearson, *The Life, Letters and Labours of Francis Galton*, 3 vols (1914)

Genius H. Young (ed.), *The Genius of Wedgwood*, Victoria and Albert Museum exhibition catalogue (London, 1995)

Gibbs Frederick William Gibbs, *Joseph Priestley* (1965)

Gignilliat George Warren Gignilliat, *The Author of Sandford and Merton, a Life of Thomas Day Esq.* (New York, 1932)

Golinski Jan Golinski, *Science as Public Culture: Chemistry and Enlightenment in Britain, 1760–1820* (1992)

Griffiths John Griffiths, *The Third Man: The Life and Times of William Murdoch, 1754–1839* (1992)

Harris J. R. Harris, *Industrial Espionage and Technology Transfer; Britain and France in the Eighteenth Century* (1998)

Heilbron J. L. Heilbron, *Electricity in the 17th and 18th Centuries* (Berkeley, Ca.,1979)

Hist. Sci. *History of Science*

Hutton *History* William Hutton, *History of Birmingham*, 3rd edn (1795)

Hutton *Life* *The Life of William Hutton* (1816), edited by Carl Chinn (Birmingham, 1998)

Inquiry John Whitehurst, *Inquiry into the Original State and Formation of the Earth* (1778)

Isis *International Review of History of Science*

Israel Jonathan Israel, *Radical Enlightenment: Philosophy and the Making of Modernity 1650–1750* (Oxford, 2001)

Jacob Margaret C. Jacob, *Scientific Culture and the Making of the Industrial West* (Oxford, 1997)

Keir *Day* [James Keir], *An Account of the Life and Writings of Thomas Day, Esq.* (1791)

Keir *Dictionary* [James Keir, translation of Pierre Joseph Macquer], *Dictionary of Chemistry*, 2nd edn, 1777

King-Hele Desmond King-Hele, *Erasmus Darwin: A Life of Unequalled Achievement* (1999)

Klonk Charlotte Klonk, *Science and the Perception of Nature; British Landscape Art in the late eighteenth and early nineteenth century* (Cambridge, 1992)

Langford Paul Langford, *A Polite and Commercial People, England 1727–1783* (Oxford, 1989)

McNeil Maureen McNeil, *Under the Banner of Science; Erasmus Darwin and his Age* (Manchester, 1987)

Marsden Ben Marsden, *Watt's Wonderful Engine* (Cambridge, 2002)

Meteyard Eliza Meteyard, *The Life of Josiah Wedgwood*, 2 vols (1865); edited by R. W. Lightbown (1970)

Misc. Tracts William Withering junior, *Miscellaneous Tracts of the Late William Withering, M.D., F.R.S.* (1822)

Moilliet and Smith J. L. Moilliet and Barbara Smith, *A Mighty Chemist: James Keir of the Lunar Society* (privately printed, 1982)

Mokyr Joel Mokyr, *The Gift of Athena: Historical Origins of the Knowledge Economy* (Princeton, 2002)

Money John Money, *Experience and Identity: Birmingham and the West Midlands, 1760–1800* (Manchester, 1977)

Morrell Jack Morrell, *Science, Culture and Politics in Britain, 1750–1870* (Aldershot, 1997)

Muirhead *Life* James Patrick Muirhead, *The Life of James Watt with Selections from his Correspondence* (1858)

Muirhead *Mechanical Inventions* James Patrick Muirhead, *The Origins and Progress of the Mechanical Inventions of James Watt*, 2 vols (1854)

Musson and Robinson A. E. Musson and Eric Robinson, *Science and Technology in the Industrial Revolution* (Manchester, 1969)

Ormolu Nicholas Goodison, *Ormolu: The Work of Matthew Boulton* (1974; new edition 2002)

Pardoe F. E. Pardoe, *John Baskerville of Birmingham: Letter-Founder & Printer* (1975)

Partington J. R. Partington, *A History of Chemistry*, vol. III (1962)

Partners in Science E. Robinson and D. McKie, *Partners in Science: The Letters of James Watt and Joseph Black* (1970)

Peck T. W. Peck and K. D. Wilkinson, *William Withering of Birmingham* (Bristol, 1950)

Phil. Trans. *The Philosophical Transactions of the Royal Society*

Phytologia Erasmus Darwin, *Phytologia; or the Philosophy of Agriculture and Gardening* (1800)

Plan Erasmus Darwin, *A Plan for the Conduct of Female Education in Boarding Schools* (Derby, 1797)

Porter *Beddoes* Roy Porter, *Doctor of Society: Thomas Beddoes and the Sick Trade in Late Enlightenment England* (1991)

Porter *Geology* Roy Porter, *The Making of Geology: Earth Science in Britain, 1660–1815* (Cambridge, 1977)

Priestley *Autobiography* *The Autobiography of Joseph Priestley* (1806), edited by Jack Lindsay (1970)

Reilly Robin Reilly, *Josiah Wedgwood, 1730–1795* (1992)

Reilly and Savage Robin Reilly and George Savage, *The Dictionary of Wedgwood* (1980)

Richards Sarah Richards, *Eighteenth-century Ceramics: Products for a Civilised Society* (Manchester, 1999)

Robinson and Musson Eric Robinson and A. E. Musson (eds), *James Watt and the Steam Revolution: A Documentary History* (1969)

Roll Eric Roll, *An Early Experiment in Industrial Organisation* (1930)

Rolt L. T. C. Rolt, *James Watt* (1962)

Rowland Peter Rowland, *The Life and Times of Thomas Day, 1748–89* (Lewiston, 1996)

Rutt *Life* J. T. Rutt (ed.), *Life and Correspondence of Joseph Priestley*, 2 vols (1831–2)

Rutt *Works* J. T. Rutt (ed.), *The Theological and Miscellaneous Works of Joseph Priestley* (1817–32)

Sandford and Merton Thomas Day, *The History of Sandford and Merton*, 3 vols (1783–89)

SchimmelPenninck Christiana C. Hankin (ed.), *The Life of Mary Anne SchimmelPenninck*, 3rd edn, (1860)

Schofield Robert E. Schofield, *The Lunar Society* (Oxford, 1963)

Schofield *Priestley* Robert E. Schofield, *The Enlightenment of Joseph Priestley* (Pennsylvania, 1997)

Scientific Autobiography Robert E. Schofield (ed.), *A Scientific Autobiography of Joseph Priestley (1733–1804)* (Cambridge, Mass.,1966)

Scientific Correspondence H. C. Bolton, *Scientific Correspondence of Joseph Priestley* (1891)

Seward Anna Seward, *Memoirs of the Life of Dr Darwin* (1804)

Seward *Letters* A. Constable (ed.), *Letters of Anna Seward* (Edinburgh, 1811)

Seward *Poetical Works* Walter Scott (ed.), *Poetical Works of Anna Seward* (1810)

SHPS *Studies in History and Philosophy of Science*

Sketch [A. Moilliet], *A Sketch of the Life of James Keir* (1859)

SL George Savage and Ann Finer (eds), *The Selected Letters of Josiah Wedgwood* (1965)

Smiles Samuel Smiles, *Lives of Boulton and Watt* (1865)

Spectator *The Spectator*, edited by Donald F. Bond, 5 vols (Oxford, 1965)

Tann Jennifer Tann (ed.), *The Selected Papers of Boulton & Watt*, I *The Engine Partnership 1775–1825* (1981)

Temple of Nature Erasmus Darwin, *The Temple of Nature, or, The Origin of Society* (1803)

UBHJ *University of Birmingham Historical Journal*

Vases and Volcanoes Ian Jenkins and Kim Sloan, *Vases and Volcanoes: Sir William Hamilton and his Collection*, British Museum exhibition catalogue (1996)

Weather Journals *The Weather Journals of a Rutland Squire: Thomas Barker of Lyndon Hall*, edited by John Kington (Oakland, 1988)

Wedgwood Circle Barbara and Hensleigh Wedgwood, *The Wedgwood Circle 1730–1897* (1980)

Whitehurst *Works* C. Hutton (ed.), *The Works of John Whitehurst, F.R.S.* (1792)

Wright of Derby Judy Egerton, *Wright of Derby*, Tate Gallery exhibition catalogue (1990)

Young Hilary Young, *English Porcelain, 1745–95: Its Makers, Design, Marketing and Consumption* (1999)

Zoonomia Erasmus Darwin, *Zoonomia; or, The Laws of Organic Life*, part I (1794); part II (1796); 3rd rev. edn (1801)

NOTES

Unless otherwise stated, all letters from Erasmus Darwin are quoted from *EDL*, whose illuminating notes are also invaluable. Desmond King-Hele is now embarking on a revised edition to include the many letters which have come to light in recent years.

The Wedgwood–Bentley correspondence can be found in Farrer but in a highly edited form. I have therefore quoted the originals from the Wedgwood Archives at Keele (W.); many, however, can also be found in *SL*. References to the Boulton & Watt and other papers in Birmingham City Archives (BCA) are given in the form used in 2001. The whole labyrinthine archive is currently being re-catalogued to make it easier for future researchers and new references will be adopted, but cross-references will be made from the old to the new system.

Spelling and punctuation follow that of the sources.

Waxing

PROLOGUE: 'SURPRISE THE WORLD'

1 JP, *Experiments on the Generation of Air from Water* (1793), Dedication. Membership of the Lunar society is uncertain, sometimes including Baskerville, sometimes excluding Wedgwood, although he was central to their work. See, for example, Eric Robinson, 'The Lunar society: its membership and organisation', *Trans. of the Newcomen Society*, XXXV (1962–63).
2 ED to MB 11 March 1766.
3 John Woodward, *Essay Towards a Natural History of the Earth* (1695).
4 Rosamond Wolff Purcell and Stephen Jay Gould, *Finders, Keepers: Eight Collectors* (1992) 17.
5 Ibid.
6 R. Porter, 'Science, provincial culture and public opinion in Enlightenment England', *BJECS*, 3 (1980). See also Golinski, and Larry Stewart, *The Rise of Public Science: Rhetoric, Technology and Natural Philosophy in Newtonian Britain, 1660–1750* (Cambridge, 1992). For a different perspective on the 'Industrial Enlightenment' and birth of the knowledge economy, see Mokyr; and for a reading of the 'Radical Enlightenment', see Israel.
7 JWe to TB 31 October 1768, W. E25–18212.
8 *BG*: *Ec. Veg.* 1. 529 note. The suggestion was first made by William Small in correspondence with James Watt in the early 1770s.
9 See D. Read, *The English Provinces,* c. *1760–90* (1964) and P. Borsay, *The English Urban Renaissance: Culture and Society in the Provincial Town, 1660–1770* (Oxford, 1989).
10 See John Styles, 'Manufacturing, consumption and design in eighteenth-century England', in John Brewer and Roy Porter (eds), *Consumption and the World of Goods* (1993) 536–8.
11 Stella Tillyard, *Aristocrats: Caroline, Emily, Louisa and Sarah Lennox, 1740–1832* (1995) 171. See also Maxine Berg and Helen Clifford (eds), *Consumers and Luxury: Consumer Culture in Europe 1650–1850* (Manchester, 1998).
12 See *Enlightenment*, and David Spadaforda, *The Idea of Progress in Eighteenth-Century Britain* (Newhaven and London, 1990).
13 *Weather Journals* 24.
14 David Souden (ed.), *Byng's Tours: The Journals of the Hon. John Byng 1781–1792* (1991) 23.

First Quarter

1 EARTH, ELSTON & ELECTRICITY

1 *Weather Journals* 59.
2 Robert Darwin, 'old notebooks before I burnt them in 1800', DAR 227.5.12.
3 Darwin 6, and ED Commonplace Book.
4 Darwin 2. After the Civil Wars Erasmus's Royalist great-grandfather trained as a barrister and married the daughter of Commonwealth diplomat Erasmus Earle. Their eldest son married a Nottinghamshire heiress, Ann Waring, who inherited the manor of Elston.
5 ED to Thomas Okes [23?] November 1754.
6 Darwin 4. Portrait by Jonathan Richardson, 1717.
7 Darwin 4, *Phil. Trans.* April and May 1719.
8 See W. Moore, *The Gentlemen's Society at Spalding* (1851); D. M. Owen (ed.), *The Minute Books of the Spalding Gentleman's Society, 1712–1755*, Lincoln Record Society, 73 (Lincoln, 1981).

9 ED to Thomas Okes [23?] November 1754.

10 King-Hele 5. For portraits of Robert and Elizabeth Darwin see *Galton*, plate VII.

11 Hutton *Life* 29.

12 *Aris's Birmingham Gazette*, 17 April 1746.

13 ED to Susannah Darwin, March 1749, and UCL 578,34, *EDL* 4, n. 1.

14 Susannah Darwin to ED, 20 February 1749, DAR 227.3.1.

15 ED to Susannah Darwin, March 1749.

16 ED to Samuel Pegge, 28 December 1749.

17 Quoted in Porter *Beddoes* 24.

18 Darwin 6. A recent deposit by Christopher Darwin of over sixty notebooks in Cambridge University Library, contains numerous verse poems by Erasmus and his brothers, including ten of his 'Enigmas' and many juvenile poems.

19 *Spectator* IV 442.

20 *Spectator* III 575. See Patricia Fara, *Isaac Newton: The Making of Genius* (2002).

21 Alexander Pope, 'Epitaph, Intended for Sir Isaac Newton', *Pope: Poetical Works*, edited by Herbert Davis (Oxford, 1966) 651.

22 J. T. Desaguliers, quoted in Daniels 38. See also *Newton Demands the Muse: Newton's Optics and the Eighteenth Century Poets* (Princeton, 1946) 37.

23 See Paul Elliott, 'The birth of public science in the English provinces: Natural Philosophy in Derby, c.1690–1760', *Annals* 57 (January 2000) 61–100.

24 *Gentleman's Magazine*, 1732, see Craven 17.

25 Daniels 48.

26 Edgeworth *Memoirs* 26.

27 With thanks to Patricia Fara for describing this. For a concise, lively account of these early discoveries, see her *An Entertainment for Angels: Electricity and Enlightenment* (2002).

28 Heilbron 247.

29 Henry Baker to Henry Miles, 29 April 1745, quoted in W. D. Hackman, *Electricity from Glass: The History of the Frictional Electrical Machine* (Alphen aan den Rijn, 1978) 105. See Darwin's version of this in *BG Ec. Veg.* I 349–56.

30 See Hackman, *Electricity from Glass*, 90–103, and Heilbron 309–23.

31 Musschenbroek to Reaumur, 20 January 1746, Heilbron 313–14.

32 John Neale, *Directions for Gentlemen who have Electrical Machines* (1747); quoted in Simon Schaffer, 'The consuming flame: electrical showmen and Tory mystics in the world of goods', in John Brewer and Roy Porter (eds), *Consumption and the World of Goods* (1993) 491. See also Maurice Daumas, *Scientific Instruments of the 17th and 18th Centuries and their Makers* (1989).

33 W. Henly to J. Canton, n.d., Heilbron 317.

34 See Schaffer, 'The consuming flame', 497, and his 'Natural Philosophy and Public Spectacle in the Eighteenth Century', *Hist of Sci*, 21 (1983) 1–43.

35 Isaac Newton, *Opticks* (1704; 1952 edition) 404.

2 TOYS

1 For Midlands lectures, see Heilbron, and Musson and Robinson 381–2.

2 Darwin 17.

3 Notebook, MBP 290/2.

4 Ibid.

5 *Sketchley's Birmingham Directory*, 1767, 56.
6 *A Blacksmith's Shop* (1771), *An Iron Forge* (1772), *An Iron Forge viewed from without* (1773). See Daniels 50–53, and *Wright of Derby* 101–4.
7 John Leland, *Itinerary* (1538), in Henry Hamilton, *The English Brass and Copper Industries to 1800* (1926) 122–3.
8 Alexander Missen, in Hutton *Life* xii. See Berg *Manufactures* 287–313, M. J. Wise (ed.), *Birmingham and its Regional Setting. A Scientific Survey* (Birmingham, 1950), and P. Hudson, *Regions and Industries* (Cambridge, 1989). Early works include S. Timmins (ed.), *The Resources, Products and Industrial History of Birmingham and the Midland Hardware District* (1866), and W. H. B. Court, *The Rise of the Midlands Industries 1600–1838* (1938).
9 The Clarendon Code included the Corporation Act (1661); the Act of Uniformity (1662); the Conventicle Act (1664); the Five Mile Act (1665), banning Nonconformist ministers from living within five miles of their old parish, and the Test Acts (1673).
10 For a comparison of Birmingham and Sheffield, see Maxine Berg, 'Small Producer Capitalism in Eighteenth-Century England', *Business History*, 35, 1 (1993) 17–39.
11 Hutton *History* 90–91.
12 See Carl Chinn, *Birmingham: The Great Working City* (Birmingham, 1994) and Eric Hopkins, *Birmingham, The First Manufacturing Town in the World, 1760–1840* (1989).
13 Hutton *History* 55.
14 See William Bennett, *John Baskerville* (1937) and uncorrected galleys (BCA; MS 705/4).
15 William Hutton, *European Magazine*, November 1785; Pardoe 19.
16 Pardoe 20.
17 Will of John Baskerville, January 1773, Pardoe 126.
18 See Eric Robinson, 'Matthew Boulton's Birthplace and his home at Snow Hill: A problem in detection', *Transactions of the Birmingham Archaeological Association*, 1957, vol. 75, 88–89.
19 Hutton *History* 379.
20 *Aris's Birmingham Gazette*, 29 December 1746.
21 Dent 90–92.
22 JW Memorandum (1809) 1.

3 SCOTLAND

1 For a survey see Gerard L'E. Turner, 'Scientific Instruments', in Pietro Corsi and Paul Weidling (eds), *Information Sources in the History of Science and Medicine* (1983) 243–58.
2 John Wallis, in Heilbron, 10; see her careful introductory survey.
3 For Thomas Watt, see Muirhead *Life* 4–9, Dickinson *Watt* 15; for Watt's uncle John, see Jacob 100–105.
4 George Williamson, *Memorials of the Lineage, Early Life, Education and Development of the Genius of James Watt* (1856), quoted in Dickinson *Watt* 19.
5 JWP 4/161.
6 In Harry Grey Graham, *The Social Life of Scotland in the Eighteenth Century* (Edinburgh, 1901) 142.
7 See Jane Rendall, *Origins of the Scottish Enlightenment* (1978) 17, and A. L. Brown and Michael Moss, *The University of Glasgow: 1451–2001* (Edinburgh, 2001).
8 Chambers, *Traditions of Edinburgh*, I 105; see also John Thomson, *An Account of the Life, Lectures and Writings of William Cullen, M.D.*, 2 vols (Edinburgh, 1859).

9 JW to James Watt senior, 1 July 1755, JWP 6/46.
10 'The improvement of the Sand glass for the True Measuring of Time in order to find Longitude'; G. L'E. Turner, *Scientific Instruments*, XIV 19.
11 JW to James Watt senior, 21 July 1755, JWP 6/46.
12 JW to James Watt senior, 31 March 1756, JWP 6/46; Muirhead *Life* 39.
13 JW to James Watt senior, 20 April 1756, JWP 6/46.
14 JWP April 1756, 19 June 1756, JWP 6/46.
15 JW to James Watt senior, 2 October 1756, JWP 4/11.2.
16 'Robison's Narrative' (1796); Robinson and Musson 24.
17 Henry Cockburn, *Memorials of his Time* (1910 edn), 46. For Black, see R. G. W. Anderson, 'Joseph Black', in D. Daiches et al. (eds), *A Hotbed of Genius* (Edinburgh, 1986), and A. D. C. Simpson (ed.), *Joseph Black, 1728–1799* (Edinburgh, 1982).
18 Introduction to Black's *Lectures* (1803), in James Sambrook, *The Eighteenth Century* (1986) 15.
19 Adam Hart-Davis, 'James Watt and the Lunaticks of Birmingham', *Science*, 6 April 2001, 56.
20 Smiles 33.
21 David Hume, *A Treatise of Human Nature*, edited by Ernest Rhys, 2 vols (1911) 5.
22 Lord Charlemont, quoted in Alistair Smart, *Allan Ramsay, 1713–1784* (Edinburgh, 1992).
23 See Christopher Lawrence, 'The Nervous System and Society in the Scottish Enlightenment', in Barry Barnes and Steven Shapin (eds), *Natural Order: Historical Studies of Scientific Culture* (1979).

4 THE DOCTOR'S BAG

1 Darwin 12.
2 Darwin learned from the *Brachygraphy* of Thomas Gurney; his notes were on lectures by Dr George Baker at King's College: Wellcome WMS 2043; see King-Hele 13–14.
3 King-Hele 15.
4 See Roy Porter, 'Medical lecturing in Georgian London', *BJHS*, 28 (March 1995), 91–100, and W. F. Bynum and Roy Porter (eds), *William Hunter and the Eighteenth-Century Medical World* (Cambridge, 1985), Porter, *The Greatest Benefit to Mankind* (1997) 245–305.
5 *The Collected Letters of Oliver Goldsmith*, edited by Katharine C. Balderston (Cambridge, 1928) 3.
6 *SchimmelPenninck* 37.
7 *Phytologia* 208; King-Hele 18.
8 Quoted in D. Daiches et al. (eds), *A Hotbed of Genius* (Edinburgh, 1986) 1.
9 See Christopher Lawrence, 'Ornate physicians and learned artisans: Edinburgh medical men, 1726–1776', in Bynum and Porter, 153–76; and R. G. W. Anderson and A. D. C. Simpson (eds), *The Early Years of the Edinburgh Medical School* (Edinburgh, 1976).
10 See Guenter B. Risse, *Hospital Life in Enlightenment Scotland. Care and Teaching at the Royal Infirmary of Edinburgh* (Cambridge, 1986).
11 *Pharmacopoeia Edinburgensis; or New Edinburgh Dispensatory*, 4th edn (1744).
12 JK to Robert Darwin, 1802; King-Hele 17.
13 The Leyden-trained Haller published his physiological textbook, *Primae lineae physiologiae*, in 1747, followed by *Elementa physiologiae corporis humani* (8 vols, 1757–66).
14 For Cullen's ideas and impact, see Golinski 11–40.
15 Robert Whytt, *On the Vital and Other Involuntary Motions of Animals* (1751); discussed in Klonk 18; Porter, *The Greatest Benefit*, 251.

16 ED to Thomas Okes, [23?] November 1754.
17 ED to Albert Reimarus, 9 September 1756, Darwin 18.
18 DAR 227: 1.4; King-Hele 20.
19 Boswell *Johnson* II 464.
20 Seward 2–3. For an interesting discussion of Seward, see the chapter in Brewer, 'Queen Muse of Britain', 573–612.
21 ED Commonplace Book.
22 DAR 227: 1.14; King-Hele 29.
23 ED to MH, 24 December 1757.
24 Ibid.
25 King-Hele 32, 34.
26 ED to MD, 18 May 1758, DAR 227: 1.16.
27 Seward 15.
28 Seward *Poetical Works* vii.
29 Seward 65–8.
30 *Aris's Birmingham Gazette*, 25 October 1762.
31 Heilbron 342.
32 The paper was read on 5 May and published in *Phil. Trans.* I (1757) 250–54.
33 Seward 16. A tutor at Queen's College, Michell had lectured on Hebrew, Greek, arithmetic, geometry and philosophy.

5 POTS

1 'Burslem in 1750 based on a plan by Enoch Wood', F. Falkner, *The Wood family of Burslem* (1912). But for the unreliability of Burslem maps of this date, see Lorna Weatherill, *The Pottery Trade and North Staffordshire, 1660–1760* (Manchester, 1971) 155–6.
2 JWe Experiment Book 1759–74, W. E26–19115.
3 See 'The Earliest Entrepreneurs', Young 33–53.
4 JWe Memorandum, 'Pot-works in Burslem about the year 1710–1715' (1776), *SL* 24.
5 Simeon Shaw, *History of the Staffordshire Potteries* (Hanley, 1829) 110–12.
6 Ibid., 97–8.
7 JWe to TB, 3 April 1765, W. E25–18072.
8 Meteyard I 200.
9 Ibid., 104–5. See also Enoch Wood, quoted in Reilly 5.
10 Seward 2.
11 Indenture, Wedgwood Archives, signed by his mother, and his uncles Abner Wedgwood and Samuel Astbury; Reilly 3–4.
12 JWe Notebook, *c.*1749, W. E26–19115.
13 Reilly 18.
14 Thomas Whieldon's notebook, Weatherill, *The Pottery Trade*, 104–8.
15 *Aris's Birmingham Gazette*, 17 June 1754.
16 Weatherill, *The Pottery Trade*, 87.
17 For Wedgwood's technical experiments, see N. McKendrick, 'The role of science in the Industrial Revolution, a study of Josiah Wedgwood as a scientist and industrial chemist', in M. Teich and R. Young (eds), *Changing Perspectives in the History of Science* (Dordrecht, 1973).
18 JWe Experiment Book 1759–74, W. E26–19115.

19 See Lilian Beard, 'Unitarianism in the Potteries from 1812', *Transactions of the Unitarial Historical Society* VI (1935–8), 14–16.
20 *Wedgwood Circle* 11.
21 See *Genius* 63–7.
22 See Richards 53–5, and 82, nn. 62–5.
23 See Richard Bentley, *Thomas Bentley* (Guildford, 1927).
24 JWe to TB, 15 May 1762, W. E25–18048.
25 See Neil McKendrick, 'Josiah Wedgwood and Thomas Bentley: an Inventor–Entrepreneur Partnership in the Industrial Revolution', *Transactions of the Royal Historical Society*, 5th ser., XIV (1960), 1–33, although this occasionally overstates Bentley's role.
26 JWe to TB, 16 September 1769, W. E25–18254.
27 JWe to TB, 26 October 1762, W. E25–18049.

6 HEADING FOR SOHO

1 MB to Benjamin Huntsman, 19 January 1757, MBP Letter Book 1.
2 Notebook 1, 1751 (later dates inside), MBP 376.
3 I am indebted to Shena Mason for this detail.
4 ED to MB, [30? October 1762].
5 JK Memorandum 1809, MBP 290/112.
6 MBP 366/1. See the memoir prefacing Whitehurst *Works*; also W. Douglas White, 'The Whitehurst family', *Derbyshire Miscellany Supplement*, March 1958.
7 BF to Joseph Galloway, 6 September 1758, *Franklin Papers* VIII 146.
8 BF to Deborah Franklin, 6 September 1758, ibid.
9 Ibid.
10 John Michell to MB, 5 July 1758; Schofield 24.
11 William Shenstone to MB, 19 July 1758, Marjorie Williams (ed.), *The Letters of William Shenstone* (Oxford, 1939) 484.
12 Ibid., 489.
13 Ibid., 490–93.
14 Since Matthew paid tribute to Mary 'bearing many children', there may have been more who died at birth, or twins as in her own family. See MBP 252/60, Dorothy Robinson's Memoranda, 'account of my children': twins, 1725 (John, d.1728, and one stillborn); twins, 1727 (Mary, and one stillborn); son, 1728 (stillborn); twins, 1731 (Luke, and William, d.8 weeks); daughter, 1733 (Anne).
15 Notes, MBP 290/6.
16 'Upon seeing the Corps of my Dear Wife many Excellent Qualitys of Hers arose to my Mind which I could not then forbare acknowledging Extempory with my pen & depositing it in her Coffin', MBP 291/72.
17 MB to Dorothy Robinson, n.d., MBP 252/62. The letters in MBP 252 and 279 are undated, but the sequence can be established from internal evidence.
18 MB to AR, n.d., MPB 279/8.
19 MB to Dorothy Robinson, n.d., MBP 252/62.
20 MB to Dorothy Robinson, n.d., MBP 252/61. For his irreverence, see MBP 279/4 on offending Mr Barker by casual jokes about working on a Sunday.
21 MB to AR, n.d. [December 1759, not 'April 1763' as marked], MBP 279/8.

22 ED to MD, 12–13 June 1759, DAR 227: 1.17.
23 MB to AR, n.d., MBP 279/4.
24 *Journal of the House of Commons* XXVIII (1757–61) 785–901.
25 MB to AR, n.d., MBP 252/8.
26 John Fry, *The case of marriage between near kindred* (1756), published at Boyle's Head in Fleet Street. Boulton's order is dated 22 April 1760: Dickinson *Boulton* 34–5.
27 MB to AR, n.d., MBP 279/9.
28 Dressmaker's bill, Elizabeth and Hannah Concher, 17 July 1760. Soho House.
29 MB to RLE, 20 November 1780 (MBP 143, Letter Book K, 58); and RLE to MB, 2 October 1780 (MBP 230/101).
30 Dressmaker's bill , 14 May 1763; paid 28 September 1768. Soho House.
31 John Taylor, testimony to House of Commons, 1759, quoted in *Genius* 100.
32 See Harris 114.
33 See D. G. C. Allan, *William Shipley: Founder of the Royal Society of Arts: A Biography with Documents* (1968) 169–88.
34 MB to Lord Hawkesbury, 17 April 1790, MBP 237.
35 MB to Timothy Hollis, 3 February 1761, Schofield 26.
36 His first architect was Joseph Pickford (who also built a house for Whitehurst), Craven 65.
37 Notes, MBP 308/9.
38 See E. Robinson, 'Boulton and Fothergill 1762–68 and the Birmingham Export of hardware', *University of Birmingham Historical Journal* VII, no. 1 (1959). The partners would share profits equally after receiving interest on the amount invested. Boulton's capital investment was £6,206 17s 9d, including buildings, tools and materials; Fothergill's £5,394 16s 0d.
39 JF to MB, 7 May 1762, *Ormolu* 10.
40 Robinson notebook, 31 August 1764, MBP 252/68. The dispute centred on a codicil signed with his mark, leaving a farm to cousins if Anne should die without issue.
41 Kenneth Quickenden, 'Boulton & Fothergill's Silversmiths', *Silver Society Journal* VII (autumn 1995) 342.
42 Ibid., 348–54: a fascinating picture of workshop organization. For extracts from the Inventory, taken by Scale after Fothergill's death, see *Ormolu* Appendix V, 269–77.
43 JF to John Lewis Baumgartner, 13 July 1765, Letter Book B, 158; MB to James Adam, 1 October 1770, Letter Book D, 29–30. See Quickenden, 'Silversmiths', 347.
44 See Robinson, 'Boulton and Fothergill', and *Ormolu* 15–18, and J. E. Cule, *The Financial History of Matthew Boulton, 1759–1800*, unpublished thesis, University of Birmingham, 1935.
45 Notes on dispute with Fothergill, MBP 308/54.
46 Ibid.

7 INGENIOUS PHILOSOPHERS

1 MBP 367/1 Notebook 1751–59, 376. Boulton used the English translation of 1750, *The art of hatching and bringing up domestic fowls of all kinds at any time of the year, either by means of the heat of hotbed or that of the common fire*. See also MBP 290/7.
2 Ibid., 21.
3 Mark Akenside, *Hymn to Science* (1744).
4 Priestley *Autobiography* 92.
5 JP, *Familiar Letters Addressed to the Inhabitants of the Town of Birmingham, by the Revd. Mr Madan*

(1790–92), Letter 4, 6. See *Enlightenment* 398.

6 Priestley *Autobiography* 71.

7 The Reverend George Haggerstone had studied at Edinburgh under Maclaurin; the books included Isaac Watts's *Logic*, Locke's *Essay on Human Understanding* and 'sGravesande's *Elements of Natural Philosophy*. See Rutt *Life* I 13.

8 Rutt *Life* I 62.

9 Priestley *Autobiography* 85.

10 Anna Laetitia Aikin to Betsy Belsham, quoted in Betsy Rodgers, *A Georgian Chronicle: Mrs Barbauld and her Family* (1958) 52.

11 Rutt *Life* I 50.

12 Priestley *Autobiography* 71.

13 See *Enlightenment* 180–83, and Richard C. Allen, *David Hartley on Human Nature* (New York, 1999).

14 Quoted by Roy Porter in 'Matrix of Modernity', Royal Historical Society Gresham Lecture, November 2000, reprinted in *History Today*, April 2001, 24–31.

15 Adam Smith, *Theory of Moral Sentiments* (1759), edited by D. D. Raphael and A. L. Macfie (Oxford, 1976) 166, part III, ch. 5.

16 Joseph Priestley, *History of the Corruptions of Christianity* (1782). See Rutt *Works* IV, Introduction, 7.

17 JP, *Disquisitions relating to Matter and Spirit*, 2nd edn (Birmingham, 1782) I vii.

18 In his *Naturalis Philosophiae Theoria*, the Jesuit Roger Joseph Boscovitch (1711–87) put forward the idea of forces and '*puncta*'.

19 Priestley *Autobiography* 86.

20 Ibid., 87.

21 JP to John Canton, 14 February 1766; *Scientific Autobiography* 15.

22 Gibbs 27.

23 Ibid., 28.

24 That the force of attraction between two charged bodies varies inversely with the square of the distance between them. Demonstrated by Coulomb in 1784–85.

25 *Experiments and Observations* I xi, in Golinski 81; see 51–90 for Priestley's writing.

26 See Jan Golinski, *Making Natural Knowledge: Constructivism and the History of Science* (Cambridge,1998) 2–4.

27 *Experiments and Observations* I xi.

28 JWe to TB, 9 October 1766, W. E25–18130.

29 Schofield 28.

30 'Franklin's "glass harmonica" used glass basins of graduated size (and so pitch) half-submerged in water, mounted on a spindle. When the spindle was rotated the glasses revolved and the friction of the player's fingers produced the sound.' Shena Mason, unpublished paper on music at Soho, 1998.

31 Lectures noted in *Aris's Birmingham Gazette* from 1750 include William Griffiths, 1755; Joseph Hornblower, 1757; James Ferguson, 1761 and 1771; John Arden, 1765, 1767 and 1771; John Warltire, annually from 1776 to 1782, 1784 and 1789. See Money 151.

32 See Berg *Manufactures* 295–301. For patents, see Christine McCleod, *Inventing the Industrial Revolution: The English Patent System, 1660–1800* (Cambridge, 1988), and for a nineteenth-century view, see Richard Prosser, *Birmingham Inventors and Inventions* (1881).

33 JR to John Seddon (probably pre-1762), Schofield 29.

34 T. C. Hansard, *Typographia* (1825) 717–18, 311.
35 See J. M. Robinson, *The Wyatts: An Architectural Dynasty* (Oxford, 1979).
36 *Aris's Birmingham Gazette*, 25 October 1762.
37 ED to MB, [30] October 1762.
38 'An uncommon case of a Haemoptysis', *Phil. Trans.* LI (1760) 526–9. For ED's election to the Royal Society, see King-Hele 45–6.
39 *Aris's Birmingham Gazette*, 17 January 1763; Schofield 31.
40 ED to MB, 1 July 1763.
41 King-Hele 50 points out that the ideal gas law is credited to J. A. C. Charles (1787) and partial pressures to John Dalton (1801).
42 JWe to TB, 31 March 1763, W. E25–18052.
43 BF to MB, 22 May 1765, MBP 233/119.
44 Merrill D. Perriman (ed.), Thomas Jefferson, *Writings* (New York, 1984), 'Autobiography', 4.
45 See Gillian Hull, 'William Small, 1734–1775: no publication, much influence', Journal of the Royal Society of Medicine, 90 (February 1997), 102.
46 Small's advice, in a letter from Stephen Hawtrey to New Hawtrey, 26 March 1765. Swem Library Archives, College of William and Mary, Williamsburg, Virginia. Herbert Ganter Collection, Box III, Folder 21. Cited by Martin McClaggett, unpublished paper.
47 Ibid.
48 Both were friends of the postmaster William Hunter, and in April 1763 Franklin went to Virginia to deal with Hunter's will, in which he and Small were mentioned.
49 For Franklin's connections with these clubs, see Carl van Doren (ed.), *Benjamin Franklin's Autobiographical Writings* (New York, 1945) 141–2, and for Lunar links, see E. Robinson, 'R. E. Raspe, Franklin's "Club of Thirteen" and the Lunar Society', *Annals* II, 2 June 1955.
50 John Baskerville to MB, 9 December 1765, MBP 219/201.
51 Keir *Day* 29–30.
52 Sending envoys and setting up cells was an acknowledged aspect of Franklin's proselytizing technique.
53 ED to MB, 11 March 1766.

8 REACHING OUT

1 JWe to TB, 9 January 1764, W. E25–18055.
2 JWe to TB, 23 January 1764, W. E25–18056.
3 Ibid.
4 JWe to TB, 28 May 1764, W. E25–18057.
5 JWe to JnWe, 11 March 1765, W. E18071–25.
6 JWe to JnWe, 2 February 1765, W. E25–18059.
7 Zaccheus Walker married Mary Boulton and was a clerk at Soho from 1760 and later accountant; Thomas Mynd married Boulton's sister, set up on his own in 1773 but took Soho commissions; John Bentley, a second cousin, whose father Richard also worked for Boulton and Fothergill, went independent in 1776, but continued to supply goods.
8 JWe to TB, 6 March 1765, W. E25–18070.
9 *Arts and Sciences United* 12; see generally, D. Towner, *Creamware* (1978).
10 Supplemented by a £500 loan from her brother John. Josiah C. Wedgwood and Joshua G. E. Wedgwood, *Wedgwood Pedigrees* (Kendal, 1925) 174, n. 3; Reilly 36.

11 JWe to JnWe, [summer 1765], *SL* 34.
12 JWe to JnWe, [17 June 1765], W. E25–18073.
13 *Wedgwood Circle* 20.
14 JWe to JnWe, 7 August 1865, W. E25–18089.
15 JWe to TB, [after 8 September 1767], W. E25–18167.
16 MB to AB, February/March 1760, MBP 279/14–20.
17 JWe to JnWe, 7 August 1765, W. E25–18089.
18 See *Genius*, 162–9.
19 JWe to TB, 3 September 1770; *SL* 96.
20 JWe to TB, 26 July 1767, W. E25–18160.
21 See MBP Letter Book A (1757–65). Fothergill spoke both French and German; see Eric Robinson, 'Boulton and Fothergill, 1762–82 and the Birmingham export of Hardware', *UBHJ* VII 1 (1959).
22 1763. Shena Mason, talk to the Society of Jewellery Historians, 10 April 2000.
23 See Robinson, 'Boulton and Fothergill', 67–8, and Musson and Robinson 218.
24 See Dickinson *Boulton* 49–50; Musson and Robinson 224–7 and Daniel Solander correspondence, MBP 254/S2/269: there were strict laws against the emigration of skilled workmen.
25 MB to AB, 18 November 1765, MBP 279/25.
26 MB to AB, 24 November 1765, MBP 279/27.
27 William Wyatt to MB, 26 January 1766, MBP 375/228.
28 Journal of the House of Commons, 16 February 1763, Petition to parliament for a turnpike from Lawton to Stoke; partially quoted in *SL* 24.
29 Arthur Young, 1771, in *English Canals* I (Lingfield, 1967).

9 STEAM

1 ED to MB, [1764?].
2 See L. T. C. Rolt, *The Horseless Carriage* (1950) 16.
3 ED to MB, 11 March 1766.
4 ED to MB, 12 December 1765.
5 Muirhead *Mechanical Inventions* II, 204; for Robison's idea see the *Universal Magazine of Knowledge and Pleasure* XXII (1757) 229–31.
6 References in this paragraph are to letters between JW and James Watt senior, and to John Watt, 1757–58, JWP 4/11.7–23.
7 *Glasgow Journal*, 1 December 1763, in Rolt 21.
8 An article by Michael Wright in *New Scientist* (March 2002) describes the flute-making tools found in material from Watt's workshop, now in the Science Museum, and also a steel stamp marked '*TLot*', the mark of the great Paris flute-maker Thomas Lot.
9 For Watt's organ, see 'Robison's Narrative' in Robinson and Musson 38.
10 'Robison's Narrative', Robinson and Musson 24.
11 'Letter to Dr Brewster from Mr Watt', in John Robison, *A System of Mechanical Philosophy*, edited by David Brewster (Edinburgh, 1822) II, 116n.
12 Robison, *Mechanical Philosophy* II, ix.
13 Recounted by JW in 1817 to the Glasgow engineer Robert Hart: 'Reminiscences of James Watt', *Transactions of Glasgow Archaeological Society*, 1859. (The wording varies.)
14 Joseph Black, 'History of Mr Watt's Improvement of the Steam-Engine' [1796–7],

Muirhead *Life* 58.

15 JW to James Lind, 29 April 1765, JWP C1/15.

16 Robinson and Musson 27.

17 Muirhead *Life* 95.

Second Quarter

10 THEY BUILD CANALS

1 For the canal era, the authority is still Charles Hadfield, *British Canals* (1959) and *The Canals of the West Midlands* (Newton Abbot, 1966; 3rd edn, 1985); see also Jean Lindsay's excellent *The Trent and Mersey Canal* (1979).

2 See H. Bode, *James Brindley* (1973).

3 By a second Act of 1760, the canal was to end at Manchester, not Salford, while a 1762 Act authorized the extension to Runcorn.

4 See *EDL* 18–27.

5 See *EDL* 19–20 n. 5.

6 JWe to TB, 2 January 1765, W. E25–18058.

7 JWe to JnWe, 11 March 1765, W. E18071–25.

8 JWe to ED, 3 April 1765, DAR 227:3:4.

9 JWe to TB, 2 January 1765, W. E25–10859.

10 JWe to ED, 5 June 1765, DAR 227.3.11.

11 See Darwin's correspondence with Wedgwood, and canal pamphlet drafts and notes in Cambridge: DAR 227, and UCL 10.

12 UCL 190A: King-Hele 57.

13 JWe to TB, 26 September 1765, W. E25–18091.

14 JWe to TB, 27 September 1765: W. E25–18092.

15 JWe to TB, 7 October 1765, W. E25–18093.

16 JWe to ED, DAR 227.3:14 and 15.

17 JWe to TB, 7 October 1765, W. E25–18093.

18 JWe to TB, 18 November 1765, W. E25–18099.

19 JWe to TB, 7 October 1765, W. E25–18098.

20 JWe to TB, 12–15 December 1765, W. E25–18015.

21 JWe to TB, [November 1765], W. E25–18012.

22 DAR 220:3:20, King-Hele 57.

23 *Aris's Birmingham Gazette*, 28 April 1766.

24 Meteyard I 455–6.

25 JWe to TB, [December 1767], W. E25–18176.

26 *The History of Inland Navigations*, 2nd edn (1769), quoted in Langford 416.

27 JWe to TB, 2 March 1767, W. E25–18139.

28 Reilly 58.

29 ED to MB, 12 December 1765.

30 MB to JWe, 7 October 1766, MBP 298/97. (Boulton is mentioned re the navigation petition, 'Lichfield, Jany 17, 1766', signed by John Barker: MBP 298/93.)

31 See correspondence, MBP 298/119–25.

32 Notes on petition, Birmingham Canal Company vs Charles Colmore, MBP 298/11.

33 WS to M. Meredith, [January 1771?], MBP 298/13.

34 See correspondence with M. Meredith, MBP 298/14–24.

35 MB to the Earl of Warwick, [1772], Schofield 42.

36 Birmingham Navigation account with Boulton & Watt, 1784: MBP 298/9.

37 See WS to JW, MBP 340/11–32 (December 1769 to June 1773); many of these letters are printed in Muirhead *Life* 161–250.

38 JW to WS, 9 September 1770, JWP 4/59, Muirhead *Life* 204.

39 King-Hele 136–7. The first lift was opened in the late 1780s in Saxony: lifts operating on similar principles to Darwin's were built on the Dorset and Somerset Canal in 1800, and at Northwich in Cheshire in 1875.

11 PAINTING THE LIGHT

1 For Burdett, Ferrers and Whitehurst, see David Fraser's article in *Wright of Derby* 16–19. King-Hele 97 identifies Darwin as the man with the watch, seated foreground left, and suggests the boys may be Charles and Erasmus; one boy, however, has been identified as the nephew of Earl Ferrers, see *Wright of Derby* 54.

2 *A Philosopher giving that lecture on the Orrery, in which a lamp is put in place of the Sun*, 1766 (Derby Museum and Art Gallery); *An Experiment on a Bird in the Air Pump*, 1768 (National Gallery, London). See *Wright of Derby* 54–61 and Daniels 32–42.

3 Elisabeth Inglis-Jones, *The Great Maria* (1959) 57. With thanks to Pauline Diamond.

4 JK to ED, 20 August 1766, *Sketch* 48.

5 *Partners in Science* 13.

6 See JW to JWe [1785], JWP C1/12.

7 ED to MB, 27 January 1782.

8 MB to AB, October 1766, Rita McLean and Val Loggie, 'The Correspondence between Josiah Wedgwood I & Matthew Boulton', *Etruria*, spring 2001, 9.

9 Edgeworth *Memoirs* 118.

12 MAGIC & MECHANICS

1 ED to MB [summer 1766].

2 Quoted in Patricia Fara, 'A treasure of hidden virtues: The attraction of magnetic marketing', *BJHS* 28 (June 1995) 28.

3 *EDL* 40: Boulton's endorsement to the letter.

4 See Schofield 51, and Fara, op.cit. Artificial magnets were made by Gowin Knight in 1745, but John Canton and John Michell both devised their own methods. Boulton was corresponding with J. H. Magellan, about a delayed shipment of magnets, 9 September 1773, MBP 244/52–64.

5 Edgeworth *Memoirs* 105.

6 Ibid., 116.

7 JW to MW, 11 April 1767, JWP 4/4/20.

8 Fara 61.

9 Edgeworth *Memoirs* 68.

10 Ibid., 102.

11 Ibid., 93–4.

12 Ibid., 106–7.
13 JWe to TB, W. E25–18432.
14 See 'Erasmus Darwin's improved steering for carriages – and cars', Notes and Records of the Royal Society, 56 (2002) 41–62.
15 ED to JWe, 14 June 1768, King-Hele 81.
16 WS to JW, 12 August 1768, MBP 340/1.
17 WS to JW, 18 April 1769, MBP 340/5.
18 JWe to TB, 23 May 1767, W. E25–18147.
19 ED to JW, 18 August 1767.
20 JW to MB, 20 October 1768, MBP 348/1.
21 JR to JW, 30 October 1768, Robinson and Musson 48.
22 MB to JW, 7 February 1769, JWP 6/19/23.
23 JWe to TB, 3 March 1768, W. E25–18191.(This was a huge undertaking, as Darwin had explained that it would require '68,000 brick, 4000 ft of inch deal boards, and a shaft 30 ft long and a foot diameter, having two sails'.) See *EDL* 54, and *Essential Writings* 121–3.
24 ED to JW, 15 March 1769.
25 ED to JW, 1771, DAR 227:1.32.
26 JW to WS, 31 January 1770, JWP 4/59. For optics see also WS to JW, 3 February and 19 October 1771, 27 January 1773, MBP 340/16, 17, 26. For Darwin, see ED Commonplace Book.
27 For Priestley and Cook, see *Scientific Autobiography* 95–8.
28 ED to BF, 18 July 1772.
29 *Temple of Nature*, Additional Note xv, 107–20.
30 With thanks to Chris Keneally for this explanation.
31 Edgeworth *Memoirs* 356.
32 Down MS. 3 September 1771, *Essential Writings*, fig. 10.

13 DERBYSHIRE EXPLORATA

1 Thomas Hobbes, *De Mirabilis Pecci: Being the Wonders of the Peak in Derbyshire* (1678) 14. Quoted in Ian Ousby, *The Englishman's England: Taste, Travel and the Rise of Tourism* (Cambridge, 1990) 132.
2 Meteyard I 210.
3 Jane Austen, *Pride and Prejudice*, edited by R. W. Chapman (Oxford, 1933) 239. Thanks also to Richard Hamblyn, '"An Absurd Curiosity": Science and the Souvenir Object', unpublished paper, Luxury Conference, University of Warwick, 1999.
4 Patty Fothergill's diary, 1793, Soho House.
5 Daniels 15. For Burdett and the Derbyshire map, see *Wright of Derby* 88.
6 See the quotations from Ferber in Faujas de St Fond, *A Journey through England and Scotland* (1784; 1907 edn) II 283–4.
7 This clay, discovered by André Duché in 1739, had already been specified as one of the ingredients in the Bow porcelain.
8 JWe to TB, 21 July 1774, W. E25–18548.
9 JWe to TB, 16 February 1767, W. E25–18186.
10 JWe to TB, 8 September 1767, W. E25–18166.
11 ED to JWe, 2 July 1767. See King-Hele's informative n. 1, *EDL* 43–4; he suggests the

bones are from the Pleistocene, and the fossils from the Carboniferous era. For Wedgwood's response to the Harecastle fossils, see Meteyard I 500–502.

12 ED to JWe, 2 July 1767.

13 W. Hunter, 'Observations on the bones, commonly supposed to be elephants bones, which have been found near the River Ohio in America', *Phil. Trans.* LVIII (1768) 34–5.

14 ED to JWe, 18 August 1767.

15 ED to JWe, 2 July 1767.

16 ED to MB, 29 July 1767.

17 Ibid.

18 WS to JW [September 1768], MBP 340/3. For Watt on Keir, see JW to James Lind, 25 November 1768, JWP C1/15/4.

19 In Charles's student thesis (1780), 127–8, King-Hele 73.

20 See Trevor D. Ford, *The Castleton Area, Derbyshire*, Geologists Association Guide, no. 56 (1996) 23; see also Ford's *Treak Cliff Cavern and the Story of Blue John Stone* (Castleton, 1992), and his 'Blue John Fluorspar', *Proceedings of the Yorkshire Geological Society* (1955).

21 John Fleming, *Robert Adam and his Circle* (1962) 246–7.

22 *Ormolu* 26.

23 E.g. Robert Wood and James Dawkin, *The Ruins of Palmyra* (1753); the Comte de Caylus, *Recueil d'Antiquités Egyptiennes, Etrusques, Grecques et Romaines* (1752–67); J. F. Neufforges, *Recueil Elementaire d'Architecture* (1757–68): see *Ormolu* 49, 51.

24 *Vases and Volcanoes* 51. See also Constantine 32–45. (The vases were in fact not Etruscan but Greek imports.) The collection was bought by the British Museum in 1772.

25 JWe to TB, 16 February 1767, W. E25–18163.

26 [Sir William Hamilton], D'Hancarville, *Collection of Etruscan, Greek and Roman Antiquities* (1766–76) I viii, quoted in *Ormolu* 55.

27 P. J. Wendler to MB, and MB to Wendler, 1767, MBP 261/35–9.

28 This was infringed in 1770 by a local rival, Palmer of Hanley.

29 For Boulton's enquiries, see Musson and Robinson 218–20.

30 MB to JWh, 28 December 1768.

31 Receipt, 2 March 1769, to John Platt, MBP 249/32. See J. D. Potts, *Platt of Rotherham, Mason Architects, 1700–1810* (Sheffield, 1959).

32 JW to TB, 6 June 1772, W. E25–18376.

33 See *Phil. Trans.* LVII (1768) 192–200, and LVIII (1769) 1–12.

34 Porter *Geology* 109.

35 Hutton's paper was read in Edinburgh in 7 March and 4 April 1785. His 'Theory of the Earth', in the *Transactions of the Royal Society of Edinburgh* (1788) 209–304, was elaborated in *The Theory of the Earth, with proofs and illustrations*, 2 vols (Edinburgh, 1795). For the Scottish school, see Porter *Geology* 149–56, 184–96.

36 John Michell, 'Conjectures concerning the cause and observations upon the phenomena of earthquakes', *Phil. Trans.* LI (1760) 566–634. See also Susanne B. Keller: 'Sections and views; visual representations in eighteenth-century earthquake studies', *BJHS* 31 (June 1999) 129–60.

37 R. E. Raspe studied basalts in Hesse and came independently to the same conclusion in 1769, and in *Phil. Trans.* LXI (1771) 580–83. For background, see Rachel Laudan, *From Mineralogy to Geology, the Foundations of a Science, 1650–1830* (1967).

38 John Whitehurst, *Inquiry*, 1st edn, 49, quoted in David Fraser, 'Fields of radiance: the scientific and industrial scenes of Joseph Wright', in Denis Cosgrove and Stephen Daniels (eds),

The Iconography of Landscape (Cambridge, 1988), 125.

39 JWr to Richard Wright, 11 November 1778. Quoted in William Bemrose, *The Life and Work of Joseph Wright, ARA* (London and Derby, 1885) 34–5.

40 JWr to his sister Nancy, 4 May 1775, *Wright of Derby* 175.

41 Unpublished talk by Helen Langdon, British Academy, 14 March 2001. See drawings of caves (*Wright of Derby* 87, 95, 96), and paintings (97, 98, 99).

42 JWe to TB, n.d. [autumn 1767], W. E25–18188.

43 Ford, *The Castleton Area*, 7–13.

44 King-Hele 89.

45 *Vases and Volcanoes* 235.

46 ED to Thomas Gifford, 4 September 1968, DAR 268.1.

14 CHEMICAL REACTIONS

1 Schofield 86 notes that Watt's recipe is quite close to the 'soft-paste porcelain' of Chelsea, Derby and Worcester.

2 WS to JW, 12 February 1769, MBP 340/4.

3 JWh to MB, 13 April 1769, MBP 366/3.

4 Schofield 100–101; J. J. Ferber, *Versuch einer Oryctographie von Derbyshire in England* (Mittau, 1776; translated by J. Pinkerton, 1808).

5 *Sketch* 8.

6 Ibid., 14.

7 Edgeworth *Memoirs* 184–5.

8 Keir *Dictionary* Preface, quoted in Moilliet and Smith 14.

9 P. J. Macquer, *Dictionary of Chemistry*, translated by J. Keir, 3rd edn (1771) 1 9.

10 See Clow and Clow 203.

11 Derby Art Gallery, first exhibited 1771: see *Wright of Derby* 84–6.

12 Paul Strathern, *The Search for the Elements* (2000) 188.

13 See Daniels 27–31.

14 JWe to TB, 31 November 1771, *SL* 117.

15 Macquer article on 'Fluor', Schofield 94.

16 *Sketch* 51.

17 These included Gellert's *Chymie Metallurgique*, 2 vols; Swedenborg's *Principia Rerum Naturalium*, 3 vols, incorporating his 'Treatise on Iron'; Schluer's *Traité de Mines*, 2 vols.

18 ED to JWe, 14 June 1768.

19 WS to JW, 26 July 1769, MBP 340/7.

20 Peter Woulfe to MB, 3 September 1769, MBP 262/223.

21 Moillet and Smith 22.

22 For Watt's involvement, see 'James Watt and Early Experiments in Alkali Manufacture', Musson and Robinson 352–71.

23 WS to JW, 10 October 1769, MBP 340/8.

24 JK to JW, 27 February 1772 (see also 14 December 1771); *Partners in Science* 27.

25 *Journals of the House of Commons*, 37, 915; see Clow and Clow 92–100; Schofield 78.

26 Moillet and Smith 25.

27 For the parliamentary petitions 1780–81, see Musson and Robinson 363–5.

28 Moillet and Smith 16. See D. R. Guttery, *From Broad-glass to Cut Crystal. A History of the*

Stourbridge Glass Industry (1956).
29 WS to JW, 19 October 1771, MBP 340/17.
30 Clow and Clow 95.
31 JK to MB, 22 October 1772, MBP 319/3.
32 Ibid.
33 *Phil. Trans.* LXVI (1776) 530–42.
34 *Sketch* 75–6.
35 For the industry, see 'Soap', in Clow and Clow 116–29.
36 K. T. Rowland, *Eighteenth-Century Inventions* (Newton Abbot, 1974) 83.
37 ED to RLE, 20 February 1788.
38 ED to RWD, 21 February 1788, King-Hele 223.
39 Simeon Shaw, *History of the Staffordshire Potteries* (Hanley, 1829) 136–7; Moillet and Smith 30–31.

15 TRIALS OF LIFE

1 JWe to TB, 20 May 1767, W. E25–18147.
2 Money 162–6. In October 1765, the *Birmingham Gazette* announced that American orders had fallen by £600,000 over eighteen months.
3 Money 65.
4 Marginalia in *An Inquiry*, an anonymous pamphlet. *Franklin Papers* 17, 341–2.
5 JWe to TB, 16 March 1768, W. E25–18194. Reilly 253.
6 JP, *Appeal to the Professors of Christianity* (1770), *The Doctrine of Philosophical Necessity Illustrated* (1777), *A History of the Corruptions of Christianity* (1782), *History of Early Opinions Concerning Jesus* (1786). See J. A. Passmore, *Priestley's Writings on Philosophy, Science and Politics* (1965).
7 JP, *An Essay on the First Principles of Government* (1768; 2nd edn, 1771).
8 Ibid., 12–13.
9 Priestley's individualism has been seen as anticipating later liberalism. See D. O. Thomas, 'Progress, Liberty and Utility: The Political Philosophy of Joseph Priestley', in Anderson and Lawrence 73–9.
10 JP to BF, 14 February 1769, *Franklin Papers* 16, 41–2.
11 Priestley *Autobiography* 92.
12 Langford 430.
13 Darwin 42.
14 Ibid., 35; Keir Memorandum (1802).
15 Seward 3–4.
16 Ibid., 108–15.
17 ED [to unknown], 7 February 1767.
18 Darwin 40.
19 Ibid., 43.
20 Ibid., 41. Johnson's edition of Shakespeare appeared in 1765.
21 *Aris's Birmingham Gazette*, 29 September 1766.
22 ED to RWD, 19 April 1789.
23 DAR 227.7.9, King-Hele 74.
24 DAR 227.7.9, King-Hele 75.
25 Ibid.

26 Seward 64.
27 King-Hele 76.
28 WS to JW, 17 September 1771, MBP 340/14.
29 MB to AB [April 1772], MBP 349/38.
30 ED to MB, 9 June 1769.
31 JWe to TB, 2 March 1767, W. E25–18139.
32 JWe to TB, 31 December 1767, W. E25–18182.
33 JWe to TB, 6 July 1767, W. E25–18160.
34 Peter Swift to William Cox, 31 May 1770, Reilly 69.
35 JWe to TB, 20 June 678, W. E25–18200.
36 JWe to TB, 19 September 1772,W. E25–18402.
37 JWe to TB, 14 November 1772, W. E25–18419.
38 JWe to TB, 12 October 1772, W. E25–18412.
39 ED to JWe, 14 June 1768.
40 King-Hele 82.
41 Seward 62.
42 WS to JW, February 1769, MBP 340/4.
43 Dorothy Porter, *Patient's Progress* (1989) 150.
44 DAR 227.1.142, King-Hele 94.
45 Ibid.
46 DAR 227.2:7, King-Hele 92.
47 Darwin College, Cambridge, *Wright of Derby* 220.
48 For Lucy Swift, see King-Hele 105.

16 ROUSSEAU & ROMANCE

1 Darwin 26.
2 Edgeworth *Memoirs* 111–12.
3 Darwin 36.
4 TD to RLE [1769], Edgeworth *Memoirs* 142.
5 Ibid., 113.
6 Ibid., 124.
7 Seward 27; for Day's inheritance see Gignilliat 2–3.
8 Seward 35.
9 Hume challenged this ideal in *Treatise of Human Nature* (1739–40): see *Enlightenment* 247–51. See also Elizabeth Rawson, *The Spartan Tradition in European Thought* (1969).
10 Edgeworth *Memoirs* 135.
11 Seward 36.
12 Edgeworth *Memoirs* 137.
13 TD to RLE, November 1769, Edgeworth *Memoirs* 143.
14 Ibid., 139–40.
15 Ibid., 143.
16 Seward 18.
17 National Portrait Gallery, London. Anna Seward claimed, wrongly, that the book was Hampden's speech in defiance of the tyrant Charles I. Wright produced two versions; the second came to Edgeworth, now in Manchester City Art Gallery. See *Wright of Derby* 118.

18 Seward 39.
19 Ibid., 40.
20 Thanks to Anne Secord for this observation.
21 Seward 52.
22 Ibid., 41.
23 See poems to Honora in Seward *Poetical Works* I 65–103.
24 'Monody of the Death of Major André', Seward *Poetical Works* II 68–88.
25 Edgeworth *Memoirs* 115.
26 Ibid., 158–9.
27 Ibid., 161.
28 Ibid., 162.
29 TD to AS, 31 August 1771, JBM ms 2001.71.60.
30 TD to AS, 18 December 1771, Mary Alden Hopkins, *Dr Johnson's Lichfield* (1956) 155.
31 Seward 44.
32 TD to AS, 14 March 1772, JBM ms 2001.71.16.
33 Hopkins, *Dr Johnson's Lichfield*, 116–17.
34 Edgeworth *Memoirs* 174.
35 Ibid., 207.
36 WS to ED, 27 August 1773, DAR 227.6.1; Edgeworth *Memoirs* 209.

17 VASES, ORMOLU, SILVER & FROGS

1 Keir Memorandum (1809) 9.
2 Elizabeth Montagu to MB, 31 October 1771, MBP 330/1.
3 JWe to TB, 15 March 1768, W. E25–18193. (Wedgwood reminded Bentley that they had seen these at Lord Bolingbroke's, and 'two or three old China bowles, for want of better things, stuct rim to rim which had no bad effect but look whimsical and droll enough'.)
4 William Cox, SWe and Catherine Willett, joint letter to TB, 29 November 1768, W. M2.
5 JWe to SWe, 15 February 1769, Reilly 38.
6 William Cox, SWe and Catherine Willett, joint letter to TB, 29 November 1768, W. M2.
7 JWe to TB, 21 November 1768, W. E25–18215.
8 MB to JWe, 28 December 1768, 17 January 1769, MBP 136, Letter Book D.
9 JWe to TB, 27 September [1769], W. E25–18258.
10 JWe to TB, 23 February 1769, W. E25–18231.
11 JW to TB [13 September 1769], W. E25–18252.
12 Elizabeth Montagu to MB, 31 October 1771, MBP 330/1.
13 MB to Elizabeth Montagu, 16 January 1772, draft, MBP 330/3.
14 MB to AB, [1770], MBP 279/38.
15 See *Ormolu* 53–4 for Boulton's use of Stuart's designs: the 2002 edition contains more information on Stuart.
16 JWe to TB, 24 December 1770, W. E25–18334.
17 ED to MB, 4 January 1771.
18 JWe to MB, 19 February 1771, MBP 361/8.
19 JWe to TB, 7 September 1771, *SL* 114. In 1770 Boulton bought a lion, a ram's head, a deer's head, a sleeping Bacchus and a 'Group of Hercules and Atlas', *Ormolu* 57–8.
20 MB to AB, 6 March 1770, see also 3 March 1770, MBP 279.

21 For this as a general issue in ceramics, see Young 94–154.

22 See G. Blake-Roberts, 'The London Decorating Studio', in *Genius* 92–101.

23 See JWe to TB, 25 June 1769, W. E25–18245; 17 September 1769, W. E25–18255.

24 MB to James Adam, 1 October 1770, MBP 136, Letter Book (1768–73) 29. For drawing and design generally, see Charles Saumarez-Smith, 'Eighteenth-century man', *Designer* (March 1987) 19–21. For Birmingham, see Young 100.

25 For a clear account of all the processes involved, see *Ormolu*, 'Manufacture'.

26 MB to J. H. Ebbinghaus, 20 October 1767, MBP 135, Letter Book (1767).

27 These recipes and others, from Boulton's Notebook (1768–75) are cited in *Ormolu* 71–3.

28 On espionage, see MB to the Paris merchant, Solomon Hymen, 17 June 1772, MBP Letter Book (1771–73), and 13 March 1775, Letter Book (1774–77).

29 JWe to TB, 21–2 April 1771, *SL* 106.

30 JWe to TB, 23 August 1772; W. E25–18392.

31 Constantine 65.

32 Ibid., 72.

33 Kenneth Crisp Jones (ed.), *The Silversmiths of Birmingham and their Marks, 1750–1980* (Fakenham, 1981) 25–6. See Kenneth Quickenden, 'Boulton & Fothergill's Silversmiths', *The Silver Society Journal* VII (autumn 1995) 343.

34 For Boulton's silver, see Quickenden, 'Boulton & Fothergill's Silversmiths'; Delieb (sometimes unreliable on design and dating); Robert Rowe, *Adam Silver* (1965); David Kynaston, 'The History of Birmingham Silversmiths', in Crisp Jones (ed.), *The Silversmiths of Birmingham.*

35 MB to Shelburne. For his campaign, see Arthur Ryland, 'The Birmingham Assay Office' in Samuel Timmins (ed.), *The Resources, Products and Industrial History of Birmingham and the Midland Hardware District* (1866) 499–510, and Jennifer Tann, *Birmingham Assay Office 1773–1993* (Birmingham, 1993) 13–27.

36 MB Diary, MBP 381 (1773). Targets included Shelburne, the Dukes of Portland and Richmond, and the Marquess of Bute. See E. Robinson, 'Matthew Boulton and the art of Parliamentary lobbying', *Historical Journal* VII (1964).

37 WS to MB, 28 January 1773, MBP 340/27.

38 See the detailed study of the 1780 Boulton & Fothergill Day Book, 1779–81, by Kenneth Quickenden, in 'Silver, plated and silvered products from the Soho Manufactory, 1780', *The Silver Society Journal* X (autumn 1998), 77–95.

39 For a survey of clockcases see *Ormolu* 104–29.

40 See *Ormolu* 109 and plate 85.

41 MB to Lord Cathcart, 30 October 1771, in *Ormolu* 121.

42 MB to JWh [11 April 1772], MBP 279/38.

43 MB to John Scale, 31 March 1772, MBP 337/7.

44 For James Cox, goldsmith, jeweller and entrepreneur, see Clare Le Corbeiller, 'James Cox: a biographical review', *Burlington Magazine* CXII (1970) 351–6, Catherine Pagani, 'The Clocks of James Cox', *Apollo*, January 1995, 15–22, and Roger Smith, 'James Cox (*c.*1723–1800): a revised biography', *Burlington Magazine*, June 2000, 353–6.

45 MB to AB, [11 April 1772], MBP 279/38.

46 William Porter to MB, 1776, *Ormolu* 123.

47 See the wonderful illustrated catalogue, Michael Raeburn, Ludmila Voronikhina and Andrew Nurnberg (eds), *The Green Frog Service* (London and St Petersburg, 1995).

48 Ibid., 11.

49 See M. Baker, 'A Rage for Exhibitions: The Display and Viewing of Wedgwood's Frog Service', in *Genius*.
50 *BG Ec. Veg.* II 291–6.

18 RUNNING THE SHOW

1 JWe to TB, 13 September 1769, W. E25–18252.
2 JWe to TB, 1 October 1769, W. E25–18264.
3 JWe to TB, 19 November 1769, W. E25–18269.
4 JWe to MB, 19 September 1771, MBP 361/8.
5 Keir Memorandum (1809).
6 Watt Memorandum (1809).
7 See Reilly 74.
8 See JWe to TB, 7 September 1771, *SL* 113–15.
9 MB to London agent, J. L. Baumgarten, August 1767: MBP 135, Letter Book C (1767). For Boulton and display, see also Peter Jones, 'Matthew Boulton's "Enchanted Castle": Visions of the Enlightenment in the English Midlands *c.*1765–1800' in Roland Mortimer (ed.), *Visualisations* (Berlin, 1999).
10 EM to MB, 31 October 1771, MBP 330/1. (There are 63 letters between 1771 and 1786.)
11 Lord Edward Fitzmaurice, *Life of William, Earl of Shelburne* (1875–76) I 277.
12 M. L. Mare and W. H. Quarrel, *Lichtenberg's Visits to England* (Oxford, 1938) 97; in *Ormolu* 13.
13 Berg *Manufactures* 293.
14 See *Ormolu*, 'Manufacture', 62–77.
15 JWe to TB, 24 December 1770, W. E25–18334.
16 JWe to TB, 9 September 1769, W. E25–18265.
17 Young 23.
18 See Reilly, 'Management and Security', 133–48, for a succinct outline.
19 JWe to TB, 3 September 1770, *SL* 94–7.
20 JWe to TB, 16 February 1771, *SL* 103–4.
21 See Neil McKendrick, 'Josiah Wedgwood and Cost Accounting in the Industrial Revolution', *EHR* XXIII (1970) 45–67.
22 JWe to TB, 22 July 1772, W. E25–18381.
23 'Proposal to Boulton & Fothergill by John Scale' [1773]. MBP [Scale] 14.
24 JWe to TB, 20 September 1772, W. E25–18407.
25 MB to Peter Bottom, 30 March 1768, in John Lord, *Capital and Steam Power, 1750–1800* (1966) 58.
26 See John Hodges to Dr Thomas Percival, 16 November 1782, by which time the 'Soho Club' had been running for some years: MBP 147, Letter Book N 48.
27 McNeil 142.
28 For the administration, see Roll 225–36.
29 Boswell *Johnson* II 459.
30 Quoted in Anna E. Meredith, *Matthew Boulton* (Birmingham, 1972) 3:1.
31 For factory discipline generally in Britain and France, see Paul Mantoux, *The Industrial Revolution in the Eighteenth-century* (1983 edn) 375–6. For Wedgwood, see N. McKendrick, 'Josiah Wedgwood and Factory Discipline', *Historical Journal* IV (1961) 30–55.
32 Minutes of the Evidence, Committee of House of Lords, 1785 252, 253, in Witt Bowden,

Industrial Society towards the end of the Eighteenth Century (1924; 1965 edn) 275.

33 William Blake, 'Jerusalem: The Emanation of the Giant Albio' (1804–20), McNeil 43.

34 See E. Robinson, 'Eighteenth-century commerce and fashion: Matthew Boulton's marketing techniques', *EHR* XVI (1963–64) 39–60; and Hilary Young, 'From the Potteries to St Petersburg: Wedgwood and the Making and Selling of Ceramics', in *Genius* 9–20.

35 Richard Cumberland, *The Observer* I (1786), Essay 25, 233–4; with thanks to Elizabeth Eger.

36 MB to AB, 6 March 1770.

37 JWe to TB, 31 May 1767, W. E25–18141.

38 See Barker-Benfield 211–12.

39 See James Adam to MB, 14 August 1770, 5 November 1770 and 28 January 1771, and MB to Adams, 1 October 1770, MBP 218/4–6. For Wedgwood, Boulton and the Adelphi, see W. E25–18334. See also later discussions: John Wyatt to MB, 27 February 1776; to B&F, 8 March 1776, MBP 375/72 and 76, and *Ormolu* 87–95.

40 JK to MB, April 1772, MBP 319/2.

41 JWe to TB, 11 April 1772, W. E25–18365.

42 JWe to TB, 24 March 1768, W. E25–18196. See also JWe to MB, 19 March 1768, MBP 361/6.

43 Ornamental catalogues appeared in 1774, 1777, 1779 and 1787, and a separate catalogue of gems (also in a French edition). For advertising generally see Maxine Berg and Helen Clifford, 'Commerce and Commodity: Graphic Display and Selling New Consumer Goods' in M. Nash and D. Ormrod (eds), *Markets for Art* (1999) 187–99.

19 EDDIES

1 Notebook 6 (1771) 88, MBP 376/6.

2 ED to JWe, 30 September 1772.

3 *BG Ec. Veg.* III 329–36, see *EDL* 66.

4 JW to WS, 7 November 1772, Muirhead *Life* 207.

5 Edgeworth *Memoirs* 219–20.

6 Gibbs 87.

7 Will of John Baskerville, Pardoe 125; other quotations from accounts in Pardoe 148–157.

Third Quarter

20 EXPERIMENTS ON AIR

1 *Experiments on Air* II 33.

2 Ibid.

3 Ibid., 29.

4 Ibid., 34.

5 Ibid., 43.

6 Becher (whose *Physica Subterranea* was edited by Stahl), refined this triad into three 'earths' found in all minerals: *terra mercurialis*, *terra lapida* and *terra pinguis*, the inflammable principle: when a metal was calcined this rose into the air leaving a metal *calx*.

7 For the long sequence of experiments, see *Experiments on Air* II 29–103.

8 Ibid.,102.

9 For Cavendish, see Christa Jungnickel and Russel McCormmach, *Cavendish, The Experimental Life* (1999).
10 Cited in James Keir, *Treatise on Gases* (1777) 73–4.
11 *Experiments on Air* I 36. Also JP to Richard Price, 19 October 1771, *Scientific Autobiography* 90.
12 For Priestley's links with medicine, see Golinski 105–17, 'Airs and their uses'; Anderson and Lawrence, *passim*; Schofield *Priestley* 256–9.
13 For the controversy, correspondence and JP's replies, see *Experiments on Air* II 293–399. For apparatus, see Anderson and Lawrence, cat. 29, 103.
14 J. J. Schweppe & Co. to MB, 4 August 1802, MBP 253/6 (with thanks to Shena Mason). For Priestley's response to competitors, see Golinski 114–17.
15 Tobias Smollett, *The Expedition of Humphry Clinker* (1771; 1985 edn) 95–6, quoted in Richards 115.
16 *Electricity* xxii, in Martin Fitzpatrick, 'Joseph Priestley and the Millennium', Anderson and Lawrence 31.
17 JW to JB, 3 February 1783, *Partners in Science* 118.
18 *Experiments on Air* I 87–9.
19 Journal of Jonathan Williams junior, [May 1771], *Franklin Papers* 18, 115.
20 [July 1772] *Franklin Papers* 19, 215–16.
21 BF to JP, 10 April 1774, *Franklin Papers* 21, 188–9 and *Experiments on Air* I 321–3.
22 Ibid.
23 ED to BF, 18 July 1772.
24 BF to ED, 1 August 1772, in *EDL* 64, n. 7.
25 'Observations on Different Kinds of Air', *Phil. Trans.* LXII (1772) 147–264.
26 WS to JW, 15 March 1773, MBP 340/29.
27 Sir John Pringle, *A Discourse on the Different Kinds of Air* (1774) 26–7.
28 JWe to TB, 4 October 1772, W. E25–18411.
29 For Bowood, see Simon Schaffer, 'States of Mind: Enlightenment and Natural Philosophy', in G. S. Rousseau (ed.), *The Languages of Psyche, Mind and Body in Enlightenment Thought* (Berkeley, 1990) 250–52.
30 JP, *An Examination of Dr Reid's Enquiry into the Human Mind on the Principles of Common Sense* (1774), and *Hartley's Theory of the Human Mind* (1775). See the clear argument in Klonk 15–18, also Isaac Kramnick, *Republicanism and Bourgeois Radicalism; Political Ideology in Late Eighteenth-Century England and America* (New York, 1990) esp. 80–91.
31 *Experiments on Air* I 10.
32 Boswell *Johnson* IV 238, and Golinski 93.
33 JWe Commonplace Book I, Schofield 93 (see also 184–5).
34 *Experiments on Air* II 36.
35 For Lavoisier, see Jean-Pierre Poirier, *Lavoisier: Chemist, Biologist, Economist* (Philadelphia, 1996) and Partington 363–494.
36 Moilliet and Smith 1.
37 *Experiments on Air* II 276: Golinski 114.
38 *Philosophical Empiricism* (1775) 166–7.

21 'WHAT ALL THE WORLD DESIRES'

1 WS to JW, 5 February 1769, MBP 340/4. See Robinson and Musson 56–61, and C. McCleod, *Inventing the Industrial Revolution: The English Patent System, 1660–1800* (1988).
2 JB to JW, 28 January 1769, *Partners in Science* 19.
3 JW to WS, 28 April 1769, MBP 348/8.
4 MW to JW, 9 August 1768, JWP 4/4.15.
5 MW to JW, 9 August 1768, 2 March 1770, 30 August 1770, JWP 4/4.15, 14, 10.
6 WS to JW, [1772], MBP 348.
7 JW to WS, 18 April 1769, MBP 348/7.
8 JW to WS, 20 September 1771, MBP 348/20.
9 WS to JW, 27 January 1773, MBP 340/26.
10 Quoted in Langford 570.
11 JB to JW, 1 October 1772, *Partners in Science* 33.
12 For Boulton's and Small's share dealings with Day, see Rowland 94–5. The loan to Small was £2,400 against eight shares.
13 17 May 1773, Muirhead *Life* 244. For Roebuck's offer on 17 May, and the signed agreement of 29 May, see Tann 33–5.
14 JW to WS, 25 July 1773, MBP 348/38.
15 Although his plans were shelved, he was consulted by Thomas Telford, who built the Caledonian Canal much along Watt's line.
16 MW to JW, 14 September 1773, JWP 4/4/15.
17 Watt's diary, 1773, B&W/MI/20, 76–7.
18 Ibid.
19 JW to WS, September 1773, MBP 348/38.
20 WS to JW, 8 October 1773, see also 27 October, MBP 340/35–7.
21 JW to WS, 11 December 1773, MBP 348/38.
22 Ibid.
23 JR to WS, 14 March 1774, MBP 334/21.
24 JR to WS, 28 January 1774, MBP 334/28.
25 JW to WS, 29 April 1774, MBP 348/42.
26 JR to JW, 12 November 1774, JWP 4/57.
27 JW to James Watt senior, 11 December 1774, JWP 4/60.
28 King-Hele 121: DAR 227.7:19.
29 James Hutton to JW, [1774], JWP 4/76.
30 WS to JW, autumn 1773, MBP 340.
31 WS to JW, Muirhead *Life* 251.
32 MB to JW, 25 February 1775, JWP 4/59.
33 MB to JW, 28 February 1775, JWP 4/59.
34 MB to Robert Small, MBP 340.
35 Muirhead *Life* 254.
36 JW to MB, February 1775, MBP 348/46.
37 Robison passed on an invitation to become Master Founder of Iron Ordnance to the Imperial Court, 22 April 1771: *Partners in Science* 24.
38 MB to JW, spring 1775, MBP 6/26.
39 ED to JW, 29 March 1775.

40 JW to James Watt senior, 8 May 1775, JWP 4/60.
41 MB to JW, August/September 1775, Muirhead *Life* 257.
42 *Aris's Birmingham Gazette*, 11 March 1776. The engine is in the Thinktank, Birmingham's science and technology museum. (Thanks to Shena Mason's memory of the dinosaur.)
43 MB to JW, summer 1776, Muirhead *Life* 257.
44 JW to MB, July 1776, Dickinson *Watt* 92–3. For the partnership and financing of the engine business, see also James Lord, *Capital and Steam Power, 1750–1800* (1923).
45 MB to JW, 3 July 1776, MBP 348.
46 Boswell *Johnson* II 459.

22 'BANDY'D LIKE A SHUTTLECOCK'

1 Thomas Jefferson to WS, 7 May 1775, MBP 340/42.
2 Bicknell's and Day's source was Michel Adanson's *Voyage to Senegal* (1759), and A. Benezet, *A Caution to Great Britain and her Colonies* (1766), *A Historical Account of Guinea* (1771).
3 For the shift see Roger Anstey, *The Atlantic Slave Trade and British Abolition, 1760–1810* (1975) and Robin Blackburn *The Overthrow of Colonial Slavery, 1776–1848* (1988).
4 *The Dying Negro* (1775 edn), dedication.
5 *A Fragment of an original letter on the Slavery of the Negroes.* David R. Chesnutt et al., *The Papers of Henry Laurens*, XI, quoted in Rowland 85.
6 *Ode to the New Year, The Devoted Legions* (1776) and *The Desolation of America* (1777).
7 TD to MB, 24 March 1777, 18 August 1777, MBP 305/11, 14. Boulton partially repaid Day, while Keir took over most of the loan.
8 Also backed by Roebuck, anxious about Carron. See Money 59–60, 101–2; also Brian D. Barger, 'Matthew Boulton and the Birmingham Petition of 1775', *William and Mary Quarterly* XIII (1956) 26–30.
9 JWe to TB, 8 August and 22 December 1777, W. E25–18805 and 18776. See Reilly 235.
10 JWe to TB, 25 February 1778, W. E25–18814.
11 JWe to TB, 19 March 1778, W. E25–18818.
12 ED to Charles Greville, 12 December 1778.
13 JP to MB, 6 November 1775, *Scientific Autobiography* 153.
14 See Eric Robinson, 'R. E. Raspe, Franklin's "Club of 13" and the Lunar Society', *Annals* XI (June 1955).
15 JWe to TB, 4 May 1776, W. E25–18666.
16 King-Hele 131.
17 JWe to TB, 15 January 1775, W. E25–18584.
18 'Josiah Wedgwood's Journey into Cornwall', Part 2, edited by G. Wills, *Proceedings of the Wedgwood Society* II (1957) 98.
19 Ibid., 87.
20 Seward 68.
21 Ibid., 75.
22 ED to WW, 25 February 1775.
23 Quoted by Nicholas Spice, review of Brian Robins (ed.), *The John Marsh Journals*, in *London Review of Books*, 18 May 2000, 3.
24 MB to JK, 1 March 1776, *Sketch* 56–7 (misdated 1777; see Schofield 142 n. 1).
25 MB to John Wyatt, 24 February 1776, B&F Letter Book G, MBP 137.

26 JWh to MB, 14 August 1776, MBP 366/17.

27 Papers included JR, 'Experiments on Ignited Bodies' and JWh, 'Experiments on Ignited Substances', *Phil. Trans.* LXVI (1776), Schofield 167–8. For the meeting, see MB to JW, 25 July 1776, MBP 348.

28 ED to MB, 5 April 1778.

23 PLANTS & PASSIONS

1 A. G. Morton, *History of Botanical Science* (1981) 257.

2 For Linnaeus see Wilfrid Blunt, *The Compleat Naturalist. A Life of Linnaeus* (1971); Lisbet Koerner, *Linnaeus: Nature and Nation* (Cambridge, Mass., 2000), Koerner's 'Linnaeus in his Time and Place', *Cultures in Natural History*, 145–62.

3 Other works included his *Genera plantarum* (1737) and *Species Plantarum* (1753).

4 This involves only seven parts: four of the flower – calyx, corolla, stamen and pistil – and three of the fruit – ovary, receptacle and seed.

5 The 44 vols. of *Histoire Naturelle* appeared from 1749, the last 8 after Buffon's death.

6 Quoted in Brian L. Silver, *The Ascent of Science* (1998) 259.

7 King-Hele 124.

8 Garden Committee of the Society of Apothecaries, 8 February 1775; in Sue Minter *The Apothecaries' Garden: A History of Chelsea Physic Garden* (2000) 47.

9 *Misc. Tracts* I 19.

10 'Experiments upon the different kinds of Marle found in Staffordshire', *Phil. Trans.* LXIII (1773) 161. (All Withering's articles are in *Misc. Tracts.*)

11 Schofield 123.

12 ED to WW, 13 May 1775.

13 Buffon, *Histoire Naturelle* III (1753), in Morton, *History of Botanical Science*, 291.

14 ED to WW, 13 May 1775. Linnaeus's emphasis on sexuality was attacked by British writers such as William Smellie, principal compiler of *Encyclopaedia Britannica*, 1771, while popularizers played down his sexual imagery.

15 Seward *Letters* III 154; King-Hele 132.

16 ED to Elizabeth Pole, [1775?].

17 Seward 104–5.

18 See *Zoonomia* II 52, 379.

19 King-Hele 139.

20 Ibid. Seward (104–5) describes Elizabeth Pole visiting in spring 1778, bringing the children who were ill after being wrongly dosed for whooping cough.

21 Ibid., 106.

22 Quotations from Seward 125–8.

23 ED to Joseph Cradock, 21 November 1775.

24 Seward 81–2.(The woman is not named in the *Memorials*, but Seward identifies herself in correspondence with Hayley.)

25 Peck, 70.

26 Ibid., 75: 'Effects, Rules, Cautions'.

27 Ibid., 72.

28 Ibid., 73.

29 ED Commonplace Book, 120.

30 Ibid., 8–9.
31 WW to Dr Hall Jackson, New Hampshire, 27 October 1786; WW to J. C. Lettsom [1788], WWP 1/19, 5/111. (Lettsom had recently proposed WW as a corresponding member of the London Medical Society.)
32 JK to ChasD, 2 May 1776, *Sketch* 53.
33 ED to EDjnr, 12 May 1787.
34 See King-Hele 144.
35 *Experiments establishing a Criterion between Mucaginous and Purulent Matter* (1780).
36 ED, 'A new case in Squinting', *Phil. Trans.* LXVIII (1778) 86–96; WW, *An Account of the Scarlet Fever and Sore Throat, or Scarlatina Anginosa particularly as it appeared in Birmingham in 1778* (1779). For Duncan's description see *Medical Commentaries* VII (1780) 193–203.

24 CONQUERING CORNWALL

1 J. Hornblower senior to JW, 23 December 1776, JWP 4/36.5.
2 JW to MB, 14 August 1777, MBP 348/76.
3 AW to AB, 1 September 1777, MBP 357/13.
4 JW to MB, August 1777, MBP 348.
5 JW to John Smeaton, 4 April 1778, in John Lord, *Capital and Steam Power 1750–1800* (1923) 199, which also quotes Watt's 1769 letter about Smeaton's coinage of 'engineer'.
6 JW to MB, 9 August 1777, MBP 348/75.
7 JW to MB, *c.* September 1777, MBP 348/120.
8 JW to MB, 1779, Griffiths 91.
9 JW to MB, 2 February 1778, MBP 348/84.
10 JW to MB, 6 September 1778, MBP 348/108.
11 Orders in 1778/79 included a 40-inch cylinder engine for Hallamanning, two 63-inch engines for Poldice and Wheal Chance; two 58-inch for United; in 1780, five large engines for Wheal Virgin; engines for Dolcoath, Wheal Treasury and Wheal Crenver.
12 See JW to MB, 2 and 8 July 1778, MBP 348/90, 93.
13 JF to MB, July 1778; Dickinson *Boulton* 110.
14 See MB to JW, 25 July 1778, B&W Box 20, Bdl. 5, and Watt Memorandum (1809).
15 Thus Poldice agreed £1,500 for two engines; Consolidated Mines £2,500 for five.
16 Roll 98–101.
17 JW to JB, 12 December 1778, *Partners in Science* 43.
18 For the complicated history see 'The Steam Engine in France', Harris 287–322, and Jacob 174–7.
19 Rolt 76.
20 MB to JW, 1 September 1778, B&W Box 20, Bdl. 5.
21 JW to MB, 31 June 1779, MBP 348/143.
22 JW to MB, May 1779, MBP 348.
23 Rolt 78.
24 AW to MB, 28 June 1781, MBP 357/20.
25 JW to MB, 23 January 1782, Griffiths 88.
26 Smiles 198–9.
27 See JK to MB, 16 January 1777, MBP 319/10. For the agreement, see Statement by William Matthews, 29 May 1775, MBP 334/31.

28 Keir Memorandum (1809).
29 Ibid. For Keir's suggestions, see JK to MB, 14 October 1778, 1 February 1779, MBP 319/27, 39.
30 *Sketch* 66–75, Schofield 158–9. Patent lodged 10 December 1779; the proportions were similar to Muntz metal, a copper sheathing alloy developed in the mid-nineteenth century.
31 Keir Memorandum (1809).
32 These were prints, hand-coloured, then varnished to resemble oils, for wall-paintings and furniture decoration. Despite clients like Mrs Montagu, and some export, they made no profit.
33 See Griffiths 29.
34 Smiles 210.
35 MB to AB, 14 September 1780, MBP, unnumbered.
36 MB to JW, 12 November 1782, MBP Parcel D/4, Folder 4.
37 MB to AnneB, 6 and 17 August 1785, MBP 291/10–11.
38 MB to JW, 25 and 30 September 1780, MBP 143, Letter Book J 16/18.
39 Boulton & Watt took one sixteenth share in Wheal Union and one thirty-second in Pool Mine in 1780, and shares in Consolidated (with John Wilkinson), Poldice, Polgooth, Wheal Jewel and United. In 1784, with Wilkinson, they held a major interest in Chacewater. Boulton transferred a £1,000 share in Polgooth to Wedgwood, October 1782.
40 JW to MB, 31 October 1780, MBP 349/22.

25 DULL EARTH & SHINING STONES

1 JW to JB, 8 January 1779, *Partners in Science* 44.
2 JW to JB, 13 January 1779, *Partners in Science* 50.
3 JB to JW, 7 September 1782, *Partners in Science* 116.
4 JW to John Williams, 16 March 1777, Dickinson *Watt* 177. See John Williams, *An Account of Some Remarkable Ancient Ruins, Lately discovered in the Highlands and Northern Parts of Scotland* (1777). Williams also wrote *The Natural History of the Mineral Kingdom* (1789).
5 JWe, 'Notes & Experiments', W. E26–19109–13.
6 JWe Notebook, 1779, W. E26–32346A.
7 Notes to *BG Ec. Veg.* II 134, 143, 215, 228.
8 ED to JWe, 7 April 1779, King-Hele 158.
9 Ibid.
10 Science Museum cat. 17.
11 JWe to TB, 21 May 1776, W. E25–18643.
12 JWe to TB, 3 November 1777, W. E25–18790.
13 JWe to TB, 14 February 1776, W. E25–18654.
14 *Dictionary of Chemistry* (2nd edn, 1777) III; Schofield 172–3; see also R. E. Schofield, 'Josiah Wedgwood and the Technology of Glass Manufacturing', *Technology and Culture* III (1962) 285–97.
15 'JW Notes and Experiments', [1779], W. E26–19113, 'Attempt to discover Causes of Cords and Wavings in Flint Glass'.
16 The idea of achieving homogeneity through stirring is credited to Louis Guinand, 1798.
17 JK, 'On the Crystallisations observed on Glass', RS Paper, 23 May 1776, Schofield 174–5.
18 JWe Experiment Book, W. E26–19117.315, trial 2956.
19 James Hutton, 'Theory of the Earth', *Transactions of the Royal Society of Edinburgh* I (1788) 304, in Klonk 77. (See chapter 13, n. 35.) See also Porter *Geology* 149–65.

20 JWe to TB, 24 October 1778, W. E26–18857.
21 JWe to TB, 4 November 1778, W. E26–18859.
22 *Inquiry* 191.
23 See Porter *Geology* 125–6.
24 *Inquiry*, 2nd edn, 1786.
25 TD to MB, 29 October 1780, MBP 305/27.
26 JW to MB, 18 October 1781, MBP 349/101.
27 WW, 'Experiments and Observations on the Terra ponderosa, &c.' *Phil. Trans.* LXXIV (1784) 293–311. At first said to come from the lead mine at Alston Moor, Cumberland, but Withering later gave a more likely source, the lead mines in Furness, Lancs, also a source of 'cawk' for Wedgwood. See Iain A. Williamson, 'The Anglezarke Lead Mines', *Mining Magazine*, 108, (March 1963), 133–9.
28 Peck 141: he found a new salt at Nevil Holt, near Uppingham, and translated Bergman on spa waters. For rock samples, see WW, 'An Analysis of two Mineral Substances &c.', *Phil. Trans.* LXXII (1782) 327–36, and JP, *Experiments and Observations* III 215–22.
29 All were credited in Richard Kirwan's *Elements of Mineralogy* (1784), and J. H. Magellan's new edition of Cronstedt's *Mineralogy* (1788), Schofield 306.

26 CREATIVE COPYING

1 *Zoonomia* I 376; *Enlightenment* 603, n. 75.
2 *A catalogue of cameos, intaglios, medals, busts, small statues, and bas reliefs; with a general account of vases and other ornaments, after the antique, made by Wedgwood & Bentley* (5th edn, 1779).
3 ED Commonplace Book, 14–15; also JB to JW, 11 March 1779, *Partners in Science* 61.
4 ED Commonplace Book 33, 38; for guided missiles, see *Essential Writings* 125.
5 The electrical doubler (Commonplace Book 79), inspired by Volta, was a generator with charged central disc of metal between two glass sheets and brass discs that could be winched in and out, passing the charge through a conductor to a Leyden jar.
6 ED Commonplace Book 53, 69, 78.
7 JW to ED, 12 May 1779, King-Hele 155.
8 JW to MB, 28 June 1779, MBP 348.
9 MB to JW, 14 May 1780, B&W Box 20, Bdl. 8.
10 Ibid.

27 SONS & DAUGHTERS

1 JWe to TB, [December 1779], *SL* 248.
2 JWe to TB, 17 July 1778, W. E25–18848.
3 JWe to TB, 1 and 19 August 1778, W. E25–18843 and 18845.
4 JWe to TB, 19 August 1778, W. E25–18845.
5 Ibid.
6 JWe to TB, 24 August 1778, W. E25–18846.
7 ED to JWe, August 1779, DAR 221:1:56.
8 *Wedgwood Circle* 89.
9 ChasD to RWD, 23 September 1777, DAR 227.6.4.
10 JK to ED, [nd], *Sketch* 66.

11 JWe to TB, 7 March 1779, W. E26–18881.
12 JWe to TB, 9 October 1779, W. E26–18929.
13 JWe to TB, 19 December 1779, W. E26–18946.
14 The Reverend Henry Pickering to MB, 5 December 1779, in Eric Robinson, 'Training Captains of Industry: The Education of Matthew Robinson Boulton [1770–1842] and the Younger James Watt [1769–1848]'. *Annals of Science* x (December 1954), 301–13.
15 ED to JWe, December 1779, *EDL* 99
16 JWe to TB, 19 December 1779, W. E26–18946.
17 SuWe to JWe, 13 March 1783, DAR 246.238.
18 See, for example, JW to AW on bleaching, 20 October 1787, JWP Letter Book 1.
19 *Schimmelpenninck* 6.
20 JWh to MB, 30 October 1775, MBP 366/15.
21 Hannah More, *Essays on Various Subjects* (1791) 9.
22 ED to Richard Dixon, 30 March 1791.
23 *Plan* 11.
24 Ibid., 46.
25 Keir *Day* 25–6.
26 SuWe to JWe, W. L95/17561A.232.
27 Letter to the author from Marilyn Butler, October 2001.
28 RLE and ME, *Practical Education* (2nd edn, 1801) I 95–6.
29 JWe to TB, 18 December 1779, W. E26–18945.
30 Published as *Hartley's Theory of the Human Mind, or the Principle of the Association of Ideas* (1775). See Butler 59–67.
31 RLE and ME, *Practical Education* (1798) II 734.
32 RLE to JP, [early 1780], Butler 63.
33 Edgeworth *Memoirs* 237.
34 Ibid., 238.
35 RLE to JWe, in JWe to TB, 10 May 1780, W. LHP.
36 Edgeworth *Memoirs* 243.
37 ED to AS, 6 November 1780, and Seward 133–45.
38 Seward 148–9.
39 Darwin–Pole children: Edward, 31 January 1782; Violetta, 23 April 1783; Emma, 24 August 1784; Francis Sacheverel, 17 June 1786; John, 5 September 1787; Henry, 10 April 1789, d. 1790; Harriot, 5 July 1790.
40 Memoir of John Rylands, Gibbs 139–40.
41 Ibid., 140.
42 For Esther's father Richard (1705–1757) see Thomas Lowndes, *Select Miscellaneous Productions of Mrs Day and Thomas Day, Esq.* (1805). For the Wakefield connections, see J. Hunter, *Familiae Minorum Gentium*, edited by J. W. Clay (1894) I 49–57 (with thanks to Deborah Scriven of Wakefield Library). Also see *Wright of Derby* 69.
43 The wife of John Milnes of Wakefield was a Shore, whose relations were banking partners of the Roebucks. With thanks to Steve Uglow.
44 TD to EDjnr, 29 January 1779. BL Add. MSS 29, 300, f55.
45 EDjnr to RWD, [1776], DAR 227.6.2.
46 ME, *Letters for Literary Ladies* (1795) 22–31.
47 Butler 204–5.

48 Keir *Day* 7.
49 See Gignilliat 264.
50 On Day's 'softened Stoicism' see Barker-Benfield 148–53.

28 BRINGING ON THE ARTISTS

1 JWe to TB, 12 November 1780, *SL* 261.
2 Ralph Griffiths to JWe, 25 November 1780, W. E30–22420.
3 Reilly and Savage 335.
4 Christie & Ansell sale catalogue, 1781, in *Etruria* 5, no. 3 (Spring 2002), 24.
5 JWe to Charles James Fox, 23 July 1785, *SL* 284.
6 JW to Lady Templetown (draft), 27 June 1783, E26–18958.
7 See Carola Hicks, *Improper Pursuits: Lady Di Beauclerk, Woman and Artist* (2001).
8 John Flaxman to JWe, 8 July 1782, W. E2/30187.
9 John Flaxman to JWe, 22 August 1782, Samuel Smiles, *Josiah Wedgwood* (1894) 210.
10 JWe to ED, 28 June 1779, W. E26–19001.
11 David Irwin, *John Flaxman 1755–1826* (1979) 20–21.
12 John Flaxman to JWe, 5 February 1784, W. E2/30188.
13 A later story suggested Stubbs saw this in North Africa: Constance-Anne Parker, *Mr Stubbs the Horse Painter* (1971), 72. He was more likely inspired by an antique sculpture in Rome. See Francis Haskell and Nicholas Penny, *Taste and the Antique* (1981) 251, and *Stubbs* 90.
14 Unfortunately this mix melts at the temperature required to separate the dirty varnish from old canvases: Robert Shepherd, 'Stubbs: a Conservator's View', *Stubbs*, 20–21.
15 Paul Mellon Collection, *Stubbs* 123: the impressed mark 'Wedgwood' can be seen at the edge.
16 JWe to TB, 4 November 1777: see also 17 October 1778, W. E26–18856.
17 JWe to TB, 30 May 1779, Bruce Tattersall, *Stubbs and Wedgwood* (1974), 19. The early small plaques were about 22 inches by 15 inches.
18 JWe to TB, May 1779, W. E26–18890.
19 JWe to TB, 21 October 1780, Tattersall, *Stubbs and Wedgwood*, 21.
20 JWe to TB, 25 September 1780, Reilly 300.
21 For Stubbs and 'gentlemen', see Judy Egerton in *Stubbs* 14–19.
22 JWe to TB, 13 August 1780, Farrer II 475.
23 See Louise Lippincott, 'Expanding on portraiture. The market, the public and the hierarchy of genres', in Ann Bermingham and John Brewer (eds), *The Consumption of Culture 1600–1800: Image, Object, Text* (1995) 85.
24 See JWr to JWe, 29 May 1783 and 23 October 1783: *Wright of Derby* 131.
25 JWr to JWe, 31 December 1783, Benedict Nicolson, *Joseph Wright of Derby, Painter of Light* (1968) I 147.
26 Ibid.
27 *The Corinthian Maid*, National Gallery of Art, Washington. See *Wright of Derby* 132–3. See Anne Bermingham, 'The Origin of Painting and the Ends of Art: Wright of Derby's *Corinthian Maid*', in John Barrell (ed.), *Painting and the Politics of Culture* (1992) 135–66.
28 William Hayley, *An Essay on Painting* (1778); *Wright of Derby* 133.
29 JWe to TB, 5 May 1778, Daniels 55.
30 JWe to JWr, 29 April 1784, W. E26–18966.
31 See Daniels 56–8.

32 They never met at Etruria, although the sleeping greyhound in *The Corinthian Maid* echoes the enamel plaque of *Labourers* made for Wedgwood in 1781.

29 FROM THE NATION TO THE LAND

1 *Speech of Thomas Day Esq., As delivered to the Freeholders of Essex at Chelmsford* ([24 January] 1780).
2 See Roy Foster, *Modern Ireland, 1600–1972*, 241–58.
3 JWe to TB, 13 October 1779, W. E26–18930.
4 See John Goodchild, 'James Milnes; An eighteenth-century Merchant Prince', *Wakefield Historical Society Journal* 10 (1983), n.p.
5 Key texts were Rousseau, *Discours sur les sciences et les arts* (1750) and Adam Ferguson, *Essay on the History of Civil Society* (1767). See Butler *Romantics* 23.
6 *Speech of Day at Chelmsford.*
7 *Two Speeches of Thomas Day, Esq., at the General Meetings of the Counties of Cambridge and Essex* ([25 March, 25 April] 1780). See also Gignilliat 181.
8 JW to JB, 9 June 1780, *Partners in Science* 94.
9 Reginald Blunt, *Mrs Montagu, 'Queen of the Blues': Her Letters and Friendships from 1762–1802* (1923) II 96–7.
10 JWe to TB, 9 September 1780, *SL* 256.
11 TD to MB, 3 May 1782, MBP 305/35.
12 JWe to RLE, 25 January 1784, *SL* 274.
13 RLE to ED, 7 September 1794, Edgeworth *Memoirs* 337–8.
14 ED to RLE, 20 February 1788.
15 Published 1776. Other Edinburgh works included Francis Home, *Principles of Agriculture and Education* (1757); William Fordyce, *Elements of Agriculture and Vegetation* (1765); Alexander Hunter, *Georgical Papers*, 4 vols (1770–72); James Anderson, *Essays relating to Agriculture* (1775).
16 TD to RLE, [early 1787]: misdated 1789 in Edgeworth *Memoirs* 306; RLE's reply in NLI is 5 February 1787.
17 *Sandford and Merton* II 14.
18 TD to RLE, [1784], Rowland 261.
19 *The Black Book of Edgeworthstown* 251–2, in Butler 82.
20 Edgeworth *Memoirs* 260–61.
21 Butler 87.

Fourth Quarter

30 FIRE

1 JWe to MB, 10 March 1781, MBP 361/9.
2 WW to Mr Rathbone junior, 16 February 1781. WWP 10.
3 MB to LH, 6 September 1781; Smiles 373.
4 Muirhead *Mechanical Inventions* II 123.
5 ED to JW, 6 January 1781.
6 Darwin returned to this when writing to Day about *Divine Benevolence asserted* (1781) by Thomas Balguy, Bishop of Winchester, which 'makes it evident that every thing was made by the Lord, with the design of producing happiness'. ED to TD, 16 May 1781.

7 ED to RLE, 20 February 1788.

8 ED to MB, 27 January 1782.

9 *SchimmelPenninck* 251.

10 Schofield 270–75. Priestley sent Galton's paper to the RS, April 1781. The colour wheel is usually credited to Thomas Young. RWD used SG's findings in 'New Experiments in the Ocular Spectra of Light', *Phil. Trans.* LXXVI (1786) 313–48, reproduced in *Zoonomia* I (1794) 534–66; his own paper was not published until 1799, in the *Monthly Magazine.*

11 *SchimmelPenninck* 36–7.

12 A. M. Broadley and Lewis Melville (eds), *The Beautiful Lady Craven, the Original Memoirs of Elizabeth, Baroness Craven, afterwards Margravine of Aspach* [1913] I 46; Schofield 228.

13 JW to MB, 26 October 1782, MB 350/59.

14 Addition, 30 April 1783, in H. Moyes to WW, RSM MS534.62–3. The note in Seward *Letters* I (1785) 97, calls Mary Knowles 'the celebrated Quaker lady who worked the King's picture so admirably in worsted'. Thanks to Norma Clarke.

15 C. Wright to WW, 30 January 1785, RSM MS534.69.

16 Catherine Wright to WW, 11 August 1785, RSM MS534.79. For Wright, see Patricia Fara and Anne Secord in M. Ogilvie and J. Harvey (eds), *The Biographical Dictionary of Women in Science* (2000) 1404–5.

17 Faujas de St Fond, *Journey through England and Scotland in 1784* (1797; 1907 edn) II 357.

18 JP, *Experiments and Observations Relating to the Analysis of Atmospherical Air &c.* (Philadelphia, 1796), quoted by John McEvoy in 'Continuity and Discontinuity in the Chemical Revolution', in *Osiris* 4 (1988) 195.

19 See John McEvoy, 'Causes and Laws, Powers and Principles: The Metaphysical Foundations of Priestley's Concept of Phlogiston', in Anderson and Lawrence 55–71.

20 MB to JWe, 30 March 1782, MBP 146, Letter Book M, 271.

21 JWe to MB, March 1782, MBP 361/12. See *Scientific Correspondence* 97.

22 See B. Bensarde-Vincent and F. Abbi (eds), *Lavoisier in European Context: Negotiating a New Language for Chemistry* (Nantucket, 1995).

23 Lissa Roberts, 'The Death of the Sensuous Chemist: The "New" Chemistry and the Transformation of Sensuous Techonology', *SHPS* 26, no. 4 (December 1995), 503–29.

24 Muirhead *Mechanical Inventions* II 225–6.

25 For the language debate, see Golinski 130–52, and summary in *Making Natural Knowledge: Constructivism and the History of Science* (Cambridge, 1998), 118–19.

26 Quoted in E. Robinson, 'Eighteenth-century Commerce and Fashion: Matthew Boulton's Marketing Techniques', *EHR* XVI (1963) 49.

27 JW to Richard Kirwan, 14 November 1783; John Whitehurst, *An Attempt toward Obtaining Invariable measures of Length, capacity and Weight, from the Measuration of Time, independent of the Mechanical Operations Requisite to ascertain the Centre of Oscillation, or the True Length of Pendulums* (1787); Schofield 256–61.

28 For twenty years, Chisholm was secretary to William Lewis, chemist to the RSA, author of *Philosophical Commerce of Arts* (1763–65) and *Chemical and Mineral History of Iron* (1767–70).

29 See Science Museum cat. 18–19, 34–37. Wedgwood's papers included, 'An attempt to make a thermometer for measuring the higher Degrees of Heat', *Phil. Trans.* LXXII (1782) 305–26; 'Attempt to compare and connect the Thermometer for Strong Fire . . . with the common Mercurial Ones', *Phil. Trans.* LXXIV (1784) 358–84; 'Additional Observations', *Phil. Trans.* LXXVI (1786) 390–408.

30 JP to JWe, 21 March 1782, *Scientific Autobiography* 206.

31 'Frigorific Experiments on the mechanical Expansion of Air, explaining the Cause of the great Degree of Cold on the Summits of high Mountains, the sudden Condensation of Aerial vapour, and of the perpetual Mutability of atmospheric Heat', *Phil. Trans.* LXXVIII (1788) 43–52. See *Essential Writings* 181–8.

32 Arthur Donovan, *Antoine Lavoisier: Science, Administration and Revolution* (Oxford, 1993) 154.

33 ED to JW, 6 January 1781.

34 Gibbs 155.

35 Described in Warltire's letter to JP, 18 April 1781, Appendix to *Experiments and Observations* II (1781) 395; Muirhead *Life* 315–16.

36 JW to MB, 10 December 1782, MBP 350/72. For Watt's agitation, see letters to Jean de Luc between December 1782 and December 1783; to his brother-in-law Gilbert Hamilton, 26 March 1783, and to JP, 2 May (in Muirhead *Life*); to JB, 21 April 1783, 12 April 1784 (*Partners in Science*), and *Phil. Trans.* (1784), 331–3.

37 JP to JWe, 8 January 1783, Gibbs 157.

38 JP to Joseph Banks, 14 December 1783, *Scientific Autobiography* 229.

39 JW to Joseph Fry,15 May 1784, Smiles 305–6.

40 Donovan, *Lavoisier*, 159–65.

41 JK to ED, 15 March 1790, *Sketch* 108.

31 BOULTON'S BLAZING SKIES

1 ED to JWe, [summer 1782], King-Hele 183.

2 ED to MB, 26 December 1782.

3 ED to MB, 4 March 1783.

4 MB to JW, 21 June 1781, B&W D/3, Folder 2.

5 MB to JW, 7 August 1781, B&W D/3, Folder 3.

6 MB to JW, 7 December 1782, B&W Parcel D/4, Folder 5.

7 For patents cf. 1781, 1782, 1784 and 1785, see Musson and Robinson 89–125. The little 'planet', a fixed cog wheel on the end of the pump-rod, circled round the 'sun', a larger rotating cog attached to a drive-shaft, turning twice or more at each stroke.

8 The 1781 patent also covered a double engine, toothed piston rack and steam-wheel.

9 JW to MB, 30 June 1784, MBP 350/113.

10 MB to WM, 19 June 1782, MBP 328. Calculations by their foreman, Zaccheus Walker, showed that on the original £20,000 investment in Soho, the loss stood at £11,000.

11 MB to WM, 28 June 1781, MBP 146, Letter Book M.

12 WM to MB, 2 April 1782, MBP 328/67.

13 MB to JW, 2 November 1782, MBP 146, Letter Book M, 396.

14 John Southern to MB, 30 July 1782, Fitton 57.

15 ED to MB, 4 March 1783.

16 John Wesley, *Journal*, July 1782.

17 Gilbert White, *Natural History of Selborne* (1789), Letter LXV.

18 *Weather Journals* 88.

19 MB to Elizabeth Fothergill, 2 August 1783, MBP 307/296.

20 SG to Mrs Barker, 13 July 1783, MBP 308/75. See also Elizabeth Fothergill to MB, 18 July 1783, MBP 307/304.

21 MB to AB, 9 November 1780, MBP, unnumbered.
22 MB to AB, 3 May 1781, MBP 279/101. (For trivial details, see MBP 279/66–91.)
23 MB to AnneB, 17 August 1785, MBP 291/11.
24 MB to JW, 28 October 1783, Fitton 65.
25 MB to JW, September 1783, B&W Box 20, Bdl.10.
26 For ED's notes on winds, see ED *Commonplace Book* 46–7, 121–3, 137–8, 147–8.
27 Schofield 250. He included a jovially coarse paper on gases to send to Priestley, 'who is apt to give himself airs (i.e. fixed, dephlogisticated , &c, &c) and has a kind of right in everything his friends produce upon that subject'.
28 JW to JB, 25 September 1783, *Partners in Science* 128.
29 ED to MB, 17 January 1784.
30 The Charles calculations were published in the *Journal de Paris*, December 1783; the Faujas de St Fond book, *Descriptions des expériences de la Machine Aérostatique de MM de Montgolfier*, was sent to Boulton in July 1784 by a M. Pradeux.
31 Faujas de St Fond, *Journey through England and Scotland in 1784* II (1797; 1907 edn) 351.
32 For fun, see the account of this on the website 'printsgeorge.com/ArtEccles_Aeronauts 4.htm'. In fact Mrs Sage, at fourteen stone, was so heavy that Lunardi had to get out, and during the flight of 13 miles to Harrow she knelt on the barometer seen in Biggins's hand, and broke it.
33 Gilbert White to Ann Barker, 19 October 1784, Richard Mabey, *Gilbert White* (1986) 195.
34 RLE to ME, 12 February 1784, Butler 92.
35 Smiles 270.
36 JW to James Lind, 26 December 1784, JWP C1/15.
37 *Aris's Birmingham Gazette*, 3 January 1785.
38 Ibid., 10 January 1785.
39 *BG Plants* II 41–58.
40 'Directions for Working Rotative Engines *c.*1784', Dickinson and Jenkins, Appendix III, 399–400.
41 For Argand, see John J. Wolfe, *Brandy, Balloons and Lamps: Ami Argand, 1750–1803* (Carbondale, 1999).
42 MB to JW, 24 February 1786, B&W Box 20, Bdl 12, Folder 1.
43 March 1786. Quoted in Peter Matthias, 'An Industrial Revolution in Brewing, 1700–1830', *The Transformation of England*, 225. Joseph Delafield of Whitbread told his brother that the old wheel needed six horses: they had ordered an engine equal to the power of ten, by Watt's calculation, 'and the work it does we think is equal to 14 horses'. The Whitbread engine was shipped to Australia in 1888, and is demonstrated daily at the Powerhouse Museum, Sydney.
44 JW to John Robison, 20 January 1784, *Partners in Science* 133.

32 THE LINNAEAN ROW

1 Jedediah Strutt took out a patent taken in 1758 for making ribbed stockings by machine. William Strutt succeeded Darwin as President of Derby Philosophical Society in 1802.
2 See 'The Derby Philosophical Society', Musson and Robinson 190–99.
3 SuWe to JWe, 13 March 1783, DAR 264:238.
4 See King-Hele 197–8. Annual subscription was one guinea, and the fines were substantial.
5 King-Hele 191, and *EDL* 129–31.

6 'An Account of an artificial Spring of Water', *Phil. Trans.* LXXV (1785) 1–7.
7 *Zoonomia* IV 90.
8 ED to Carl Linnaeus junior, 14 September 1781.
9 See notes to *EDL*, 114–15.
10 ED to Sir Joseph Banks, 29 September 1781.
11 [ED], Botanical Society of Lichfield, *A System of Vegetables* (Lichfield 1783), Preface.
12 [Samuel Goodenough], *Monthly Review* LXXII 401–10 and LXIII, 1–13.
13 WW to Ralph Griffiths, RSM MS534.75. By the mid 1790s Goodenough was exchanging plant specimens with Withering: see MS534/166–73.
14 Stokes's early writings include Botanical Commentaries to *Fothergill Catalogue* (1782); *de Aere Dephlogisticato* (Edinburgh, 1782), 'On the Nomenclature of Fossils', Edinburgh University Natural History Society minutes (1782) and *Hortus Uptonensis: a Catalogue of Stove and Green-House Plants in Dr Fothergill's Garden at Upton* (*c.* 1785).: Schofield 223. For Lunar members at RS meetings, see *Journal Book of the Royal Society* (1780–82).
15 JS to Carl Linnaeus, June–21 August 1783, Schofield 226.
16 His later work included *A Botanical Materia Medica* (1812) and *Botanical Commentaries* (1830).
17 *SchimmelPennick* 43.
18 ED Notebook, DAR 227.2.11.
19 ED Commonplace Book 109.
20 Ibid., 90.
21 ED to Joseph Banks, 23 February 1782. The books requested were H. L. Duhamel's *La Physique des Arbres* (1758) and Charles Bonnet's *Recherches sur l'Usage des Feuilles des Plantes* (1754).
22 *Phytologia* 45.
23 ED Commonplace Book 72.
24 *BG* 'Proem', vii.
25 *BG Plants* I 51–6. For poetry related to natural philosophy, see William Powell Jones, *The Rhetoric of Science; A Study of Scientific Ideas and Imagery in Eighteenth-Century English Poetry* (1966).
26 *BG Plants* I 457–61.
27 Ibid., note to *Rubia*, I 373.
28 Ibid., note to *Curcuma*, I 65.
29 ED to Joseph Banks, 24 October 1781.
30 ED to Joseph Johnson, 23 May 1784.

33 PROTECTING OUR INTERESTS

1 TWe to JWe, 11 March 1783, *SL* 268.
2 *An Address to the Young Inhabitants of the Pottery* (Newcastle, 1783).
3 See Harris 481.
4 MB to Thomas Wilson, 16 December 1784, Dickinson *Boulton* 130.
5 MB to Samuel Garbett, [late 1784], MBP 305.
6 JWe to MB, 21 February 1785, W. E26–19098.
7 JWe to RLE, 3 October 1785, *SL* 285.
8 Hornblower claimed that he had thought of this in 1776, and that the second cylinder was not a 'separate condenser'. See Hugh Torrens, 'Jonathan Hornblower (1753–1815) and the steam engine: A historiographic analysis', in Denis Smith (ed.) *Perceptions of Great Engineers: Fact and Fantasy* (1994).

9 MB to JW, 7 July 1781, MBP 146 Letter Book M, 95.
10 JW to MB, 28 September 1782, Dickinson *Watt* 133.
11 JW to MB, late 1782, Smiles 256.
12 Dickinson and Jenkins 327. See Fitton 96.
13 Thomas Carlyle, 'Chartism', quoted in *Wright of Derby* 197–8.
14 MB to JW, 7 August 1781, Fitton 98.
15 *BG Plants* II 97–104.
16 George Canning et al., 'The Loves of the Triangles', Lloyd Sanders (ed.), *Selections from the Anti-Jacobin* (1904) 101.
17 ED to MB, 25 January 1785.
18 Fitton 138.
19 See E. Robinson, 'James Watt and the Law of Patents', *Technology and Culture* 13 (1972) 127–8; Watt's 'Thoughts upon Patents' is in Robinson and Musson 214–28.
20 MB to JW, 2 September 1786, MBP. For Murdoch's steam carriage see Griffiths, 142–68.
21 JWe to JW, 30 January 1784, *SL* 279. In April he drafted a memo suggesting that Committees of Commerce be encouraged to correspond and work together.
22 MB to Joseph Banks, 10 February 1787, Tann I 157.
23 Soho Visitors' Book (1781–83), MBP 204/206.
24 EM to MB, 28 September 1780, MBP 330/33 (see correspondence 1780–82, MBP 330/29–60).
25 MB to Mary Linwood, MBP 148, Letter Book O, 66–7.
26 JW to MB, 10 March and 17 April 1786, MBP 351/52 and 64.
27 MB to JW, 18 July 1785.
28 MB to JW, 26 July 1785.
29 MB to AnneB, 6 August 1785, MBP 291/10.
30 JWe to RLE, 13 February 1786, *SL* 291.
31 House of Commons *Journals* XLII 266–74, quoted in Witt Boden, *Industrial Society in England towards the End of the Eighteenth Century*, 2nd edn (1965).
32 *Birmingham Life* I 329.
33 MB to AnneB, 18 November 1786, MBP 291/16.
34 MB to C. L. Berthollet, [March] 1788, MBP. See also Clow and Clow, 189–90, and 'The Introduction of Chlorine Bleaching', Musson and Robinson 251–337.
35 JWe to JW, 27 March 1787, MBP Letter Books (Office) 1786–88, 159, in Boden, 191. For manufacturers' views on the General Chamber, see Money 44.
36 William Eden to Lord Sheffield, February 1787, Reilly 276.
37 MBP, Letter Book (Office), 1786–87.
38 TD, *A Letter to Arthur Young, Esq., on the Bill now depending in Parliament to prevent the exportation of Wool* (1788).
39 Ibid.
40 Ibid.

34 FAMILY & FEELING

1 ED to MB, 12 March 1789.
2 ED to JWe, 13 March 1784.
3 SuWe to JWe, 1 February 1785, DAR 264:240.

4 ED to JWe, [?23 March 1784].
5 Eliza Meteyard, *A Group of Englishmen (1795–1815)* 35.
6 See JW to Margaret Miller, 25 April 1791, JWP LB2/25.
7 JW to JWjnr, 16 January 1784, and AW to JWjnr, 17 July 1785; JWP, LB1.
8 JWjnr to MB, 26 March 1789, MBP 353/1.
9 MB to MRB, 19 December 1787, in Musson and Robinson 110–11.
10 Dent 221.
11 *Aris's Birmingham Gazette*, 14 June 1781. Within a year, the stock was 500 titles; by 1786 it reached 1,600 and 3,400 by 1790. Later a subscription was raised to build a proper library (on land leased from Withering), opened in 1797.
12 Gibbs 174.
13 MB Diary, MBP 381 (1786).
14 JWe to RLE, 24 December 1786, *SL* 301.
15 Joseph Spence, *Grand Repository of the English Tongue* (1775). The first vol. of Horne Tooke's *Diversions of Purley* appeared in 1786 (2nd edn, 1798; vol. 2, 1805): see Darwin, *Temple of Nature*, n. XIV. See Olivia Smith, *The Politics of Language, 1791–1819* (Oxford,1984).
16 On sensibility and reform see Langford 461–518, and Barker-Benfield 215–86.
17 Anna Laetitia Barbauld, in Langford 504.
18 See Butler *Romantics* 31–8.
19 Francis D. Klingender, *Art and the Industrial Revolution* (1947) 72–3.
20 *Enlightenment* 361. See J. Walvin, *Slavery and British Society, 1776–1848* (1982).
21 JWe to JW, 1788, JWP C1/8.
22 JWe to AS, February 1788, *SL* 319–10.
23 JWe to BF, 29 February 1788, *SL* 311.
24 Neil McKendrick, 'Commercialization of Fashion', cited in Barker-Benfield 213.
25 King-Hele, 231. For Wedgwood's wooden leg, see JWe to ED, 27 June 1788, *SL* 314.
26 *BG Plants* II 421–30. See David Turley, *The Culture of English Anti-Slavery, 1780–1860* (1991).
27 *BG Plants* III 439–42.
28 ED to JWe, 13 April 1789, DAR 227.1.112. King-Hele 232.
29 Quoted in Edward Royle and James Walvin, *English Radicals and Reformers, 1760–1848* (Brighton, 1982) 36.
30 *Petition against the Abolition of the Slave Trade* (1789): Birmingham City Library.
31 MB to JW, 19 April 1783, B&W Box 20 Bdl 10.
32 John Dawson to B&W, 9 November 1790, B&W Box 4.3.12.
33 SuWe to JWe, April 1777, W. L95/17651A244.

35 GRAND PROJECTS

1 For a detailed study see R. Doty, *The Soho Mint and the Industrialisation of Money* (1998), and also J. O. Sweeney, *A Numismatic History of the Birmingham Mint* (Birmingham, 1981).
2 MB to Lord Hawkesbury, 14 April 1789, MBP 237/13.
3 For his evidence in January 1788 and later, see the PRO 'Papers relating to copper 1798–1802', Dickinson *Boulton* 136–56 *passim*.
4 Thanks to George Demidowicz for details on the mint: the information will be in the forthcoming book, George Demidowicz (ed.), *Matthew Boulton and Soho: House, Manufactory, Mint and Landscape*.

5 *London Magazine* (1787), in R. C. Bell, *Commercial Coins, 1787–1804* (Newcastle, 1963), 145. Bell illustrates numerous tokens, with detailed annotation.
6 Patent no. 1757, 8 July 1790.
7 *BG Ec. Veg* I 283–8.
8 See Reilly and Savage, 276; W. Mankowitz, *The Portland Vase and the Wedgwood Copies* (1952); D. E. L. Haynes, *The Portland Vase*, 2nd edn (1975), and A. Dawson, *Masterpieces of Wedgwood* (1984; rev. edn, 1995) 112–25.
9 Quoted in *Vases and Volcanoes* 187.
10 JWe to Sir William Hamilton, 24 June 1786, W. E26–18976.
11 JWe to Sir William Hamilton, 16 June 1787, *SL* 307.
12 ED to JWe, October 1789, DAR 221.1:117, King-Hele 243.
13 See J. Wedgwood, *Account of the Barberini, now Portland vase, With various explications of its bas-reliefs that have been given by various authors* (1788); also JWe to Sir William Hamilton, [before 29 April 1788], *SL* 313.
14 Martin Priestman, *Romantic Atheism: Poetry and Freethought, 1780–1830* (1999) 72–3.
15 Reynolds [Sci. Mus. 61]. Tom Byerley and Jos Wedgwood took a copy to The Hague, where William Eden (now Lord Auckland) was Ambassador, and then to Germany. The original was returned to the family and given to the British Museum in 1810: in 1845 it was smashed by a deranged visitor, and painstakingly restored.
16 'Appeal to the Faculty Concerning the case of Mrs Houlston, February 1789'. See Peck 108–22.
17 *SchimmelPenninck* 177.
18 MB to ED, 4 January 1790, *EDL* 198, n. 1.
19 JW to ED, 24 November 1789, JWP LB2.
20 ED to JWe, 9 July 1791, DAR 227.3.39.
21 King-Hele 236.
22 JK to ED, 15 March 1790, *Sketch* 111.
23 *BG Ec. Veg.* IV 207n.
24 *Phytologia* 103.
25 Strongly argued in McNeil 16–21. But Francis D. Klingender, *Art and the Industrial Revolution* (1947) 10–20, shows the convention dated back to Yalden (1710) and Dalton (1755).
26 *BG Ec.Veg.* I 101–2. See Butler *Romantics* 78.
27 For Darwin's deism, see Priestman, *Romantic Atheism*, 62–80, and Priestman's excellent chapter, 'Masters of the Universe', for the influence of Lucretius and d'Holbach's *Système de la nature* (1770) on William Jones, Richard Payne Knight and Darwin.
28 *Zoonomia* III, v–vi.
29 For complex influences on *Zoonomia*, see McNeil 92–105.
30 King-Hele 298.
31 *Zoonomia* II 505 (and in *Essential Writings* 87).
32 Galvani's first experiment was done in 1781; his researches were published in 1796.
33 The 'animalculae' from paste are also in d'Holbach; Priestman, *Romantic Atheism*, 16.
34 Mary Shelley, Preface to *Frankenstein, or The Modern Prometheus: The 1818 Text*, edited by Marilyn Butler (Oxford, 1994).
35 *Zoonomia* II 237.
36 *Temple of Nature* IV 451–4. For Darwin and 'Progress' see *Enlightenment* 437–45.

36 TURNING

1 *SchimmelPenninck* 241; *Enlightenment* 438.
2 RLE to JWe, 15 May 1788, W. EB1.474.
3 ED to RLE, 20 February 1788, *EDL* 179.
4 *SchimmelPenninck* 125–6.
5 ED to JW, 18 November 1788. *Elements of Natural History and Chemistry* (4 vols) by A. F. de Fourcroy was published in 1788.
6 *Sketch* 94.
7 Ibid., 93–4.
8 ED to MB, 18 January 1789.
9 JWe to ED, 15 March 1789, *SL* 316.
10 Dent 197.
11 *SchimmelPenninck* 118.
12 Gignilliat 349.
13 Edgeworth *Memoirs* 311.
14 ED to RWD [October 1789].

37 RIOTS

1 Coleridge, 'The Fall of the Bastille', Richard Holmes, *Coleridge: Early Visions* (1989) 33.
2 Conclusion to *Marriage of Heaven and Hell* (1790–92). See also *Visions of the Daughters of Albion* (1791–93), *The French Revolution* (1790–91) and *America* (1791–93).
3 JWe to ED, [July 1789], Farrer III 92–3.
4 *BG Ec. Veg.* II 377–94.
5 ED to JW, 19 January 1790.
6 JW to JWjnr, 11 October 1789, JWP W/6.
7 *Sketch* 118.
8 *Enlightenment* 448.
9 Dent 224.
10 Alexander Armstrong, Whip-maker, and Abel Sharp, Spur-maker, *Very Familiar Letters to John Nott, Button Burnisher* (Birmingham, 1790), letter VI, 12–13, quoted in Money 147.
11 JP to JWe, 26 February 1791, *Scientific Autobiography* 255.
12 *Letters to the Right Honourable Edmund Burke*, 2nd edn (1791),154. Anderson and Lawrence 31.
13 Gibbs 195–6.
14 For these issues and Hutton's position, see Money 11–22.
15 JK, letters of 20 and 24 July, appended to some editions of *An Authentic Account of the dreadful Riots in Birmingham* (Birmingham, August 1791).
16 *Authentic Account* 5.
17 Hutton *Life* 107–8.
18 *An Appeal to the Public on the Subject of the Riots in Birmingham* (1792).
19 Dent 230.
20 The Reverend W. Jesse to the Earl of Dartmouth; Gibbs 202.
21 *The Times*, Sunday 17 July 1791, Dent 243.
22 WW to Sir Henry Gough Calthorpe, 21 September 1791, Peck 192.
23 *Authentic Account* 21.

24 George Rude, *The Crowd in History* (1985 edn) 145.
25 MB to Charles Dumergue, 18 August 1791, MBP 229/203.
26 MB to F. Swediaur, 21 July 1791, MBP 346/64.
27 Published 19 July 1791, Richard Godfrey, *James Gillray: The Art of Caricature* (2001), 95.
28 B. Dobree (ed.), *The Letters of King George III* (1935) 212.
29 See *The Report of the Trials of the Rioters; taken in shorthand, 11 August 1791* (1791).
30 JWe to JP, July 1791; *Scientific Correspondence* 112.
31 SG to JP, 5 September 1791; *Scientific Correspondence* 118.
32 ED to JWe, 25 July 1791, DAR 227.1:138, King-Hele, 256.
33 *EDL* 215–16.
34 JW to JP, [September 1791], Muirhead *Mechanical Inventions* II 243–4.
35 Not all the witnesses were needed; in the end Galton and Johnson appeared. Priestley's claim was £4,083 10s 3d; his apparatus was valued at £602 2s; books £432 15s 6d; papers and manuscripts £420 15s; household goods £1,307 8s; building damage £1,267 4s. He was allowed damages of £2,096 10s raised through the rates, eventually paid two years later. See also Dent 250–56, and Gibbs 212–13.
36 JP to JW, *Scientific Correspondence* 135; see also letters relating to the riots, 120–38.
37 *A Letter from Timothy Sobersides, Extinguisher-Maker, at Wolverhampton, to Jonathan Blast, Bellows-Maker, at Birmingham* (1792), extensively quoted in Moillet and Smith, Appendix B, 66–72.
38 Dent 229.
39 Money 147. The 'cabal' of Banks and the RS had already been satirized by John Wolcott ('Peter Pindar') and Francis Garden: see W. P. Jones, *The Rhetoric of Science* (1966) 195–9.
40 *Thoughts on French Affairs* (1791), quoted in Maurice Crosland, 'The Image of Science as a threat: Burke versus Priestley and the "Philosophic Revolution"', *BJHS* (1987) 20, 281.
41 *Letters to a Noble Lord*, Crosland, 'The Image of Science', 294.
42 *Dictionary of Chemistry* (1789) Preface.
43 Edmund Burke, *Reflections on the Revolution in France*, edited by Conor Cruise O'Brien (Harmondsworth, 1981) 93–4.
44 Ibid., 90.
45 Thomas Belsham, *Memoirs of the late Theophilus Lindsey* (1812) 375, in Martin Fitzpatrick, 'Joseph Priestley and the Millennium', Anderson and Lawrence 36, n. 30.
46 Money 148.

38 HANDING ON

1 JW to JB, 23 November 1791, *Partners in Science* 185.
2 For the von Breda paintings see MBP 223/28–39. For Wright, see *Wright of Derby* 221–2.
3 JWejnr to JWe, 13 April 1791, *SL* 338–9.
4 'Observations on The Production of Light from Different Bodies, by Heat and by Attrition', *Phil. Trans.* LXXII (1792) 28–47, 270–82.
5 TWe and H. Davy, 'An Account of a Method of Copying Paintings upon Glass, and of Making Profiles by the Agency of Light upon Nitrate of Silver', *Journal of the Royal Institution* I (1802) 70.
6 ED to TWe, 10 August 1794.
7 For this episode, see Peter M. Jones, 'Living the Enlightenment and the French revolution: James Watt, Matthew Boulton, and their sons', *Historical Journal* XLII (1999), 157–82. For

radical groups generally, and the role of Cooper and Walker, see Edward Royle and James Walvin, *English Radicals and Reformers, 1760–1848* (Brighton, 1983) 48–79.

8 AW to JW, in Jones, 172.
9 JW to JWjnr, 27 May 1792, JWP LB2.
10 JW to AW, 28 February 1792, MBP MII 4.4.28. quoted in Jacob 118.
11 *The Times*, 14 June 1792, Gibbs 215.
12 TWe to JWe, 7 July 1792.
13 JW to AW, 16 September 1792, MBP MII 4.4.
14 Priestley *Autobiography* 130.
15 JP to WW, 2 November 1792, WWP 5/124.
16 JP to Antoine Lavoisier, 2 June 1792, *Scientific Autobiography* 264.
17 JWjnr to JW, 17 October 1792, Jones, 'Living the Enlightenment', 176.
18 JW to JWjnr, 17 July 1793, Jones, 'Living the Enlightenment', 177.
19 Quoted in McNeil 59.
20 ED to Richard Dixon, 25 October 1792.
21 RAJ to WW, 23 March 1793, WWP 5/95.
22 JP to WW, 15 April 1793, WWP 5/125.
23 JWjnr to L. D. Huichelbose van Liender, 5 April 1794. JWP, LB 7, Jones 182.
24 Marianne Sykes, King-Hele 274.
25 ED to Richard Dixon, 25 October 1792.
26 Patty Fothergill's diary, 27 August 1793, Soho House.
27 Ibid., 30 August 1793.
28 King-Hele 257.
29 JP, *Experiments on the Generation of Air from Water* (1793), Dedication.
30 Priestley *Autobiography* 120.
31 In 1792 he published a long poem with notes, *Alexander's Expedition*, in open imitation of the *Botanic Garden*.
32 For the influence of Day's ideas, see Dorothy A. Stansfield, *Thomas Beddoes M.D., 1760–1808: Chemist, Physician, Democrat* (Dordrecht, 1984).
33 RLE, note added to Maria's letter to her aunt Mrs Ruxton, 21 July 1793, Butler 110.
34 ED to JW, 1 September 1794, *EDL* 94. For the joint work on airs, see Dorothy A. Stansfield and Ronald G. Stansfield, 'Dr Thomas Beddoes and James Watt: Preparatory Work 1794–6 for the Bristol Pneumatic Institute', in W. D. A. Smith, *Under the Influence: A History of Nitrous Oxide and Oxygen Anaesthesia* (1982).
35 *Letters from Dr Withering, Dr Ewart, Dr Thornton and Dr Biggs . . . on Asthma, Consumption, Fever and other Diseases* (1794).
36 ED to JW, 21 November 1795.
37 ED to JW, 1 September 1794.
38 ED to RLE, 15 March 1795.
39 ED to JWe, 9 December 1794, DAR 227.1:155, King-Hele 294.
40 EDjnr to RWD, 12 January 1795: DAR 227.6:42.
41 JWejnr to RWD, 13 January 1795, DAR 227.6.
42 For Tom Byerley's account, see Reilly 340. Reilly gives no credence to the second version, recorded in Farrer 1 vii, and repeated in *Wedgwood Circle* 104.

39 'FIRE ENGINES & SUNDRY WORKS'

1 ED to JW, 21 June 1796, JWP W/9.
2 MB speech, *Aris's Birmingham Gazette*, 1 February 1796.
3 MB to JW, 23 March 1791, B&W Box 20, Bdl 20, Folder 1.
4 *Rules of the Soho Insurance Company*, 'Explanation of the Plate', MBP
5 Watt Memorandum (1809).
6 See handwritten test of speech, MBP 290/19.
7 *Aris's Birmingham Gazette*, 1 February 1796.
8 John Southern to JWjnr, 2 February 1796, JWP/LB 1, fol.65r.
9 JW to MB, 25 January 1799, MBP 352/98.
10 JWjnr quoted in Dickinson *Watt* 177; see also JWjnr to MRB, 25 January 1799, B&W Folder 1.
11 FRS witnesses included de Luc, Herschel, Lind, Robert Mylne and Jessie Ramsden; employees included Murdoch, Southern, John Rennie and Richard Mitchell. See David Philip Miller, 'The Usefulness of Natural Philosophy: the Royal Society and the culture of practical utility in the later eighteenth century', *BJHS* 32 (June 1999) 193–7.
12 Heathfield was demolished in 1926, and houses built in the grounds.
13 AW to Gregory Watt, 10 November 1795, JWP6/14/45.
14 Gregory Watt to William Creighton, undated [1796–1800], JWP C/24 and 6/10.
15 JW to Gregory Watt, 24 August 1800, JWP6/14.
16 For the two plans see Samuel Wyatt, 'Soho House: Elevation of the Front', 1787, and John Rawsthorne, 'Soho House, Plan and Elevation', 1788: Christie's catalogue, *Fine Architectural Drawings and Watercolours*, 16 December 1986, 32, 33.
17 James Newton, 1797–98, MBP Miscellaneous papers, Box 1; Thos. Downing to MB, 7 February 1799. MBP 229, Box D2, 69.
18 Humphrey Repton to MB, 6 December 1795. MBP: MRB, General Correspondence.
19 MB to Lord Lansdown, 12 November, 1 December 1794, MBP 242/88, 91. WW to MB, 24 November 1796, WWP 5/36. See also Robertson Buchanan, *Essays on the Economy of Fuel and Management of Heat* (1810) Preface.
20 MB to Thomas Wilson, 14 February 1792, MBP 369/123.
21 MB to Thomas Wilson, 26 February 1792, MBP 369/126.
22 Sarah Withering to William Withering junior, 16 March 1798, WWP, 5/177.
23 Quoted in Anna E. Meredith, *Matthew Boulton* (Birmingham, 1972) 5.
24 Robbers, MBP 1799.
25 Charlotte Matthews to MB, 1794, MBP 325/119.
26 MB to Charlotte Matthews, 7 August 1799, MBP 323/163. Quoted in unpublished talk by Shena Mason, June 1999, 'Hark I hear Musick! Music, and the Boultons of Soho'.
27 MB to ED, 7 January 1793, MBP 304/40.
28 *A Chemical Analysis of the Water at Caldas da Rainha* (Lisbon, 1795).
29 WW to George Robinson, 10 November 1794, WWP, 1.
30 'A New method of preserving Fungi', *Transactions of the Linnean Society* II (1794) 263–6.
31 As n. 27.
32 JW to William Roebuck, 19 August 1795, JWP LB 2.
33 'Mineralogy of the South West part of Staffordshire' (article dated 14 June 1798), in Stebbing Shaw, *History and Antiquities of Staffordshire* (1798–1801).

34 See ED to MB, 8 October 1794, and n. 1, *EDL* 263.
35 WW to Dr Edward Ash, 12 May 1798, WWP 8.

40 'TIME IS! TIME WAS! . . .'

1 Byron's Journal at Ravenna, 19 January 1821, in Butler 230.
2 RLE to ED, 11 December 1794, Edgeworth *Memoirs* 338–9. See also RLE, 'An Essay on the Art of Conveying Secret and Swift Intelligence', *Transactions of the Royal Irish Academy* VI (1797) 95–139, and Supplement, 313–17.
3 ED to RLE, 15 March 1795.
4 Edgeworth *Memoirs* 338.
5 Ibid., 378–9.
6 RLE to ED, 31 March 1800, ibid., 389.
7 Frances Edgeworth to Louisa Beaufort, 27 July 1799, Butler 143.
8 *Zoonomia* IV 152; *Essential Writings* 76.
9 *Zoonomia* IV 155.
10 R. M. Wardle (ed.), *Godwin and Mary* (1967) 99.
11 Samuel Taylor Coleridge to Josiah Wade, 27 January 1796, King-Hele 301.
12 *Wedgwood Circle* 112.
13 Humphry Davy to Davies Giddy, 22 February 1799, J. A. Paris, *Life of Humphrey Davy* (1839–40) I 171.
14 John Davy (ed.), *The Collected Works of Humphry Davy* (1839–40) III 308; Golinski 168.
15 For Mesmerism in this context, see Fara 201–7.
16 Samuel Taylor Coleridge to TWe, [*c.* February 1803], *Wedgwood Circle* 127.
17 Ibid.
18 'Observations on Basalt, and on the Transition from vitreous to the stony Texture', *Phil. Trans.* XCIV (1804) 279–314.
19 Quoted in Rolt 133.
20 Darwin 105–6.
21 Thomas Brown, *Observations on the Zoonomia of Erasmus Darwin* (1798). See *EDL* 298–303.
22 Lloyd Sanders (ed.), *Selections from the Anti-Jacobin* (1904) 92–3.
23 *Enlightenment* 462. See Desmond King-Hele, *Erasmus Darwin and the Romantic Poets* (1986).
24 Archibald Cochrane, Earl of Dundonald, used new French nomenclature in his *Treatise, shewing the Intimate Connexion which subsists between Agriculture and Chemistry*. While Richard Kirwan published *On the manures most advantageously applied to various Sorts of Soils* (1796).
25 Jeremy Bentham, *Introduction to the Principles of Morals and Legislation* (1789). On *Phytologia*, see McNeil 106–15.
26 RLE to TD, 5 February 1787, Edgeworth NLI, Ms10, 166.
27 *Phytologia* 526; *Essential Writings* 118.
28 ED jnr to RWD, [Summer 1791], DAR 227.6.32.
29 Darwin, manuscript of *Life*, King-Hele 326.
30 Ibid.
31 Mrs Clive to unnamed correspondent, DAR 227.6.53.
32 ED to Georgiana, Duchess of Devonshire, [November 1800]. Volta's finding that the contact of dissimilar metals would generate a current was published early in 1800.
33 ED to RLE, 17 April 1802, Edgeworth *Memoirs* 396 and *EDL* 338.

34 See letter from the solicitor, N. Edwards, partner of EDjnr, to RWD, 18 April 1802: DAR 227.6.71.

35 Edgeworth *Memoirs* 397.

36 DAR 210.14:16, in the pamphlet by Nicholas Redman, 'Dr Erasmus Darwin and Breadsall Priory, Derbyshire' (2001).

Waning

EPILOGUE: 'FIDDLEDUM, DIDDLEDUM'

1 Gibbs 248.

2 Information from Tim Procter of the Soho Archive team, who have uncovered the existence of this banking company, 'M. R. Boulton, J. Watt & Co.'

3 MBP 342, Dickinson *Boulton* 74.

4 JK to MB, 1807, Moillet and Smith 59.

5 MB to AnneB, 11 March 1809. Dickinson *Boulton* 192.

6 Moilliet and Smith 54.

7 In David Philip Miller, '"Puffing Jamie": the commercial and ideological importance of being a "Philosopher"' *Hist. Sci.* XXVIII (March 2000) 2. See also Christine McCleod, 'James Watt, heroic invention and the idea of the industrial revolution', in Maxine Berg and Kristine Bruland (eds), *Technological Revolution in Europe: Historical Perspectives* (Cheltenham,1998).

8 Edgeworth's later publications included *Essays on Professional Education* (1809) and *An Essay on the Construction of Roads and Carriages* (1813).

9 L. Sanders (ed.), *Selections from the Anti-Jacobin* (1904) 29; *Enlightenment* 466.

10 Hellen Shelley, quoted in Thomas Jefferson Hogg, *The Life of Percy Bysshe Shelley* [1858], edited by H. Wolfe, 2 vols (1933).

11 Estimates differ: Musson and Robinson, *James Watt and the Steam Revolution* (1969), suggest 500 engines erected by 1800; Jennifer Tann, *Boulton & Watt's organisation of Steam Engine Production* (1977), suggests 451, with an output of 11,251 horsepower.

12 Ebenezer Elliott, in Griffiths 81.

LIST OF ILLUSTRATIONS

Plate Section I

Plate Section II

ILLUSTRATIONS IN THE TEXT

LIST OF ILLUSTRATIONS

The author and publishers would like to thank the following individuals and institutions for their kind permission to reproduce illustrations:

Plate sections Art Gallery of Birmingham, Alabama, Dwight and Lucille Beeson Collection 15; Birmingham Museums & Art Gallery 2, 23, 24, 37, 45, 47; British Museum 44; Michael Butler 34; Canterbury Cathedral Library 20; David Craig 5; Darwin College, Cambridge 1; Derby City Art Gallery 17, 18, 21; The Viscount Dunluce 16; Trustees of the Fitzwilliam Museum, Cambridge 19; Institution of Mechanical Engineers, London 38; The Board of Trustees of the National Museums and Galleries on Merseyside 30; National Gallery, London 22; National Gallery of Art, Washington, Paul Mellon Foundation 28; National Gallery of Ireland, Dublin 35; National Portrait Gallery, London 8, 10, 29, 46; Nottingham Castle Museum 25; Faculty of Medicine, University of Birmingham 39; Victoria and Albert Museum 9; Images by courtesy of The Wedgwood Museum Trust Limited, Barlaston, Staffordshire, England 11, 12, 13, 14, 31, 32, 33, 36

Illustrations in the text Birmingham City Archives pp. 68, 93, 203, 277, 472; Birmingham Museums & Art Gallery pp. 57, 464; The British Library p. 486, 498; Derby City Art Gallery pp. 18, 390; Darwin Commonplace Book, Erasmus Darwin House, Lichfield, on loan from Down House, © English Heritage (photographs by Bob Booth, Lichfield) pp. 304, 385; Fitzwilliam Museum, Cambridge, p. 493; Metropolitan Museum of New York p. 432; National Portrait Gallery, London, p. 181; Images by courtesy of The Wedgwood Museum Trust Limited, Barlaston, Staffordshire, England pp. 55, 85, 209, 328, 329, 415; © The Royal Society p. 164; The Library and Information Centre of the Royal Society of Chemistry p. xvi; The Wellcome Trust pp. xiii, 20, 126, 154, 163, 211, 266, 398

INDEX

Figures in *italics* refer to captions. 'ED' indicates Erasmus Darwin, 'JP' Joseph Priestley, 'JWe' Josiah Wedgwood, 'JW' James Watt, and 'MB' Matthew Boulton.

INDEX

INDEX